GENERAL GENETICS

*The whole subject of inheritance
is wonderful*

(*Darwin, An. and Pl., I, 146*)

GENERAL GENETICS

BY

M. J. SIRKS

Professor of Genetics, Government University
Groningen (Netherlands)

With 5 coloured plates and 238 illustrations

from the fifth dutch edition translated by

Dr JAN WEIJER
Groningen

and

D. WEIJER-TOLMIE, B. Sc.
London

SPRINGER-SCIENCE+BUSINESS MEDIA, B.V.

In 1922, the Centenary year of GREGOR MENDEL
and FRANCIS GALTON the first Dutch edition of this book was
dedicated as a tribute to their pioneering initiative

ISBN 978-94-017-0043-6   ISBN 978-94-015-7587-4 (eBook)
DOI 10.1007/978-94-015-7587-4

PREFACE

The year 1922 was a landmark in the life of Genetics. In that year the Centenary of both MENDEL's and GALTON's birth was celebrated. In that year also, Genetics came of age.

By lucky chance the first edition of this book appeared in the same year; after a few revisions the fifth edition left the press in 1951. At present Genetics has no reason to complain of lack of textbooks. Its contents, its theories and its value have been written for us in many languages.

Genetics is a rich Science. During the last fifteen years particularly several new avenues have been opened up. Its development since 1940 has been almost unbelievable in scope, and often in rapidity.

However, there is also some danger in this feverish development. Modern textbooks seem to suggest, that for Genetics "life begins at forty". This implies a dangerous overestimating of the value and reliability of recent results and their theoretical considerations, which are not always well-founded or predestined to a long life. It also implies an underestimation of the work carried out by our pioneering colleagues, whose activity ended before the second world war.

In one of the recent textbooks, and a very good one at that, no mention is made of the work done by BATESON, BAUR, CUÉNOT, HAECKER, Oscar HERTWIG, LANG, PLATE, RENNER, STRASBURGER, VAVILOV, WEISMANN or WETTSTEIN, while the names of others, such as CORRENS, NILSSON-EHLE, SHULL are recorded only incidentally. These scientists however constructed the canvas upon which the pattern of modern Genetics is embroidered. They built classical Genetics.

This book tries to establish an equilibrium between Genetics as it was built up before 1940, and Genetics as it has developed since. Lest we forget.

My thanks are due to the Publisher, Messrs Martinus NYHOFF at The Hague, who took the initiative for this translation, and to Dr Jan WEIJER and Mrs D. WEIJER-TOLMIE for the careful way in which they have performed this laborious task.

Groningen, September 1955.   M. J. SIRKS

CONTENTS

COLOURED PLATES

CHAPTER I

INTRODUCTION, CONCEPTS AND METHODS

Heredity is a fact of experience. Every day we will encounter many phenomena, which point to the existence of resemblances in external appearance between organisms and their offspring.

Each day we can observe, that parents and children resemble each other in certain characteristic features, and that the resemblance between relations in a direct family-line is, in general, greater than that between two individuals at random.

This resemblance expresses itself especially in specific parts of the body or in a complex of its properties: striking features and prominent forms of a single organ in many families. One of the most remarkable is present in the famous *Hapsburger lower jaw*: a peculiar jaw form, by which so many of the numerous members of the Royal House of Hapsburg were characterized. From the investigations of GALIPPE (1905), HAECKER (1912) and STROHMAYER (1937) it has been brought to light, that this type of face appears continuously for hundreds of years within the Hapsburg family and that certainly many members of this family from the time of Ernst the Iron, who reigned circa 1400, have shown this characteristic.

Also several such characteristics can go together; comparing the different human races, some characteristic differences will easily be found after a little reflection. That

Fig. 1. Ernest the Iron
(after HAECKER 1912).

1

is to say we can distinguish negroes, malayans, mongolians and white people one from the other by complexes of properties which in each of these races are passed from parent to child and hence are hereditary.

Side by side with this experience goes the admission, that resemblance must have a basic cause, and by virtue of this the phenomenon of heredity becomes a problem and a subject of scientific research and discussion.

By virtue of the penetrating influence, which heredity wields in human society and in the life of man as an individual, and the wide sphere over which its influence can be felt, its problems and their answers, evoked general interest as soon as man attempted to form an idea of causes and consequences present in Nature. For a long time therefore, heredity has been a subject of discussion of the most divergent character. Contemplations from the point of view of the purely philosophical and judicial-sociological aspects of the problems of heredity have been made, all of which have their own individual merit, but which are still greatly lacking in their objective value. This is due to the fact, that in such speculations, which are based on a purely subjective point of view one thing is forgotten and this is most important, namely that mankind, as do all organisms around him, forms a great part of Nature and is subject to the laws, which govern all living things and that also man and his heredity is an object of biology. This is even more relevant when considering plants and animals. The problem of heredity is a biological-one, its study is a subject of investigation for the biologist. The purely scientific character of heredity compels our respect; it exists and it is not possible to argue it away.

It induces us into a critical examination of its handling, that is to say: the natural scientific character of heredity demands natural scientific methods.

If we will now try to examine this problem as accurately as possible, one thing must be factually ascertained viz., what heredity is. Not everything which is labelled as hereditary in every day life is in fact so. Many false ideas are held concerning the concept of heredity, which we have to put right at once because they can give rise to misunderstandings.

There are within the phenomena, which are usually described as heritable, those which bear this name incorrectly. All things which are only caused by the presence of other *symbiotic organisms* in the body of parents and children are excluded from heredity proper. The penetration of other organisms into the reproductive cells (e.g., egg-cells, spermatozoon, pollen-grains) or into the young embryo inside the body of the mother can lead to a change in the young life which is developing. This imitation of inheritance is found to occur occasionally in only a few individuals but in some cases it occurs regularly in all organisms of a particular species. A mother infected by tuberculosis can also very easily infect her children during their embryonic lives. In this case the children will be born with tuberculosis affections. Through the investigations of PASTEUR (1870) we know of an analogous phenomenon in the pebrine disease of the silkworm. This is an infectious disease caused by the parasite *Nosema bombycis* which penetrates into the eggs and causes infection in the offspring which develop. Besides these cases in which this supposed heredity occurs in some of the offspring there also occur examples of a transference of contamination from parents to all offspring. For example *Lolium temulentum* is a species of grass in which the seeds systematically accomodate a parasite which is passed on from parent to children through each and every generation. In order to make plants free from parasites definite precautions must be carried out as was pointed out by HANNING (1907). But we must be on our guard that such cases of transference of infections in the reproductive cells or transference during the period of embryonic growth are not mistaken for heredity. This phenomenon has nothing at all to do with heredity: it is an early, *pre-natal contamination*, a *pseudo-heredity*. However pseudo-heredity does not necessarily need to be a consequence of the direct presence of parasitic organisms: for example poisoning of the reproductive cells by some means or others (e.g., alcoholism) can bring about aberrant phenomena in the offspring which were also observed in the parents and because of this may be wrongly thought of as heritable abnormalities. To what extent such damage can in fact invoke a change of the hereditary capacities (so called *germ-damage*) is of extreme importance, but it is also a very difficult question. This question will be discussed more fully later on (chapter XIX).

In the first place a strict distinction must be made between such cases of pseudo-heredity and real heritable properties. The first is a subject for pathology, the second is one for genetics. If it can be pointed out that the resemblance in external appearance between parents and children is not the direct result of their both having present parasitic organisms, or of poisoning of the reproductive cell, then we can try to explain this resemblance by means of genetical methods. An observable resemblance in external appearance does not necessarily prove the existence of a genetical resemblance. If we try to define real heredity, we must put it into these words: heredity is the phenomenon involved when organisms and their descendants show the same outward observable characteristics. But this description of heredity introduces two new conceptions which we have to define, if we wish to have a clear cut conception of genetics before us. It must be admitted that this given definition of heredity is not precisely correct.

The first of the two new conceptions in the above definition requiring elucidation, is that of *"descendants* or *offspring"*. An organism can produce new organisms in two ways: asexually or sexually. The difference between these methods of reproduction can easily be understood. In the case of asexual reproduction a more or less arbitrary part of the organism develops independently as a whole unit and becomes a new organism. In sexual reproduction two cells (reproductive cells or gametes) fuse and form a new organism. This fundamental difference in the manner of their origin must be acknowledged as being of primary significance in experiments on heredity. Wherever one speaks of heredity it must only be in connection with offspring formed by a sexual process. Those offspring developed by an asexual process are not descendants in the strict sense of the word. For this reason WEBBER (1903) suggested the term: *"clone"* for this type of descendants and this was supported by SHULL (1912). It appears acceptable that those clones which originate from one and the same organism, and which thus properly form together one single individual, also obtain all the properties of this original individual. If such is really the case is quite another question, which will be discussed later on (chapter XVI), but at the moment it must remain fixed that descendants are only formed in a sexual way.

It is not so easy to define the term *characteristic* in the previous definition of heredity. If we examine a given organism as accurately as possible, we will observe a number of external properties — colour, form and measurements — which together give a picture of this organism. The individual is characterized by these external properties.

These external properties are all characteristics of the organism we examined, and together they build up the external appearance of the organism. JOHANNSEN defined this as the *"phenotype"* of the individual (1909; 1926 p. 162). However, this phenotype is a deceitful guide. The external appearance of an organism is the product of two different groups of influences: inherited properties (the *"genotype"*)

Fig. 2. Two clones from the same plant *Helianthemum vulgare* (after BONNIER 1895). On the left: Lowland type. On the right: Alpine type.

and the circumstances of its life (*environment* or *peristase*). In the beginning of its life, that is, at the moment of fertilization from which it will arise, every organism gets a number of hereditary characteristics from its parents. Since from this moment it is also subjected to the influence of environment (light, temperature, food, humidity and other factors) which give to the extant hereditary properties the opportunity to unfold; one property can be favoured, the other opposed. Therefore it is possible that two organisms which had the exactly same hereditary properties in the beginning, which will persist throughout their whole lives, will still exhibit a different external appearance when they are fullgrown. It would be very easy to give a great number of examples. The most classic example which illustrates this very clearly are those of BONNIER (1895). His investigations dealt with the influence of the effect of the Alpine climate as an environment compared with that of the Paris basin.

In order to compare these influences as accurately as possible BONNIER took as experimental material plants with as close an heritable disposition as possible, i.e. parts of one individual obtained by asexual multiplication. Through the deep influence, which the environment wields in the Alps (lower temperature, slight humidity, more intensive insolation) the parts of plants grown there finally had a totally different external appearance to those planted near Paris.

A picture of both forms of *Helianthemum vulgare* shows this clearly (figure 2).

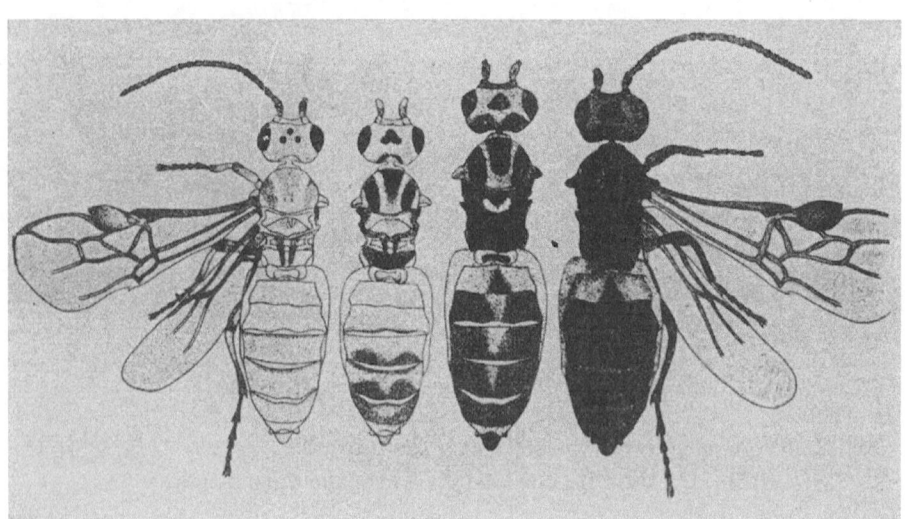

Fig. 3. Temperature modifications in Ichneumon fly, *Habrobracon juglandis*. From the left to right, the flies were reared at 35°, 30°, 20° and 16° C. (after SCHLOTTKE 1926).

Such differences in external appearance are purely a consequence of difference in environment which we can find everywhere in wild nature and also under bad cultivation, if we take the pains to observe it.

The whole external appearance of organisms can change solely as a result of the environment, but this change in external appearance may be limited to one property only. When the ichneumon fly, *Habro-bracon juglandis*, is reared under different temperatures, the colour of the body reacts to the temperature: i.e., flies reared at 35°C will be light, at 16°C darker, and between these temperatures there are various transitions (fig. 3). However in another race of this species which has the same colour as that which the changeable race develops at 16° C, there is no change of colour with temperature. This is to say, that the latter mentioned race shows no change in external appearance and is thus always darker in colour. When examined from the point of view of low temperature (16°) we would like to give to both races similar inheritable properties on account of this resemblance in external appearance (phenotype) but from experiments conducted at higher temperatures it has been shown that a difference exists in their heritable disposition.

Therefore in genetical investigations external characteristics have no great validity, but on the other hand the way in which organisms react to environment has great validity. This last characteristic of reaction to environment is really important; that is to say the ability to react to similar circumstances in a similar way. This definition of the conception of "*property*" (character) was first given very clear by BAUR (1911; 1930 p. 8) and it is to his definition that we must adhere. To cover changes in external appearance BAUR introduced the term "*modification*". This phenomenon has caused many mistakes — for only the manner of reaction to circumstances of life is heritable.

This misleading character of modification is much increased by the possibility that a long period of time may pass between the instant of reaction to some definite environment and the moment at which the effect is visible. Frequently it occurs that the organ in question on which a certain property is observable is only open to external influences during a short *susceptible* (*sensible*) *period* of its development; whereas it is only much later on (when full-grown) that the result is observable.

Modifications can lead us astray if we only take account of observable characteristics in external appearance. Therefore we have to exclude these from an exact definition of the concept of heredity. Therefore the previous definition must be replaced by the following: "heredity is the phenomenon whereby organisms and their descendants have the ability to react to similar external environment in a similar way". Heredity, so defined, is the subject matter of genetics and its investigation must be carried out by means of natural science as mentioned earlier.

Methods in genetics have undergone many changes over the years and to a large extent these changes are still observable. The most primitive and most obvious method is that of pure speculation which builds up a more or less clearly defined hypothesis without taking any facts into consideration. This was followed by other procedures which did take facts into account as bases for theoretical views. This succession is not a mere replacement. At the present time problems in genetics are still approached in all imaginable ways. The reason for this lies in the interest which this subject holds for every one.

Philosophers, writers and jurists have supposed that they have the same rights to discuss problems of heredity as indeed have men of medicine and biology.

In this way a whole library of purely philosophical literature concerning genetics originated. There are a great many deductive views concerning heredity; almost as many views as publications. Possibly there is a great value in the rich content of what is mainly original thought, but to the scientist, who takes up heredity as a biological problem, these philosophical contemplations are only enjoyable from an aesthetical or an ethical point of view. However, from the point of view of pure science these contemplations are too speculative.

Philosophical line of thought alone, concerning a biological problem, will not bring us to a resolution, completely satisfactory to future generations. Such a resolution of the biological problem of inheritance has to be sought out along the paths of pure biological science, that is to say, it is of the first order of importance that the philosophical way of thinking is transferred from the beginning of the problem to the end. It must not be entirely lacking and hence it is never excluded from any publications, because every conclusion and every premise to which scientific research leads is discovered along the path of logic. Should this appear lacking then in many cases it is due to difficulties of recognition, but it is still there. If it is really missing then the character of the publication is dubious. The philosophical way of thought must play its rôle, but this rôle commences only after all the other aspects have been considered. The expressions of the logical way of thought must always be controlled by scientific arguments.

Other approaches preceed the philosophical in scientific genetics. First are the *observations*, a very obvious and important method for those people who like to look round them in the world. Under this method, phenomena become our starting point. At first only unrelated processes in Nature are observed. Accumulation of such observations can lead to the discovery of an accordance between some of the observed phenomena. At this point the faculty of thought comes to the fore. It is acceptable that those accordances have a causality as a common foundation which is derived from external factors.

LAMARCK was the first to frame uncontrolled observed phenomena with logical arguments. This was, in his time, a great and significant improvement on the philosophical contemplations of Nature which reigned supreme in his days in Europe (first part of the 19th century). It is not so much the question of heredity itself which caught LAMARCK's attention but more the closely connected question of the variability, i.e., of imperfect inheritance. This is placed before us by the apparent precarious deviations in external appearances which descendants often have when compared with their parents.

This lead to the seeking of the cause of differences between children of the same parents and the cause of the slow changes in flora and fauna during the history of the world as the natural consequence of this imperfect inheritance. LAMARCK was looking for facts to explain this precariousness of heredity. Upon these facts he built his rules, which could explain their occurrence.

The conclusions which LAMARCK drew from the collection of observations will be considered later (Chapter XIX) in connection with other questions of inheritance. For, in the light of todays' knowledge they are wrong, or else unproven. This is not the fault of the logical way of thought of LAMARCK, but of the untrustworthiness of the observed phenomena which could not be verified. In the time of LAMARCK, his methods of observations could have satisfied us. Indeed, the enormous flourishing of biology in the mid nineteenth century and the work of DARWIN and his circle, is the best evidence as to the value of accumulated and critically judged observations. We may not go so far in our appreciation of modern methods as to deny all significance to his method of "observation". Negation of the value of them would be a sign that we overrate our own procedure and its value. Today however, genetics makes higher demands in the matter of technique;

stray observations are no longer sufficient. Indeed genetics followed the same trend as all the exact sciences, which try to sway and to know everything.

But observations still play a very important rôle in different branches of genetical methodology. First, in the *direct method of cytology*, which tries to determine the presence of heritable characters by means of research into the cell and its compounds which are the units from which all bodies of plants and animals are built up. It is common knowledge that in general a specific part of the cell, the nucleus, can be considered as the storage place of the heritable characters. This idea was started by OSCAR HERTWIG (1884), STRASBURGER (1884) and WEISMANN (1885) and is based on the following points: since the reproductive cells (egg-cells and spermatozoa and pollen) form the only link between parents and offspring, it is thus a logical necessity to suppose, that that which is responsible for the heritable characters is located in these cells. In cytological investigations it is remarkable that during the process of cell-division the nucleus particularly (as well as the rest of the cell) is split up very precisely into two similar parts. Within the nucleus, the chromosomes appear to be of most importance on account of the care with which they are longitudinally split. For instance we may consider the nucleus of the banana fly, *Drosophila melanogaster*, one of the pre-eminently suited animals.

Through the works of MORGAN and his school this has become a classic animal of genetics. This is evident from MULLER's (1939) and HERSKOWITZ' (1953) "Bibliography on the Genetics of Drosophila" in which more than 6800 publications are listed.

In the normal body

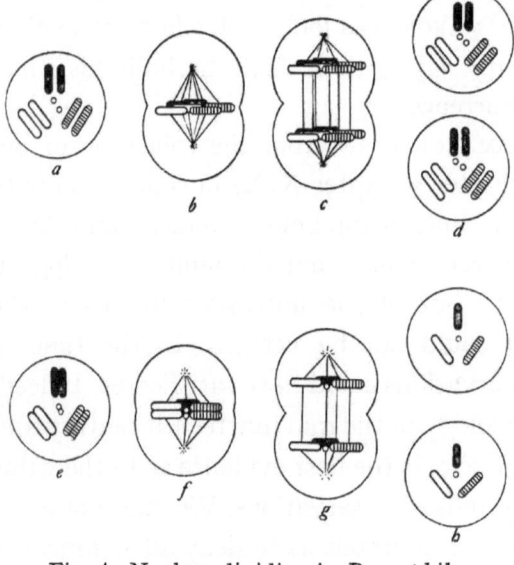

Fig. 4. Nuclear division in *Drosophila*.
a — d in body cells, e — h formation
of the reproductive cells
(after MORGAN 1916).

cells of the individuals of this species, we always find eight or rather four pairs of chromosomes (fig. 4a). These are divided by longitudinal splitting (fig. 4b, 4c) during the normal somatic cell-division, and are distributed between both the daughter cells (fig. 4d) which originate from the mother-cell. In the formation of the reproductive cells, it takes place in another way. There is no splitting of the eight chromosomes, but from each of the four chromosome-pairs (which may be readily distinguished) one chromosome goes to one daughter-cell, the other to the second daughter-cell. This is the reason why in each reproductive cell there are four chromosomes (figs 4e-h).

The regular course of this somatic cell-division and of the cell-division by which the reproductive cells originate was a revelation to the biologists during the 30 years before 1900. Indeed, it was undoubtedly a happy thought of HERTWIG, STRASBURGER, and WEISMANN to seek a relation between these processes, and the presence of heritable characters in the cell. Truly a fine piece of work was done by these research workers in this direct method of cytology. Their technique was based on making visible all the complicated structural details within the living cell and this was really a remarkable step forward. Their observations could only be obtained after patient devotion. Their conclusions are often expressions of sharpness which cannot be other than enviable. The whole synthesis of the facts and observations was a study which seems to be fully worthy of our confidence.

There exists a great danger to which a solely cytological worker exposes himself. He must feel the limitations of the data which his technique can procure for him for he may try to convince himself that he can solve the questions concerning the nature of heritable characteristics alone. The material basis of the heritable characters cannot be analysed by using a microscope only — it must be stressed that cytology is incomplete when it is standing on its own as an observing science. Although it does not entitle one to explain the problems of inheritance, it must be pointed out that cytology can give excellent services and can give first rate indications if it collaborates with other methods.

Fortunately it can be said that cytologists understand their task very well. Cytologists are busily engaged in changing over from the science of strict observations to experimental science. Much experimental work has been done, and through this it has been shown that the investigations into the cell can have a great bearing on the final conclusions regarding the problem of heredity. Many opportunities· will arise in this book to observe their influence.

There is also another method which deals with observations. This latter method does not take the heritable characters as the basis of research, but considers only their "end product", that is to say, that the objects of study are the external, visible properties of the organisms and this method compares these properties in groups of related individuals. This is the school of biometrics or the method of statistical comparison. Convinced of the value of earlier data and of theoretical conclusions derived from these data FRANCIS GALTON (fig. 5) founded his own school of investigators with their own working methods, their own viewpoint and their own review. This took place at about 1900 under the impulse of the revived interest in the problems of genetics. GALTON tried to apply the modern method of statistics to the problems of biology, especially to genetics.

"Those methods", he says, "deal comprehensively with entire species and with entire groups of influences, just as if they were single entities, and express the relations between them in an equally compendious manner" (1901, p. 7). This is the method with which GALTON's successors (WELDON, PEARSON, YULE, and others) worked particularly in trying to derive the laws of heredity. They attempted this by means of detailed studies of the external morphological properties of organisms with the help of measuring, counting and weighing.

Fig. 5. FRANCIS GALTON
(1822–1911).

The fully-grown or growing individuals which resemble one another morphologically are grouped together and reduced to a statistical basis. These leaders of the biometrical school were so convinced of the exactitude and the authority of their methods, that PEARSON could write in a discussion to

BATESON: "Biometricians have not only to collect material, analyse it, and see its bearing on vital phenomena, but they have still to convince the great body of biological workers that their methods are the only logical methods for solving, not necessarily every problem, but certainly many problems in the evolution of life" (1901, p. 321). This statement is still the most striking piece of evidence for the peculiar mental condition of a number of advocates of the biometric school. They have obstinately persisted in regarding genetics as mathematics and refused to recognise fundamental biological errors inherent in their working methods. The great majority of biologists in 1901 were not convinced that the biometric was the only logical method, and they are still not convinced.

Unquestionably the mathematically trained geneticist renders many good services to-day, but there are some cardinal objections which are very important, so that the usefulness of biometrics on its own is limited. The significance of this method, supposedly fundamental, has been reduced to one of a more complementary nature.

The weak places in the biometric school are suspiciously many and include the manner in which the methods of biometrics were and sometimes still are applied by their followers.

This does not inexorably imply that the method of statistical work is useless, in this respect it can be compared to the cytological method: standing on its own without realising the limitation of its abilities biometrics overrates its own importance and works with material insufficiently tested.

But if statistics are used in twin-research (p. 43), or for consequences of prolonged inbreeding (p. 46), then they can give a reliable basis for genetical conclusions. Recently there has developed a school of research in which the premeditated experiment is woven together with the mathematical exactitude of statistics (genetics of populations, see chapter XX). Methods which are too exclusively based on observations alone, carry with them the insuperable difficulty, that it is not possible to verify the reliability of data. This means that not everything can be controlled by the investigator. For this reason geneticists who discover laws and like to work as exactly as possible could not be content with this method and hence there was the necessity to find another method whereby the field of interest could be more fully controlled; that is to say, genetics started to work *experimentally*.

If we want to be on the track of laws and to perfect our knowledge of the properties of heredity then we have to try to explain not only heredity but also its imperfections. Since we have learned that also in these imperfections there lie definite laws, that is to say, that these imperfections also follow definite lines, we also know that these rules (by which genetics is governed in its perfections and imperfections) can only be discovered by an exact study of all the descendants of any one parental pair, and by judgment of the heritable value of an individual on the basis of its descendants. These descendants can only be seen as a whole if we cultivate plants (or breed animals) and observe them ourselves. In other words: the rules of genetics can only be discovered by specially arranged, i.e. systematic, experiments.

Fig. 6. GREGOR MENDEL (1822–1884).

Everything hangs on the planning of these experiments. At the time when no basic rule was yet known, it was important to extract something conclusive from the mass of existing information. It was GREGOR MENDEL (fig. 6) who saw this problem with extraordinary sharpness. It was he who comprehended that the resolution of this simple looking problem could be had only after much seeking. His contemporaries, NAUDIN and WICHURA for instance, carried out investigations in the same sphere but they could not come to an agreement. Their material was too complex and because of this their results were too complicated to enable them to discover the leading thread which ran through it.

The genius of MENDEL drove him to ask very easy, clear cut questions and he found the answers through his exact work. To be sure, there was also a great deal of good fortune responsible for the excellent results of his work. But we may still honour him as the first investigator into Nature, who realised that the question of inheritance must be gone into with experimental precision. MENDEL (1865) propounded rules, which are still held as fundamental in the eyes of modern geneticists. More than this, he introduced and built up with exactitude a new methodology which was one of "the controlled experiment", which helped found our genetics.

The majority of investigators — who have as their field of interest the correctness, incorrectness or general validity of MENDEL's basic laws — nowadays work with this experimental method.

Much is still achieved today with this experiment, but it is not solely due to MENDEL's work that modern geneticists have accepted the experimental method. Undoubtedly without the strong initiative of HUGO DE VRIES this method would not have found such a wide acceptance so quickly, and would not be so important in the flourishing of genetics. One has to keep an eye on this because experiments must make high demands, and their reliability and accuracy must be ascertained before we judge the value of the results they achieved. Too many experiments overlook the most simple precautions.

The experiments performed by MENDEL, to which we will return later on, were based on the following considerations. If we intend to find out something about an existing or non-existant regularity in the process of genetics we must not start with individuals about whose descent nothing is known, but with pure races which have been cultivated, or bred, as the case may be, for years. We will only obtain true views concerning heredity, if we exclude all differences in environment, that is to say, if we cultivate all our research material under similar conditions. Moreover, it is necessary to choose material, in which a single individual can give an offspring, hence the choice must be a self-pollinating plant-species. On these grounds MENDEL chose varieties of peas for his experiments. These peas had very little modifiable properties and hence they gave a good picture of the heritable tendency. In order to trace the course of inheritance, MENDEL started with artificial crossings between two individuals of different varieties of peas, hence he produced artificial hybrids of which the parents were known and studied the offspring of each hybrid separately. This is the experimental method of artificial hybridization to establish the inheritance of the characters in which the original varieties differ.

By means of this method a tremendous amount of work has been carried out since MENDEL's time and still goes on today. It was in this way that the basic rules of modern genetics were discovered. The question now arises, can everything pertaining to the subject of genetics be found in this way?

At the present time it is frequently considered almost heresy to doubt that there can be another experimental way other than that of MENDEL. But it is a mistaken worship of MENDEL's genius to believe that his method is the only one which can bring to light all the solutions of problems related to genetics. His methodology was original and brilliant, this we cannot repeat too often. This method has uncovered the basis of the genetical problems and still gives us brilliant services every day. Nevertheless it is not the only method which has a raison d'être.

MENDEL's methodology deals only with clear cut differences between individuals, which can be crossed with one and another or with a third individual respectively. Therefore the sphere of work in genetics is limited in a fixed manner. Hence the necessary result is that up to now we can only demonstrate differences in heritable characters between two plants or animals, if we can make a cross between them (or with a third one which serves, one might say, as a reagent) and provided we can compare the offspring. Nevertheless a danger exists in this method that we confuse our ideas, because our experimental methods will not bring us further than tracing the inheritance of points of difference only. If we do not succeed in investigating the similarities between organisms then this is due to our faulty methods and not to the process of inheritance.

Neither chemistry, in its early days occupied itself in studying that which is present inside the molecule and inside the atom, nor did genetics occupy itself until now in studying that part of the heritable character in which animals, plants or even groups of plants or animals agree. All vertebrates are jointly characterized in a number of points of similarities. But genetics, thus far cannot give the solution of the problem as to which heritable characters underlie these similarities because it is still impossible to cross a vertebrate with an invertebrate. Nevertheless some things have also come to light in this direction.

Between the heritable properties in the gametes and the externally visible characters of the growing organism which develops from them, is a wide distance which is itself also connected with the question of inheritance and must thus also be studied.

This development of heritable tendency to a point where the external characteristics show, can be followed in two directions: we can either start with the reproductive cells during fertilization and investigate the behaviour of that which we consider to be the basis of inheritance during the further development of the egg to the full-grown stage (by this means we can influence the development by experimental treatment). This is SCHAXEL's method of *cytomorphosis* (1915). Or, the other way is to start with the phenotype and try to explain the visible characteristics by a tracing backwards in the development till we track down the heritable characters which underlie the visible ones (*phenogenetics* of HAECKER, 1918, 1925).

Following the work of these two scientists, a method was developed and genetics became linked with embryology. This method tries to estimate the background of embryonic differentiation by experiments and the chemical processes which take place between the fertilized egg stage and the phenotype. Hence the experimental field of developmental genetics was opened up (see chapter XXI).

Summarizing, we can start by clearly distinguishing from one another, *pure heredity* and *pseudo-heredity*. The latter is caused by early infection or poisoning and only the first mentioned is a subject of research in genetics. Next, inheritance must be described as "the phenomenon by which organisms and their sexual offspring concur in their ability to react to similar external life circumstances (environments) in similar ways".

This form of inheritance can be studied by means of:

1st. Direct observation of the reproductive cells (cytology). This method is insufficient as a solely observing procedure but has given very important contributions to genetics on an experimental basis, and will still continue to do so. This method tries to locate the heritable characters in the chromosomes which are situated in the cell nucleus.

2nd. Indirectly by statistical studies, which on their own give very little certainty and must be reviewed critically or preferably combined with investigations into the derivation of the material studied.

3rd. Indirectly by experimental crosses, which is a limited method because of the impossibility of crossing between animals or plants which differ too greatly.

4th. Indirectly by means of experiments on the path of development lying between the heritable properties in the reproductive cells and the

external characteristics of the fully developed organism in a constructive (*cytomorphological*) or destructive (*phenogenetical*) way.

For the present, experimental hybridization must still remain in the
front as the pioneer, always busy clearing a way towards broadening
new views, to explore new avenues. A close contact must always exist
between experimental cytology (which is rapidly developing today),
cytomorphosis and phenogenetics. Finally, in special cases the statistical study can be used to explore spheres opened up by the experimental
method. Nevertheless, this method may not work vertically but only
horizontally. It can however help in breaking up the field already
opened up by experiments.

The last words concerning problems in genetics are still unspoken
and it is always a point of discussion between the different methods
which one will win the right of way to solve them. It seems as if the
experimental analysis in itself has reached its highest point. This means
that this method undoubtedly needs the indispensable help of experimental cytology and embryology (including the biochemical side of
developmental processes).

CHAPTER II

THE STATISTICAL METHODS AND THEIR RESULTS

In spite of the important and basic points in which the conceptions of the biometric school differ from those of modern genetics, it will be clear to everyone that the very precise nature of statistical methods can be an example to genetics. The exactitude associated with mathematics is highly desirable, and it would be a good thing to introduce it to our experiments if this were possible. As already mentioned, statistics can give us valuable assistance in the area where the experiment fails, and therefore it is necessary to become acquainted with the results of this method.

The first investigator who attempted an application of the *statistical method* to biological problems was the Belgian Anthropologist QuÉTE-LET (1871) whose studies (measurements concerning the shape and size of the human body) gave striking results in that these measurements showed a rather strong regularity. The tallest and the shortest people seem to occur in nearly equal, but low percentages; those who were 5 cms. shorter than the tallest ones were less rare and occurred in numbers nearly equal to those who were 5 cms. taller than the shortest. The group of people showed a certain definite regularity in their "*variability*". The complete group of individuals measured gave a distribution which takes the form of the well known binomial theorem of Newton. That is to say, they agree with the coefficients of the factors which are obtained by expansion of the binomial formula $(a + b)^n$. For the successive values of $n = 1, 2, 3, 4, 5$, etc. and for $a = b = 1$, this development gives the following results:

$$(a + b)^1 = 1 + 1$$
$$(a + b)^2 = 1+2+1$$
$$(a + b)^3 = 1+3+3+1$$
$$(a + b)^4 = 1+4+6+4+1$$
$$(a + b)^5 = 1+5+10+10+5+1$$
$$(a + b)^6 = 1+6+15+20+15+6+1$$
$$(a + b)^7 = 1+7+21+35+35+21+7+1$$
$$(a + b)^8 = 1+8+28+56+70+56+28+8+1$$
$$(a + b)^9 = 1+9+36+84+126+126+84+36+9+1$$
$$(a + b)^{10} = 1+10+45+120+210+252+210+120+45+10+1$$

In general form these coefficients can be expressed by the following formula:

$$1, \frac{n}{1}, \frac{n(n-1)}{1.2}, \frac{n(n-1)(n-2)}{1.2.3}, \frac{n(n-1)(n-2)(n-3)}{1.2.3.4}, \text{etc.}$$

It follows immediately that there is a symetric distribution of these numbers on both sides of a central dividing line. These numbers decline slowly at first, then more quickly, then decline slowly again.

GALTON (1876, 1877) followed soon after QUÉTELET, and applied measurements and countings to the most divergent organisms. It was his work that illustrated the importance of what is generally known today as the *'frequency distribution curve"* or *"GALTON-curve"*. The origin of this curve is very typically shown in fig. 7. A total of 175 students at Connecticut Agricultural College were classified into groups according to their heights. As a standard for this classification into groups a unit of difference of 1″ was taken, so that students who were 4 ft. 10″ or a little taller, but shorter than 4′11″, were taken as one group. Similarly, students who were 4′11″ or slightly more, but less than 5′, were grouped together. Only one student belonged to the group which included the shortest individuals (4′10″ to 4′11″). In the next groups, i.e. 4′11″ to 5′ and 5′ to 5′1″ there was no representative. Only one person belonged to the 5′1″ to 5′2″ group and so on, so that the whole classification took place as follows:

Height	4′10″ 4′11″	4′11″ 5′0″	5′0″ 5′1″	5′1″ 5′2″	5′2″ 5′3″	5′3″ 5′4″	5′4″ 5′5″	5′5″ 5′6″	5′6″ 5′7″
Number	1	0	0	1	5	7	7	22	25

Height	5′7″ 5′8″	5′8″ 5′9″	5′9″ 5′10″	5′10″ 5′11″	5′11″ 6′0″	6′0″ 6′1″	6′1″ 6′2″	6′2″ 6′3″
Number	26	27	17	11	17	4	4	1

Total number = 175

By placing students belonging to one group behind each other with equal intervening spaces, the graph given in fig. 7 can be obtained. This figure gives a very clear picture of the typical classification of the body lengths of the students. If the back persons of each column held fast to a rope, then this rope would lie in the shape of the curved line of a definite form which is called a *"frequency distribution-curve"* in statistical biology. It is evident from its frequency of occurrence in nature that this curve is of general importance. Some examples follow which are selected from a few of the very divergent cases which occur. Recalling in the first instance the study of PETERSEN we can count the fin-rays which appear in the caudal fin of a flounder (see JOHANNSEN 1909, 1926, p. 11). We will find that the smallest number is 47, present

Fig. 7. Frequency curve according to body height (after BLAKESLEE, 1914).

in 5 fish, and the greatest is 61 shown by one individual only. The numbers are presented in the following manner:

numbers of rays	47	48	49	50	51	52	53	54	55	56	57	58	59	60	61
number of fish	5	2	13	23	58	96	134	127	111	74	37	16	4	2	1

To represent these numbers by a frequency distribution curve we must first set equidistant points along a horizontal line. Along these points we write successively the numbers 46, 47, 48, 49, 50, ..., 61, 62. Through these points we draw vertical lines, the length of which is determined by the number of the fish belonging to the respective group. By connecting the tops of these perpendicular lines we will obtain a frequency-curve (fig. 8) which has nearly the same shape as that for the case of the American students previously descri-

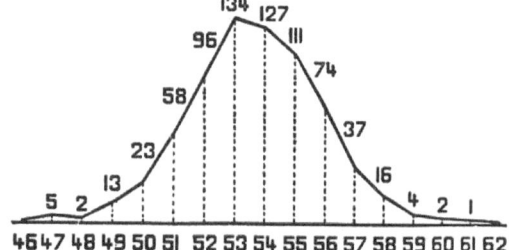

Fig. 8. Frequency distribution-curve of the number of fin-rays in tail fins of flounders (after JOHANNSEN, 1909–1926).

bed, but now it is represented graphically instead of realistically.

No transition occurs between the individuals which are brought together into groups. There exist no flounders with $46\frac{1}{2}$ or $54\frac{1}{4}$ fin rays. Because of this we will denote this type of variability by the term *"discontinuous variability"*.

But if we work with material that we cannot count, that is to say if we have to fall back on weighing and measuring, then it will obviously not always be possible to have whole number groups. For instance it is not possible to carry out exact measurements of the lengths of seeds of a scarlet runner bean, and moreover, between the whole numbers we will obtain all the transitions when measuring any quantity of beans. We call this variability *"continuous"*.

This compels us to classify individuals investigated into groups divided by borders which are chosen arbitrarily. JOHANNSEN measured the lengths of 558 seeds of scarlet runners. The smallest were slightly more than 17 mm in length and the largest slightly less than 33 mm. From these measurements also, we can make a diagram. The points along the horizontal line show the dividing points between the groups (17—18 mm., 18—19, 19—20, 20—21 mm., etc.) and on every part of the line which had the same length, a rectangle was constructed. The height of the rectangle was determined by the number of seeds belonging to each definite group. In fig. 9, the sixteen groups are represented by 16 rectangles (histogram). Thus we form a notion of the classification of scarlet runners according to their seed lengths. This picture can also be shown in a practical way. For instance if we take one kilogram of commercial kidney beans as our experimental material and we

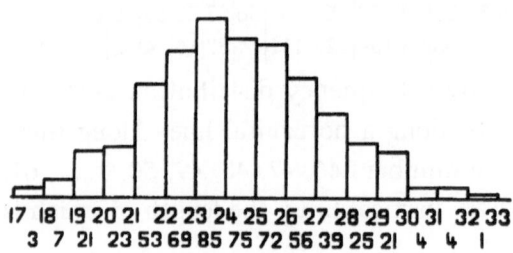

Fig. 9. Frequency distribution-curve of the lengths of seeds of scarlet runners; class intervals of 1 mm (after JOHANNSEN 1909; 1926).

spare no pains in measuring the lengths of all the seeds present in the kilogram it will be found that all these measurements are situated between the limits 6 and 16 mm. A classification into ten groups will obviously be 6—7, 7—8. 8—9 mm. etc. Fig. 10 represents a box which contains beans with the smallest measurements in the left hand compartments and the longest beans in the right hand compartments. The beans in the various compartments form columns. There exists a very obvious correlation between the length of the beans and their volumes. Therefore the shorter beans have the smallest volume for the same number and the larger beans the greatest volume for the same number of individuals. Hence the beans in the left hand compartment form shorter columns than do those in the right hand columns. Thus the curve is rather too high to the right side of the box and too low to the left, but this incorrectness is of relatively minor significance.

There are three concepts that should be considered in any discussion of the distribution curve:

1st, the *class interval* or *class range* which embodies the distance between the two successive class limits.

2nd, the *class value* which is expressed as the number or quantity of the property measured, counted or weighed.

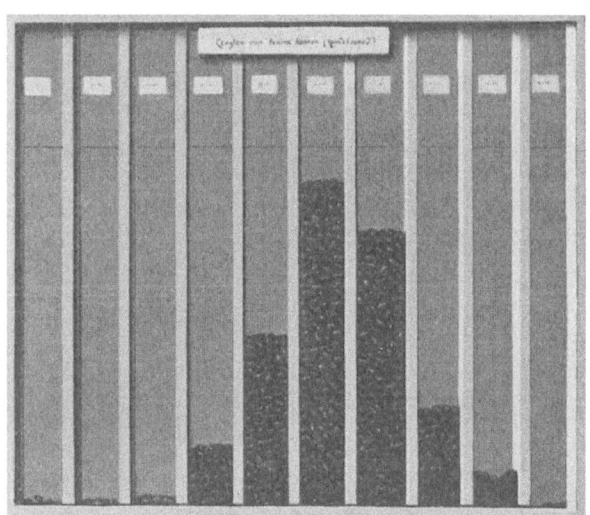

Fig. 10. Frequency distribution-curve "in natura" of the lengths of a random sample of beans; class intervals of 1 mm.

3rd, the *class number* expresses the number of individuals which belong to a definite class.

For an exact investigation into the "variability" of the material, the preparation of a distribution curve is not sufficient. For an exact determination of the variability a mathematical approach is necessary, which can be one of two methods:

1st, the *method of the ogive* including the Median and the Quartile.

2nd, the *frequency distribution curve* with its average in the sense of an Arithmetic Mean and Standard Deviation.

We will not enter into the mathematical basis of these statistical methods. We can refer for this to the books of JOHANNSEN (1909; 1926) written for biologists, and of LANG (1914, p. 204–464) and to the more mathematically coloured discussions of BERNSTEIN (1929), CHARLIER (1920), DAVENPORT and EKAS (1936), FISHER (1948) and WEBER (1935).

We will show here only in a very short and simple way how it is possible to determine the mathematical constants of a series of data by both methods.

GALTON (1889) set both methods side by side in his classical work "Natural Inheritance". In the operation of his material he restricted himself to the above mentioned *ogive-method*. The difference between the two methods can be shown in the simplest way by means of a single example. It must be assumed that these methods differ in procedure whether we are dealing with whole numbers obtained by counting the properties, or values obtained by measuring or weighing; the only point to notice is that in the case of counting, the class limits are fixed by whole numbers but in the case of weights or measures, we have to choose our class limits (i.e. our class limits will be arbitrary).

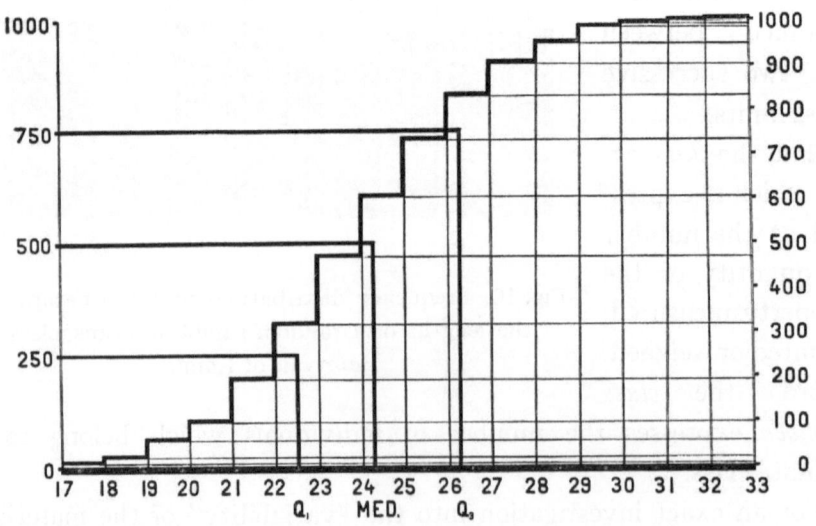

Fig. 11. Ogive of the scarlet runner seeds of figure 9.

We shall take as our example the data summarized in fig. 9, which is concerned with the lengths of the 558 seeds (JOHANNSEN, 1926). In the construction of the ogive (fig. 11), the individuals in progressive classes are successively added together. For simplification we will convert the numbers present in each class on a basis of a total of 1000 beans as opposed to the actual 558.

The classification will be as follows:

Class limits	17	18	19	20	21	22	23	24	25	26	27	28	29	30	31	32	33
Number out of 558		3	7	21	23	53	69	85	75	72	56	39	25	21	4	4	1
Number out of 1000		5	13	38	41	95	124	152	134	129	100	70	45	38	7	7	2
Sum of successive classes		5	18	56	97	192	316	468	602	731	831	901	946	984	991	998	1000

From the "*sum of classes*" it is evident that 5 seeds per 1000 are not longer than 18 mm., 18 not longer than 19 mm., 56 not longer than 20 mm., etc. In the diagram these numbers compose a characteristic distribution curve, which is known as an "*ogive*" (fig. 11). In order to indicate the course of this line by a few figures the position of three points are determined (as pointed out by GALTON). Firstly the measurement which is not exceeded by 250 promille i.e. per thousand individuals (*quartile*–1 or q_1), secondly the measurement which is not exceeded by 500 promille individuals (*median*, Med.) and thirdly, the measurement which is not exceeded by 750 promille individuals (*quartile*–3 or q_3). The three measurements can be found by interpolation; for example 192 individuals reach the 22 mm in length but 316 individuals will not pass the 23 mm. in length so that the point q_1 can be calculated by the formula:

$$q_1 = 22 + \frac{250-192}{316-192} \times 1 = 22 + \frac{58}{124} \times 1 = 22 \cdot 47$$

Similarly it can be found that

$$\text{Med.} = 24 + \frac{500-468}{602-468} \times 1 = 24 + \frac{32}{134} \times 1 = 24 \cdot 24 \text{ and}$$

$$q_3 = 26 + \frac{750-731}{831-731} \times 1 = 26 + \frac{19}{100} \times 1 = 26 \cdot 19$$

The three limits q_1, Med., and q_3 divide the distribution curve into 4 parts in each of which are situated 250 promille individuals.

If we consider the Median (24·24) to be the principal point then it is obvious that q_1 is situated at 24·24–22·47 = 1·77 mm. below it, and q_3, 26·19–24·24 = 1·95 mm., above it. From both these numbers the mean average can be derived which then indicates Q (*Quartile*) $= \dfrac{1 \cdot 77 + 1 \cdot 95}{2}$

= 1·86 which together with the Median (24·24) form characteristic points in the classification into classes.

These numbers derive their significance from the fact that half the total number of individuals is situated between the limits Med. + Q and Med. – Q so that the Quartile indicates the „*probable deviation*''. This means that you can forecast two to one (50% probability) that a distinct individual from this material is situated between the named limits. However, the ogive method has one objection which must not be underrated: it did not fix the variability outside the limits q_1 and q_3. This becomes clear if one bears in mind that the constants do not change if in the example discussed all individuals from classes 17–18 up to and including 21–22 were grouped together in class 21–22, and likewise all individuals over 27 mm, grouped in class 27–28. In this way we would find the same values for q_1, Med. and q_3, but in spite of this the variability of the material was absolutely different.

The second method of handling the data by means of the *frequency distribution curve* is that shown in figures 8 and 9. At first sight it would appear that the starting-point (*minimum*), the terminal point (*maximum*) and the top (*modus*) should be important in these curves but further consideration soon show that neither the minimum nor the modus nor the maximum has any significance. A correct interpretation leads to the determination of the arithmetical mean $\overline{x}$ and the standard deviation σ or sigma.

This method is rather more cumbersome than that of the ogive but these figures have preference over the Median and Quartile.

The average ($\overline{x}$) can be calculated in the usual manner by multiplying the number of individuals by the class values, e.g. for class 17–18 multiply 3 by the number 17·5; for 18–19 multiply 7 by the number 18·5 and so on.

The sum of these products is divided by the total number of individuals and the quotient represents the average. But this working-procedure entails a lot of multiplications. It is because of this, that in practice a simplified method of calculation is recommended. This method is as follows:

It would appear from the figures that the mean will lie between 23 and 24, because this class is represented by the greatest number of individuals. Because of this the „*provisional average*'' (A) can be fixed as 23·5. To calculate the exact average ($\overline{x}$) we consider that the beans of class 22–23 are just as far from the provisional average as the beans in class 24–25, but the first group is in a negative, the second in a positive direction. Their deviation from the provisional average can thus be expressed in the class-interval (in this case equal to 1 mm) by

—1 and +1 respectively. In the same way the deviation for classes 21–22 and 25–26 can be expressed as —2, +2 respectively.

Thus, on shifting the provisional average to real average a bean from the class 21–22 wields the same influence as a bean from the class 25–26, only in an opposite sense. If we work in this way we can set the numbers of beans which raised the provisional average against the beans which reduce the provisional average by the same amount.

So we obtain the following classification:

Class	Deviation — or +	Indiv. —	Indiv. +	more —	more +	Value of the deviation —	Value of the deviation +
23–24	0	85		—		—	
22–23 24–25	1	69	75	—	6	—	6 × 1 = 6
21–22 25–26	2	53	72	—	19	—	19 × 2 = 38
20–21 26–27	3	23	56	—	33	—	33 × 3 = 99
19–20 27–28	4	21	39	—	18	—	18 × 4 = 72
18–19 28–29	5	7	25	—	18	—	18 × 5 = 90
17–18 29–30	6	3	21	—	18	—	18 × 6 = 108
30–31	7	—	4	—	4	—	4 × 7 = 28
31–32	8	—	4	—	4	—	4 × 8 = 32
32–33	9	—	1	—	1	—	1 × 9 = 9
I	II	IIIa	IIIb	IVa	IVb		Total — 0 + 482

Total sum of the deviations + 482 — 0 = + 482.

In this table the class-intervals by which the various classes are characterized are successively indicated (I); the value of the deviations per individual from the provisional assumed average of 23·5 (II); the number of individuals which show this deviation in a negative and a positive direction (IIIA and IIIB respectively); the number of individuals which are more frequent in the negative than in the positive direction (IVa), and the converse (IVb), and the value of these deviations as the product II × IVa and II × IVb. The sum of these positive values (here = + 482) is decreased by the sum of the negative values (here = 0) and so 482 is obtained in this case as the total sum of deviations. For this purpose all individuals, i.e. 558 have played their part and the exact average $\bar{x}$ of the data amount to: the provisional average (A = 23·5) + mean deviation or

23·5 + (482 : 558) × 1 = 23·5 + 0·8638 = 24·3638 mm. $(\bar{x})$

This figure thus indicates the *real arithmetic mean* of all data. However a complete characterization of the material is not given by this average alone: for instance a number of beans, which lie only in the classes 22–23, 23–24 and 24–25, can also have a average of 24·3638 as the afore mentioned ones, which differ between 17–33. The measurements of the beans can be spread over the classes 22–23, 23–24, and 24–25 only, and still give the same average as that of the example in which beans are spread over classes 17–33. Therefore the statement of the average must be accompanied by the ,,*standard deviation*" which originated from GAUSS' theory of mathematical error. In the calculation of the standard deviation all individuals exert an influence, especially those individuals with great deviations from the average. It is characteristic of the average that the sum of the squares of all deviations from the mean is the smallest one, whereas the sum of the squares of all deviations with regard to any other number has a greater value. In other words the ,,standard deviation" (σ) can be defined as the "root of the mean square of the deviation", that means that firstly, the deviation must be fixed for every class, then the product of the number of individuals belonging to that class together with the square of the deviation must be calculated, and after this, the average of all these products must be fixed, and finally, the root of this number must be extracted. The general formula for this standard deviation is:

$$\sigma = \pm \sqrt{\frac{\Sigma p\alpha^2}{n}} = \pm \sqrt{\frac{\Sigma pa^2}{n} - b^2}$$

where:

σ = the standard deviation

p = number of individuals with the same positive or negative deviation

α = the value of the deviation of the calculated average

a = the value of the deviation of the provisional average

n = total number of individuals measured

b = the real difference between the provisional average and the calculated average (in this case 0.8638).

Grouping the data according to these principles gives this table:

Classes		Deviation	−	+	sum = p	a²	pxa²
	23–24	0		85	85	0	85 × 0 = 0
22–23	24–25	1	69	75	144	1	144 × 1 = 144
21–22	25–26	2	53	72	125	4	125 × 4 = 500
20–21	26–27	3	23	56	79	9	79 × 9 = 711
19–20	27–28	4	21	39	60	16	60 × 16 = 690
18–19	28–29	5	7	25	32	25	32 × 25 = 800
17–18	29–30	6	3	21	24	36	24 × 36 = 864
	30–31	7	—	4	4	49	4 × 49 = 196
	31–32	8	—	4	4	64	4 × 64 = 256
	32–33	9	—	1	1	81	1 × 81 = 81

$$\Sigma \ pa^2 \qquad = 4512$$

Total of individuals $\quad = 558$

$$\frac{\Sigma pa^2}{n} = \frac{4512}{558} = 8\cdot08602$$

We already found that $b = 0\cdot8638$

$$\text{thus} \frac{b^2 = 0\cdot746150}{\sigma^2 = 7\cdot339871}$$

It follows that for the standard deviation, a value of $\pm 2\cdot7095$ can be calculated.

The variability of the lengths of the seeds of the scarlet runner in the material of JOHANSNEN was characterized by both of these data $\bar{x} = 24\cdot3638$ mm and $\sigma = \pm 2\cdot7095$. Such values can be calculated for all types of material investigated.

Just as previously for the distribution of the variants between the limits Med $\pm$ Q it was found, that 50% of the variants were situated between these limits and 50% outside, so a relationship between the distribution of variates and the standard deviation can be calculated by the use of mathematics. This is possible because a ratio exists between Q and σ as given in the formula $Q : \sigma = 0\cdot6745 : 1$.

From this it can be pointed out at once that 50% of the variates are between the limits $\bar{x} = \pm 0.6745 \, \sigma$. Further calculations have now lead to the conclusion that in the ideal or binomial frequency distribution, the distribution of the variates is as follows:

Limits	Percentages of the variates between the limits	Percentage of the variates outside the limits
$\bar{x} \pm 0.5 \, \sigma$	38·30	61·70
$\bar{x} \pm 1.0 \, \sigma$	68·26	31·74
$\bar{x} \pm 1.5 \, \sigma$	86·64	13·36
$\bar{x} \pm 2.0 \, \sigma$	95·46	4·54
$\bar{x} \pm 2.5 \, \sigma$	98·76	1·24
$\bar{x} \pm 3.0 \, \sigma$	99·74	0·26
$\bar{x} \pm 3.5 \, \sigma$	99·96	0·04

Thus practically all variates are situated between the limits $\bar{x} \pm 3$ which is evident from figure 12, in which an ideal frequency distribution curve is given, in which $\bar{x} \pm \sigma, \bar{x} \pm 2 \, \sigma$, and $\bar{x} \pm 3 \, \sigma$ are marked. With regard to operating material such as this and determining the average and the standard deviation, two questions must still be considered.

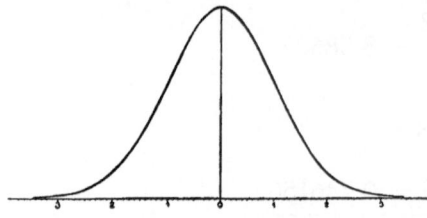

Fig. 12. Scheme of the ideal frequency distribution-curve.

Firstly, the scarlet runners were classified in classes of 1 mm, and this will not always be the case, especially if the extreme types diverge very much and one has then to recourse to classes with another interval. The given formula for the standard deviation does not hold for this new class interval. Thus, for all classes in which the class-interval is not equal to one unit, a correction must be introduced in which the calculated standard deviation is expressed in the material-unit. If, for instance the class-interval is 2 mm, the calculated standard deviation is expressed in the unit "2 mm" and subsequently the real standard deviation expressed in mm, will have twice the value of the calculated one.

Secondly, all frequency distributions in which absolute values are dealt with, e.g. those concerning the number of fin rays on flounders, can be calculated in the same way as the measurements of the scarlet runners. It is only necessary that the absolute variates are considered to be class-variates, of which the class-limits are situated between these absolute values, but only one value is possible and this is situated between the limits. For example in the frequency distribution of flounders (figure 8) we can choose as class-limits the numbers 46·5, 47·5, 48·5 etc., but it is obvious that only the values 47, 48, which are situated between the limits, really exist. Considered in this way, the operation of the material will not present any obstacles. A value of $b = + 0·671$ is found if we take the value 53 (class 52·5 — 53·5) as the provisional average.

Hence an absolute average $\bar{x} = 53·671$ and the standard deviation $\sigma = \pm 2·131$ can be calculated. It must be stressed that the choice of the provisional average cannot influence the final result. By whatever means this provisional average is fixed, the values of $\bar{x}$ and σ will always be the same.

To sum up, there are two important points to be noted when dealing with frequency-distribution:

Firstly: the Median (Med) and the Quartile (Q) are in some cases sufficient and may be preferred in view of their simplicity in calculations. The "variability" is not however sufficiently expressed by these constants.

Secondly: the average $\bar{x}$ and the standard deviation σ together give an absolutely reliable view of the nature of the data.

That which has been discussed up to now was necessary for mathematical characterization of all the data available, for further details we must refer to the more specialized books mentioned previously. But it is noteworthy that not all frequency distributions found by variability studies are constructed on the same basis. The original view of QUÉTELET was, that the frequencies of the different groups agreed with the coefficients of the binomial theorem of Newton, and that thus an absolute symmetry of distribution of individuals on both sides of the mean would be found. Indeed in many cases this exists and for all such ideal distributions the curves can be classified into a general scheme i.e. the ideal or normal frequency-distribution. However, from a later investigation in the path of GALTON and QUÉTELET it was clear that these ideal, absolutely symmetrical frequency distribution curves are not at all common and that the majority of living material gives rise to a graphical representation, which does not agree with the symmetrical distribution.

There are all kinds all possibilities of deviations: 1stly, that in which the slope of the curves is different on either side of the top, so that the top and average are not situated in the same class; in this case we call the distribution curve, *asymmetric, skew* or *parabinomial*.

2ndly: Those curves in which the distribution is truly symmetrical on both sides of the average, but the numbers of individuals in the classes close to the mean are too high, and the rest are too low in comparison to those of the ideal distribution. This aberrant curve is called *high-topped* or *hyperbinomial*.

3rdly: Those curves in which the distribution is *one-sided* (half-distribution), that is to say, from the top, the curve descends only to the right or only to the left, while on the remaining side of the top, no individuals occur in the remaining classes.

4thly: Those curves in which the distribution is a *double-half* i.e. when there are two tops, of which one is situated close to the minimum, the other close to the maximum. The number of individuals in the classes which are situated between these two tops is less than those of the minimum and maximum.

5thly: Those curves in which the distribution has two or more tops, giving the impression that this distribution is composed of two (or more) binomial curves (fig. 13). This curve is called *multi-topped*.

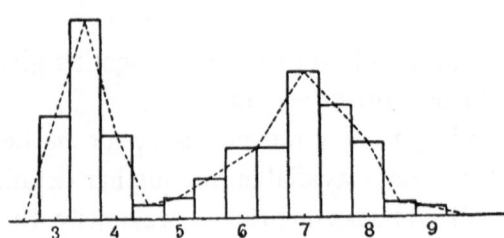

Fig. 13. Double-topped curve originating from the lengths of the pincers of earwigs *Forficula* (after BATESON 1894).

For none of these forms is the normal calculation of the average and the standard deviation valid. These distributions demand a special operation for which one must refer to handbooks on the statistics of variability. The multi-topped distribution is especially important from the point of view of the phenomena of genetics.

In addition to the variability of one distinct characteristic, we can try to find out, also by statistics, the relation which exists between two definite characteristics. That is to say, that by exact measurement of the lengths and widths of a great number of marginal flowers of *Chrysanthemum Leucanthemum*, for example, it will be possible to eventually find out the connection existing between the lengths and widths, i.e. find a *correlation*. We attain this as follows. Taking both characteristics separately from which the possible existence of a correlation must be shown, the two frequencies of the different classes are determined independently. For each of these characteristics (i.e. length- and width-distribution) a frequency will be obtained and also both the averages and standard deviations can be calculated. Every group is now subdivided again. For instance we calculate the distribution of widths for each length. In *Chrysanthemum leucanthemum* (SIRKS and BIJHOUWER 1919) the following frequencies were found by measuring all marginal flowers of an inflorescence:

Length	18	19	20	21	22	23	24	25	26	27	28	29	
Width 4	1												1
4·5		1		1									2
5						1	1						2
5·5						4	3	1					8
6						1	1	2					4
6·5							1	1	2	2			6
7										5	7	3	15
	1	1		1		6	6	4	2	7	7	3	38

In this case, the total number of marginal flowers was too small to give a really nice distribution of frequencies, but the main result is still obvious, i.e. that there exists a clear correlation between length and width of the marginal flowers in the same capitulum, that is to say that the smallest lengths are accompanied by the smallest widths and the greatest lengths by the greatest widths. This finds expression in the fact that all the numbers of the frequencies are ranged along a diagonal in the rectangle which runs from the top left to the bottom right. Such a correlation where small is correlated with small, and great with great, is called positive. If the frequency numbers lie completely on the diagonal, then the *positive correlation* is an *absolute* one but if they are more or less arranged beside the diagonal although still in the general direction of the diagonal, then we call the correlation *gradually positive*. Conversely, one can imagine a correlation in a negative sense. This is found many times. For instance where this correlation exists, the smallest lengths occur with the greatest widths and the converse (the longest lengths with the smallest widths), all individuals are found to lie on or beside the diagonal which runs from top right to bottom left, and the *correlation* is in this case a *negative* one.

A third possibility is that the numbers of individuals in the classes involved in the supposed correlation are spread over the whole area of classes with the highest numbers in that class which contains both the averages of the curves dealt with, and the number of individuals falls gradually to both sides of this class. In such a case, no correlation exists between the characteristics investigated.

If we wish to apply more exactly this method of determining the correlation, we can find distinct degrees of correlation (a correlation coefficient) with the aid of mathematics. The formula for this *correlation-coefficient* (r) is:

$$r = \frac{\Sigma p a_x a_y - n b_x b_y}{n \sigma_x \sigma_y}$$

where the letters have the same significance as previously (p. 28). But one must bear in mind that each class of the one frequency distribution (x) is now distributed over a number of classes of the other frequency distribution (y). The correlation-classes are thus characterized by two values of a (the deviation of the selected provisional average) for both the frequency distributions and therefor by the product $a_x a_y$. In each correlation-class, there exist p individuals, and the sum total of the aberrations of both the selected provisional averages is thus $\Sigma (p \times a_x \times a_y)$. Again, b_x and b_y represent the differences between the provisional averages and the true calculated average of both the frequency distributions x and y, likewise σ_x and σ_y the standard deviations. For the calculation refer to the work of JOHANNSEN (1909, 1926, p. 352 and on).

Thus the value of the correlation-coefficient expresses the degree of the correlation, $+ 1$ means an absolutely positive correlation, $+ 0.80$ a strong positive, $+ 0.20$ a very weak positive, 0 means no correlation, $- 0.80$ is a strong negative and $- 1$ an absolutely negative.

Hence the "variability" of a definite characteristic in a number of individuals can be investigated statistically in this way, and can be represented by a drawing. Also, the relation (correlation) between two properties in their variability can be recorded in a correlation scheme.

But let us now examine the biological background of this phenomenon of variability and its correlation, the data of which can be operated so precisely by mathematics. This is really the purpose of the investigation into the phenomena of heredity, that is to say not the greater or lesser importance of the data from a mathematical point of view, but the effects of the conclusions of the mathematical operations on the material on the biological field. These effects are of great biological significance.

On their own, such variability and correlation studies are nearly always insufficient to justify conclusions with respect to inheritance, because from material investigated we really do not know anything concerning the mutual background of descent. To be sure, the investigations of variability enable us to define a rather vague property by means of its variable nature and to define it by numbers, but they will not give us an insight into the heritable nature of the character. Determinations of correlation teach us, fundamentally nothing other, than, that in a number of individuals, definite measurements or numbers are related, whereas in other individuals, other measures and numbers occur simultaneously. With respect to the case where a property, as defined by its characteristics, is handed down to the off-spring, these calculations cannot teach us anything. Therefore detecting a correlation between two properties will not give us absolute evidence for existence of heredity in this property.

It is true that by mathematical methods, with the aid of the standard deviation, it can appear probable that either, the differences in averages which occur in two collections of organisms can only be due to causes which are situated outside of the organism, or there exist differences between the organisms of both groups which are of a fundamental heritable nature. But these mathematical contemplations cannot give more than an indication of probability. Therefore, if we wish to examine the biological significance concerning the heredity of such properties more closely, the statistical study of the variability or of the correlation must go hand in hand with the *scientific-genealogical* data. This means not only must we collect data from material which was procured by good luck, but also from the relations of the organisms under examination.

This can be done in two ways, preferably by deliberate breeding or cultivation, but should this be quite impossible, then by study of the genealogical tree in the full sense of the word.

The nature of the material with which we can work carries with it the choice of this method. Premeditated investigation into the heredity of plants and animals, which can be bred or cultivated intentionally in large numbers required by statistics is possible, but investigations into the heritable characters of man and the larger domestic animals must of necessity take place through the study of the genealogical tree. It follows that there are two different methods according to the material type: a mixed administrative experimental way, in which the data is obtained from individuals which were bred with deliberation and a purely administrative way, which only works with data collected at random. In those cases where we are able to choose, the material with which we have to carry out the statistical investigations is bred in such a way that the influence of the environment on all individuals is the same. This is only possible under controlled experiments and is not the case in nature. It is for this reason that experiments on wild plants and animals are so difficult, and such work must be attended by experimental breeding, if this is at all possible.

In the beginning of statistical research, in the time of Quételet and Galton, all differences between individuals which were classified in the different groups of a normal frequency-distribution curve, were put down to nutritional influences. Indeed, we can obtain a good frequency distribution if we subject material of the same heritable tendency to a different food supply. These influences of nourishment including climatological circumstances, were considered by Galton to be factor of *coincidence* (*chance*). Therefore in general, frequency distribution curves were given the name of *curves of coincidence*. But if this was correct and only differences in nourishment were considered as the cause of the origin of such distributions then the problem holds no further interest from the point of view of genetics. Then it would only be a problem of physiology to explain the binomial frequency curve or the other forms of frequency-distribution curves that occur apart from the ideal. This, however, is not the case, because by numerous experiments it turned out that the variability in the binomial as well as in the deviated distribution is partly caused by the inequality of the heritable characters of the material. Therefore we distinguish this form of variability which is called *diversity* from the variability caused by nutritional differences (*modification*). In order to illustrate the influence of the study of heredity on the development of the use of frequency-distribution curves, an investigation into the origin of the multiple topped curve undertaken by DE VRIES (1899) may be cited. He obtained a frequency-distribution curve for the number of marginal flowers of the primary capitula of *Chrysanthemum segetum* grown from commercial seed. Primary capitula are those which develop first on each plant. The following result was obtained:

No. of marginal flowers	12	13	14	15	16	17	18	19	20	21	22
No. of individuals		1	14	13	4	6	9	7	10	12	20	1

Thus there exists a definite double-number on which the tops are situated at the points 13 and 21. This means that the capitula with 13 and 21 marginal flowers are the most frequent.

From this material plants were selected which had 13 and 21 marginal flowers. The seeds of these plants gave an offspring which we can classify according to the number of marginal flowers on the capitula once again:

No of marg. flowers:	9	10	11	12		13	14	15	16	17	18	19	20		21	22	23	24	25	26	27
descendants from 13:	1	3	8	32	221	50	8	5	4	3	1	2	—	—	—	—	—	—	—	—
descendants from 21:	—	—	—	—	—	1	3	0	3	7	14	43	142	43	21	11	5	3	1		

It is clear from this, that fundamental differences exist between the plants with 13 and those with 21 marginal flowers in the original seed, and that these differences did not bear any relation to the nutritive circumstances. The difference was one of a heritable nature; the initial two-topped curve was the consequence of the fact that the seed was a mixture of plants which differed individually in heritable tendency in the number of their marginal flowers.

This was not only found to be the case with a two-topped curve, obtained from commercial seed of a cultivated species. Hence it was evident that in many other cases the cause of the two-and more-topped curves, which were obtained from wild growing plants or in animals occurring in nature, could be found to lie in the fact that the groups of organisms investigated were heterogeneous in a heritable sense.

It is now clear, that to a large extent, differences in genotype (heritable tendency) are responsible for many-topped curves. But how far does this hold for other forms of the frequency distributions? High-topped and half-curves are in general of a physiological nature. The development of the organs concerned, sometimes entails that the precise measurement or a definite number occurs frequently. It is known that the number of marginal flowers of the Compositae almost agree with one of the numbers of the progression of Fibonacci (1, 2, 3, 5, 8, 13, 21, 34, 55 etc.). Skew-curves can be caused by technical errors, and also by physiological circumstances, for instance, if the material measured is not fully grown. On the other hand, in such curves, differences in heritable tendency between the investigated individuals can operate. Even the most regular ideal frequency distribution is often misleading in this respect. As was already mentioned above, it was formerly generally accepted, that the differences between the individuals which together gave such a normal distribution curve was due only to nutritional influences. We are indebted to JOHANNSEN (1903) for proof that this is not absolutely the case. In 1900 he began cultivation experiments with kidney beans from which he made a determination of the frequencies of the different classes by weighing in milligrams, and by measuring the lengths and widths in millimetres. Starting with 8 kilograms of kidney beans (nearly 16000 seeds), the average weight per bean was first determined (495 mg). Next a number of beans was selected out of this, e.g. 25 very small ones and 25 very big ones, which were sown in 1901 and the seeds separately harvested from these plants after self-fertilization were again measured individually and weighed and again sown. This was continued in 1902 and then it was evident that the families descended respectively from one single bean selected in 1900 individually differed considerably in average and in frequency of the different classes, whereas the total of all these families counted together, still yielded an absolutely regular frequency distribution. This is obvious from the table borrowed from the work of JOHANNSEN (1903 p. 25). The italicized printed numbers indicate the groups in which the average is situated.

Thus all the families counted together, yield an absolutely normal frequency distribution and also every family considered separately yields a normal frequency distribution. Differences in heritable tendency, with respect to the weight of the seeds, has been proved to exist between families. By comparison in the table, family A for instance has a maximum in the class 60–65 centigrams with an average of 641·9; family O on the other hand has its highest point in the group 35–40 centigrams with an average of 351·3.

Groups concerning the weight

	10	15	20	25	30	35	40	45	50	55	60	65	70	75	80	85	90	Average
Family A	—	—	—	—	—	2	5	9	14	21	22	24	23	17	6	2	—	641·9
B	—	—	—	1	6	19	32	·66	88	100	90	50	19	1	3	—	—	557·9
C	—	—	—	—	—	5	14	50	76	58	44	29	5	1	—	—	—	554·3
D	—	—	—	5	2	9	21	38	68	77	62	22	3	—	—	—	—	547·6
E	—	—	—	4	1	12	29	62	65	57	19	6	—	—	—	—	—	511·9
F	—	—	—	2	8	21	46	74	46	28	14	1	1	—	—	—	—	481·8
G	—	—	3	9	28	51	111	174	101	44	6	—	1	5	—	—	—	465·1
H	—	—	1	6	20	60	106	114	75	33	3	—	—	—	—	—	—	455·3
J	—	1	2	14	38	104	172	179	140	53	9	—	—	—	—	—	—	454·4
K	—	—	1	2	6	31	55	55	28	6	4	—	—	—	—	—	—	449·5
L	—	—	1	5	15	37	88	76	33	13	4	1	—	—	—	—	—	446·2
M			4	9	26	56	82	76	32	9	1	—	—	—	—	—	—	428·4
N	1	3	11	22	29	72	120	69	23	5	2	—	—	—	—	—	—	407·8
O	4	4	5	19	69	69	44	5	—	—	—	—	—	—	—	—	—	351·3
P	—	—	—	3	1	18	35	27	13	3	4	2	—	—	—	—	—	452·8
Q	—	—	1	2	7	16	44	93	80	52	10	—	—	—	—	—	—	492·0
R	—	—	—	2	3	12	17	27	19	3	—	—	—	—	—	—	—	455·1
S	—	—	1	2	3	8	27	47	37	30	4	—	—	—	—	—	—	488·8
T	—	—	—	—	1	6	20	37	39	30	8	—	—	—	—	—	—	506·2
All families together	5	8	30	107	263	608	1068	1278	977	622	306	135	52	24	9	2		

Parallel results appear in other experiments of JOHANNSEN concerning the length of the seeds. The curves of four families A–D were considered individually and are shown in figure 14 (borrowed from JOHANNSEN 1915, p. 609). In the bottom figure, the total of these four families is given. The mixture i.e. total of A–D of seeds in which some differ genotypically from others as well as each family A, B, C and D, obtained by selection, show an ideal frequency distribution. The mixture is a composition of the other four and thus still consists of individuals which have different genotypical tendencies, although the form of the curve agrees with the ideal and does not seem to justify any presumption of heterogeneity of the material.

With that goes a second possibility in which an ideal frequency-distribution curve can be found. That is to say, by the summation of a number of smaller frequency distribution curves a binomial curve can be obtained. Apart from this possibility there is another which we called "coincidence" caused by the combination of factors of the environment. We will discuss a third possibility later on.

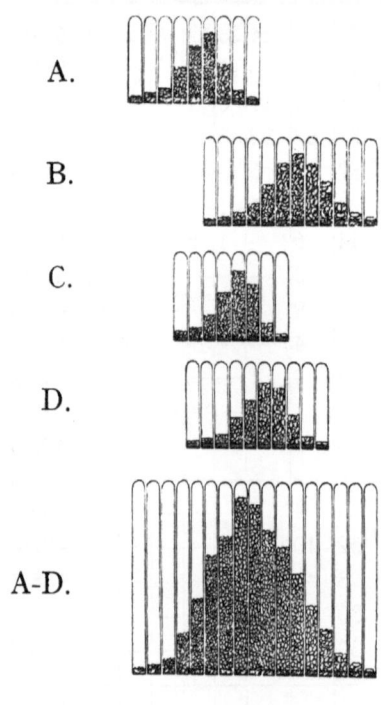

A.

B.

C.

D.

A-D.

Fig. 14. (after JOHANN-SEN, 1915).

The mixture of different types which usually occur in commercial seed and certainly in wild groups of organisms, are called *populations* by JOHANNSEN, and the families selected from these, which consist of individuals with a similar tendency, are called *pure lines* (reine Linien). Of course, he has given different definitions of this last conception over the course of the years. In 1903 (p. 9) he wrote: "As a pure line I take together those individuals which originate from only one self-fertilized individual". In 1915 (p. 606): "A pure line is the whole of all the descendants of one absolutely self-fertilized individual which does not have a hybrid nature". This however will not influence the fundamental importance of the "*pure line*" as opposed the "*population*".

It is now certain that similarity and ideal form of a frequency-distribution curve originating from one of the populations, may have an apparent simplicity, however even such an ideal curve may still be a summation of a number of different frequency-distribution curves.

After the statistical method we come to a purely administrative one which is a necessary evil; necessary because the material under investigation admits of no other operation and an evil, because the data on hand is always more or less indefinite and incomplete. Moreover, it is impossible to control the data since it is borrowed from ancestors which are dead. Nevertheless, this method, providing it is applied well, is of great use and will be used to a wider extent in the future provided that a start is made now in the administration of the data in a systematic manner. The research into the heredity of distinct observable characteristics can be carried out in two ways: firstly, by consideration of one or more organisms in relation to all their descendants. From this latter, a *table of progeny* will result. Secondly, by considering the organism as a descendant from a number of ancestors, so that a *table of ancestry* can be obtained. The *progeny table* shows us the relationship between a definite individual or pair of individuals and all their descendants: the procedure by which this is usually carried out is complete in only two respects. Too often mention is made in the pedigree of the descent in the direct and masculine line only, and this is not admissible from a biological point of view (HAECKER 1917), since heritable characters are transmitted to descendants by both parents. Every individual originating by sexual reproduction, owes its constitution of heritable characters to two individuals, father and mother (with the exception of self-fertile plants only). Hence for an accurate pedigree the female line must also be inserted. The second incompleteness is that, although a certain characteristic (as will be shown later on) can appear to be heritable, nevertheless, it is possible that this characteristic can be missing in a number of generations of direct descendants and that these characters can still be present in the material under investigation. This presence reveals itself occasionally in the brothers or sisters of the direct ancestors i.e. uncles, aunts, great uncles, great aunts etc. Therefore the genealogical tree (the pedigree) must extend in this direction if this method is to be useful in scientific research. VAN BEMMELEN (1939) used this method of research with considerable success in cases where the character could be observed in successive generations.

The first objection to the pedigree does not apply to the *table of ancestry* because all direct ancestors (including the female ones) are included. This method shows all these forebears equally to better advantage. The second point (that is missing out of indirect relatives) is however applicable to the table of ancestry as used in genealogy. There exists one further objection to this type of table. If we did not know better, it would seem plausible that, everyone who worked with a table of ancestry as material for an investigation into heredity, would be inclined to equalize the heritable tendency of individuals, which have the same ancestry (i.e. all brothers and sisters). That this conception, which originates from GALTON, is incompatible with the present day view, which is based on MENDEL's work, will be fully explained in Chapter III.

Thus we see that both the table of progeny and the table of ancestry in their original naive form are useless for research into heredity. Therefore they must be adjusted in order to reach the agreement which led to a combination of both these research methods which are now united into one method of procedure: pedigree research. Figure 15 gives a picture of a *pedigree*, as it must be for scientific use, according to the International Congresses of Eugenics. Above all it is desirable to have a uniform and exact symbolism for the successive generations and indi- viduals which belong to these generations. This was proposed by TAMMES (1930). We can see from this, that the progeny table is no longer in opposition to the table of ancestry in their more elaborate form, but are now woven together.

SCHEME OF A PEDIGREE

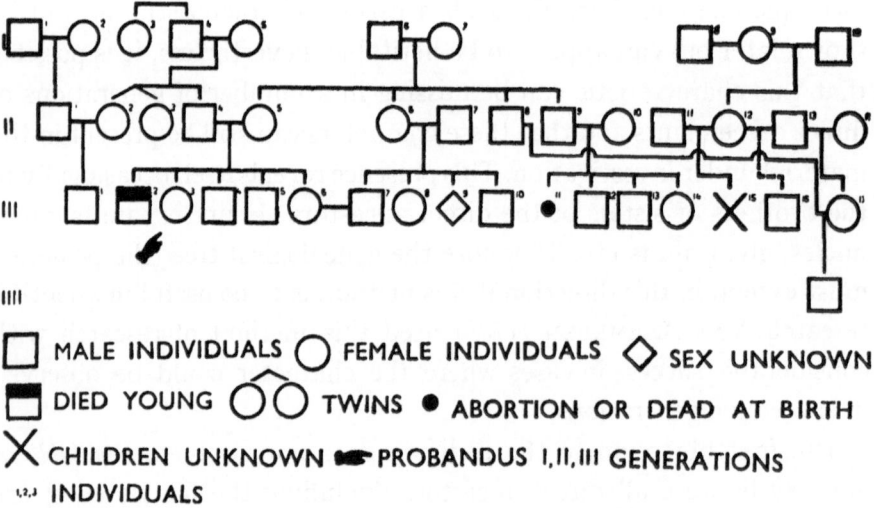

MALE INDIVIDUALS ○ FEMALE INDIVIDUALS ◇ SEX UNKNOWN

DIED YOUNG ◯◯ TWINS ● ABORTION OR DEAD AT BIRTH

X CHILDREN UNKNOWN ➤ PROBANDUS I, II, III GENERATIONS

ᵘ·ʲ INDIVIDUALS

Fig. 15. (simplified after VAN HERWERDEN and LAUGHLIN, 1927).

Despite the conscientiousness with which the administration of these genealogical studies is carried out, and the objectivity with which the judgment and description of the individuals is formed, objections still arise in connection with the study of the pedigree, which can be of great importance. Because of this we have to accept the results with some reserve, not so much with respect to the question as to whether or no, one property has an underlying heritable basis, but more a question of how the inheritance takes place. The data concerning ancestors is in most cases hard to obtain, and even then it is unreliable. It would be possible to rectify this eventually if, from now on, records were made and kept for us in the form of archives over a long period of time.

The impossibility of obtaining very accurate data is more serious (that is to say, data excluding all modifications). Material which is not intentionally bred always gives phenotypes which are unreliable as a measure of judgement for the genotypes which are fundamental to these. This objection to modification is of less importance in the two other methods of approach which came to the front in recent years, as a completion of the pedigree research method, viz. the study of identical twins and the comparison between special sections of population, in which many consanguinous marriages had taken place, and the population as a whole.

Identical twins (one egg twins) owe their origin to a splitting of either egg (or embryo) into two parts, soon after the process of fertilization. Each of these parts then continues through a normal development until it forms a fully-grown individual. These twins thus originate from one egg and from the point of view of heredity, these twins are considered to be identical, that is to say, there are good reasons that the postulation can be held that the heritable tendencies of both these twins is absolutely similar. Identical twins are either two boys or two girls. We will describe this phenomena more extensively later (Chapter XVIII). Such identical twins usually are referred to by IT. By far the larger number of twins are *fraternal twins, (two-egg-twins)* who owe their origin to the fact, that two separate eggs were fertilized at the same time, each by one spermatozoon. Fraternal twins can be of the same sex, and are then considered as fraternal twins in the limited sense of the word (FT). They can however be of a different sex (boy and girl) and are then called *couple twins* (CT).

These three types open up possibilities for investigating the phenomena of genetics. The possibility has two sides. The first, which was pointed out by SIEMENS (1924) and DAHLBERG (1926) and later specially developed by VON VERSCHUER (1932, 1937) and now put into active practice by many investigators, is of a statistical nature (see LOTZE, 1937; LUXENBURGER, 1940). The starting point is a number of identical twins who are compared with the same number of FT's. Comparison with couple-twins could perhaps give an unjust conclusion as a result of sex-differences. If a distinct character is investigated in a hundred cases, and we find *concordance* in ninety twins between the twin partners, and in ten cases there is *discordance*, whereas from the hundred FT's there are perhaps ten concordant and ninety discordant. In this case, the strong concordance between the IT's and the slender concordance between the FT's for this character gives the right to consider the character as being highly heritable. Moreover, we can gain some idea from the percentages of concordance and discordance, of the influencing power of the environment. The greater the concordance between the IT's, the stronger the heritable element which results in the character; the smaller the concordance the more important the environmental or peristatic influence.

This proportion between the influence of the heritable tendency and the influence of the environment can be even better investigated along the second path into which the research of twins turned, i.e. the comparison of identical twins, who have grown up under similar circumstances, with identical twins who had a different education and secondly with non-identical (FT) pairs, who have had the same environment in youth. From the data obtained by this, a ratio can be fixed for a number of body characteristics. This ratio of environmental influences: heritable influences, is called the *weighing up*. For instance, VON VERSCHUER gives a valuation of the ratio in which the environment is always fixed as 1), for body weight 1 : 2, for chest width 1 : 2 · 4, for body length 1 : 10 · 4, and for cranium length 1 : 5·6.

The cases of identical twins who are subjected to different environments are extremely important. It is possible to trace the influence of gymnastics on one twin, from a pair in which one practices athletics and the other does not; or the influence of a profession in a case of two IT brothers who are butcher and carpenter respectively. But it is also possible to fix in this way, the influence of the environment on characters which are difficult to analyze, but which are still extremely important, such as intellect and nature.

Pioneer work was carried out in relation to this by NEWMAN, FREEMAN and HOLZINGER (1937), who examined nineteen pairs of identical twins who had been raised separately. Their conclusion was that the difference in environment was so important, that the heritable tendencies could be repressed to a great extent. The extreme difficulty of

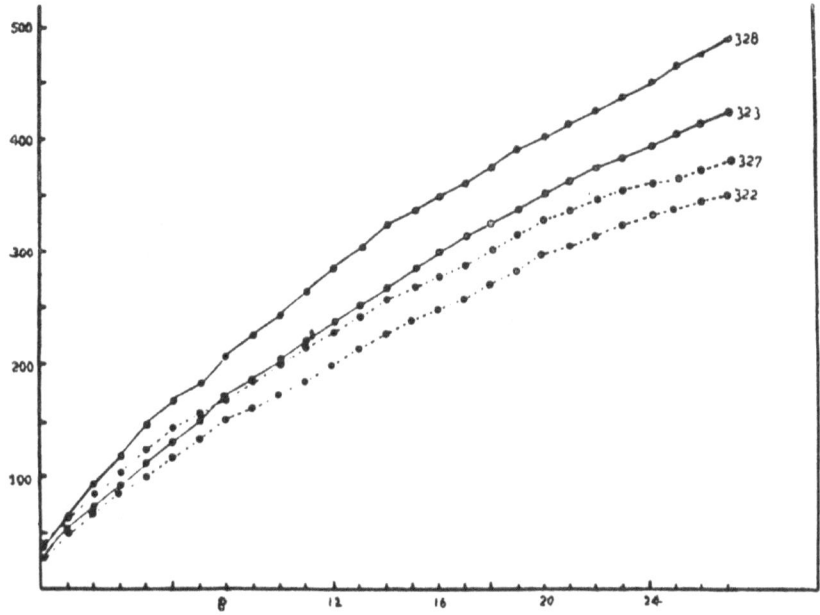

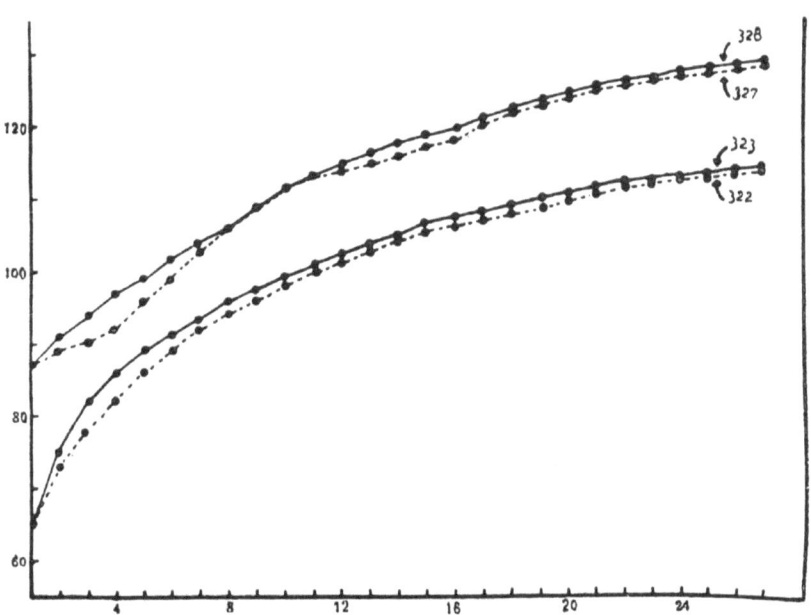

Fig. 16 The influence of nourishment upon body weight and shoulder height in twin calves (after BONNIER and HANSSON, 1948).

making an objective and unprejudiced conclusion in such subjects is shown in this work. NEWMAN who as a biologist, allows himself to be affected by FREEMAN and HOLZINGER, who were both pedagogues, in such a way, that some years later he changed his point of view (GARD-NER and NEWMAN, 1940) absolutely and states that a great rôle is played by heredity. Indeed strong objections to the mathematical calculations and their interpretation from the data of 1937 can be made. But this does not alter the fact, that the work of NEWMAN and his colleagues is a nice example. The principle of their method of investigation is of great significance, but their manner of operating the data is far from faultless. The future will undoubtedly show that the method introduced by these investigators will lead to an excellent result.

The significance of twin research was very clearly revealed in the studies of G. BONNIER and A. HANSSON (1948) concerning cattle twins (figure 16). In two identical twins (322 and 323; 327 and 328) the numbers 322 and 327 were poorly fed, their partners 323 and 328 were fed on rich diet. The result regarding the body weight, was that number 323 was much heavier than his twin-partner 322 and similarly, 328 was much heavier than 327 (solely due to the influence of the food), whereas there was a recognizable difference between 328 and 323 on the one hand and 327 and 322 on the other (figure 16a). For the height of the shoulders also, the heritable differences came clearly to light, but no difference caused by good and bad diets was observed (figure 16b).

The other path along which statistical research is led today is one of comparison of the frequency of a character in an affected family belonging to a well described portion of the population, in which many *blood-related weddings*, or *consanguineous marriages* (biologically called: *inbreeding*) have taken place, with the frequency of the character in a large population. The theoretical basis of this method was laid down by HULTKRANZ and DAHLBERG (1927). It is a matter of course, that the frequency of a heritable character in a family in which these characters appear will be greater than that of the same character in the whole population of the country. But it is not so obvious that these frequencies can differ in different groups of the population. In every population there are known areas in which consanguineous marriages are more frequent than in the average population. A definite heritable characteristic usually comes very strongly to the front and the frequency of this characteristic increases in this limited section of the population as can be calculated mathematically (see Chapter XX).

Conversely it is also permissible to conclude that the frequent appearance of a character in an inbred group of individuals, compared with the presence of the same character within the whole population is evidence that these characteristics have a heritable basis. An investigation of POLMAN (1950) concerning *anencephaly* has shown very clearly indeed the significance of this method. For further details concerning the application of the statistical methods in genetical experiments we may refer to the literature of GEPPERT and KOLLER 1938, KEMP 1942, KOLLER 1940 a and b, SCHULZ 1936, SANDERS 1941, in which human heredity is discussed.

From the biological point of view we can summarize the significance of statistics for the genetical investigation as follows: Statistical methods enable us to give a description of experimental material by means of the mathematical calculation applied to this material. The description can be made by means of frequency-distribution curves, which form a diagrammatic representation. This is the limit of such statistical experiments into variability. It is a good thing that we do not overrate the exactitude of these methods. Their results are only valuable to pure genetics if the statistical investigations are combined with methods of biological origin, such as descent investigations, investigations into identical and fraternal twins, or with studies concerning the biological composition of inbred populations and the average population. Only by this will statistical work attain significance. However, it will always be the basis of heritable contemplations and is always necessary in order to draw any conclusions regarding the material. It will also always appear inferior compared with experiments in which organisms and their offspring, observed and examined accurately, can be judged sharply for their heredity and their response to environmental influences, through deliberate breeding experiments.

CHAPTER III

PREMENDELIAN THEORIES OF GENETICS

Genetics, which is a branch of biology, developed out of physiology and is a science which observes phenomena in the lives of plants and animals, especially the phenomena of reproductive physiology. Out of the conceptions formed, arose the first notions concerning heredity. It is clear that the physiologists' point of view formed the basis for the ideas concerning genetics. Therefore, revolutions and evolutions which the science of reproduction and fertilization passed through were also shocks and crises for the development of genetics. Hence we find a parallel in both developments; the sharp contrast between preformation and epigenetical theories, which were, in the time of LEEUWENHOEK, in the foreground of the reproductive theories, seem to be driven back, only to reappear again with new energy. Through the centuries we can follow like a red thread, the same contrast between *preformation and epigenetic theories* through all the concepts of heredity, though they take a rather different form.

The advocates of these theories concerning the physiology of ferti-lization fought against another through the centuries (see SIRKS 1942a). The preformatist supposed that he saw, in the material from which the new organism develops after fertilization, all the organs and all the parts of the new organism, already (i.e. previously) formed, that is to say, preformed. This implies that the whole development from the moment of fertilization onwards becomes a real "development" i.e. an unfolding, an evolution of already long determined forms and organs. This preformation could be situated in the material, handed over by the male parental organism. Among the many *"animalculists"* who advocated this we will name only LEEUWENHOEK and BOERHAVE (Dutch scientists) who held a place of importance, whereas SWAMMER-DAM and other famous investigators belonged to the *"ovists"*, who believed that the preformed state must be found in the egg, that is to say, in a part of the mother.

Both antagonistic groups carried this problem to an extreme, the conceptions of the preformatists led to foolishness; for example, the pictures of spermatozoa in old books, showed a whole human organism of microscopical measurement within the spermatozoa, or the calculation of MALEBRANCHE as to how many human ova must have been present in the ovary of Eve, illustrate the ridiculous lengths to which a defence of a point of view can go.

The reaction against these conceptions, which were carried too far by the preformatists, was introduced by BUFFON and reached their culminating point through the investigations of C. F. WOLFF, who tried, in his Theoria Generationis (1759) to replace the theory of the preformatists by a theory of epigenesis. According to WOLFF, nothing is preformed in the reproductive cells. At the moment of fertilization nothing has any structure. Firstly, this inorganic material gradually changes into organic material. This is brought about by the development of vesicles and blood vessels. The cause of this translation into organic material was due to a *"vis essentialis"* of which no description was offered.

A greater contrast than that which existed between these original preformation and epigenetical schools in the science of reproduction is unimaginable, and it stands to reason that in their extreme form, both are inaccurate. The modern fertilization-physiology lies between them. It is clear that one investigator leans more to one side, the other to the opposite side.

We will also meet such antipodes of conceptions in genetics. In the most primitive conceptions we will already find expression by ARISTOTELES and HIPPOKRATES, who represented the most extreme points of view.

It was also of great merit to JOHANNSEN, that he pointed out that two fundamental different theories, which still have their advocates today and their opponents, can be traced back in primitive Greek literature (see JOHANNSEN 1917 a and b). HIPPOKRATES believed that it would be possible to explain heredity of body properties by supposing that every body-cell, down to the smallest, gives off a germule, and all these germules together congregate into one central body, viz., that which is handed over by the parents. ARISTOTELES visualized the

4

course of events in another way. The body will be formed from a central "mass of seed" (in which all parts of the body are represented to be in the course of construction) which can grow by means of uptake and translation of food from the outside.

From the construction of the parts of this body, not all the "seed" available is necessary, so that a remnant remains and this remnant is in its turn, the origin of the next generation. Just as a painter uses a little of all his paints on his pallet for painting a picture, so ARISTOTELES believes that he can explain the cause of similarity between parents and children as both originating from the same material which is handed over unchanged from generation to generation.

The fundamental difference between both conceptions is clearly demonstrated by a picture borrowed from JOHANNSEN (figure 17).

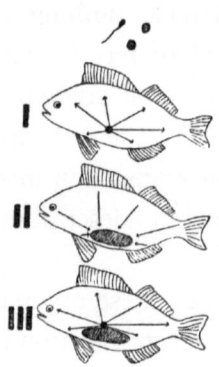

Fig. 17.
I. Development of an organism from the fertilized egg.
II. The view of HIPPOKRATES.
III. The conception of Aristoteles. (from JOHANNSEN, 1917b).

Every living organism which has sexual reproduction originates from only one ovum, which is penetrated by a spermatozoon. Both cells fuse wholly together and from this fertilized egg, the whole body of animal and plant develops. This is represented in figure 17[1], where the fertilized egg gives rise to the body of a fish. Supposing that this fish makes its own reproductive cells (spermatozoa or ova) then we have to accept the fact that the heritable characters which are handed over from father and mother to child are represented in one or another form in the reproductive cells of the child and hence must have been introduced there in some way or other, or must have been present in the beginning. The first conception we could call epigenetic however, this conception is still not quite comparable to the epigenetical point of view in the fertilization-physiology. It is the Hippocratical one, which is represented in figure 17[11]; all parts of the body give off germules which . accumulate in the blood stream and are transported by the blood to the reproductive organs. There exists another opposing point of view, viz., that from the beginning, all heritable characters are present in the reproductive organs. The reproductive cells of all successive generations form a continuous series and the body develops anew from a part of this heritable material in each generation. In this material the characters are preformed. This representation is Aristotelian or preformatic (fig. 17[111]). Then for centuries this problem of inheritance remained in the background. Other problems took its place until in the medical literature of the early nineteenth century, the interest in the phenomena of inheritance re-awakened. In his handbook of human anatomy (1812), J. F. MECKEL gives an enumeration of aberrations in the human body, which occurred in different successive generations. HOFACKER compiled numerous heritable characteristics known to him in man and animals in 1828, with an emphasis on horse-breeding. In 1847, PROSPER LUCAS published a detailed study concerning the physiological and psychological aspects of the problems of inheritance, with special detail given to nervous affections. By means of these investigations, a lot of material was compiled but a classification of the views concerning the process of heredity was not made and a theory based on these facts was not forthcoming.

As in many other cases, the tide of interest turned towards the biological sciences. As the first of a series of authors, HERBERT SPENCER gave his opinion in his "Principles of Biology" (1864) over the nature of heredity based on observation. SPENCER also tried to include all groups of phenomena of biology as a part of a system of synthetic philosophy.

One of these groups is thus heredity, the cause of which is sought by him in the following direction. In every organism, the total of heritable properties is determined by distinct units. The question now becomes: "What is the nature of these units". To say they are "chemical units" is not a helpful explanation. It is not possible to explain the variation of organisms by this means. "Morphological units" (cells) may not be taken into consideration because the multi-cell organisms take their specific structure from the cooperation of all the cells which take part in the building up of the organism.

That is to say, that a single cell which does not cooperate with another, will not cause a specific structure. SPENCER supposed the "physiological unit" to be the only acceptable unit in which heritable characters of an organism could be located. This unit is defined as follows: "There seems to be no alternative but to suppose that the chemical units combine into units immensely more complex than themselves, complex as they are; and that in each organism the physiological units produced by this further compounding of highly compound molecules, have a more or less distinctive character. We must conclude that in each case some difference of composition in the units, or of arrangement in their components, leading to some difference in their mutual play of forces, produces a difference in the form which the aggregate of them assumes" (1864, 1898, p. 226).

Thus every organism is characterized by a complex of chemicals which form together a physiological unit. This physiological unit is responsible for the heritable character of the organism. This unit is handed over from the parental organisms to the offspring by the reproductive cells, which only act in the role of carrier: "And here the assumption to which we seem to be driven by the ensemble of evidence, is, that sperm-cells and germ-cells are essentially nothing more than vehicles in which are contained, small groups of the physiological units in a fit state for obeying their proclivities towards the structural arrangement of the species they belong to" (1864; 1898, p. 317).

The question now arises as to whether this physiological unit (according to SPENCER) is an invariable one, a whole one, or whether changes which the outside of the organism undergoes, as a result of the effect of the environment, react on the physiological unit. In other words, is the theory of the physiological unit strictly preformatic or is there still room for more or less epigenetically coloured contemplations. Without any doubt, SPENCER's conception is one of a strict epigenetical nature and according to his opinion, the physiological unit varies under the influence of the environment. In his own words: "Or, bringing the question to its ultimate and simplest form, we may say that as, on the one hand, physiological units will, because of their special polarities, build themselves into an organism of a special structure; so, on the other hand, if the structure of this organism is modified by modified function, it will impress some corresponding modification on the structure and polarities of its units. The units and the aggregate must act and re-act on each other. If nothing prevents it the units will mould the aggregate into a form in equilibrium with their pre-existing polarities. If, contrariwise, the aggregate is made by incident actions to take a new form, its forces must tend to re-mould the units into harmony with this new form" (1864; 1898 p. 319).

Thus it was SPENCER who supposed that there exists in the very beginning of the development, a set of physiological units which under normal circumstances are able to build up a body which is of predetermined shape. Under abnormal circumstances, the body will obtain an other form than that previously determined. The form will be embodied in the physiological units and will be transmitted to the offspring. The theory is thus one of partial preformation i.e. that the heritable tendency of an organism can still change later on. Some years later, DARWIN drew up in his "Variation of Animals and Plants" a more elaborate theory of heredity based on observations in nature and in cultures which were carefully evaluated according to the standards of his day. From the data compiled, he drew up five "Laws of heredity" which he had already put forward vaguely (p. 12) in his "Origin of Species" (1859), but which he now formulated positively, summarized as follows: "Firstly, a tendency in every character, new and old, to be transmitted by seminal and bud generation, though often counteracted by various known and unknown causes. Secondly, reversion or atavism which depends on transmission and development being distinct powers: it acts in various degrees and manners through both seminal and bud generation. Thirdly: prepotency of transmission, which may be confined to one sex, or be common to both sexes (1868: of the prepotent form). Fourthly, transmission, (1888: as) limited by sex, generally to the same sex in which the inherited character first appeared; (1888: and this in many, probably most cases depends on the new character having first appeared at a rather late period of life). Fifthly, inheritance at corresponding periods of life, with some tendency to the earlier development of the inherited character" (1868, II, p. 84; 1888, II, p. 61).

To DARWIN these were fixed rules, to which heredity was subjected and he tried to find an explanation for these rules by means of his well-known *hypothesis of the pangenesis*, which he offered to his co-investigators in the same publication for discussion, but he still considered his hypothesis as an interim-one. He also acknowledges this hypothesis to be composed of a number of suppositions: "It is (edition of 1868: almost) universally admitted, that (edition of 1888: the) cells or (1868: the) units of the body (1868: propagate themselves, 1888: increase) by self-division or proliferation, retaining the same nature, and (1868: ultimately becoming; 1888: that they ultimately become) converted into the various tissues and substances of the body. But besides this means of increase I assume that (1868: cells, before their conversion into completely passive or "formed material"; 1888: the units) throw off minute granules (1868: or atoms) which (1868: circulate freely; 1888 are dispersed) throughout the (1888: whole) system; (1868: and; 1888: that these) when supplied with proper nutriment, multiply by self-division, (1868: subsequently becoming; 1888: and are ultimately) developed into (1868: cells; 1888: units) like those from which they were (1888: originally) derived. These granules (1868: for the sake of distinctness) may be called (1868: cell-gemmules, or as the cellular theory is not fully established, simply) gemmules. (1868: They are supposed to be transmitted from the parents to the offspring, and are generally developed in the generation which immediately succeeds; 1888: They are collected from all parts of the system to constitute the sexual elements and their development in the next generation forms a new being); but (1868: are often transmitted; 1888: they are likewise capable of transmission) in a dormant state (1868: during many; 1888: to future) generations and (1868: are then; 1888: may then be) developed. Their development (1868: is supposed to depend; 1888: depends) on their union with other partially developed (1888: or nascent) cells (1868: or gemmules) which precede them in the regular course of growth. Why I use the term union, will be seen when we discuss the direct action of pollen on the tissues of the mother-plant. Gemmules are supposed to be thrown off by every (1868: cell or) unit, not only during the adult state, but during (1868: all the stages; 1888 each stage) of development (1888: of every organism; but not necessarily during the continued existence of the same unit). Lastly, I assume that the gemmules in their dormant state, have a mutual affinity for each other, leading to their aggregation (1868: either) into buds or into sexual elements. Hence, (1868: speaking strictly), it is not the reproductive (1868: elements, nor the; 1888: organs or) buds which generate new organisms, but (1868: the cells themselves throughout the body; 1888: the units of which each individual is composed)" (1868 II p. 374; 1888 II p. 369-370).

Hence this pangenesis-theory is a whole construction based on suppositions and it is only due to the known facts of DARWIN's time that DARWIN was not very fortunate in his conclusions. In two respects his conclusions differ from the views of SPENCER; physiological units were supposed to be of the same sort, whereas DARWIN's *gemmules* (or *gemmulae*) were considered to be very different in origin according to the organ of the body which handed them over (see SPENCER 1898 p. 356). Besides this, in SPENCER's theory there is no discussion over the transport of these units from the body-cells to the reproductive organs. For the same reason this is also the weakest point in DARWIN's complex of hypotheses. His supposition did not find favour in the eyes of his successors.

DARWIN considers nothing to be preformed in the reproductive cell. Everything comes from the outside and is stored solely in the reproductive cell ready to be transmitted to later generations. Of all the premendelian contemplations concerning heredity, this theory of DARWIN has the greatest tendency towards the epigenetical side. It is in the supposition of the transport-possibility of gemmulae that there is room for all possible kinds of fantasy.

Nevertheless, the *pangenesis-hypothesis* of DARWIN has been of a greater historical importance than the theory of SPENCER, because it formed the first working out of the suppositions which could be traced in a more indefinite form in the works of HIPPOKRATES and more especially because this theory has attracted more general attention and hence became the starting point for other investigations which introduced changes of varying degrees in DARWIN's explanation.

It was possible to obtain improvements in DARWIN's pangenesis-hypothesis by analysing it into different parts as a separate problem. Immediately one of these points attracted attention, this point was that of the transportability of the pangenes. The majority took sides against this because in this the main point against the pangenesis-theory was rightly sought. Not every attack was equally strong. In the years following 1870, the principal fight was led by DE VRIES (1889), GALTON (1878, 1889) and WEISMANN (1883 etc.).

These scientists expressed practically the whole range of different standpoints, which existed between DARWIN and the afore-mentioned investigators, but they ranged themselves totally on the side of the preformatists. In 1889, DE VRIES gave an *intra-cellular* instead of DARWIN's *inter-cellular pangenesis*. DARWIN supposed that the transport of gemmulae took place from the body-cell to the reproductive cells; DE VRIES denies this absolutely, but he believes that the pangenes can move only from the nucleus to the protoplasm and that they are able to maintain themselves in this medium. Hence there does not exist any possibility of handing over changes in the body-cells, which originate from a changed environment, to the offspring. The intracellular pangenesis of DE VRIES is more a hypothesis, explaining the differentiation of the body-cells than a theory of heredity. The more so, since DE VRIES ranged himself on the side of WEISMANN, according to heredity.

More fundamental theories of heredity than the intracellular pangenesis had already been given by GALTON years previously, more or less contemporaneously with DARWIN, hence he was also the first important opponent to the darwinistic pangenesis-hypothesis. GALTON at once saw the great difficulties which the transport hypothesis produces. He started to arrange some experiments which had to show whether or not the supposition was tenable. He spoke over this in his autobiography as follows: "According to DARWIN's theory, every element of the body throws off gemmules, each of which can reproduce itself, and a combination of these gemmules forms a sexual element. If so, I argued, the blood which conveys these gemmules to the place where they are developed, whether to repair an injured part or to the sexual organs, must be full of them. They would presumably live in the blood for a considerable time. Therefore, if the blood of an animal of one species were largely replaced by that of another, some effect ought to be produced on its subsequent offspring. For example, the dash of bulldog tenacity that is now given to a breed of greyhounds by a single cross with a bull-dog, the first generation corresponding to a mulatto, the second to a quadroon, the third to an octoroon, and so on, might be given at once by transfusion. Bleeding is the simplest of operations, and I knew that transfusion had been performed on a large scale; therefore I set about making minute inquiries".

"These took a long time, and required much consideration. At length I determined upon trying the experiment on the well-known breed of rabbits called silver-greys, of which pure breeds were obtainable and to exchange much of their blood for that of the common lop-eared rabbit; afterwards to breed from pairs of silver-greys in each of which alien blood had been largely transfused. This was done in 1871 on a considerable scale. I soon succeeded in establishing a vigorous cross-circulation that lasted several minutes between rabbits of different breeds, as described in the Proceedings of the Royal Society, 1871. The experiments were thorough, and misfortunes very rare. It was astonishing to see how quickly the rabbits recovered after the effect of the anaesthetic had passed away. It often happened that their spirits and sexual aptitudes were in no way dashed by an operation which only a few minutes before had changed nearly half of the blood that was in their bodies. Out of stock of three silver grey bucks and four silver grey does, whose blood had been largely adulterated, and of three common bucks and four common does whose blood has similarly altered, I bred eighty-eight rabbits in thirteen litters without any evidence of alteration of breed. All this is described in detail in the Memoir" (1908, p. 296-7).

This quotation is not so much important because of the results of the experiments, but more because it proves that GALTON had at that time a good point of view concerning the methods which alone give a well-founded survey of a hypothesis. We would not wonder today that GALTON's experiments dealing with transportation could not find support. In spite of the rather weak points which his experiments showed, they still had a conclusive significance for him, because GALTON could use them to arrive at his own opinion, which differed all the world from the darwinistic one.

GALTON was already a leading man in genetics, when DARWIN's theory of pangenesis was published. It is true that he had at that time only a small paper concerning heredity of talent and character published, but he was already busy with more extensive studies of the same material, later published as "Hereditary genius" (1869), shortly after DARWIN's "Variation". Herein GALTON declares, while still under the first impression that the genius of the pangenesis-hypothesis made on him, "In short, the theory of pangenesis brings all the influences that bear on heredity into a form that is appropriate for the grasp of mathemathical analysis" (1869 p. 358). He does not say more over this question. He assumes its exactitude, but meanwhile holds that further evidence must be produced to support it. Seven years later he fixed his standpoint in accordance with the pangenesis-hypothesis. In a publication of classical value, "A Theory of Heredity" (1876) he leans upon treatments, from which the results were negative and he suggests another theory of heredity in place of the pangenesis. GALTON will readily accept, that definite "organic units" underlie all heritable characters, but he returns to the view, that a relationship between these cannot exist. They together form a whole, a totality, which he calls *stirp*. "Before proceeding, I beg permission to use, in a special sense, the short word "stirp", derived from the Latin stirpes, a root, to express the sum-total of the germs, gemmules, or whatever they may be called, which are found, according to every theory of organic units, in the newly fertilized ovum" (1876, p. 330).

By the introduction of this expression, GALTON took an important step backwards from the conception of the absolutely independent gemmulae as accepted by DARWIN, which was that all gemmulae belong to each other and together form a whole.

Notwithstanding, GALTON asserts the existence of these units in his contemplations: "We will begin with a statement of the four postulates that seem to be almost necessarily implied by any hypothesis of organic units, and which are included in that of Pangenesis. The first is, that each of the enormous number of quasi-independent units of which the body consists, has a separate origin, or germ. The second is, that the stirp contains a host of germs, much greater in number and variety than the organic units of the bodily structure that is about to be derived from them; so that comparatively few individuals out of the host of germs, achieve development. Thirdly, that the undeveloped germs retain their vitality: that they propagate themselves while still in a latent state, and contribute to form the stirps of the offspring. Fourthly, that organisation wholly depends on the mutual affinities and repulsions of the separate germs; first in their earlier stirpal stage and subsequently during all the processes of their development" (1876, p. 331).

The third theorem is quite new and was never expressed as clearly by any of his predecessors; the gemmulae multiply themselves, while they are in a latent condition, and contribute to form the "stirp" of the offspring. The same idea can be said even more explicitly put in these words: "We have thus far dealt with three agents- (1) the stirp, which is an organised aggregate of a host of germs; (2) the personal structure, developed out of a small portion of those germs; and (3) the sexual elements, generated by the residuum of the stirp" (1876, p. 343). There is now no question of collecting gemmulae in the reproductive cells, only one of a continuous chain, which the reproductive cells of an individual form with those of his offspring. In GALTON's principal work dealing with heredity in general, "Natural inheritance" he expresses his views in this way: "It appears that there is no direct heritable relationship between the personal parents and the personal child, except perhaps through little-known channels of secondary importance, but that the main line of hereditary connection united the sets of elements out of which the personal child was evolved. The main line may be rudely likened to the chain of a necklace, and the personalities to pendants attached to its links". (1889, p. 19).

The fundamental difference between both these conceptions: the pangenes transport conception of DARWIN, and the stirp conception of GALTON can be explained once more by means of a figure. You will find this scheme in figure 18. From the germ cells (or the reproductive cells, KC), the body of the organism is formed (L). According to DARWIN this new body will give rise to newly-formed germ-cells, whereas, according to GALTON, the germ-cells, from which the individuals of the next generation originate, owe their origin directly to the remnant of the heritable substance.

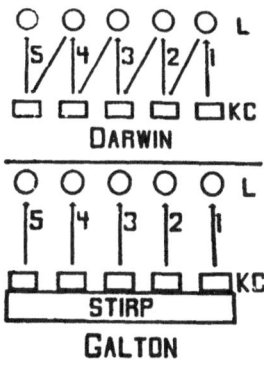

Fig. 18. The gemmules of DARWIN and the stirp-theory of GALTON.

In spite of his absolutely preformatistic point of view, GALTON could not withdraw completely from the epigenetical influences. There were still in his time always plenty of arguments to bring forward in support of a "heredity of acquired characters". Thus it amounts to a change of the stirp, that is to say, of the heritable character of the organism by means of the influence of the environment, either indirectly by way of the body cells, or directly. His conclusion is finally: "The variability of germs under changed conditions, and that of their progeny, may be small, but the change is indubitable; absolute uniformity being scarcely conceivable in the condition of growth, and therefore, in the reproduction of any organism. The law of heredity goes no further than to say, that like tends to produce like; the tendency may be very strong, but cannot be absolute" (1876 p. 338). Another fact that GAL-TON has to face in connection with his stirp-theory, was that of the relationship of *both* parents with regard to their offspring. If an organism descends from another single organism, then it was easy to accept a stirp of an unbreakable chain between all the reproductive cells of the successive generations, but if an individual was born of two parents, then this necessitated another explanation. GALTON went into this subject in a dissertation over "The average contribution of each of several ancestors to the total heritage of the offspring" (1897 b).

Here and in a small publication in "Nature", he puts forward a new law of heredity: "It is that the two parents contribute on the average, one-half, (or 0·5) of the total heritage of the offspring; the four grandparents contribute one quarter $(0·5)^2$; the eight greatgrandparents one eighth or $(0·5)^3$ and so on. Thus the sum of the ancestral contributions is expressed by the series $(0·5) + (0·5)^2 + (0·5)^3$ etc., which sum being equal to one, accounts for the whole heritage" (1897 b, p. 402).

The next year GALTON suggested his schematic views (1898) con-

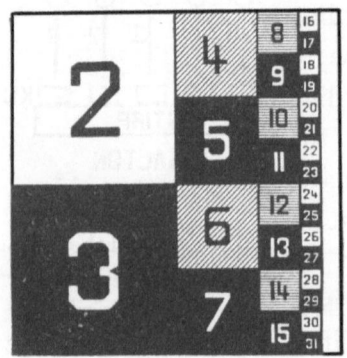

cerning the amount which the ancestors should contribute to the heritable structure of the offspring, by the picture in figure 19. 2 and 3 repeat the parents and 4, 5, 6 and 7 the grandparents, and 8, 9, 10, 11, 12, 13, 14 and 15 the great-grandparents etc., and the whole square contains the heritable tendency of the individual.

Fig. 19. Scheme of the composition of an individual from ancestral elements (after GALTON, 1898).

This was the stirp-theory in its extreme consequence, and if this was correct, then the differences between individuals, which descend from the same pair of parents, should be only the result of environment. Such individuals form a *"fraternity"* and GALTON says concerning the different fraternities: "Circumstance comprises all the additional accidents, and all the peculiarities of nurture both before and after birth, and every influence that may conduce to make the characteristics of one brother differ from those of another" (1889, p. 195).

It is a pity that GALTON, who showed here and there in his work that he attached great value to experiments to prove a definite point of view, was not known to have set a higher value to the experiments of MENDEL.

He knew the work of MENDEL and held the view, that "MENDEL clearly showed that there were such things as alternative atomic characters of equal potency in descent" (1908, p. 308). But what he did not understand was that the simple results of MENDEL's work, threw his "law of ancestral inheritance" overboard. The simple fact shown by MENDEL, that the reproductive cells formed by an organism can vary among themselves, is a flagrant contradiction to this law. It is a pity, since the stirp-theory was very good progress in the preformatic direction, and this progress became eclipsed by being overrated and hence led to incorrect deductions. It is clear that an overrating of the preformatic contemplations, such as the one indicated by GALTON must lead to a dead end. This was acknowledged by the last of the premendelian theorists, AUGUST WEISMANN who had been the most persistent of all the preformatists. WEISMANN's position in relation to the problem of inheritance was totally different from that of GALTON. WEISMANN stood in the middle of an animated and cheerful sphere of treatments and contemplations, which characterized natural science in Germany in the late nineteenth century. He even made himself answerable for a great part of the work and published many divergent zoological papers, and was the most respected and among the most prominent scientists in Germany. In contrast GALTON was entirely dependent on himself and stood alone without wide knowledge of biological data. WEISMANN on the other hand had at his disposal more than sufficient material and had close contacts with the great minds on the subjects of fertilization and reproductive sciences. He was a contemporary and compatriot and also a kin of O. HERTWIG and STRASBURGER. Thus WEISMANN could give his contemplations a stronger foundation than could GALTON, who was really a man of brilliant intuition.

BÜTSCHLI (1876), JÄGER (1878) and NUSSBAUM (1880) had already proclaimed views analogous to the stirp-theory of GALTON, when WEISMANN after some more or less vague introductory contemplations came out with his lecture dealing with the continuity of the germplasm in the "Naturforscherversammlung" in Straszburg. Soon after this, he placed the facts before natural scientists in detail in a dissertation (1885).

JÄGER in his works had put forward the opinion that the body of a higher organism should be composed of types of cells (ontogenetical and phylogenetical ones) from which the latter, i.e. the reproductive cells are not formed as a product of the first (the body-cells) but they should originate directly from the reproductive cells of the parents.

He considered it as practically proved that the "formation of the reproductive cells takes place in the first stages of embryonic life" Later, this was the basis upon which WEISMANN built his theory on the continuity of germ-plasm. He also believed, (and in some cases pointed out by his own experiments) that in an early stage of the embryonic life, definite parts were separated, which later on would form the origin of the future reproductive cells. These parts were then cut off from the circuit of "fertilized ovum to full grown body" and kept until the moment of maturity of that body had arrived and should be ready for reproduction.

As already mentioned, WEISMANN had a greater quantity of sound data at his disposal than had his principal predecessor GALTON. This was the reason why he could found his theory more satisfactorily, and also develop it further. Cases in which the formation of the future reproductive cells was observed to take place directly out of the fertilized ovum were seldom found (Diptera). Therefore WEISMANN took another step backwards from the stirp-theory. He also believed that germ-plasm was given to some somatic cells now and then which contains all the heritable characters of the organism. This however does not interfere with the theory of the continuity. The continuous chain is still there, and that is the fundamental point of his theory.

In 1866 HAECKEL put forward a theorem with regard to the composition and localization of the stirp. GALTON had previously spoken of this very indefinitely. HAECKEL's theorem was that the cell nucleus ought to provide for the handing over of the heritable characters. On the other hand, the cell-plasm ought to provide for the function of feeding and adaptation to external circumstances (1866, 1, p. 287). WEISMANN could go further than this. It was established as acceptable by the work and contemplations of O. HERTWIG, STRASBURGER and WEISMANN himself, that the nucleus in the germ-cell was the depository of this germ-plasm. The internal structure of the nucleus, the chromosomes with their adherents, began to be brought to light at this time.

It is to the credit of WEISMANN, that he was able to recognise the significance of this data. This made it possible for him to draft a theory and to carry it through in detail. He published his theory for the first time in 1892, and it was a very detailed work, exclusively consecrated to the germ-plasm and its significance for inheritance. Briefly we may summarize it as follows: According to WEISMANN, NAEGELI's (1884) so-called idioplasm, (the substrate of all heritable characters in the reproductive cell) is located in the chromosomes. He had proposed earlier that these chromosomes should be known as Idants (1891, p. 39). Each idant should be composed of a number of spherical bodies (or ides) situated next to each other. The ides are small globular bodies which occur in many animal species and build up the separate idants. Hence "the so-called microsomes of the cytologically well-known animal *Ascaris megalocephala* can be thought of as being the globular bodies (ides) which are responsible for the formation of separate idants" (1892, p. 9). Indeed, these ides are units; "a life-unit which exists in itself, with a definite architecture", but they are complex units. Their parts are naturally of a more limited significance. Every cell or cell-group of a fully grown body is represented in the ide by a determinant, a small isolated body comparable with the gemmulae of DARWIN, but fixed in the ide and not transportable. Just as every cell or cell group has its distinct characteristics, so there must be a basic sub-unit in the germ-plasm responsible for each of these characteristics. It is clear that the conception of the determinant is also not a simple one, and we have to suppose that the determinant is in its turn composed of a group of definite units of life, which were called bio-phores by WEISMANN; summarising, each idant (each chromosome) is built up of ides, each ide contains determinants, which are responsible for distinct cell groups and each determinant is again composed of biophores, which all represent a distinct characteristic.

In the realm of the architecture of the heritable characters, WEIS-MANN's theory is worked out to a greater degree than the theory of GALTON. In another realm also, the theory of WEISMANN is different from that of GALTON. GALTON has supposed the heritable constitution of each organism to be made up of contributions from both parents and the four grandparents and so on, according to the statistical material available. No trace of this line of thought can be found in WEISMANN. He had the advantage of knowing the cytological processes of fertilization and of recognizing their significance in heredity. "The fundamental process of sexual reproduction lies in the mixture of two tendencies of heredity (which are individually different from a materialistic point of view), in the combination of two heritable materials which will give rise to one individual" (1892, p. 308). Such a supposition could still be in agreement with GALTON, but WEISMANN realized and rightly so, that not only the amphimixis, but also the so-called reduction division (by which the reproductive cells are formed and by which, as is known, the number of chromosomes is reduced by half) is related to the phenomena of heredity.

"If idants do not change basically during the ontogenesis (that is to say from the fertilized ovum to the germ-cells of the new organism), then we may conclude from definite phenomena of heredity, that the reduction of the ides by half does not separate one from another in advance, a definite group, and not always the same group of ides; sometimes it separates one group and then another time it separates another group. The consequence of this must be that the germ-cells of one individual contain totally different combinations of ides, and thus also a totally different mixture of heritable tendency, which is represented in the germ-plasma of the parents" (1892, p. 321).

This is thus in fundamental opposition to the similarity in heritable tendency of the reproductive cells formed by an organism as accepted by GALTON. WEISMANN accepted the dissimilarity of these reproductive cells in consequence of the reduction division. It speaks well for the brilliant logical view of AUGUST WEISMANN that eight years before the crisis in genetics, (i.e. before the crucial year 1900) in which year the work of MENDEL was rediscovered, he already recognized the fundaments of MENDEL's laws. A more brilliant affirmation could not be desired by WEISMANN.

Naturally, not all the details of WEISMANN's germ-plasm theory and its development are correct and also not all his views over the construction of the chromosomes can be accepted, but still it is a good thing to point out that apparently despite the earlier mocking of his "fantastic views", there are now many arguments which the theory support. Let us remember the results of MORGAN's work over the localisation of genes (see Chapter XI). It stands to reason, that WEISMANN's germ-plasm theorem in relation to the significance which he attached to fertilization, does not leave any room for epigenetical contemplation.

In the time before 1900 there was no more stern opponent to the variability of heritable characters under the influence of environment, than WEISMANN himself. He doesn't admit to any evidence of "heredity of acquired characters" as being sound. He absolutely denies this kind of heredity and puts forward with this the heritable tendency

of a given organism as being unchangeable during life and only interrupted by combinations ("Kombinationen") of the chromosomes that
is to say, by reduction division and by amphimixis. This standpoint of
WEISMANN was for a long time accepted as unassailable, and only in
recent years was experimental data obtained which indicated that
WEISMANN's conclusions are in general valid. The results however of
very extreme external circumstances (X-ray irradiation) can make
it an exceptional case (see Chapter XIX).

With the crisis of centuries, a crisis cradled in the manner of the
handling of questions in genetics, from the period of the theoretical
contemplations, based on observations, genetics has changed to a
period of experiments which are possible to verify. This has been the
cause of the end of the time in which HAECKEL could write: "All
opinions, which are given and find expression in the next theories of
heredity, are based on nothing more than suspicions and are considered
— strictly considered — nothing more than speculative metaphysical
contemplations" (1911, p. 197). For that we have to thank MENDEL.

CHAPTER IV

THE RISE OF THE EXPERIMENTAL METHOD

Genetics is, in its theories as well as in its experimental method, a branch of reproductive physiology; its technique is that of hybridization, i.e. the pairing of two organisms which differ in heritable tendency. Hence, the practicability of this technique is closely connected with the possibility of hybridization. It is obvious that the admission of this possibility is dependent on the theories and experimental results concerning the process of reproduction. Both in the botanical and in the zoological field, the process of reproduction was the basis for experiments on cross fertilization. The actual process of reproduction was questionable for centuries, and hence the basis of the experiments was not very certain.

So long as such peculiar false notions existed concerning the reproduction of animals (such as the fight between the "*animalculisis*" and "*ovists*") a clear conception over the process of fertilization was out of the question. The animalculists saw a diminutive creature in the spermatozoon, but on the other hand, the ovists saw the whole individual together with all its organs represented in the microscopical egg.

The *epigenesis-theory* of C. F. WOLFF, though a sane reaction, did not bring a chance for the better. Hence the hybridization research in the zoological sphere has been very barren. Only scattered data concerning the appearance of hybrids between animal species are given in the literature. There has been no record of systematic breeding of animal hybrids, and still less of systematic inbreeding of later generations, which are both necessary for tracing the heredity of the characters of both original parents. Moreover, animal hybrids most frequently described were those which were considered to be completely sterile. Because of this, it seemed to be an investigation doomed to failure. This was true even when the views on the sexuality of animals were more enlightened.

Good summaries of early data concerning animal hybrids published in different periodicals are given by HOFACKER (1828, p. 81–92), ACKERMAN (1897–1898) and recently by GRAY (1954).

So far as the existence of sexuality in the animal kingdom was concerned the fact was already recognized and ancient use made of it in cross-breeding of domestic animals. This practical breeding did not however have any scientific value from the point of view of the problem of hybridization.

The fertilization-process in plants had still to pass through a longer period of doubt to come to full acceptance. The first condition for acceptance was naturally the recognition of pollen as the necessary material for obtaining good seeds and we will find that, already in the mediaeval literature, there were some indefinite indications on behalf of this, but in general these were not very convincing. The opinion of

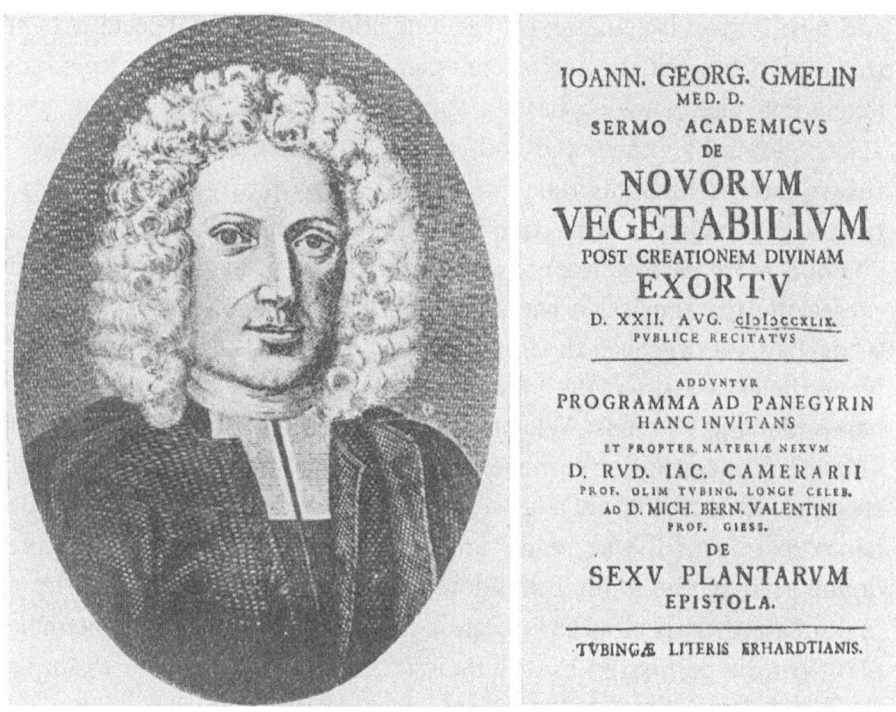

Fig. 20. R. J. CAMERARIUS (1665–1721).

IOANN. GEORG. GMELIN
MED. D.
SERMO ACADEMICVS
DE
NOVORVM
VEGETABILIVM
POST CREATIONEM DIVINAM
EXORTV
D. XXII. AVG. cIↄIↄCCXLIX.
PVBLICE RECITATVS

ADDVNTVR
PROGRAMMA AD PANEGYRIN
HANC INVITANS
ET PROPTER MATERIÆ NEXVM
D. RVD. IAC. CAMERARII
PROF. OLIM TVBING. LONGE CELEB.
AD D. MICH. BERN. VALENTINI
PROF. GIESS.
DE
SEXV PLANTARVM
EPISTOLA.

TVBINGÆ LITERIS ERHARDTIANIS.

Fig. 21. Title-page of CAMERARIUS' De sexu plantarum epistola (Ed. 1749).

the day, which considered the stamens as organs which secreted harmful material was not shaken at all.

This concept began to waver for the first time in the last part of the seventeenth century, when RUDOLPH JACOB CAMERARIUS (figure 20) published his experiments in this sphere, in a letter to VALENTIN. This letter became one of the classics of biological science (figure 21). In this letter dated August 25th 1694, CAMERARIUS shows his great interest in the whole question. He reviewed the literature concerning this subject very carefully and recognized that only experiments (as opposed to loosely compiled data) could give decisions. He speaks with certain pleasure of experimental science which he likes very much. Extensive experimental work was carried out by him, particularly with dioecious plants, such as *Mercurialis* and hemp, and also monoecious plants such as *Mays* and *Ricinus* offered him the necessary material. From *Ricinus* he saw, for instance, that seed-formation was absolutely excluded if the male inflorescences had been removed before the stamens had opened and that he could prevent pollination and fertilization by cutting off the long stigma. All his experiments in this direction were absolutely convincing. "In the vegetable kingdom (excluding plants which have a supposedly spontaneous origin, and also excluding all other methods of reproduction, such as tree buds, tubers, etc., which may be considered as partly reproductive and thus secondary) no fertilization is brought about by seeds (seeds are a gift of a more perfect Nature and form the most common method of preserving the species) if the stamen (which appears earlier than the seeds) had not prepared the flower of the plant for it" (Ed. 1899, p. 25). Nevertheless CAMERARIUS found exceptions where an isolated female hemp plant gave seeds. What the explanation of this is he does not know, but he raised the surmise, "that in the garden, plants of another species which flower in a great number had pollinated the female hemp plant which was ready for fertilization. Nobody would have doubt, that in the animal kingdom a female could be fertilized by a male of another species. The question now arises as to whether a female plant can be pollinated by the male of another species. For example the female hemp by male hops, or the *Ricinus* (from which the spherical stamen flowers are removed) by the pollen of the turkish corn (maize) etc., or, how far does an altered embryo originate" (Ed. 1899, p. 49). With this, CAMERARIUS began the long chain of people who became interested in this problem of hybridization and its consequences.

Hence he was a pioneer in two fields, the field of flower-biology (concerning the life-cycle of flowers, pollination and fertilization) and the school of hybridization. In the first field, much war was still waged after his work was published. Partly by the experiments of MILLER (1731) (with tulips from which the stamens had been artificially removed and in which the stigmas were found to be covered with pollen after the visitation of bees), and partly also by the more systematic studies of flowers which SPRENGEL (1793) gave us, the signficance of flower parts in the life-cycle of flowers was finally recognized.

The second group of questions all of which concern hybridization were touched only very lightly by CAMERARIUS. He could not prove the possibility of hybridization. It was thus the work of his successors to investigate this possibility in an experiment. It seems that an English grower FAIRCHILD grew an artificial hybrid between two species of *Dianthus* for the first time (BRADLEY 1719, p. 16). The next was LINNAEUS, who was not particularly engaged on hybridization, but whose wide general interest in botanical phenomena led to his choice of this problem. Apart from some crossings which failed, a successful hybrid was obtained by him between the yellow flowering *Tragopogon pratensis* and the purple-flowering *T.porrifolius*. In 1757 he made this crossing and in 1759 he obtained the first hybrids of both these species. These plants gave red flowers which had yellow bases. Concerning these experiments, a communication was given by him in several of his publications (1759, p. 13; 1760, p. 27; 1762, p. 14). But for a great part his task lay in another field, systematic hybridization was too comprehensive to treat as a lesser piece of investigation, and so the significance of LINNAEUS's work does not lie in the field of experimental hybridization but more in pure systematics.

But this task was soon taken over by a younger man, by J. C. KOELREUTER (fig. 22) and performed in a brilliant way. In 1760, the crossing of *Nicotiana rustica* ♀ and *N. paniculata* ♂ was made and he followed the growth of the plants obtained from this seed with a great interest. "I observed with pleasure, that not only did these plants lie between both natural species in the ramification and in the position and the colour of the flowers, but in particular, the flowers, with the exception of the anthers showed an almost geometric proportion when compared

with the same parts of the two natural forms". This he said in 1761
(p. 40; Ed. 1893 p. 30).

The hybrids were also interesting for the manner in which they
formed their seeds. Under the magnifying glass it could be seen that

Fig. 22. J. G. KOELREUTER
(1733–1806).

Fig. 23. Title-page of the third
part of KOELREUTER's work.

the pollen was completely shrivelled and this caused complete sterility
after self-pollination. On the other hand, normal fruits with good seed
capable of germination developed either with pollen of *N. rustica* or
with *N. paniculata*, only in much smaller quantities. In the sequel to
his first book "Vorläufige Nachricht von einigen das Geschlecht der
Pflanzen betreffenden Versuchen und Beobachtungen" (fig. 23) he
gives more details of the later generations, which were obtained by
back-crossing with the original mother plants. This led him to the
following conclusion concerning the back cross (*Nicotiana rustica* ♀ ×
N. paniculata ♂) ♀ × *N. rustica* ♂. "In general all these plants resemble their mother (the *rustica* type), some bear more resemblance,

others less. Some of these plants bear, as well as a general resemblance to the mother type, also some specific likeness to *rustica*. But on the other hand, other plants, instead of inclining to *rustica* appear to diverge not only from the *rustica* type but also from the hybrid" (1761, p. 14; Ed. 1893, p. 47). Such conclusions he drew also from the back-cross with *N. paniculata* as pollinator. By this he broke the first barrier in scientific hybridization research concerning the problem as to whether or no artificially made hybrids breed true to type. He follow-ed the experiments with tobacco by a great number of other crosses between species of the genera: *Aquilegia, Matthiola, Dianthus, Melan-drium, Linum, Malva, Lavatera, Lobelia, Datura, Lycium, Verbascum, Digitalis*, and *Mirabilis* of which he published the results partly in sequels to the Vorl. Nachricht and also partly in the Proceedings of the Imperial Academy of St. Petersburg (from 1778 to 1806). In so far as the sterility of the experimentally produced hybrid plants did not prevent further experiment, there were also indications that the hybrids were not constant, that is to say the offspring of a hybrid segregated into different types.

KOELREUTER's work caught attention of others, and a number of investigators were convinced by him of the existence of the sexu-ality of plants, but the question which fascinates us today, that is to say, the behaviour of the off-spring of the hybrids, still remained in the background. Notwithstanding this, his work was predestined to be the basis of a more comprehensive study of hybridization in plants (see FOCKE, 1881; ROBERTS, 1929).

Towards the end of the eight-eenth century, interest in hybri-dization grew. First, in England where T. A. KNIGHT (figure 24, one of the founders of the famous Horticultural Society of London) began to busy himself with this

Fig. 24. T. A. KNIGHT
(1759–1833).

question (1799–1841). In contrast with KOELREUTER, whose experimental material was taken from wild plants, KNIGHT worked with cultivated plants. He was a practical horticulturist with scientific tendencies, and because of that he chose his demonstration plants from the cultivated races. Fruit trees and peas gave him the material for his crosses, and among his investigations with peas, there is one that especially demands our attention. In 1823 he described a hybrid made by himself between a pea with white flowers with uncoloured seed coat and green stems; and a pea in which the flowers and stems were violet and the seed coat was grey. The seed obtained from this cross absolutely resembled the colourless mother plant, but from this seed hybrid plants grew up which, in some respects had more of the external appearance of the father plant. The hybrids were again hybridized with a plant of the white type, which resulted in a certain number of white plants in the offspring of this back-cross. KNIGHT does not state the ratio between these white plants and the total, but he did establish that the first hybridization of the white with the purple plant gave only a purple flowered offspring.

Practically in the same year as this communication of KNIGHT, (which is to us today, the most important of his publications), two other practical scientists, JOHN GOSS (1822) and ALEXANDER SETON (1822) reported on hybridization between peas. From their publication a very important conclusion can be drawn. Goss had pollinated flowers of the variety "Blue Prussian" with pollen from "Dwarf Spanish". The first was a type with green cotyledons; in the second those were yellow. On opening, the ripe pods of the Blue Prussian plant, Goss saw to his astonishment, that the seeds inside the pods were yellow and not blue-green. The plants from these yellow seeds, when they set fruits sometimes had all green peas, sometimes all yellow seeds in their pods, but in most cases green and yellow peas were mixed. The quotation from Goss (1822, p. 235) is of special importance: "Last spring I separated all the blue peas from the white and sowed each colour in separate rows; and I now find, that the blue produce only blue, while the white seeds yield some pods with all white, and some with both blue and white peas intermixed". Especially important in his work as a pioneer experimenter (and even stronger than the experiments of KOELREUTER) is the way in which he sets the hybrids (which do not breed true to type) versus the pure races, which are constant. A similar result was obtained by SETON; after crossing green seeded peas "Dwarf Imperial" with pollen of a yellow seeded type, he noticed in the pods of the hybrid plant, "On their ripening it was found that instead of their containing peas like those of either parent, or of an appearance between the two, almost every one of them had some peas of the full green colour of the Dwarf Imperial, and others of the whitish colour of that with which it had been impregnated mixed indiscriminately and in undefined numbers" (SETON, 1822, p. 237). SETON also gives a coloured drawing of these remarkable pods. Neither GOSS nor SETON notice any regularity in the numbers.

The main point in KNIGHT's investigation did not lie so much in the heritable behaviour of the characters in which the hybridized parents differed, but in the completely different problem which he considered as more important, viz., if it would be possible to obtain new vigour by means of hybridization of weakened types, and also if hybrids between individuals, which are classified systematically as being of different species, are sterile. KNIGHT denied absolutely the possibility of making fertile hybrids between species. KNIGHT's contemporary, WILLIAM HERBERT, (who occupied himself very much with the problem of hybridization) also principally restricted himself in his discussions (1819 and 1837) to the latter question. For the rest, his work still led to important novelties in horticulture, but not to exact results concerning the heredity of different characters. In spite of great interest, which the work of KNIGHT and HERBERT attracted in England, and notwithstanding the fact that many investigators followed this track, (TH. MILNE, G. SWEET, G. DON, PAXTON, LINDLEY) there still resulted nothing from all this work on hybridization, which could throw any light upon the heredity of the different characters.

In France the same state of affairs arose: SAGERET (1826) and LECOQ (1827 and 1845) did not bring much to light. The work of the first was the most important, because he could draw some definite conclusions from his crossings between varieties of melons, which should come slightly nearer to the conclusions made by MENDEL later on. Both original types of melons differed mutually in five sharp observable characters: flesh colour, seed colour, markings on the pericarp, form and number of ribs and flavour. Contrary to his expectations, he found in some respects only, that the hybrid is an intermediate between the two parent races. In relation to most of the characters studied he saw that the hybrid expressly resembles one of the parents. "It is evident to me in general that the resemblance between the hybrid and its parents is not based on a close mixture from each parents' main characteristics, but much more on an equal and unequal distribution of the same characters". In this publication SAGERET uses for the first time the expression, that one character "dominates" over another.

In Germany, the commotion concerning the question of sexuality at the beginning of the nineteenth century had the good fortune that several investigators set to work in this field.

In 1828, WIEGMANN put forward his answer to the question offered first in 1819 and again in 1822 by the Academy of Berlin for a prize; — Does hybridization exist in the vegetable kingdom? He offered a rather detailed essay in which he not only defended almost the same postulate as that of SAGERET, but also supported the view of KOEL-REUTER "that from one and the same seed that is sown which originated from one and the same fruit produced by a hybrid plant, the fully developed plants from these seeds often differ mutually, and then again they often resemble the father or the mother" (1828, p. 25).

Two years later (1830) the Holland Society of Sciences in Haarlem offered a prize for the same problem: "What do we learn from experiments concerning the origin of new species and varieties of plants by means of artificial pollination of the flowers by pollen from another species? And which new and useful (or fine) plants can be derived or multiplied in this way?"

This prize as well as the former was very successful. CARL FRIEDRICH VON GAERTNER answered this with an essay which was the most important of all essays since the publication of KOELREUTER (1838). GAERTNER had already published a small paper on fertilization experiments which he had attempted with different species, but this paper was not of much significance. HERBERT's criticism to which the paper was later subjected viz., that it was "highly misleading" is too severe. To be sure it was an immature work and did not deserve a translation into French (1827), but through his later publications GAERTNER has shown that it was the first step in the right direction.

His Dutch essay was later greatly amplified and recast (1849) and published in German. This has been and still is, one of the most important and most consulted works. The reason for this is the enormous quantity of data he presents in it. Indeed GAERTNER was the first person after KOELREUTER to carry out large scale hybridization experiments in a systematical manner, and to attempt to draw a conclusion from the data, which the experiments yielded. They both formulated definite laws concerning the behaviour of hybrids and their offspring. A great number of conclusions are put forward in his work. From the point of view of heredity, two of those have great value. "The most important and most interesting phenomenon in hybrids which were obtained from plants by crossing is the absolute resemblance of both products. (GAERTNER alludes here to what we today call reciprocal crosses; i.e. species A as father with species B as mother and conversely, species B as father and species A as mother). The seed developed from the pollination in one direction, gives absolutely similar plants to those developed from the opposite pollination, so that differences in origin and descent (even in the most careful investigation from both hybrids with respect to their shape and external appearance) do not yield the slightest difference, and even the most highly trained connoisseur of hybrids is not able to point out the origin of the hybrid (with respect to the sex of the parents)" (1849, p. 222).

We will see later on that this statement is of great value, but is not of general validity. The second result of GAERTNER's experiment as given in the following quotation will also undoubtedly retain a permanent value:

"Several extremely fertile hybrids multiply themselves like pure species (as immutable types), so that some botanists will even class these hybrids as stable species .... Other hybrids (and indeed the majority, which are fertile) from the seeds of the second and subsequent generations produce deviated forms which differ from the normal type. That is to say, varieties, which are partly or completely different from the original hybrid plant or which can differ to a greater or lesser degree from this original hybrid plant .... In many fertile hybrids, this change appears in the second and third generations not only in the flower, but also sometimes in the whole appearance, or the appearance exclusive of the flowers. The majority of individuals which originate from a single hybrid preserve the shape of the hybrid plant, a few others begin to resemble the original mother plant and still others begin to resemble the original father plant" (1849, p. 421-3).

Even GAERTNER accentuates strongly the fact that hybrids do not breed true to type. "The variability in the offspring of the hybrids is a great mark of identification for their hybrid nature". (1849, p. 518).

These statements are still indefinite and not clear cut, but they are nevertheless the first indication as to the direction along which clarity begins to dawn concerning hybridization and the results of hybridization.

GAERTNER does not consider the lack of breeding true to type as typical in hybrids, but rather that most of the hybrids are not breeding true to type, and moreover that variability is a characteristic phenomena occurring after hybridization (he is often wrongly quoted on the latter point).

Following the prize offered by the Berlin Academy (which led to the publication of WIEGMANN) and subsequent also to the prize offered by the Holland Society of Sciences (this led to the manuscript of GAERTNER), the Paris Académie des Sciences on the 30th January 1860 issued the statement that they intended to set apart the Grand Prix des Sciences Physiques for the year 1862 for a good exposé of the following question, viz; — "A study of plant hybrids with regard to their fertility and whether or not they are breeding true to type with respect to their characteristics". More precisely it was expanded as follows: "The possibility of making hybrids between plants belonging to different species of the same genus has been an established fact for a long time, but a number of accurate experiments must still be carried out, before the next questions (which are not only important for general physiology but also for the study of species and their variability) can be answered:

1stly) in which cases are hybrids between species fertile? Is this fertility of the hybrids connected in any way with their external similarity to the species from which they originate, or is the fertility an indication of a special relationship in the reproduction, as is shown in the possibility of producing artificial hybrids?

2ndly) is a defect in the pollen the cause of sterility in hybrids? Are the ovaries and the ova always susceptible for fertilization by pollen of a suitable but other species? Alternatively, could one observe occasionally, a marked incompleteness in the ova or ovaries?

3rdly) do the hybrids which are self-fertile in some cases keep unalterable characteristics during the several successive generations, or do they always throw back in the course of some generation to the form of one of their ancestors, as seems evident from some more recent observations?"

One can observe a noticeable development in these successive prizes of the Societies. The one from Berlin only asked for the possibility of hybridization, the one from Haarlem asked for the more practical application of this and the Paris one for a very detailed discussion of the scientific results, which the investigation of hybridization can yield and in fact yield. The prize of the Académie des Sciences was also answered in a brilliant manner by CHARLES NAUDIN (fig. 25), whose essay was rightly called by the committee "one of the most important, and most rich iu excellently observed facts, which are methodically presented and which are very lucidly discussed" (DUCHARTRE 1863, p. 132). The answer of D. A. GODRON (1863a) was given second place but a great distance divided the two efforts.

Fig. 25. CHARLES NAUDIN (1815–1899).

The great difference in character between these scientists, finds its expression in the rather strongly philosophical tone of GODRON's paper which contained only a few facts and these are considered in agreement with his point of view, whereas NAUDIN compiles an overwhelming body of facts and builds some conclusions upon them. NAUDIN's publication (1855-1865, especially 1865b) are of an extremely convincing nature. GODRON's essays (1844, 1863) are rather vague and they do not prove anything. The only conclusion which GODRON intended to draw was as follows: "Successive pollination by the pollen of the parents plays an indispensible role in the fertility of hybrids and also a principal role in the formation of varieties. The mutual pollination by means of pollen from these permanent fertile varieties gives origin to new variations. Finally: pollination by pollen of the parents or by the pollen of individuals which resemble one of their ancestors again gives rise to varieties which have the appearance of one of the types of the species from which the initial hybrid originated" (1863, p. 154).

In direct contrast to the contemplations of GODRON (who built partly on other people's information), NAUDIN presented data of a number of crossing experiments, which appeared to justify his conclusions. The crosses he made were between species of the genera *Papaver, Mirabilis, Primula, Datura, Nicotiana, Petunia, Digitalis, Linaria, Ribes, Luffa, Coccinia* and *Cucumis*.

It is to the merit of NAUDIN's sharp powers of observation that in spite of objections (which hybrids of species always raise, and which formed so great a barrier for many of his predecessors, that their experiments partly led to incorrect conclusions), he still came to much the same point of view as MENDEL and unconsciously endorses MENDEL's contemplations. When we compare their contributions, we will find similar thoughts in both authors. It is NAUDIN who, unconsciously endorses MENDEL's contemplations, the only difference being that MENDEL's contemplations take a more mathematical form and because of this are more sharply propounded. Fundamentally there is only a small difference between their work and their results.

In the first place it was the great uniformity in artificial hybrids which was observed by NAUDIN. He saw a fundamental difference between the artificial hybrids and their next offspring. "To vizualise the external appearance of hybrids it is fundamentally important to make a distinction between the first generation, and subsequent generations. I have always found a great uniformity in external appearance in the hybrids (which I made myself and of which the origin was known), between individuals of the first generation (which originates from one and the same cross), however great the number was". And further: "Briefly we can say, that hybrids from one and the same cross (so far as the first generation is concerned) mutually resemble each other, and equally strongly (or very nearly so), as individuals which originate from parents of the same pure "species". (1863, p. 187; 1865, p. 146).

NAUDIN further observed the great resemblance in external appearance between the so-called *reciprocal hybrids*, that is to say the resemblance between the hybrid originating from A as the mother and B as the father, and the reverse with A as pollinator, and B as the seed bearer. "I can state, that all reciprocal hybrids which I have obtained both between closely related and between more different species, showed an equally strong resemblance, as when they had originated from the same cross .... Naturally, it is possible that this will not always be found, but if this remains true then this case is a great rarity and forms an exception to the rule rather than the rule itself (1863, p. 188; 1865, p. 147).

Apart from the uniformity of the first generation hybrids and the uniformity in external appearance of the reciprocal crosses, NAUDIN gives the following as his third observation: "The external appearance of the hybrids begins to change very radically in the second generation. Usually the absolute uniformity of the first generation is succeeded by an extremely multiform generation". Thus he records the phenomena which had already been loosely stated at an earlier time by KOELREU-TER, that hybrids (and also hybrids of species) are not uniform in their offspring but segregate and produce, as a general rule, very different types. "Indeed in a great majority of cases, (and perhaps in all), a segregation of the hybrids begins to occur in the second generation of the hybrids. This phenomena was already anticipated by many investigators, and doubted by many others, but it seems to me that it is now beyond all doubt" (1863, p. 190, p. 149).

These three conclusions are borne out by NAUDIN with arguments and we shall see later, they are parallelled by some of the rules drawn up by MENDEL in his work. But the point is reached here where both views also differ. MENDEL obtained forms from his hybrids which were not recognizable as hybrids in their offspring, and behaved as pure species. NAUDIN did not succeed in obtaining from his hybrids, forms which bred true to type: "I can only assume, that none of the hybrids which I have obtained showed any tendency at all to give rise to new species" (1863, p. 197; 1865, p. 157). We need not wonder at all the difference in results which occurs here, if we bear in mind the fact that NAUDIN made hybrids between species, viz., hybrids between two plants that differ very much in characters.

NAUDIN tried to give a theoretical explanation of the rules concerning hybrid production which he discovered. According to this NAUDIN's views are closer to our present day conceptions than are those of his predecessors: "A hybrid plant is an individual in which two different tendencies are united. These tendencies each have their own pattern in the spheres of growth and life, which work against each other and are continually fighting to get loose from each other. Are these two tendencies very mingled? Do they penetrate into one another so that every small particle of the hybrid plant as small and divided as we can possibly imagine, contain these tendencies in like manner? This perhaps occurs in the young embryo and perhaps also in the first stages of development of the hybrid, but it seems to me to be more acceptable that the first stages of development and certainly the full-grown stage is a homogeneous accumulation of particles, equally or unequally distributed over both the original species and mixed in a different ratio in the different organs of the plant .... All the facts give evidence that it is in the pollen and the ova (and especially in the pollen) that the segregation into the original species takes place most energetically .... And hence this segregation can occur in pollen as well as in ova in all its degrees, so multiformity will originate from the combinations which will be born and which are governed by chance. This was ascertained from the hybrids of *Linaria* and *Petunia* starting with the second generation" (1863, p. 191; 1865, p. 150).

Comparison between this hypothesis of NAUDIN and the statement given by MENDEL which will be discussed in the next chapter, clearly brings to light that in NAUDIN, we must see a man of genius very closely related to, and one of the most important fore-runners of MENDEL.

The works of MENDEL's predecessors may not glory in results of the same vein as those of MENDEL, but they still have a right to our appreciation. As a mark of gratitude, ROBERTS (1929) and ZIRKLE (1935) wrote an attractive history of the lives of these predecessors.

CHAPTER V

THE RESULTS OF MENDEL'S EXPERIMENTS AND OF ANALOGOUS INVESTIGATIONS

As we have seen, some data concerning the behaviour of the heritable characters after hybridization was obtained by predecessors of MENDEL, but a clear view and the rules concerning this heredity had not appeared in spite of frequent attempts and the large amount of work carried out. It was left to MENDEL to disclose these laws and it was the privilege of HUGO DE VRIES, CORRENS and TSCHERMAK (1900) to introduce the same laws to the world as a universality, independently from one another and almost simultaneously.

In a modest form and in a modest corner of scientific literature GREGOR JOHANN MENDEL, the teacher in the Brunner Augustiner-Stift, published (1865) his "Detailversuche" in relation to the problem of heredity, which "in the nature of the case was limited to a small group of plants", and after a lapse of eight years was brought mainly to a close. Still it was MENDEL's view "that he had failed to frame a general rule concerning the formation and development of hybrids".

The opinion of the investigators in the years following 1900 who carried out the same experiments as MENDEL is not so negative: MENDEL's point of view concerning the experiments which must be carried out is absolutely borne out by modern experimenters in genetics.

"Everyone who overlooks the work done in this sphere of Science attains the conviction that among the numerous experiments already done, not a single one was carried out in such a range and such a manner that it would be possible to ascertain the number of different forms in which the offspring of the hybrids occur. This means, that these forms could not be arranged with certainty in the successive generations and the mutual ratio of the numbers determined. Indeed some courage is needed to undertake such a work, but it seems to me

6

to be the only right way, along which one can reach the final solution of the question (the significance of which we should not underestimate in the developmental history of organic forms)" (1865, p. 4). With admirable sharpness, MENDEL touches the quintessence of the whole problem here: that above all one must be exact. This is the first demand that must be to the forefront in all investigations dealing with heredity and variability. MENDEL's choice of his groups of plants for his experimental material was extremely felicious.

The family of papilionaceous flowers seemed to him to be suitable on all points, in particular the races of peas within that family, viz. *Pisum sativum*. MENDEL's attention was especially drawn to those characteristics of the different types which had been comprehensively described, which differed qualitatively. Quantitative characteristics (in which the distinction is based on them being "more or less" different) were judged to be less useful by him. So he succeeded in finding pairs of characteristics in which these types occurred which could be sharply distinguished: 1) smooth or wrinkled seed-form, 2) yellow and green colour of the seed-lobes (which shine through the transparent seed-coat), 3) coloured seed -coat and flowers as opposed to white seed coat and flowers, 4) a smooth pod and an indented pod, 5) green and yellow unripe pods, 6) flowers scattered along the stem, and flowers clustered at the top, 7) height of the plant: 6-7 ft, and $^3/_4$-$1\frac{1}{2}$ ft.

The principles according to which the experiments were carried out, were in MENDEL's day, absolutely new. He deliberately hybridized two plants, viz., he brought pollen from one flower of a plant to the stigma of the other type from which the anthers of the latter had been removed before they opened. The seeds yielded by this cross were sown and the plants from these seeds obtained were "hybrids". Peas belong to the most regularly self-pollinating plants, viz., the pollen from a flower lands by itself on the stigma of the same flower, so MENDEL needed only to leave these hybrids alone to get seeds from these plants. These seeds in their turn yielded the "first" generation of the hybrids. Nowadays our nomenclature has changed and we call the hybrid itself, which was made, the first generation (F_1, in which the F is an abbreviation of filius, a child). The children obtained by self-pollination of the first generation are thus the second generation (F_2), the children of F_2 are called F_3 etc. Likewise, the parents of the hybrids are called P_1 (sometimes both parents are distinguished as P_{1a} and

P_{1b}), the grandparents by P_2 and so on (P is an abbreviation of parens, parent).

The important results which were obtained by MENDEL were these: It does not matter in hybridization of two peas of a different type, which one is taken as the pollen plant, and which one as the stigma plant. The results in both cases are the same (*principle of reciprocity*). GAERTNER had also found this (1849, p. 222). All plants of the F_1 are mutually identical (*principle of uniformity*). The plants of the F_1 generation always show the same characteristic as one of the parents (*principle of dominance*). The plants of the F_2 generation are not mutually similar but show for a part the external appearance of one parental plant, the rest on the other hand exhibit that of the other parental plant and both these parts are always in the ratio of 3 : 1 (*principle of segregation*). We will give the following example: MENDEL pollinated a green seeded pea with the pollen from a yellow seeded pea and conversely, the flowers of yellow seeded plants with pollen from green seeded plants. In both cases, he obtained only yellow seeds; the yellow colour of the seed-cotyls is a characteristic of the young hybrid which is still surrounded by the transparent seed-coat. The whole F_1-generation was composed of yellow seeds. Therefore he called the character "yellow seeded" *dominant* over the character "green seeded" which he termed *recessive*. In what is today an obsolete terminology, one could also speak of a *latent factor*, but this concept also includes other possibilities besides that of recessivity. Those possibilities will be discussed later on. The F_1-plants developing from the yellow hybrid seeds were kept under self-pollination. It was already possible to observe a mixture of yellow and green peas in the seeds which were formed on these plants, since the colour of the cotyls shines through the seed-coat (figure 26). The whole F_2-generation in MENDEL's experiment was composed of 8023 seeds, from which there were 6022 yellow and 2001 green. Thus a ratio of 3·01 : 1 was obtained. Hence he found from all the seven pairs of characteristics previously given, that the first was absolutely dominant, the second recessive, so that the F_1-hybrid always showed the characteristic of the first mentioned parental plant and the segregation in the F_2-generation always took place in the ratio of 3 : 1, which is sufficiently evident from the numbers obtained by MENDEL in his seven series of experiments. He found the ratios of 5474 : 1850 or 2·96 : 1; 6022 : 2001 or 3·01 : 1; 705 : 224 or 3.15 : 1;

882 : 299 or 2·95 : 1; 428 : 152 or 2·82 : 1; 651 : 207 or 3·14 : 1 and 877 : 277 or 2·84 : 1. The average of all these experiments is 2·91 to 1, that is to say, nearly 3 : 1. The frequent occurrence of these fixed ratios necessarily drew MENDEL's attention.

The investigators who have also engaged themselves with hybrids of yellow seeded and green seeded types of peas, along the paths indicated by MENDEL, could do no other than bear out this result and this ratio of numbers in the F_2-generation. If now these F_2-plants are kept under absolutely self-pollinating conditions and if the seed of each plant is observed individually and if the seed is sown again separately, then the plants with a phenotypical recessive character produce only a recessive offspring, as GOSS and SETON already pointed out, and had already been ascertained by MENDEL. But the plants with dominant characteristics produced, for a part, an exclusively dominant

Fig. 26. An F_1-plant from the cross yellow seeded × green seeded pea, on which F_2 seeds are growing (after DARBISHIRE, 1911).

offspring, and for the rest, they segregated again into 3 dominants: 1 recessive; the true bred dominants made up $^1/_3$ of the whole number of dominant F_2-individuals. So MENDEL obtained F_2-seeds which were yielded by F_1-plants with yellow seeds. Out of 519 plants, 166 had exclusively yellow seeds, and 353 yielded yellow and green seeds in the ratio of 3 yellow: 1 green. The ratio of yellow seeded non true breds, which segregated and true breeding yellow F_2-individuals, was in this sample, 353 : 166 or 2·13 : 1. In other experiments he also found ratios which corresponded more or less with 2 : 1, viz.

372 : 193 1·93 : 1 | 64 : 36 1·72 : 1 | 71 : 29 2·45 : 1 ⎰ Total 771 : 394
125 : 75 1·67 : 1 | 67 : 33 2·01 : 1 | 72 : 28 2·5 : 1 ⎱   1·96 : 1

Though an absolute accordance with the ratio 2 : 1 was seldom reached because the numbers per group were too small, MENDEL stated that it was allowed to suppose a ratio of 2 segregating to 1 as a constant. His later work has shown that he was right.

Continuations of the experiments led to the result that the recessive green seeds always produced a green offspring (true breeding) and that the yellow offspring of the hybrids were partly $^1/_3$ true bred yellow seeded and the remaining $^2/_3$ segregated again. In every generation this will repeat in the same manner. Thus the course of such hybridization between two plants which differ from each other in only one heritable characteristic (*monohybrids*), that is to say, one plant pure dominant (DD), another plant pure recessive (RR) and a hybrid, which we can call DR, can be summarized in a table as follows:

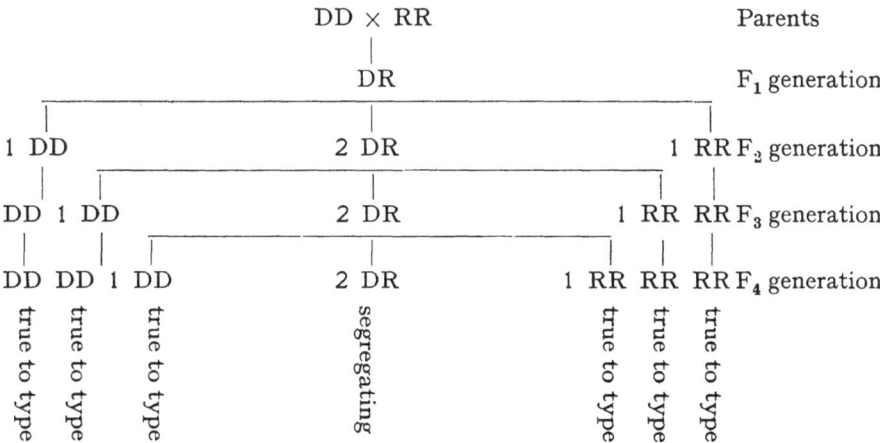

Numerous examples of such monohybrids, in which the factor (characteristic) of one of the parents is dominant have been found since 1900 by other investigators using the most divergent characters in the most different organisms.

A cross between two types of kidney beans, in which one of the parents had brown seeds, and the other one yellow (see Plate 1) (the colour of the seeds being situated in the seed-coat which belongs wholly to the mother plant) gave an F_1-generation of which the seeds were

brown in colour. This colour was not distinguishable from that of the mother type. The F_2-generation (obtained by self-pollination of the F_1-plant) was composed of 10 plants, from which 8 had brown seeds and two had yellow seeds. In the F_3-generation the offspring of both the yellow seeded F_2-individuals (total 73 plants) were again all yellow seeded. From the 8 brown seeded F_2-plants, 3 had exclusively brown seeded offspring (total 103 plants), whereas the other five plants had brown and yellow offspring in the following ratios:
a) 33 : 9. *b*) 38 : 10. *c*) 61 : 26. *d*) 33 : 12. *e*) 35 : 8. Thus totalling 200 brown: 65 yellow, i.e. 3·08 : 1.

A cross as CORRENS made between two stinging nettle plants: *Urtica pilulifera* (with indented leaves) and *U. Dodartii* (with practically entire leaves) yielded an F_1-generation in which the leaves were similar to those of *U. pilulifera*. Of the F_2-generation, according to his statement (1905, p. 6, 1912a, p. 25) 75% of the plants had indented leaves and 25% had entire leaves. These latter proved to be breeding true to type in F_3. Of the 75% $^1/_3$ bred true to type, thus 25% of the total were constant and $^2/_3$ segregated again (50%). The course of things is represented schematically in figure 27.

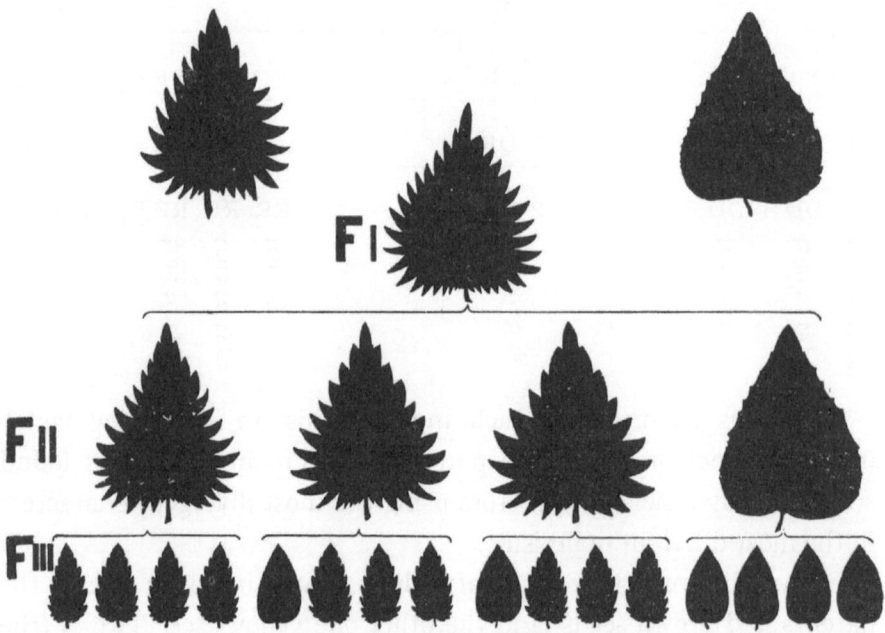

Fig. 27. A cross between *Urtica pilulifera* (left corner) and *U. Dodartii* (right corner) and next generations (after CORRENS, 1905, p. 6).

Plate I

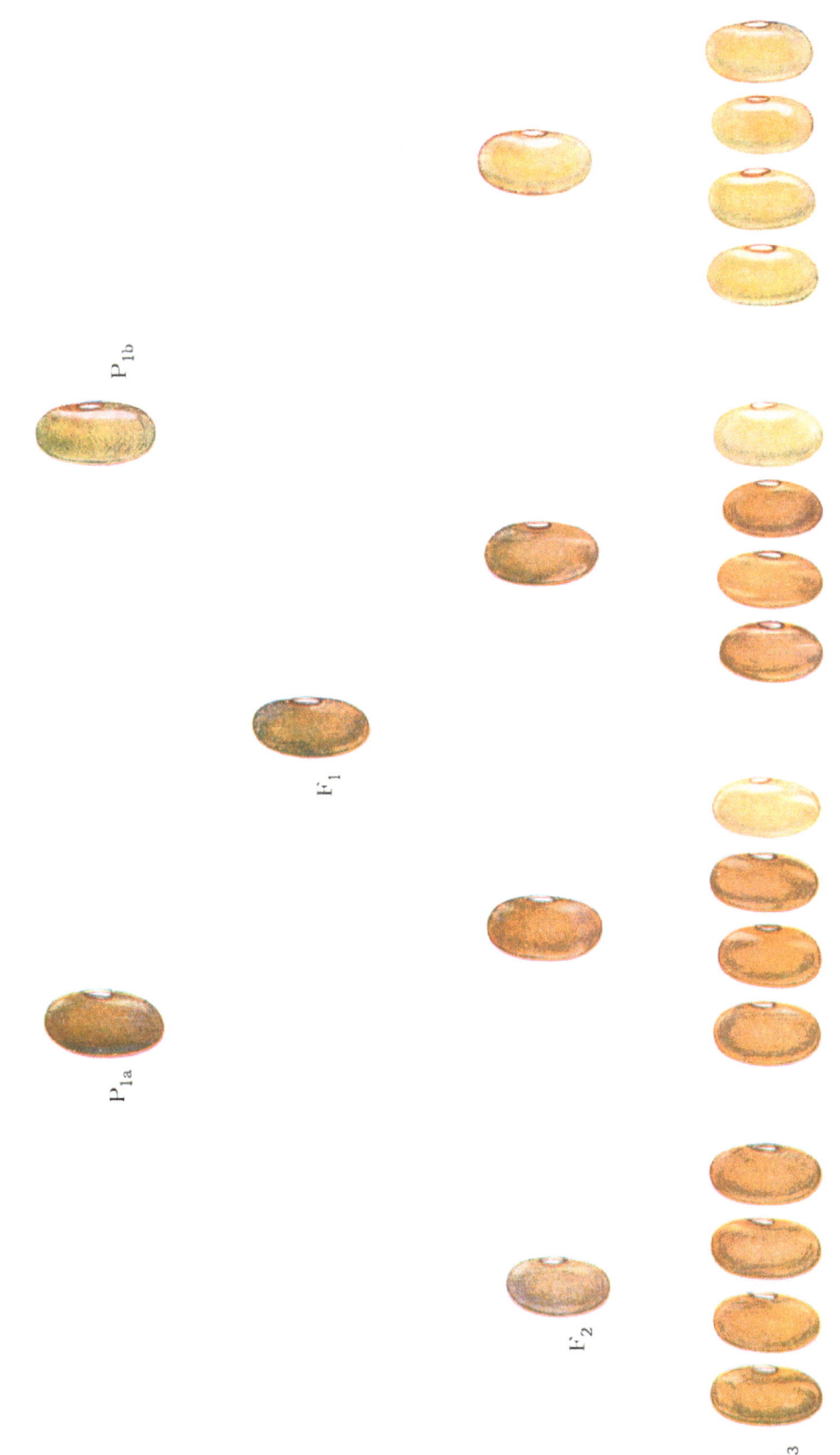

Cross between brown kidney bean and yellow kidney bean

Similar experiments were also carried out with animals. It can be seen that difficulties stand in the way of such a scheme since the self-fertilization, so often possible in plants, is out of the question in animals. By means of self-fertilization, an offspring forming an F_2-generation can be obtained from one single F_1-individual. In animals this result is only obtainable by the mating of two F_1-individuals. Still we can trace data in the literature which shows that monohybrid segregation can also occur here. For instance LANG (1904, 1905) bred several races of both the snail species *Helix (Cepaea) hortensis* and *H. nemoralis*, which are distinguishable by colour and the markings on the shells. Each race separately, was in itself a true bred so, for instance, a pair of snails with white unstriped shells never yielded anything but unstriped offspring.

Two individuals with longitudinally striped shells always yielded unexceptionally striped offspring. Mating of an unstriped with a striped animal in his experiments gave an offspring of 107 animals which were all (without exception) unstriped (1904, p. 482). Conclusion, the characteristic for unstriped shells is dominant in snails over the character for striped shells. If two of these hybrid snails with unstriped shells were mated, then an F_2-generation was obtained, which segregated into $3/_4$ unstriped individuals, and $1/_4$ striped individuals (LANG found for instance in some of his cultures, the ratio of 31 : 10).

Up to now the results of his investigations were an absolute parallel to those obtained in plants. But it is a matter of course that the breeding in the third generation did not proceed so regularly. The striped animals always yielded by mutual mating, only striped offspring. If two unstriped F_2-individuals were selected then either a wholly unstriped F_3-generation was obtained, or the offspring formed a mixed F_3-generation in which unstriped and striped individuals occurred side by side in the ratio of 3 : 1. Schematically this is represented in figure 28.

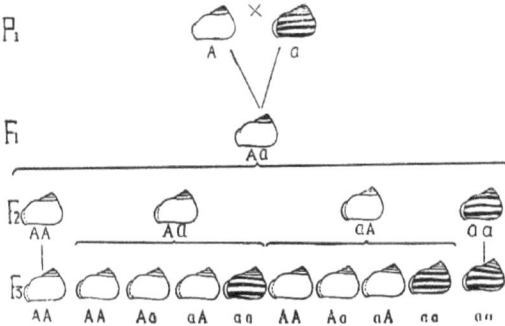

Fig. 28. Cross between a striped garden-snail and an unstriped garden-snail, F_1, F_2 and F_3 according to data from LANG (1904).

In general one can say that the previous cases confirm the conclusions made by MENDEL from his crosses of pea-types, which differed
in one heritable characteristic. But not all crosses follow this plan.
The fact is, that since the rediscovery of the results of the research of
MENDEL it has been shown more than once, that the dominance of
one characteristic over the other does not need to be complete, so that
the phenotype of the hybrid does not resemble one of the parents, but
lies between the phenotypes of both of the parents.

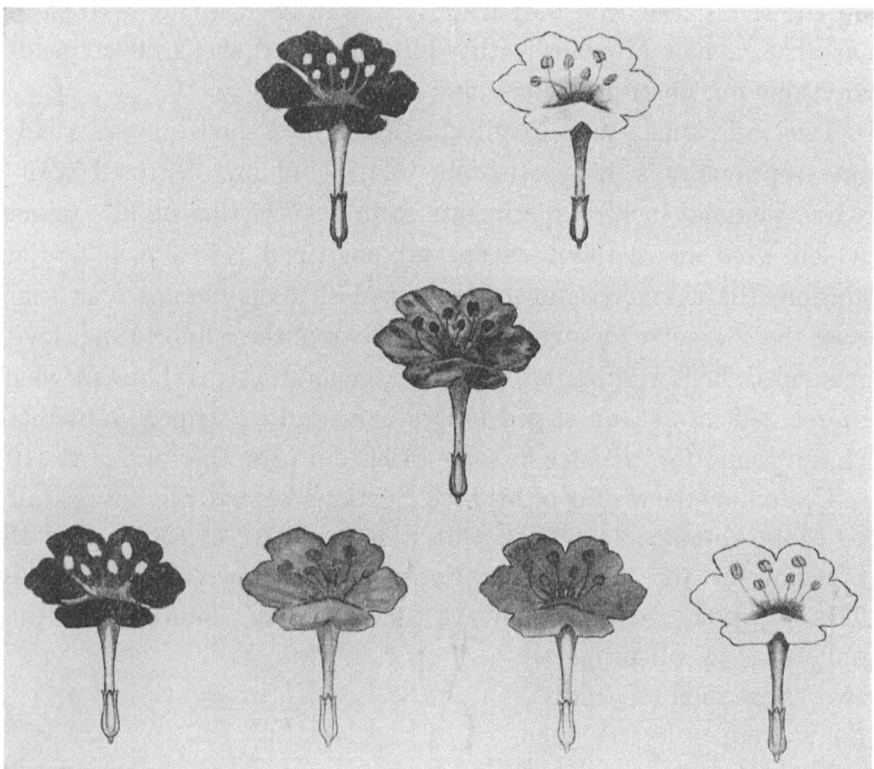

Fig. 29. A cross between *Mirabilis jalapa alba* and *Mirabilis jalapa rosea*.
Intermediate F₁ and segregating F₂ (after CORRENS, 1905).

This occurs in the hybrid made by CORRENS (1905, p. 19) between
two varieties of *Mirabilis jalapa*, viz., *alba* with white flowers and
rosea with dark pink flowers, which hybrids were, however, lighter
than those of the *rosea*-parent (figure 29). All the F₁-plants were light
pink. If the seeds of these F₁-plants were obtained by self-pollination

then the following appeared in the F_2-generation: the segregation of all individuals was 25% white-flowering, 50% light pink and 25% dark pink. Thus the segregation did not proceed as was always the case in MENDEL's own experiments, according to the scheme 3 : 1, but in the ratio of 1 : 2 : 1. This is to say, $^1/_4$ of the F_2-generation resemble the one parent, $^1/_4$ the other and $^1/_2$ had the external appearance of the F_1-plants. By continuation of the cultures, CORRENS obtained from the white flowered F_2-plants after self-pollination F_3-plants which had white flowers only; from the dark pink individuals, dark pink only, and from the light pink F_2-plants an F_3-generation of mixed composition (25% white, 50% light pink and 25% dark pink). Also the breeding of the F_4 generation proceeded in the same manner, so that white plants always gave a white offspring, and dark pink parents a dark pink one, whereas the light pink segregated into 1 : 2 : 1. Figure 30 gives a clear summary of these continued cultures. Each offspring is represented by 4 individuals and the whole offspring is connected with the parents.

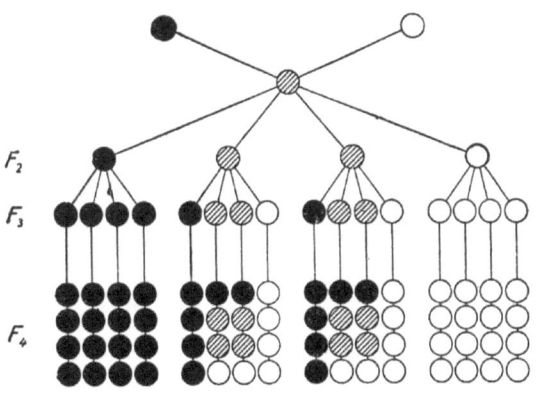

Fig. 30. Scheme of the pedigree of the cross recorded in figure 29 (after CORRENS, 1905).

The whole pedigree is pictured here, bearing in mind that the individuals belonging to the F_4 generation are not placed next to each other, but one under the other. Comparison of this pedigree with the scheme mentioned earlier (p. 85) of the cross DD × RR shows, that fundamentally no difference exists either between this cross and a cross in which dominance takes place but we deal in the cross DD × RR with hybrids DR (light pink) which is distinguishable by naked eye from the original DD parent (dark pink).

Monohybrids, viz., hybrids between two parents which differ from each other only in one characteristic can behave in two ways:

1) *The F_1-individuals are not distinguishable in external appearance from one of the parents, because the factor (the characteristic) of that parent*

is "dominant" over the other "recessive" one (according to MENDEL's
classic example indicated in the literature as Pisum-type);

2) *The F_1-individuals differ from both of the parents, but are intermediate in appearance (sometimes cited as the Mirabilis-type according to
the case of Mirabilis jalapa quoted earlier, and sometimes as the Zea-
type in other experiments of* CORRENS (1899), *preferably called intermediate monohybrids).*

*The dominant hybrids show a segregation in the F_2 into 3 individuals
having the dominant characteristic to one with the recessive. The intermediate monohybrid gives a segregation into 1 individual with the characteristic of the one parent, 2 intermediates with the phenotype of the F_1 and
1 with the character of the other parent.*

The dividing line between these two types is not always clear,
because the intermediates are in phenotype not always situated
exactly between both of the parents, but remind one sometimes very
much of one of them.

Years later, in renewed research, CORRENS pointed out, with regard
to this intermediate case that the dominance of the indented leaf-shape
of *Urtica pilulifera* over the entire shape of *U. Dodartii* is not complete.
By a very small, almost unobservable difference in dents, the F_1-
individuals are distinguishable from the pure *pilulifera's* (CORRENS
1918a, p. 222). The *dominance is thus incomplete* and it is better to
speak in such cases of *prevalence.*

It will be very easy to understand the explanation of these phenomena now we are so advanced in our knowledge of fertilization processes
and know so much more of the structure and action of the pollen and
ovules. It is unquestionable that the genius and mathematical insight
of GREGOR MENDEL was needed to find an explanation of the data at
his disposal in the year 1865. Moreover we have to admire the clarity
of MENDEL and see how he realizes the significance of his experiments
to "give information concerning the formation of the ovule and the
pollen of the hybrids", even more when we recall how primitive was
the knowledge at that time of the process of fertilization. "So far as
we can tell from our experiments, we always find that it can be proved
that a true-bred offspring can only be formed if the ovule and also the
fertilizing pollen exhibit similar tendencies. We must necessarily
accept also, that in a hybrid plant absolutely identical characteristics
cooperate in the origin of true-bred forms. Hence, the different true-

bred forms are formed on one single plant, even in the same flower and hence, the supposition must be correct, that in the ovaries of the hybrids, as many kinds of ovules and in the anthers as many kinds of pollen are formed as there are possible true-bred combinations. Also that these ovules and pollen-grains are in their internal (heritable) tendency, in accordance with the different forms" (1865, p. 23-24). With this, the root of heredity was uncovered. Later on this statement was to form the basis of all our research into heredity.

In honour of the discovery, this fundamental scientific fact is termed *"Mendelism"* and we say that a segregating hybrid plant, (which gives rise to a multiform offspring and behaves according to MENDEL's law) *"mendelizes"*. Plants which breed true to type give rise to only one kind of ovule and to one kind of pollen-grain. Hybrid plants on the other hand give rise to different kinds of ovules and pollen grains viz., in a monohybrid in which the parents differ in only one characteristic there are two kinds.

We will explain this for some of the previously described crosses: In the first place let us take the *Mirabilis* cross of CORRENS (p. 88). The parents were one plant of a white flowered constant type and one of a red flowered constant type. The first plant thus produced in its pollen-grains and in its ovules, the character (also called factor) for white

flowers. The fusion of such a pollen-grain and ovule will thus give rise to a white flowered plant. Two "red" germ-cells (a pollen-grain and ovule) give together a red flowered plant, but if we now bring together a "white" germ-cell (whether pollen-grain or ovule is immaterial) and a "red" germ-cell, then there will originate a hybrid plant with light pink flowers as we have already seen.

This is schematically represented in figure 31, where a germ-cell (Kc) or reproductive cell is pictured by a half circle, an individual by two half circles surrounded by a ring. The white P₁ give rise to "white" (r) germ-cells, the red

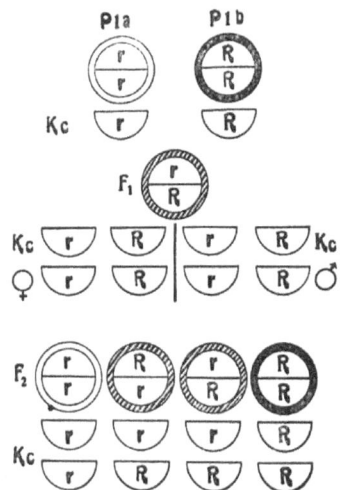

Fig. 31. Scheme of the *Mirabilis-*cross recorded in fig. 29, (Kc = germ cell, ovule or pollen grain) (after CORRENS, 1905).

P_1 to "red" germ-cells (R) only and a white germ-cell together with a red germ-cell give rise to light pink-flowering individuals. If now this pink plant goes into the production of pollen-grains and ovules, then according to the supposition of MENDEL, this plant will not make exclusively "pink" germ-cells, but "white" and "red" germ-cells, in equal numbers. If now the F_1 is self-pollinated so that an F_2 originates, then four possibilities can be realised:

a) white ovule (♀) × white pollen-grain (♂)
b) red   ovule   × white pollen-grain
c) white ovule   × red   pollen-grain
d) red   ovule   × red   pollen-grain

The possibility a) will give a white flowered plant, those of b) and c) give pink, and d) gives a red flowered plant. By supposing that these fusions have the same chance, then at the same time the striking segregation of the F_1 into 1 red: 2 pink: 1 white is explained.

The red and white plants, which originated from the germ-cells which are mutually of the same tendency, are called *homozygotes*, they are found by further self-pollination to be constant or true-breds, because they cannot give rise to other kinds of germ-cells, than those from which they originated. The pink F_1 and F_2 plants on the other hand originated from mutually different germ-cells. They are called *heterozygotes* and they in their turn give rise to germ-cells of two kinds of tendencies: "red" and "white". It is now a custom in genetics that this tendency for a definite colour of the flower is called a *character* or *factor* or more usually — a *gene*. Both *genes*, which form a pair, are indicated by the term *"allelomorphs"* or *"alleles"*. These heritable characters can be indicated by letters, viz., the character for the red flower colour by A and that for the white flower colour by a. So the homozygous red plant gets the formula AA, because this plant has inherited the gene A from the father and from the mother. The homozygous white plant will be aa and the heterozygous pink plant Aa. An AA plant gives rise only to A germ-cells, and aa plants only to a germ-cells. By fusion, of a pollen-grain A with an ovule a and the converse, there originated an Aa plant with pink flowers. This plant again gives pollengrains, 50% A and 50% a, and, likewise ovules. As a result of the self-pollination we then get the next table. Combination 1 will be a red flowered plant, 2 and 3 will be pink flowered and 4 a white flowering one. The offspring of self-pollination in number 1 cannot give flowers of a colour other than red; those of 4 when self-pollinated must always give white and those of 2 and 3 will always segregate again into 3 types as follows: 1 red: 2 pink: 1 white.

The preciseness of this view that the heterozygotes do not form pink germ-cells, but red and white can be proved by pollination of one of these pink heterozygotes with pollen of a recessive parental plant (in this case the white-flowered one) instead of with its own pollen. The white one is not exactly recessive, but is still comparable with a pure

♂ A		a
♂ A	1 AA	2 Aa
a	3 aA	4 aa

recessive, because the pink F_1 tends more to the red than to the white parent. If it is true that the pink heterozygotes form equal numbers of two kinds of ovules, red and white, and equal numbers of both kinds of pollengrains, then systematic "back-crossing" of

the heterozygote with the white parent will give rise to offspring from which half are pink-flowered and the other half are white-flowered plants. For it is the heterozygote Aa which makes 50% A ovules and 50% a ovules; the pollen grains are all a and thus the result of the hybridization will be (50% A + 50% a) × a = 50% Aa + 50% aa (or 50% pink and 50% white-flowered plants). This systematic back-crossing of a monohybrid F_1 with the recessive parental plant is shown in figure 32.

In a monohybrid, in which one of the characteristics is completely *dominant* (MENDEL's pea cross), the same reasoning is valid and the same explanation possible, but due to the fact that the heterozygote cannot be distinguished from the dominant homozygotes, these two forms are lumped together. The Aa plant does not differ

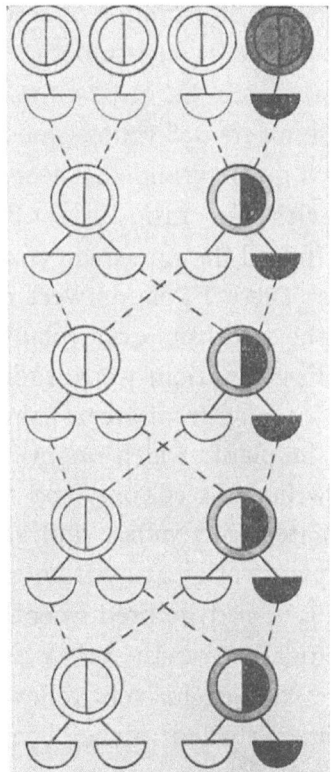

Fig. 32. Backcross of the homozygous recessive with F_1.

from the AA in phenotype and hence by investigation of the F_2-gene-
ration the AA and Aa plants and animals are counted together and
placed opposite the aa individuals (ratio 3 : 1). In the figure of the
snail cross (figure 28) these letter combinations are added to every
individual.

After MENDEL recognized the regular course and character of the
monohybrid-segregation in his experiments, he went further and
crossed peas with each other, which differ in more than one character
from each other, viz., an individual of a pea-type with yellow, smooth
seeds with a type in which the seeds were green and wrinkled. He again
found from this *dihybrid-cross*, regular and very remarkable proportions
of frequencies in the offspring. The F_1-peas were all, without exception,
yellow and smooth, which was in accordance with the previously
discussed dominance of yellow over green colour and smooth over
wrinkled seed-shape. The F_2-generation on the other hand was composed of 4 groups of seeds of different external appearance: yellow-
smooth, green-smooth, yellow-wrinkled and green-wrinkled, and
always in the constant ratio of 9 : 3 : 3 : 1. In fact, MENDEL found the
numbers 315 yellow-smooth: 108 green-smooth: 101 yellow-wrinkled:
31 green-wrinkled. Theoretically these numbers must be in accordance
with the ratio of 9 : 3 : 3 : 1 (or 312·75 : 104·25 : 104·25 : 34·75).
Indeed the agreement could not be better.

These F_2-plants were now kept under continuous self-pollination;
the seed from every plant was sown separately and the offspring (the
F_3-generation) were studied. It then became evident that the plants
obtained from green-wrinkled peas only had such seeds; one third of
the plants which originated from yellow-wrinkled seeds gave yellow-
wrinkled seeds only and so bred true to type, while the other $^2/_3$ segregated into yellow and green, but were always wrinkled. From all
plants originating from green-smooth seeds, all seeds were green, but
$^1/_3$ were true bred smooth, $^2/_3$ segregated into smooth and wrinkled.
Finally from the yellow-smooth seeds there originated plants of which
$^1/_9$ had exclusively yellow-smooth seeds, $^2/_9$ still segregated according
to the colour alone, $^2/_9$ according to the shape alone, while $^4/_9$ still
segregated into both of the characters. Hence, from the cross yellow-
smooth × green-wrinkled peas, 4 constant (i.e. breeding true to type)
descendants were obtained, viz., yellow-smooth, green-smooth, yellow-

wrinkled and green-wrinkled. The conclusion to which MENDEL came on the basis of his work was as follows: "so it has been proved that the behaviour of two different properties (or characters) in hybrids is independent of other differences between both of the parental plants" (1865, p. 22).

Analogous cases from the plant and animal world can be produced in great numbers. We will restrict ourselves to a few classical cases in which the exactitude of MENDEL's concept concerning the independent behaviour of different characters during segregation can clearly be seen. Among the maize types which are bred in agriculture, there are a number which differ in striking characters. There are races with blue grains and with yellow grains, types with smooth starchy endosperms, types with shrunken endosperms. If blue maize is crossed with yellow, then the F_1-seed is blue, from this a dominance of blue may be concluded. The cobs of the F_1-plants are composed of F_2 kernels, they are two-coloured and contain blue and yellow grains next to one another in the rate of 3 blue: 1 yellow. This colour is (as is also the reserve substance, starch or sugar) situated in the endosperm, which originates in an analogous manner as does the germ of a young plant, by fertilization of a maternal nucleus with a paternal nucleus. So an ear of a yellow-seeded plant which is pollinated exclusively and richly with pollen of a blue seedy strain, will bear nothing other than blue seeds. Likewise the smooth grain shape which contains starch is dominant over the shrunken sugary-one. If an ear of blue-sugary maize is pollinated with the pollen of a yellow-starchy one, then the seeds originating on this ear are blue and smooth, since blue is dominant over yellow and smooth over shrunken. If from these seeds, plants are bred and if the ears are pollinated with pollen of the same plant, then the result obtained will be an F_2-generation which is represented on one single ear. This F_2-generation is composed of blue-smooth, blue-shrunken, yellow-smooth and yellow-shrunken seeds and these groups are represented respectively by $9/16$, $3/16$, $3/16$ and $1/16$ of the total number of seeds; i.e. in the ratio of 9 : 3 : 3 : 1. This ratio was to be expected if the pairs of the characters (genes) blue-yellow and smooth-shrunken segregate independently from one another. Viz., from the total $3/4$ must be blue and $1/4$ yellow, from these blue ones $3/4$ are again smooth and $1/4$ shrunken and similarly for the yellow ones. From the total there must be thus $3/4 \times 3/4 = 9/16$ blue-smooth, $3/4 \times 1/4 = 3/16$

blue-shrunken, $^1/_4 \times ^3/_4 = ^3/_{16}$ yellow-smooth and $^1/_4 \times ^1/_4 = ^1/_{16}$ yellow-shrunken.

From 9 cobs, grown on F_1-plants and thus set with F_2-grains one is on plate II. The kernels of the four groups occur in the following numbers:

		1	2	3	4	5	6	7	8	9	total
numbers counted	blue-smooth	184	166	112	151	68	106	120	143	178	1226
	blue-shrunken	64	53	30	57	19	34	36	50	54	397
	yellow-smooth	59	57	32	50	23	40	41	50	51	403
	yellow-shrunken	21	17	13	15	6	14	12	18	25	141
	total	328	293	187	273	116	194	209	261	306	2167
calculated 9 : 3 : 3 : 1	blue-smooth	$184^1/_2$	$164^{13}/_{16}$	$105^3/_{16}$	$153^9/_{16}$	$65^1/_4$	$109^1/_8$	$117^9/_{16}$	$146^{13}/_{16}$	$172^1/_8$	$1218^{15}/$
	blue-shrunken	$61^1/_2$	$54^{15}/_{16}$	$35^1/_{16}$	$51^3/_{16}$	$21^3/_4$	$36^3/_8$	$39^3/_{16}$	$48^{15}/_{16}$	$57^3/_8$	$406^5/$
	yellow-smooth	$61^1/_2$	$54^{15}/_{16}$	$35^1/_{16}$	$51^3/_{16}$	$21^3/_4$	$36^3/_8$	$39^3/_{16}$	$48^{15}/_{16}$	$57^3/_8$	$406^5/$
	yellow-shrunken	$20^1/_2$	$18^5/_{16}$	$11^{11}/_{16}$	$17^1/_{16}$	$7^1/_4$	$12^1/_8$	$13^1/_{16}$	$16^5/_{16}$	$19^1/_8$	$135^7/$

The agreement between the actual numbers found and the theoretical numbers (according to the ratio 9 : 3 : 3 : 1) is such that we may conclude that an independent segregation of both characters took place.

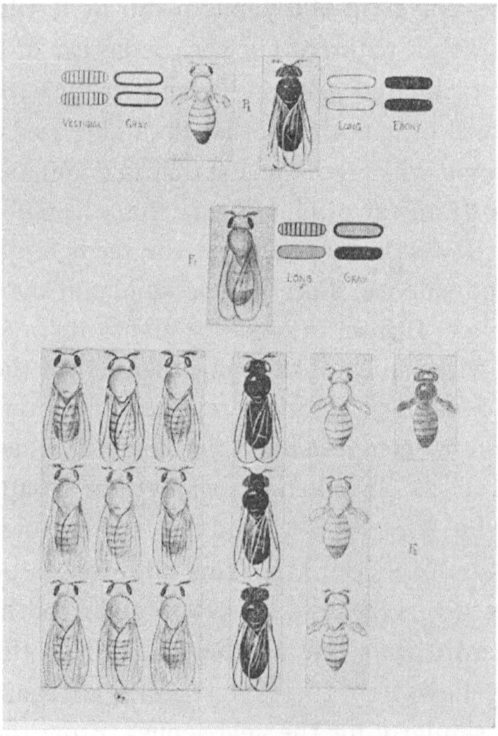

Experiments of MORGAN (1916, p. 56) have shown that in animals also, a similar segregation of characters independent from one another can be found. Among his numerous crosses with different forms of Drosophila melanogaster (the banana fly) to which we must repeatedly return, there was also a cross between a grey fly with rudimentary wings and a black one with well-developed wings. The F_1 was grey and had normal wings. From this it follows that the character grey (gray)

Fig. 33. Cross between two types of Drosophila, F_1 and F_2 (after MORGAN, 1916, p. 56).

Plate II

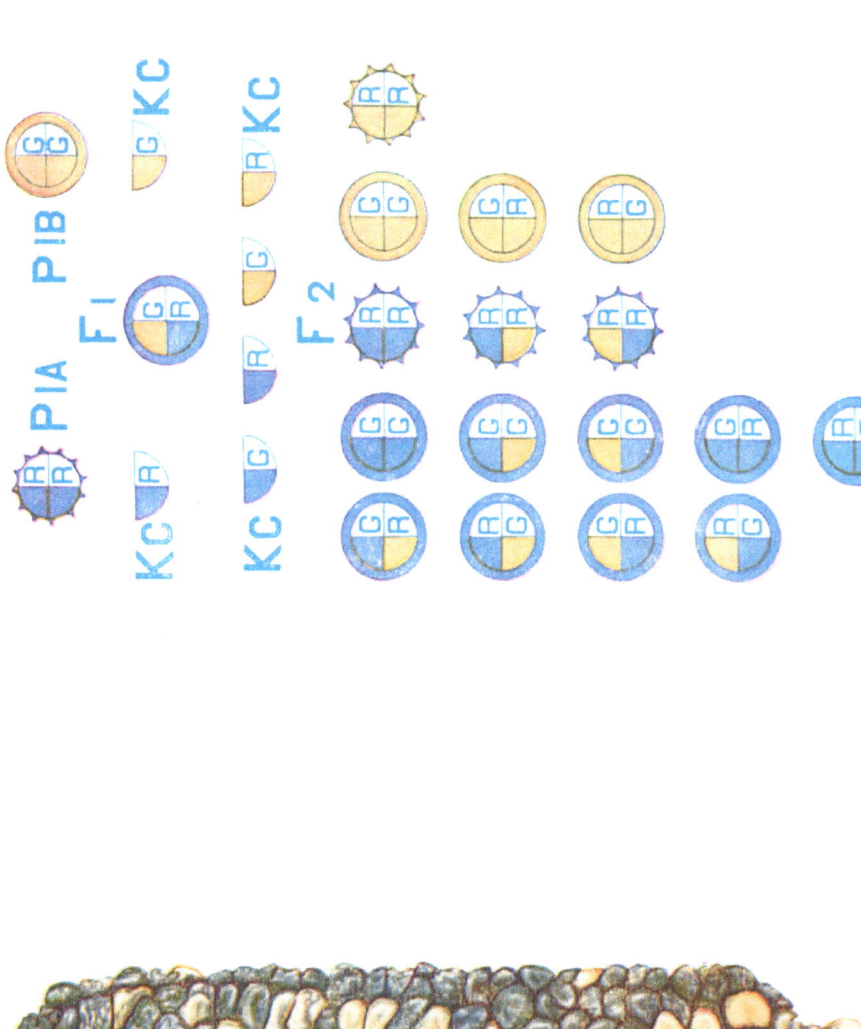

Cross between blue sugary corn and yellow starchy corn.
Cob with F₂-kernels and schematical explanation

is dominant over the character black (ebony) and the character for normal wings (long) over the rudimentary wings (vestigial). That these characters behave independently in the F_2 is evident from the ratio of the numbers: 9 grey with normal wings, 3 black with normal wings, 3 grey with vestigial wings, 1 black with vestigial wings (figure 33).

It is an obvious fact that we will only find the segregation ratio of $9 : 3 : 3 : 1$ in the F_1 if both the characters behave according to the dominance scheme. If one of the characters behaves according to intermediate patterns then the number per 16 individuals will naturally be different. This ratio is easily obtained, if the characters are considered individually. BAUR (1910, p. 82) made a cross between two types of *Antirrhinum majus* one with white bilateral symmetrical flowers (so-called zygomorphic flowers), the other with red five-sided symmetrical flowers (so-called peloric). The F_1 (figure 34) was pink and had zygomorphic flowers. Considering

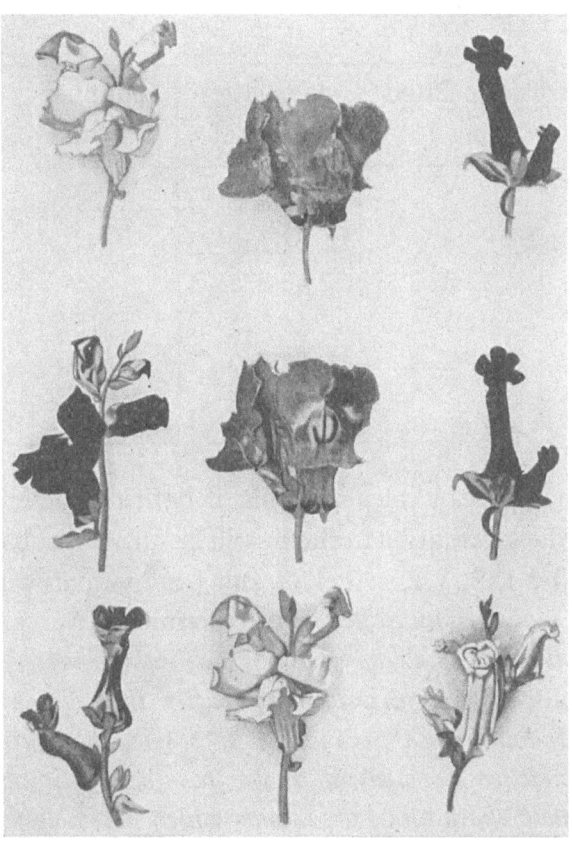

Fig. 34. Cross between white zygomorphic and red peloric *Antirrhinum majus*, F_1 and F_2 (after BAUR, 1911).

the colour alone, the hybrid behaved as an intermediary type, whereas the factor for shape behaved according to the scheme of dominance (bilateral symmetric is dominant over five-sided symmetric). The F_2 in consequence was not composed of 4, but of 6 groups: zygomorphic and peloric ones, both of the forms occurred in the colours red, pink and white. Hence, the segregation of the form had to take place according to the segregation 3 : 1. It could be expected that from the total of 16 F_2-plants, 12 would be zygomorphic and 4 peloric. The group of zygomorphic flowers must be distributed over red, pink and white flowers in the ratio 1 : 2 : 1, thus from the 12 zygomorphs 3 were red, 6 were pink and 3 white. From the 4 peloric flowers, 1 was red, 2 were pink and 1 white. Indeed BAUR found numbers which were in sufficient agreement with this (see the following table).

	calculated per 16	calculated per total	numbers found
zygomorphic red . . .	3	43·875	39
zygomorphic pink . .	6	87·750	94
zygomorphic white . .	3	43·875	45
peloric red	1	14·625	15
peloric pink	2	29·250	28
peloric white	1	14·625	13

In an analogous manner we can calculate, that if both characters give an intermediate F_1, the segregation in the F_2 will be into 9 groups, respectively represented by 1, 2, 1, 2, 4, 2, 1, 2, and 1 individuals per group. As an important conclusion from these experiments we can derive two theorems: *firstly (as was already pointed out in this chapter), that characters behaved during hybridization, and after this, during segregation with absolute independence. Secondly, it is possible to obtain as a result of this independence (by crossing of two plants or animals which differ in two heritable characters), organisms which assemble in themselves characters which originate in the different parents and which are transmitted constantly in this new combination to their offspring.* From the above-mentioned cross of MENDEL, viz., yellow-smooth with green-wrinkled peas, there originated the new combinations yellow-wrinkled and green-smooth peas, among which were those which bred true to type for these combinations. From a maize cross between blue-sugary and yellow-starchy, there originated the new combinations yellow-sugary and blue-starchy maize and also among these were individuals which bred true to type. From MORGAN's cross of a grey fly with rudimentary wings and a black fly with normal wings, there appeared in the F_2 the new combinations: black flies with rudimentary wings and grey flies with normal wings and among them were some which handed down this new combination constantly in their offspring. From BAUR's hybridization of the white zygomorphic *Antirrhinum* with the red peloric ones plants were produced with white peloric, and plants with red zygomorphic flowers, and these plants could be constant for this new combination.

Again we are indebted to MENDEL for the explanation of the course of such a dihybrid. The results of the discussed cross between the blue-sugary maize and yellow-starchy maize, where blue is dominant over yellow and starchy over sugary, can be treated under the same scheme or formula. In plate II, the course of this cross and its consequences is indicated in the same manner as is the *Mirabilis* cross in figure 31. The blue-sugary maize, belonging to a strain which breeds true to type, originates by the fusion of two germ-cells, which both contain the character blue and shrunken (sugary). In the same way, two germ-cells with the characters yellow and smooth (starchy) together give a homozygous yellow-starchy maize. Their cross, in which a "blue-shrunken" germ-cell fuses with a "yellow-smooth" one, results in a blue-smooth F_1 kernel and from this a plant grows up which in flowering time gives four sorts of reproductive cells viz., blue-smooth, blue-shrunken, yellow-smooth and yellow-shrunken. Thus the factors for blue and for yellow, form together a pair of "allelomorphs", the factor for smooth and for shrunken form another pair of allelomorphs.

If now blue is represented by A, yellow by a, smooth by B and shrunken by b (since blue and smooth are both dominant types), then the four kinds of pollen-grains (which are made by the F_1 plant in equal numbers) become AB, Ab, aB and ab. And consequently these four kinds occur in equal numbers among the ovules because the pairs Aa and Bb behave independently in segregation. So in the F_2 we obtain 4×4 or 16 possible combinations:

♂ \ ♀	AB	Ab	aB	ab
AB	1 AABB	2 AABb	3 AaBB	4 AaBb
Ab	5 AABb	6 AAbb	7 AaBb	8 Aabb
aB	9 AaBB	10 AaBb	11 aaBB	12 aaBb
ab	13 AaBb	14 Aabb	15 aaBb	16 aabb

It is obvious that all combinations of these sixteen in which A and B occur will be blue and smooth, that is, numbers 1, 2, 3, 4, 5, 7, 9, 10 and 13, i.e. a total of nine. Further there are those with A but without B and these will be blue and shrunken (6, 8 and 14), those without A but with B are yellow and smooth: 11, 12 and 15. Those without A and B (16) are yellow and shrunken. The total ratio of 9 : 3 : 3 : 1 is obvious. And it is also evident that in all 4 phenotypical groups (blue-smooth, blue-wrinkled, yellow-smooth and yellow-shrunken) there is one plant every time which is a homozygote for both of the factors. These homozygotes are successively given by the numbers 1, 6, 11 and 16. These are true breds and by selfpollination they give offspring which are absolutely identical. On plate II, all 16 combinations in the F$_2$ are pictured in colour and form; the 4 homozygotes are the last 4 of the upper row of the F$_2$.

As an explanation of MORGAN's cross with *Drosophila* (which is analogous to the latter) (figure 33) we refer to the scheme shown in figure 35, which needs no explanation after comparison with the scheme of the maize cross,

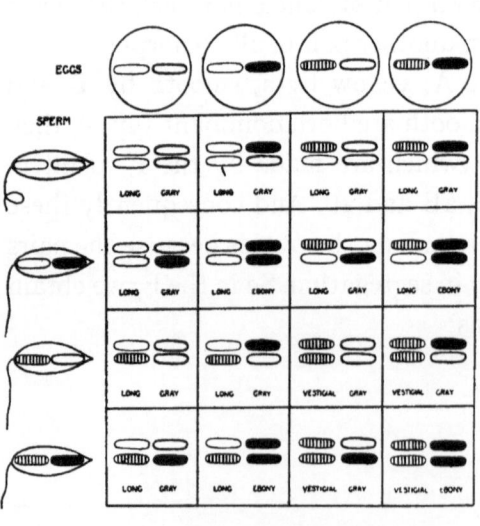

Fig. 35. Scheme of the cross recorded in fig. 33 (after MORGAN, 1916).

if we bear in mind the fact that long wings (normal) is represented by A, rudimentary (vestigial) wings by a, grey (gray) by B and black (ebony) by b. The marks, which MORGAN used for the four characteristics, are found in figure 33.

The cross of the *Antirrhinum majus* types of BAUR (figure 34) can also be explained in the same way (figure 36). As we saw already in the F_1, the bilateral symmetrical (zygomorphic) flowers are dominant over the five-sided symmetrical (peloric) ones and the F_1 is intermediate. The germ-cell from which the F_1 originates comes from a white zygomorphic plant (rrZZ) and from a red peloric (RRzz). These germ-cells are thus rZ and Rz and the hybrid RrZz. This dihybrid now makes four kinds of ovules, RZ, Rz, rZ and rz and also 4 kinds of pollen-grains. Sixteen possible kinds of combinations can also occur here in the F_2 as in the previously described AaBb segregation. From these sixteen possibilities 1, 2, and 5 are zygomorphic red, 6 is peloric red, 3, 4, 7, 9, 10 and 13 are zygomorphic pink, 8 and 14 peloric pink, 11, 12 and 15 zygomorphic white and 16 peloric white. Because they are homozygous in both characters the numbers 1 (zygomorphic red), 6 (peloric red), 11 (zygomorphic white) and 16 (peloric white) breed true to type. All the others remain heterozygous, either in one or both pairs of characters and thus give rise to more than one kind of germ-cell.

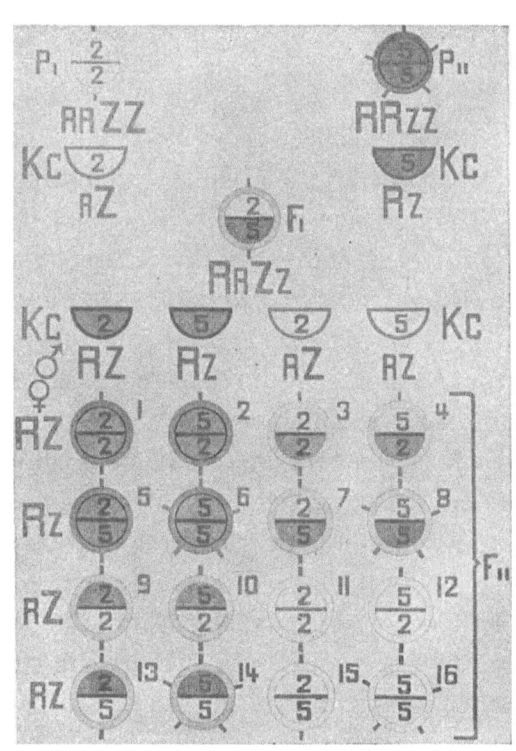

Fig. 36. Scheme of the cross recorded in fig. 34.

Comparison of figure 36 with this theoretical analysis explains the principle resemblance between the segregation into 9 : 3 : 3 : 1 and 3 : 6 : 3 : 1 : 2 : 1.

It is evident that the best way to show the "purity" of the repro-
ductive-cells in a dihybrid is by *back-crossing* this F_1 dihybrid with
an individual which is recessive for both pairs of characters.
Therefore in the *Antirrhinum* cross we can test the F_1 pink plant very
easily by crossing this F_1 plant with the peloric white form. The following result must be obtained: The F_1 plant as a double heterozygous
individual has the formula RrZz and thus gives rise to ovules of RZ,
Rz, rZ and rz; the white peloric one has the form rrzz and thus only
forms pollen-grains with the tendency rz. This back-cross must
necessarily give rise to an offspring with the composition (25% RZ +
25% Rz + 25% rZ + 25% rz) × rz = (25% RrZz + 25% rrZz +
25% Rrzz + 25% rrzz) or in words: $^1/_4$ of the total is pink zygomorphic,
$^1/_4$ pink peloric, $^1/_4$ white zygomorphic and $^1/_4$ white peloric.

Through such *back-crosses* we can test the hypothesis, concerning
similarity in number of the different sorts of germ-cells, and in many
cases a strict affirmation of MENDEL's supposition was obtained in this
way. If we go a step further and cross plants (or animals) which differ
from each other in three or more heritable characters, then naturally
all the rules, which were so far derived have a significance. Only, the
result of the experiments is more complicated and the working out
more difficult. In the case of absolute dominance we found that in
the F_2 of a monohybrid, a segregation occurred in the ratio 3 : 1. In
the dihybrid this was 9 : 3 : 3 : 1, in a trihybrid, (i.e. three differing
characters in parents) this will be 27 : 9 : 9 : 9 : 3 : 3 : 3 : 1. This is to
say that in the F_2 of a trihybrid in the case of absolute dominance
we can distinguish 8 groups of mutually similar individuals (where
external appearance or phenotype is concerned). These groups are as
follows: 27, 9, 9, 9, 3, 3, 3 and 1 individuals respectively (total number
of the F_2 equals 64).

MENDEL (1865, p. 20) made such a cross between peas of which one
had yellow smooth seeds with a brown seed-coat, the other green
wrinkled seeds with a white seed-coat. An F_2 was bred from this cross
and from the 693 individuals obtained, the eight groups theoretically
should contain 270, 90, 90, 90, 30, 30, 30, and 10 individuals respectively. In actual fact MENDEL found: 269, 98, 88, 86, 34, 30, 27 and 7
plants per group. In such complicated cross and relatively small F_2-
generation a better accordance could not be wished for. Again, as in
previous crosses, it was found that one individual from each of the

eight groups bred true to type. From these eight true breds there were 6 (2 were already used as parents) which displayed *new combinations* of the characteristics represented in the parents.

Finally we can further derive theoretically, that a trihybrid inherits its characteristics in the following manner, providing that the three characteristics are independent of one another. Suppose a plant which is homozygous for three dominant characteristics is crossed with another plant, homozygous for three recessive genes, then the following will be obtained:

$$P_{1a} \quad AABBCC \qquad\qquad aabbcc \quad P_{1b}$$
$$F_1 \qquad\qquad AaBbCc$$

Hence, germ-cells of the F_1 become ABC, ABc, AbC, aBC, Abc, aBc, abC, abc. The F_1 thus gives rise to eight sorts of pollen-grains and the same 8 sorts of ovules. These together give the following 64 combinations:

♀	ABC	ABc	AbC	aBC	Abc	aBc	abC	abc
ABC	1 AABBCC	2 AABBCc	3 AABbCC	4 AaBBCC	5 AABbCc	6 AaBBCc	7 AaBbCC	8 AaBbCc
ABc	9 AABBCc	10 AABBcc	11 AABbCc	12 AaBBCc	13 AABbcc	14 AaBBcc	15 AaBbCc	16 AaBbcc
AbC	17 AABbCC	18 AABbCc	19 AAbbCC	20 AaBbCC	21 AAbbCc	22 AaBbCc	23 AabbCC	24 AabbCc
aBC	25 AaBBCC	26 AaBBCc	27 AaBbCC	28 aaBBCC	29 AaBbCc	30 aaBBCc	31 aaBbCC	32 aaBbCc
Abc	33 AABbCc	34 AABbcc	35 AAbbCc	36 AaBbCc	37 AAbbcc	38 AaBbcc	39 AabbCc	40 Aabbcc
aBc	41 AaBBCc	42 AaBBcc	43 AaBbCc	44 aaBBCc	45 AaBbcc	46 aaBBcc	47 aaBbCc	48 aaBbcc
abC	49 AaBbCC	50 AaBbCc	51 AabbCC	52 aaBbCC	53 AabbCc	54 aaBbCc	55 aabbCC	56 aabbCc
abc	57 AaBbCc	58 AaBbcc	59 AabbCc	60 aaBbCc	61 Aabbcc	62 aaBbcc	63 aabbCc	64 aabbcc

The F_2 of a trihybrid-cross in which the three pairs of characters are independent from one another is thus composed of 64 combinations, belonging to 8 phenotypically (externally) different groups, providing that absolute dominance occurs in all three pairs of characters. In each of these eight groups one individual is represented which will be a true-bred for all of the characters, but only after self-pollination, or after fertilization with a genotypically similar individual. These 8 individuals are the combinations indicated above by the numbers 1, 10, 19, 28, 37, 46, 55 and 64. And from these 8 there are 2 which absolutely resemble the original parents in the cross, and 6 of which unite the characters of the three pairs in themselves in new and different ways.

For this reason MENDEL said very rightly "that in the ovary of the hybrids there are formed as many kinds of ovules and in the anthers as many pollen-grains as there are true breeding combinations in the F_2". This number is generally expressed by the number 2^n, if both of the original parents differ from one another in n characteristics and these characteristics behave independently from one another in segregation. By means of this we can observe in the F_2-generation of a polyhybrid an impressive diversity (see p. 37); if for example n = 30 (n equals the number of heterozygous pairs of characteristics in the F_1-individuals), then 2^{30} or more than a thousand millions different combinations, which all can be bred true to type, will be expected.

MENDEL's results concerning *monohybrids* can be summarized in rules of a mathematical nature:

1. In a cross between two plants (or animals) of different types, it does not matter which individual is chosen to be the father and which the mother (*the principle of reciprocity*).

2. All individuals of the F_1-generation are mutually similar in external appearance (*the principle of uniformity*).

3. The individuals of the F_1-generation always show the same characteristic as one of the parents (*principle of dominance*).

4. The individuals of the F_2-generation have no mutual resemblance and show in part the external appearance (*phenotype*) of the other parent. This resemblance always occurs in the proportions of 3 : 1 (*principle of segregation*).

With respect to the *di- and poly-hybrids* the following conclusion can be drawn:

5. The behaviour in segregation of each pair of characteristics in hybrids is independent of the other pairs of characteristics in which the parents can differ (*principle of independent assortment*).

6. As a result of this independent assortment it is possible to obtain, by means of crossing two plants (or animals) differing in more than one hereditary characteristic, individuals which unite characteristics in themselves, which originated from the different parents. These new combinations of characteristics are handed on as a constant to their offspring. As a result of later study of these conclusions it is clear that some and even important restrictions must necessarily be made. We will find later on in this book that these conclusions are not upheld in their entirety.

But this does not in the least decrease the great value which MEN-DEL's conclusions have for us today. This is due to their solid mathematical basis.

CHAPTER VI

THE RELIABILITY OF SEGREGATION RATIOS

Normal segregation of hybrids is accomplished only if the three following conditions are satisfied: 1) Each possible theoretical combination of genes has an equal chance to form, 2) All reproductive cells (gametes) formed must have the same chances to participate in a fertilization process, 3) The zygotes (fusion-product of two reproductive cells) must have equal possibilities of development.

No disturbing influence is theoretically allowed to affect either one of these conditions. Later on we will become acquainted with a number of instances in which the course of segregation pursues an abnormal course. The previously mentioned segregation ratio for a monohybrid (3 : 1 or 1 : 2 : 1) and that for dihybrids (9 : 3 : 3 : 1 or 3 : 6 : 3 : 1 : 2 : 1) is thus realized only if the formation and fusion of the reproductive cells is wholly controlled by chance. Even in this case it seldom happens that the observed segregation ratios are exactly in accordance with theoretical expectations: in most cases some of the phenotypical groups deviate a variable amount. The fact that the observed results are a product of chance explains the great or small differences between theory and observation.

This naturally results in a desire for a criterion with which to judge the reliability of segregation frequencies. Such a test of reliability of segregation frequencies must enable us to prove the validity of the conditions under discussion.

Again mathematics served biology very well in deriving a suitable statistical method with which to test the probability of the segregation frequencies obtained experimentally.

Thanks to its aid it is now possible to calculate a probability which borders upon certainty, if a fixed ratio obtained experimentally is to agree with a proposed mathematical scheme.

The starting-point for the calculation of this probability is the previously discussed calculation of the *Arithmetic Mean* of the *frequency distribution-curve* and the value of its *standard deviation* (σ). This last value (p. 28) has significance not only as a measure of the distribution of the frequency-curve, but also for the so-called "standard error" of one observation. Increase of the number of individuals (n) leads to decrease of the standard error. This decrease is not proportional to the number, but to the root. If the standard error is indicated by 'S.E.$\bar{x}$' then from this proportion a very important formula follows:

$$S.E.\bar{x} = \frac{\sigma}{\sqrt{n}}$$

The error-theory of GAUSS determined, that the mean of such a series of observations in the plus-direction as well as in the minus-direction can fluctuate with the values of this standard error of the Arithmetic Mean as its limits in both directions.

How can this view, which is valid for the frequency distribution-curve be applied to segregation frequencies of the F_2-generations? The answer to this question is very easy: in a segregation of a pea-hybrid, which is heterozygous in one gene (e.g. yellow and green colour of the seed cotyledons) both of the phenotypic classes into which the F_2 segregates, can be considered as two classes of a frequency distribution-curve in which the variable character behaves as an alternative one, that is to say the variable character has either the one or the other phenotype (with the class-interval = 1). In the previous example of MENDEL this segregation is thus class 0 (green seeds) represented by 2001 individuals, class 1 (yellow seeds) by 6022; hence, the total number of individuals (n) is 8023. We can draw a frequency distribution of this form

0	1
2001	6022

For this frequency distribution we can calculate the Arithmetic Mean ($\bar{x}$) in the manner given on p. 27 as well as the standard deviation (σ). Usually the individuals belonging to class 0 are indicated by p_0 and the individuals of class 1 by p_1 so that $n = p_0 + p_1$.

The Arithmetic Mean of this "variability distribution " becomes:

$$\bar{x} = \frac{p_0 \times 0 + p_1 \times 1}{p_0 + p_1} = \frac{p_1}{n}$$

If we now fix the provisional mean as 1, then the deviation (a) of the p_0 individuals from class 0 equals 1, those of the p_1 individuals from class 1 equals 0. The value of b (deviation of the provisional mean and the Arithmetic Mean) equals

$$1 - \frac{p_1}{n} = \frac{n - p_1}{n} = \frac{p_0}{n}$$

For the standard deviation σ, for which we calculated the formula:

$$\sigma = \pm \sqrt{\frac{\Sigma p a^2}{n} - b^2}$$

We now find a value of:

$$\sigma = \pm \sqrt{\frac{\Sigma p_0 \times 1 + p_1 \times 0}{n} - \frac{p_0^2}{n^2}} = \pm \sqrt{\frac{p_0}{n} - \frac{p_0^2}{n^2}}$$

$$= \pm \sqrt{\frac{n p_0 - p_0^2}{n^2}} = \pm \sqrt{\frac{(p_0 + p_1) p_0 - p_0^2}{n^2}}$$

$$\text{Hence } \sigma = \frac{\pm \sqrt{p_0 p_1}}{n}$$

As was already indicated, this standard deviation is expressed in the class-interval, which in this example equals 1. The formula for the absolute value of the standard deviation per n individuals is thus:

$$\sigma_{abs} = \pm \sqrt{p_0 p_1} \text{ (standard deviation of ratio)}$$

Between this standard deviation and the standard error the proportion was calculated : $\text{S.E.} \bar{x} = \dfrac{\sigma}{\sqrt{n}}$

Hence S.E. $\bar{x}_{abs.} = \pm \sqrt{\dfrac{p_0 p_1}{n}}$ (standard error of ratio).

In an ideal monohybrid segregation with dominance (3 : 1) the theoretical values of p_0 and p_1 will be respectively $p_0 = \frac{1}{4} n$ and $p_1 = \frac{3}{4} n$. If we substitute these values in the formula for S.E. $\bar{x}_{abs.}$ then it follows that

$$\text{S.E. } \bar{x}_{abs.\ 3\ :\ 1} = \pm \sqrt{\frac{\frac{1}{4}n \times \frac{3}{4}n}{n}} = \pm \sqrt{\frac{1n \times 3n}{16n}} = \pm \frac{1}{4} \sqrt{3n}$$

With the help of this formula $S.E.\bar{x}_{abs.\ 3:1}$ the standard error, can now be calculated for any value of n for 3 : 1 segregation.

In the following table all quadruples up to and including 400 are given and for all hundreds up to and including 2000 with the corresponding theoretical segregation ratios and the standard error found in each case.

If we have several F_2-families which can be expected to segregate according to the 3 : 1 scheme, then we can test this mathematically by dividing the deviation between the segregation frequencies observed and those calculated experimentally (the deviation D is positive in one group and negative in the other group) for a 3 : 1 ratio. The formula $D : S.E.\bar{x}_{abs.3:1}$ is of extreme importance. Hence the value of $S.E._{abs.3:1}$ is that one which goes with the n number of individuals in a 3 : 1 segregation.

n	3 : 1	$S.E.\bar{x}_{abs.3:1}$	n	3 : 1	$S.E.\bar{x}_{abs.3:1}$	n	3 : 1	$S.E.\bar{x}_{abs.3:1}$
4	3·0 : 1·0	± 0·866	136	102·0 : 34·0	± 5·050	268	201·0 : 67·0	± 7·089
8	6·0 : 2·0	± 1·224	140	105·0 : 35·0	± 5·123	272	204·0 : 68·0	± 7·141
12	9·0 : 3·0	± 1·500	144	108·0 : 36·0	± 5·196	276	207·0 : 69·0	± 7·194
16	12·0 : 4·0	± 1·732	148	111·0 : 37·0	± 5·268	280	210·0 : 70·0	± 7·246
20	15·0 : 5·0	± 1·940	152	114·0 : 38·0	± 5·339	284	213·0 : 71·0	± 7·297
24	18·0 : 6·0	± 2·118	156	117·0 : 39·0	± 5·408	288	216·0 : 72·0	± 7·348
28	21·0 : 7·0	± 2·289	160	120·0 : 40·0	± 5·477	292	219·0 : 73·0	± 7·399
32	24·0 : 8·0	± 2·448	164	123·0 : 41·0	± 5·545	296	222·0 : 74·0	± 7·450
36	27·0 : 9·0	± 2·601	168	126·0 : 42·0	± 5·612	300	225·0 : 75·0	± 7·500
40	30·0 : 10·0	± 2·740	172	129·0 : 43·0	± 5·679	304	228·0 : 76·0	± 7·550
44	33·0 : 11·0	± 2·872	176	132·0 : 44·0	± 5·745	308	231·0 : 77·0	± 7·599
48	36·0 : 12·0	± 3·000	180	135·0 : 45·0	± 5·810	312	234·0 : 78·0	± 7·648
52	39·0 : 13·0	± 3·123	184	138·0 : 46·0	± 5·874	316	237·0 : 79·0	± 7·697
56	42·0 : 14·0	± 3·241	188	141·0 : 47·0	± 5·937	320	240·0 : 80·0	± 7·746
60	45·0 : 15·0	± 3·355	192	144·0 : 48·0	± 6·000	324	243·0 : 81·0	± 7·794
64	48·0 : 16·0	± 3·464	196	147·0 : 49·0	± 6·062	328	246·0 : 82·0	± 7·842
68	51·0 : 17·0	± 3·570	200	150·0 : 50·0	± 6·124	332	249·0 : 83·0	± 7·890
72	54·0 : 18·0	± 3·674	204	153·0 : 51·0	± 6·185	336	252·0 : 84·0	± 7·937
76	57·0 : 19·0	± 3·775	208	156·0 : 52·0	± 6·245	340	255·0 : 85·0	± 7·984
80	60·0 : 20·0	± 3·874	212	159·0 : 53·0	± 6·305	344	258·0 : 86·0	± 8·031
84	63·0 : 21·0	± 3·969	216	162·0 : 54·0	± 6·364	348	261·0 : 87·0	± 8·078
88	66·0 : 22·0	± 4·061	220	165·0 : 55·0	± 6·423	352	264·0 : 88·0	± 8·124
92	69·0 : 23·0	± 4·153	224	168·0 : 56·0	± 6·481	356	267·0 : 89·0	± 8·170
96	72·0 : 24·0	± 4·243	228	171·0 : 57·0	± 6·538	360	270·0 : 90·0	± 8·216
100	75·0 : 25·0	± 4·330	232	174·0 : 58·0	± 6·595	364	273·0 : 91·0	± 8·262
104	78·0 : 26·0	± 4·416	236	177·0 : 59·0	± 6·652	368	276·0 : 92·0	± 8·307
108	81·0 : 27·0	± 4·501	240	180·0 : 60·0	± 6·708	372	279·0 : 93·0	± 8·352
112	84·0 : 28·0	± 4·584	244	183·0 : 61·0	± 6·764	376	282·0 : 94·0	± 8·397
116	87·0 : 29·0	± 4·664	248	186·0 : 62·0	± 6·819	380	285·0 : 95·0	± 8·441
120	90·0 : 30·0	± 4·743	252	189·0 : 63·0	± 6·874	384	288·0 : 96·0	± 8·485
124	93·0 : 31·0	± 4·821	256	192·0 : 64·0	± 6·928	388	291·0 : 97·0	± 8·529
128	96·0 : 32·0	± 4·899	260	195·0 : 65·0	± 6·982	392	294·0 : 98·0	± 8·573
132	99·0 : 33·0	± 4·975	264	198·0 : 66·0	± 7·036	396	297·0 : 99·0	± 8·617
						400	300·0 : 100·0	± 8·660

n	3 : 1	S.E.$\bar{x}_{abs.3:1}$	n	3 : 1	S.E.$\bar{x}_{abs.3:1}$
500	375·0 : 125·0	± 9·682	1300	975·0 : 325·0	± 15·612
600	450·0 : 150·0	± 10·606	1400	1050·0 : 350·0	± 16·202
700	525·0 : 175·0	± 11·456	1500	1125·0 : 375·0	± 16·770
800	600·0 : 200·0	± 12·247	1600	1200·0 : 400·0	± 17·320
900	675·0 : 225·0	± 12·990	1700	1275·0 : 425·0	± 17·853
1000	750·0 : 250·0	± 13·693	1800	1350·0 : 450·0	± 18·371
1100	825·0 : 275·0	± 14·361	1900	1425·0 : 475·0	± 18·875
1200	900·0 : 300·0	± 15·000	2000	1500·0 : 500·0	± 19·365

According to the rules of probability it can be pointed out that practically all deviations are situated between $+ 3$ S.E.$\bar{x}_{abs.}$ and $- 3$ S.E.$\bar{x}_{abs.}$, so that the ratio D : S.E.$\bar{x}_{abs.}$ (P.E. probable error) must always be smaller than 3. *If for this proportional number a value greater than 3 is obtained, it can be considered certain, that a disturbing influence is active.*

If we now test this rule for the segregation of monohybrids, which was mentioned in the previous chapter (peas), it is evident that the value for D : S.E.$\bar{x}_{abs.}$ is in no case more than 3. A value greater than 1 is seldom found and a value greater than 2 only occurs rarely. This is shown in the following table very clearly.

	n	found	expected	D	S.E.$\bar{x}_{abs.3:1}$	P.E.$_{3:1}$
Peas						
Mendel	7324	5474 : 1850	5493·00 : 1831·00	19·00	37·057	0·512
	8023	6022 : 2001	6017·25 : 2005·75	4·75	38·789	0·122
	929	705 : 224	696·75 : 232·25	8·25	13·198	0·625
	1181	882 : 299	885·75 : 295·25	3·75	14·881	0·251
	580	428 : 152	435·00 : 145·00	7·00	10·428	0·671
	858	651 : 207	643·50 : 214·50	7·50	12·648	0·591
	1065	788 : 277	798·75 : 266·25	10·75	14·131	0·761

In general we can say that a value of less than 1·5 for D : S.E.$\bar{x}_{abs.}$ proves an absolutely normal course of segregation. If the value for D : S.E.$\bar{x}_{abs.}$ lies between 1·5 and 3·0 it is suggested that there may be disturbing influences, and if D : S.E.$\bar{x}_{abs.}$ is greater than 3 we can accept such influence as existing.

It is evident that in the formula S.E.$\bar{x}_{abs.3:1} = {}^1/_4 \sqrt{3n}$ is only valid for the 3 : 1 segregation, and that other theoretical ratios will require other formula. If, for example, a segregation follows a true monohybrid scheme, however, we are able to distinguish the homozygous-dominant individuals from the heterozygotes then the theoretical segregation $1 : 2 : 1$ becomes valid as was shown in the last chapter in the example of CORRENS *Mirabilis*.

The previous formula is valid for both of the homozygous groups, which each amounts to $1/4$ of the total number of individuals, however, the formula is not valid for the group of heterozygotes, because the frequency of this group is 50% of the total. This latter group is in a $1 : 1$ (or $2 : 2$ for four individuals) proportion to the sum of both of the homozygous groups. In the case that $p_0 = 2$ and $p_1 = 2$ we calculated for the standard error of the ratio (S.E.$\overline{x}_{abs.}$):

$$\text{S.E. } \overline{x}_{abs \cdot 2:2} = \pm \sqrt{\frac{2/4 n \times 2/4 n}{n}} = + \sqrt{\frac{4n^2}{16n}} = \pm 1/4 \sqrt{4n}$$

Between this value for S.E.$\overline{x}_{abs2:2}$ and the previous formula S.E.$\overline{x}_{abs3:1}$ there exists a fixed relation, viz.,:

$$\text{S.E.}\overline{x}_{abs2:2} : \text{S.E.}\overline{x}_{abs \cdot 3:1} = 1/4 \sqrt{4n} : 1/4 \sqrt{3n} = \sqrt{4} : \sqrt{3} = 1 \cdot 155 : 1.$$

From the values which are given in the table for S.E.$\overline{x}_{abs \cdot 3:1}$ we can easily derive the corresponding values for S.E.$\overline{x}_{abs \cdot 2:2}$ by multiplying these by the factor $1 \cdot 155$.

In this way we can calculate from the general formula S.E.$\overline{x}_{abs \cdot} =$

$\pm \sqrt{\dfrac{p_0 \, p_1}{n}}$ the values which correspond to each value of n (for every phenotypical group and for all theoretical segregation ratios). For a dihybrid with dominance in both its characters (factors) a segregation according to $9 : 3 : 3 : 1$ can be expected. Each of these phenotypical groups has its own characteristic S.E.$\overline{x}_{abs.}$ — value dependent on the total number of individuals (n). Against the 9 individuals with both of the dominant characteristics, there are 7 which have either one or no dominant gene. In this case $p_1 = 9n$ and $p_0 = 7n$ and S.E.$\overline{x}_{abs.} =$

$\pm \sqrt{\dfrac{p_0 \, p_1}{n}}$ can be written as:

$$\text{S.E.}\overline{x}_{abs \cdot 9:7} = \pm \sqrt{\frac{9/16 n \times 7/16 n}{n}} = \pm 1/16 \sqrt{9 \times 7n} = \pm 1/16 \sqrt{63n}.$$

Each of both groups with one dominant and one recessive gene is represented by $3/16$ of the number of n individuals against $13/16$ n others. For this group the general formula becomes:

$$S.E.\bar{x}_{abs.} = \pm \sqrt{\frac{P_1\, P_0}{n}}$$

in $S.E.\bar{x}_{abs\cdot 3:13} = \pm \sqrt{\frac{{}^{3}/_{16}n \times {}^{13}/_{16}n}{n}} \pm {}^{1}/_{16}\sqrt{3 \times 13n} = \pm 16\sqrt{39n}.$

And for the group with both of the recessive genes (${}^{1}/_{16}$ of the total) this formula will be

$$S.E.\bar{x}_{abs\cdot 1:15} = \pm \sqrt{\frac{{}^{1}/_{16}n \times {}^{15}/_{16}n}{n}} = \pm {}^{1}/_{16}\sqrt{1 \times 15n} = \pm {}^{1}/_{16}\sqrt{15n}.$$

The three values of $S.E.\bar{x}_{abs.}$ for the phenotypical groups of a dihybrid segregation with dominance in both genes have a mutual proportion of $\sqrt{63} : \sqrt{39} : \sqrt{15}$, while their relation to the $S.E.\bar{x}_{abs\cdot 3:1}$ (3 : 1 segregation) can be calculated as follows:

Segregation 9 : 7: segregation 3 : 1 = ${}^{1}/_{16}\sqrt{63} : {}^{1}/_{4}\sqrt{3} = 1\cdot146 : 1$
Segregation 3 : 13: segregation 3 : 1 = ${}^{1}/_{16}\sqrt{39} : {}^{1}/_{4}\sqrt{3} = 0\cdot901 : 1$
Segregation 1 : 15: segregation 3 : 1 = ${}^{1}/_{16}\sqrt{15} : {}^{1}/_{4}\sqrt{3} = 0\cdot559 : 1$

We can thus calculate from the above mentioned table of the $S.E.\bar{x}_{abs\cdot 3:1}$ — values for a monohybrid 3 : 1 segregation the corresponding $S.E.\bar{x}$ — values for the phenotypical groups of a dihybrid 9 : 3 : 3 : 1 segregation.

Hence we are able, with help of a standard list of $S.E.\bar{x}_{abs.}$ values for 3 : 1 segregations, to derive any $S.E.\bar{x}_{abs.}$ — value, by use of a particular factor. Some of the multiplying-factors which occur most frequently are:

$S.E.\bar{x}$ abs.1:1	$= 1\cdot155 \times S.E.\bar{x}$ abs.3:1	*Backcross monohybrid recessive*
$S.E.\bar{x}$ abs.9:7	$= 1\cdot146 \times S.E.\bar{x}$ abs.3:1	*dihybrid* with segregation
$S.E.\bar{x}$ abs.3:13	$= 0\cdot901 \times S.E.\bar{x}$ abs.3:1	9 : 3 : 3 : 1.
$S.E.\bar{x}$ abs.1:15	$= 0\cdot559 \times S.E.\bar{x}$ abs.3:1	(dominance in both of the genes).
$S.E.\bar{x}$ abs.6:10	$= 1\cdot118 \times S.E.\bar{x}$ abs.3:1	*dihybrid* with the segregation
$S.E.\bar{x}$ abs.3:13	$= 0\cdot901 \times S.E.\bar{x}$ abs.3:1	6 : 3 : 3 : 2 : 1 : 1 (dominance of
$S.E.\bar{x}$ abs.2:14	$= 0\cdot704 \times S.E.\bar{x}$ abs.3:1	one gene, intermediate pheno-
$S.E.\bar{x}$ abs.1:15	$= 0\cdot559 \times S.E.\bar{x}$ abs.3:1	type of the other gene).
$S.E.\bar{x}$ abs.4:12	$= \qquad S.E.\bar{x}$ abs.3:1	*dihybrid* with the segregation-
$S.E.\bar{x}$ abs.2:14	$= 0\cdot764 \times S.E.\bar{x}$ abs.3:1	4 : 2 : 2 : 2 : 2 : 1 : 1 : 1 : 1 (in-
$S.E.\bar{x}$ abs.1:15	$= 0\cdot559 \times S.E.\bar{x}$ abs.3:1	termediate behaviour of both of the factors).
$S.E.\bar{x}$ abs.27:37	$= 1\cdot139 \times S.E.\bar{x}$ abs.3:1	*trihybrid* with dominance in all
$S.E.\bar{x}$ abs.9:55	$= 0\cdot101 \times S.E.\bar{x}$ abs.3:1	three genes (27 : 9 : 9 : 9 : 3 : 3:
$S.E.\bar{x}$ abs.3:61	$= 0\cdot488 \times S.E.\bar{x}$ abs.3:1	3 : 1).
$S.E.\bar{x}$ abs.1:63	$= 0\cdot286 \times S.E.\bar{x}$ abs.3:1	

It is possible to test the segregation frequency reported in the previous chapter dealing with dihybrids (dominance in 2 factors 9 : 3 : 3 : 1 (peas and maize), and dominance in one factor 6 : 3 : 3 : 2 : 1 : 1 (*Antirrhinum*)) and for the trihybrid we will take the pea-segregation of MENDEL (27 : 9 : 9 : 9 : 3 : 3 : 3 : 1) (table p. 114).

It is again evident from this table that the probable error (P.E.) is for most of the cases less than 1, and that only in two cases is this value slightly exceeded. At any rate this ratio is always far below the critical limit of 3.

The above mentioned procedure may be enough to illustrate the way in which probability and certainty of the segregation frequencies is assessed. Calculation of the ratio per n individuals expected, the finding of the deviation (D) from observations and theoretical number, and fixing of the $S.E.\bar{x}_{abs.}$ – value, which belongs to every phenotypical group of n-individuals, give all the values necessary for substituting in : $D : S.E.\bar{x}_{abs.}$ (P.E.).

Thus if the quotient (P.E.) is less than 1·5, we can say that the segregation took place as expected; if the quotient lies between 1·5 and 3, the possibility exists of disturbing influences, and lastly, if the quotient is greater than 3, this possibility becomes practically certain.

In this connection it must be emphatically pointed out that there exists a danger of misinterpreting the results, if the data of different experiments (with apparently equivalent material, and the same apparent characteristic) is lumped together. Even the grouping of F_2-families, originating from different F_1-individuals into one large F_2-generation, can be the cause of a mistake. It is not impossible that such a grouping of data can lead to wrong conclusions, and to different segregation frequencies in apparently analogous investigations. Indeed this is evident from the results obtained by several investigators dealing with the heredity of the formation of tendrils on the leaf of peas. Besides the normal type of leaf, which occurs in the majority of pea-stocks (in which the top-leaf in the compound leaf is changed into a tendril), there are also some types with the so-called acacia-leaves which have an ordinary top-leaf. The gene for the tendril form is dominant; the other one for acacia leaves is recessive.

On the grounds of some experiments it seems that it should be accepted that the F_2-generations segregate wholly according to the

Dihybrids.

	n	found a : b : c : d	theoretically a : b : c : d	D a	D b	D c	D d	S.E.$_{\bar{x}abs}$ a	S.E. b	S.E. c	S.E. d	P.E. a	P.E. b	P.E. c	P.E. d
Peas Mendel	555	315 : 108 : 101 : 31	312·188 : 104·062 : 104·062 : 34·688	2·812	3·938	3·062	3·688	11·690	9·191	9·191	5·701	0·240	0·428	0·333	0·647
Mays Sirks	328	184 : 64 : 59 : 21	184·500 : 61·500 : 61·500 : 20·500	0·500	2·500	2·500	0·500	8·986	7·065	7·065	4·383	0·056	0·354	0·354	0·114
	293	166 : 53 : 57 : 17	164·812 : 54·938 : 54·938 : 18·312	1·188	1·938	2·062	1·312	8·494	6·678	6·678	4·143	0·139	0·290	0·309	0·317
	187	112 : 30 : 32 : 13	105·188 : 35·062 : 35·062 : 11·688	6·812	5·062	3·062	1·312	6·785	5·334	5·334	3·310	1·004	0·949	0·574	0·396
	273	151 : 57 : 50 : 15	153·562 : 51·188 : 51·188 : 17·062	2·562	5·812	1·188	2·062	8·198	6·447	6·447	3·999	0·312	0·901	0·184	0·516
	116	68 : 19 : 23 : 6	65·250 : 21·750 : 21·750 : 7·250	2·750	2·750	1·250	1·250	5·244	4·200	4·200	2·707	0·524	0·655	0·298	0·466
	194	106 : 34 : 40 : 14	109·125 : 36·375 : 36·375 : 12·125	3·125	2·375	3·625	1·875	6·911	5·434	5·434	3·371	0·452	0·437	0·667	0·552
	209	120 : 36 : 41 : 12	117·562 : 39·188 : 39·188 : 13·062	2·438	3·188	1·812	1·062	7·174	5·645	5·645	3·503	0·340	0·565	0·321	0·303
	261	143 : 50 : 50 : 18	146·812 : 48·938 : 48·938 : 16·312	3·812	1·062	1·062	1·688	8·016	6·302	6·302	3·910	0·475	0·168	0·168	0·432
	306	176 : 54 : 51 : 25	172·125 : 57·375 : 57·375 : 19·125	3·875	3·375	6·375	5·875	8·680	6·825	6·825	4·233	0·446	0·494	0·934	1·388

Antirrhi-num

	n	found a : b/c : d/f : e/f	theoretically a : b/c : d : e/f	D a	D b/c	D d	D e/f	S.E.$_{\bar{x}abs}$ a	S.E. b/c	S.E. d	S.E. e/f	P.E. a	P.E. b/e	P.E. d	P.E. e/f
Baur	234	94 : 39 : 28 : 15 / 45 : 13	87·750 : 43·875 : 29·250 : 14·625	6·250	4·875 / 1·125	1·250 / 1·625	0·375	7·405	5·968	5·060	3·703	0·844	0·817 / 0·188	0·247	0·101 / 0·439

Trihybrids.

Peas Mendel

	n	found a : b/c/d : e/f/g : h	theoretically a : b/c/d : e/f/g : h	D a	D b/c/d	D e/f/g	D h	S.E.$_{\bar{x}abs}$ a	S.E. b/c/d	S.E. e/f/g	S.E. h	P.E. a	P.E. b/e/d	P.E. e/f/g	P.E. h
Mendel	639	269 : 98 / 88 / 86 : 34 / 30 / 27 : 7	269·573 : 89·858 : 29·953 : 9·984	0·573	8·142 / 1·858 / 3·858	4·047 / 0·047 / 2·953	2·984	12·467	8·767	5·341	3·130	0·046	0·929 / 0·212 / 0·440	0·758 / 0·009 / 0·553	0·953

theoretical ratio of 3 : 1; other experiments have produced a considerable deficit of tendril-forming plants.

Investigator	n	tendril	acacia	expectation	D.tendril	D.acacia	S.E. xabs.3:1	P.E. 3:1
BATESON (cit.a. VILMO-RIN)	279	211	68	209·25 : 69·75	+ 1·75	— 1·75	± 7·23	0·24
MEUNISSIER (1922) .	49	36	13	36·75 : 12·25	— 0·75	+ 0·75	± 3·03	0·24
PELLEW (1913) . . .	1038	768	270	778·50 : 259·50	— 10·50	+ 10·50	± 13·95	0·76
WHITE (1917)	59	45	14	44·25 : 14·75	+ 0·75	— 0·75	± 3·33	0·22
PELLEW (1913) . . .	2065	1481	584	1548·75 : 516·25	— 67·75	+ 67·75	± 19·68	3·44
VILMORIN (1910a) . .	190	118	72	142·50 : 47·50	— 24·50	+ 24·50	± 5·97	4·10
VILMORIN (1910b) . .	274	174	100	205·50 : 68·50	— 31·50	+ 31·50	± 7·17	4·39
VILMORIN and BATE-SON (1912)	449	322	127	336·75 : 112·25	— 14·75	+ 14·75	± 9·17	1·60

The above mentioned table, borrowed from WELLENSIEK (1925,
p. 439) which is given here with the values of P.E. calculated, gives a
clear picture of it. It can be concluded from the first four experiments
that the segregation of tendril: acacia is in accordance with the
theoretical scheme. The P.E.-values are all less than 1, and give, in
two cases a surplus of tendril-forming plants over the theoretical value
and in the other segregations a small surplus of acacia-leaf types. In
sharp contrast to this are the four other results, not only are the
P.E.-values too high (in one case between 1·5 and 3 and in three cases
more than 3), but it is moreover the tendril type, which now always
indicates an important deficit from the theoretical value, while on the
other hand the acacia-leaf individuals are represented by too great a
number. The cause of this surplus of acacia-leaf individuals cannot be
explained at once, but it is certain, that a disturbing influence is
active. A possible cause can be suggested: the pollen tubes which have
the recessive acacia-factor grow more quickly than the pollen tubes
which possess the tendril-forming factor, so that the latter participate
in a smaller number in the fertilization.

We will become acquainted later on (chapter XX) with this phenomenon, which is known as *certation* (competition). It will then be
evident that such a certation-factor in different, but apparently
equivalent, individuals can be present to different degrees and because
of this it is not permissible to group the F_2-families, which descend
from different F_1-individuals together, and to calculate the P.E. value
for such a total.

But whatever the explanation may be, it is evident from the discussion that the compilation of results from apparently analogous investigations can be very dangerous for the following reasons: 1) the exact course of some segregations can be lost and 2) accidental disturbances can be considered to be quite common and they hinder a more detailed study of these primary influences.

Another example of the above mentioned danger lies in combining the results of reciprocal crosses.

Whereas MENDEL's investigations (see p. 83) led to the conclusion, that it should be a matter of indifference as to which individual is chosen as father and which as mother (since the results of both of the crosses should be the same); later experiments showed some exceptions to this rule. Later on in the book (chapter X) this subject will be more fully discussed. At this stage it will be sufficient to give an example: i.e., crosses between both sub-species of the horse-bean *Vicia Faba major* and *Vicia F. minor* (SIRKS, 1931), have shown a difference between the reciprocal crosses, with respect to a number of genes. For example, in relation to the habit and also to the occurrence of a black spot in the flower. The dominant E-factor is responsible for the erect manner of growth and the recessive one causes a recumbent habit. The black spot in the flower is caused by a dominant O-gene, of which the recessive form causes the flower to be absolutely white.

Some F_2-generations, which were descended reciprocally from the same parental plants, were composed as follows:

Habit E—e mother father	n	E	e	theoretical segregation	D	P.E.$_{3:1}$
major × minor	293	212	81	219·75 : 73·25	7·75	1·05
minor × major	161	161	0	120·75 : 40·25	40·25	7·32

Black spot O—o mother father	n	O	o	theoretical segregation	D	P.E.$_{3:1}$
major × minor	1162	867	295	871·50 : 290·50	4·50	0·32
minor × major	944	944	0	708·00 : 236·00	236·00	17·68

From both these examples it is evident, that we must be very careful in presenting the results of reciprocal crosses (even if these are made with the same individuals as the parents), because these F_2-families can show a great variation in composition.

Several other causes which influence the course of segregation can be enumerated. We will have the opportunity to make our acquaintance with these later, but for the present the ones already discussed may be sufficient to make it clear that in counting together apparently analogous segregations, the great danger exists that we may hide the true result.

The improbability of some segregation frequencies is especially noticeable if, for instance, the offspring of one single parent or one pair of parents is very small in number; for example in the case of the higher domestic animals and man. This is already evident from the great value of the standard error (S.E.$\bar{x}_{abs.}$) for such a small number of offspring. Indeed the value of four descendants is $\pm 0 \cdot 866$ (see table p. 109), for eight descendants $\pm 1 \cdot 224$ and for twelve descendants $\pm 1 \cdot 500$. The admissible limits are three times these values, respectively 2·598, 3·672 and 4·500. One might at first think that the combination of a great number of segregating offspring must lead to more probable segregation frequencies, but apart from the afore-mentioned covering up the results, to which such a compilation can lead, there exists yet another objection. There is a probability that such a small offspring from one single pair of parents, which are both monohybrids (Aa $\times$ Aa), will not include recessive individuals. This probability is quite high and can be expressed by a general formula.

For example, among a total of n children, there will arise a possible total of 2^{2n} combinations within the n children.

n dom.	n-1 dom.	n-2 dom.	n-3 dom.
0 rec.	1 rec.	2 rec.	3 rec.
3^n	$3^{(n-1)} \times \dfrac{n}{1}$	$3^{(n-2)} \times \dfrac{n(n-1)}{1 \times 2}$	$3^{(n-3)} \times \dfrac{n(n-1)\,(n-2)}{1 \times 2 \times 3}$

If we calculate for the numbers n $= 1$ up to and including 7 individuals per offspring, the chances for a number of dominant characteristics (d) appearing in any one group of offspring, numbering 1-7 individuals are:

n =	1	2	3	4	5	6	7
d = 7							2187
d = 6						729	5103
d = 5					243	1458	5103
d = 4				81	405	1215	2835
d = 3			27	108	270	540	945
d = 2		9	27	54	90	135	189
d = 1	3	6	9	12	15	18	21
d = 0	1	1	1	1	1	1	1
Total 2^{2n}	4	16	64	256	1024	4096	16384

This means, that in an offspring (from Aa × Aa) composed of one individual, there are three chances that this individual has the dominant gene to one, that it carries the recessive gene.

From an offspring composed of two individuals there will be 9 (out of 16) chances that both show the dominant characteristic, to 6 (out of 16) that one of them is dominant, to 1 (out of 16) that both are recessive. In an offspring of three individuals there are 27 (out of 64) chances that all of the three are dominant, to 27 that two dominant individuals occur and one recessive, to 9 that one is dominant accompanied by two recessives and one chance (out of 64) that all of the three have the recessive phenotype.

Due to a lumping together of segregating offspring the difficulty will always exist that a certain number of families may be composed of phenotypical dominants only and that the segregation which is theoretically expected for such a family is not shown up. These families do not influence the number of phenotypical recessives, but they should be still taken into consideration.

An easy example with which to illustrate this occurrence is as follows: from 256 families, each with four children, it may be expected that for a monohybrid segregation (i.e., segregation for one character or factor) there will be approximately 256 children out of 1024 which should carry this recessive character. But for four children per family there exists a chance (in the ratio of 81 to 256) that none of the four children will show the recessive character (aa), so that in looking for these recessive individuals in an offspring of n = 4, these families cannot be taken into the research counting since we do not have any evidence that the offspring really originate from a monohybrid (Aa × Aa) as none of the children will show this recessive character. Only 175 families are therefore observed, this gives a total of 700 children, among which there are 256 recessive. According to the table of probability there are in total:

1)   81 families with 4 dom: 0 rec. or 324 dom.:   0 rec.
2) 108 families with 3 dom.: 1 rec. or 324 dom.: 108 rec.
3)   54 families with 2 dom.: 2 rec. or 108 dom.: 108 rec.
4)   12 families with 1 dom.: 3 rec. or 12 dom.: 36 rec.
5)   1 family with 0 dom.: 4 rec. or   0 dom.:   4 rec.

In reaching a total, the families under (1) will not be included, i.e., they are excluded; the others 2, 3, 4 and 5 yield a total of 175 families with 700 children, from which 444 are dominant and 256 recessive. This apparently important ratio does not agree at all with the 3 : 1 ratio expected. The deviation D is $525 - 444 = 81$, n $= 700$, the standard error $S.E.\overline{x}_{abs \cdot 3:1} = 11 \cdot 456$ and the probable error $P.E._{\cdot 3:1} =$

$$\frac{D}{S.E.\overline{x} \, abs \, 3:1} = 7,$$ which is far above the limit of 3.

Nevertheless, this is the result of a segregation for one gene. It is obvious that other methods must be found to introduce the necessary corrections. For this we may refer to the method of APERT-BERNSTEIN, the method of LENZ and the probands-method of WEINBERG.

Thusfar we saw that it is possible to determine whether the actual values for a segregating offspring compare with the calculated theoretical expectation. These methods were based on the comparison of the deviation with the calculated standard error in question. Undoubtedly, this method is a very accurate one, however, modern biometrics, based on the principles, led down by PEARSON (1900), has derived an easier method, which leads to the same result. This method is known as the Chi-Square method. We will not discuss in detail the derivation of the formula used, since the theoretical consequences are more in the field of mathematics than biology.

By Chi-Square we will understand the squared proportional deviations of each class. The expression χ^2 will give us some information concerning the possibility, that a deviation as great or greater can occur by chance.

d represents the respective deviation,

e represents the corresponding values expected.

$$\text{Hence } \chi^2 = \Sigma \frac{d^2}{e}$$

It is obvious that the number of classes to which the χ^2 refers, plays an important role. Dealing with two classes we may expect a lower value for χ^2 than in the case when we deal with 4 or 8 classes. Therefore we recognize the effect of the number of independent classes by the degrees of freedom.

In a monohybrid segregation we deal with three classes (1 : 2 : 1) and we say that there are two degrees of freedom. The value of m, designating the degree of freedom, is always one less than the number of segregating classes.

Together with the χ^2 value, we can calculate (directly from a table) the odds for a strictly chance deviation as large as or larger than the observed values (see FISHER, 1946).

If for example, in a given cross of *Neurospora crassa* between a morphological mutant "fluffy" (fl⁻) and the biochemical mutant arginineless (arg⁻) the actual segregation is

Classes	fl⁻arg⁺	fl⁻arg⁻	fl⁺arg⁺	fl⁺arg⁻
Actual segregation numbers	23	28	22	27
Expected ratio	25	25	25	25

(*Neurospora* is haploid, therefore gives a segregation into 1 : 1 : 1 : 1) hence we will derive as follow:

Actual segregation numbers	23	28	22	27
Theoretical expectation (e)	25	25	25	25
Deviation from expectation (d) . .	2	3	3	2
d^2	4	9	9	4
d^2/e	·16	·36	·36	·16

Total (sum of the d^2/e i.e. ·16 + ·36 + ·36 + ·16) = 1·04, m = 3 and P = ·75 — ·79 (from table).

In other words 75—79 per cent of similar cases will have as great or greater deviation from the 1 : 1 : 1 : 1 ratio and therefore, the present population fits the ratio very well. An independent segregation is therefore proved.

However, in another example also taken from *Neurospora crassa* investigations, we will find in a cross between "fluffy" (fl^-) and micro (pe^{m-}).

Classes	fl^-pe^{m+}	fl^-pe^{m-}	fl^+pe^{m+}	fl^+pe^{m-}
Actual segregation numbers.	29	12	14	35
Expected	22·5	22·5	22·5	22·5
d = ...	6·5	10·5	8·5	12·5
d² = ...	42·25	110·25	72·25	156·25
d²/e = ...	1·88	4·90	3·20	6·94
Total of the d²/e =	16·92	m = 3	P = ·0027	(from table)

The value of χ^2 when m = 3 corresponds to a value of P equal to ·0027. Hence, we may say that the chances of agreement between observed and calculated data as bad as, or worse than this are as small

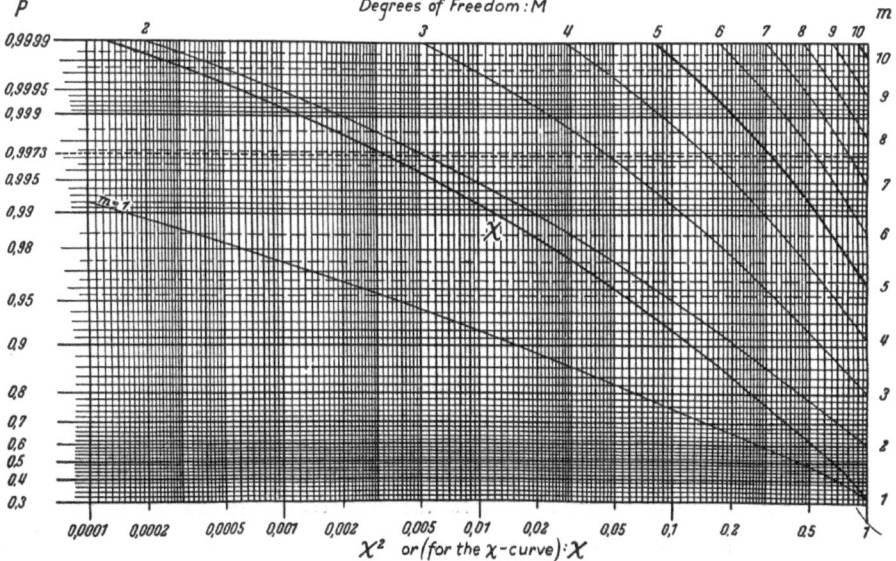

Fig. 37. After PÄTAU, 1942

or smaller than 27 per cent. Or put in other words: only ·27 per cent of similar cases with as great or with greater deviation from a 1 : 1 : 1 : 1 ratio would be found to occur.

In the case that P > ·05 there exists a small chance that the deviation from the anticipated ratio is not due to random sampling; in the case ·01 ≥ P > ·0027 there is a proved (slight) difference between the numbers expected and the numbers observed and this difference is not caused by sampling errors; in the case 0027 > P there exists a striking difference between the expected and observed ratio also no longer due to chance.

We have to be especially careful in the case that ·05 ≥ P > ·01 when deriving any conclusions, since these P-values may point to an existing difference or, to a slight agreement.

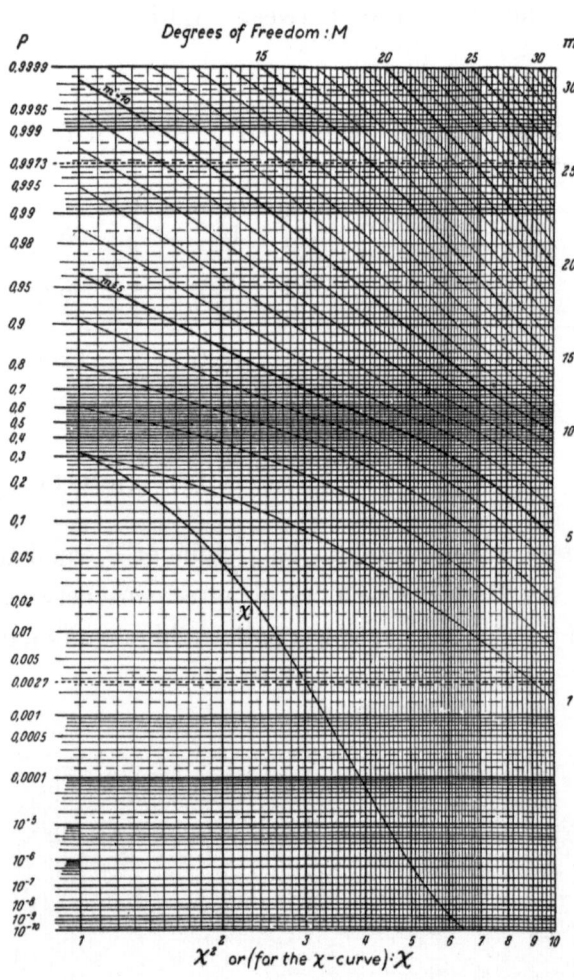

Fig. 38. After Pǎtau, 1942

Therefore, in our first example we can reasonably regard the observed deviation as a sampling or chance error and the present population fits the ratio 1 : 1 : 1 : 1 very well. In our second example no independent assortment occurred and the deviations from the expected ratios are no longer due to chance. It may be noticed, that the deviations in the second case were caused by a linkage relationship of "micro" and "fluffy" which are almost 15 Morgan units apart in the same chromosome. This phenomenon of linkage will be discussed later in this book.

For calculating the probability (P) we will refer to the book of FISHER (1946). PÄTAU (1942) gives a rather compact form of these tables in the form of graphs from which P can easily be read (see fig. 37-39).

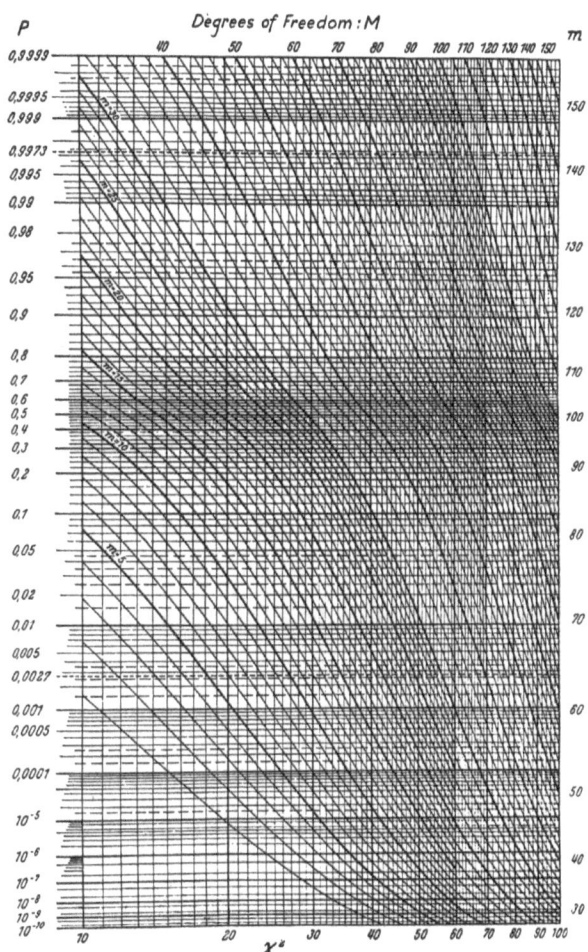

It is obvious that the Chi-Square can be used only in cases in which the class-frequency is more than 5. For reliable data we even suggest a class-frequency of at least 50. The Chi-Square method cannot be applied to percentages but to numerical frequencies only, since a percentage never refers to the total number of individuals involved in the experiment.

Fig. 39. After PÄTAU, 1942

Mathematical testing of the segregation frequencies is thus, from all points of view an absolute necessity, if we wish to derive a reliable conclusion from the data obtained.

CHAPTER VII

RELATIONS BETWEEN GENES I

(Multiple allelism, Polymery, Cryptomery,
Epistasis and Hypostasis)

It was a long time before the significant place, which MENDEL's work held in our conception of the basis of genetics, was recognized.

However, when once this milestone was reached in the course of the development of science, further unfolding could follow quickly. Therefore the twentieth century inherited a task and has shown an understanding of her duties. We are not yet at the finish of research into the problem of inheritance but we can state already that since 1900 we have obtained a general outline. Genetics became an international science with an international application and this characteristic will never be lost.

In the course of the years following 1900, supplementary investigations on the classical work of the pioneer appeared from all sides.

This supplementary work was accomplished in three ways, viz., 1st by pure repetition of MENDEL's own work; 2nd by extension of his research (which gave us a view of its wide application, and illustrated the general validity of the rules) and 3rd by virtue of unexpected results, which opened up new horizons and significantly altered conceptions already obtained.

For his investigations MENDEL had used pairs of characters and later workers also found these working-methods were the obvious ones. Through this, a great number of so-called ,,Mendelian'' pairs of characters were discovered. Among the characters that MENDEL set against one another were: smooth and wrinkled seed-forms, yellow and green coilour of the cotyledons, green and yellow unripe pods axial and termnal infslorescences and high and low stems. It wa, obvious in the first year of the new period that a start had to be mades by means of searching among such easy cases. This searching was rewarded. We are able to give a long list of characters in plants and animals,which can be set against one another in pairs. Of these a few are discussed in the previous chapter. Similarly, it was evident that these characters were not limited to colours only, but also included such factors as shape etc.

In the first place it was obviously true from the work, that the rules of MENDEL were in general valid and that a number of characters, which were of enormous importance in practice, behave according to these rules.

But on the other hand it became known from further study, that these rules could not be maintained in their absolute form and that a broad free conception was absolutely necessary. We have already seen that the rule of dominance, given by MENDEL, does not always hold and that two groups (i.e., dominant hybrids and intermediate hybrids) must be distinguished.

Closer investigation into phenotypically dominant forms showed (as in the *Urtica*-crosses of CORRENS, p. 86) that this dominance is incomplete: even in the case of MENDEL, concerning the dominance of smooth form of pea seeds over wrinkled, it appeared that the starch of homozygous-smooth seeds could be distinguished microscopically

Fig. 40. Reversion of dominance in a heterozygous *Dianthus* (after TAMMES, 1926).

from   the   starch of heterozygous-smooth   seeds   (DARBISHIRE, 1908; KAPPERT, 1920). Hence the concept of *"prevalence"* took the place of that of dominance.

Moreover the dominant gene, which expresses itself phenotypically, can be influenced by incidental circumstances. For instance, by the stage of development of the organ at which this gene operates.

A typical example of this was obtained from two heterozygous individuals originating from a cross between two *Dianthus* varieties; i.e. from a cross between a white flowered *Dianthus barbatus* and a red flowered one (from which the flower-buds are also red coloured). This cross gives rise to an F_1-generation with white flower-buds. During the development of the flowers the corolla obtains a light-pink colour at the margin at first and the rest is white. The whole corolla gradually forms anthocyane. The intensity of the colour in-creases and after ten days the fully-grown flower is as darkly red co-loured as is the homozygous parental plant (TAMMES 1926; figure 40).

Such a change in dominance also occurs in potato-plants, which are green coloured in the heterozygous condition as young seedlings. When they are fully-grown they have red stems. Hence a count of the seedlings in the F_2-generation gives a segregation of 3 green : 1 red, whereas the fully-grown F_2-generation segregates into 3 red : 1 green (SIRKS, 1929b). The dependence of age can also be observed in other characters; such as the deformed leaves on an aberrant tobacco type (*Nicotiana deformis*). HONING (1923) found that these deformations do not show in heterozygous plants until the fourth or the fifth leaf. The first leaves formed are completely normal.

This phenomenon of *reversion of dominance* is very interesting, since it gives us a better view of the nature of the factors, which will be discussed more extensively in chapter XVI.

The extension of MENDEL's work was performed soon after its rediscovery by CUÉNOT (1902–1911). In his series of hybridization experiments with mice, he established that the factor for yellow hair-colour (J, jaune) is dominant over three other factors for hair-colour: grey with white belly (G', gris à ventre blanc), grey with grey belly (G, gris or agouti) and black (N, noir). The G'-factor behaves as a dominant one over both the factors G and N and finally, the G-factor in its turn is dominant over N, so that the latter was recessive with respect to the three other types. It was evident from several and different crosses, that the F_2-generations were always comprised of both parental types and never of an animal of another colour from the afore mentioned series of colours. This means, that a cross: yellow × grey with white belly (w.b.) segregates into 3 yellow: 1 grey w.b.; yellow × grey with grey belly (g.b.) segregates into 3 yellow: 1 grey g.b.; yellow × black segregates into 3 yellow: 1 black.

grey w.b. × grey g.b. segregates into 3 grey w.b. : 1 grey g.b.

grey w.b. × black   segregates into 3 grey w.b. : 1 black.

grey g.b. × black   segregates into 3 grey g.b. : 1 black.

CUÉNOT (1911, p. 47) expresses this very clearly in the words: "The yellow mice are characterized by a factor J, which is allelomorphic with G', with G and with N and is dominant over all these"; thereupon he gives as the order of succession of dominance $J > G' > G > N$.

Some years later the same result was observed by MORGAN (1914b).

He used other symbols for indicating the factor, viz., yellow was indicated by By, grey with white belly by Bw, grey with grey belly by Bg and black by N. Independently from CUÉNOT, MORGAN recognized the principal significance of this result; he called the four factors which form this series of colours *"multiple alleles"*. There is thus no talk here of only one single pair of alleles, but of a whole series which can behave mutually as one pair of alleles.

Since this time a great number of such series of multiple alleles have been discovered; the number of members of such a series is in most cases limited to 4 or 5, but in some cases it amounts to 15. In the vegetable kingdom more of these examples have been found. We will mention here the investigations of OWEN (1928), who described a series of four alleles for the seed-colour of the soybean. In this species there exist four types of seed-

Fig. 41. Multiple allelomorphs in soybeans (after OWEN, 1928).

coats (figure 41), from which the uncoloured (light hilum, $I^h I^h$) is dominant over all the others and from which the plain-coloured one (self colour, ii) is recessive with respect to other colours. Besides the ratios which were established by other investigators between I^k and i, OWEN (1928) also observed the following segregations:

Parents	Segregation	D	S.E.$\bar{x}_{abs\cdot3:1}$	P.E.$_{3:1}$
$I^hI^h \times I^iI^i$	223 : 75	0·5	7·47	0·07
$I^hI^h \times ii$	15 : 5	0·0	1·94	0·00
$I^iI^i \times I^kI^k$	186 : 65	2·25	6·86	0·33
$I^iI^i \times ii$	250 : 92	6·50	8·01	0·82

From this it is evident, that I^h is dominant over I^i and over i; I^i over I^k and over i. The dominance of I^k over i was already known.

All these segregations agree with the theoretical expectation of 3 : 1 and hence the following series of multiple alleles can be established $I^h > I^i > I^k > i$.

There is no sense in quoting more examples, for which one can refer to the summary of JUST (1934) and of STERN (1930b). It must be emphasized that other relationships between different genes create a semblance as if multiple allelomorphism exists in such segregations. We will learn later on in this chapter for instance that the phenomenon of *epistasis* can cause analogous segregations, and also linkage between factors can be the cause of apparent multiple allelomorphism. This will be discussed in chapter VIII. The exact explanation of the segregation processes can always be obtained by specially arranged crosses.

The ratio of the numbers in MENDEL's original experiments with peas were in some cases confirmed, but deviations also existed. MEN-DEL had himself already found that in a cross of white-flowering *Phaseolus vulgaris* (common kidney-bean) with a purple-red-flowering *Ph. multiflorus* (runner-bean), the F_2-generation was composed of only 1 white-flowering plant to 30 lighter or darker red-flowering plants and on this observation he based the following interesting remark: "This enigmatic phenomenon can be also explained according to the rules which are valid for *Pisum* (peas), if we first accept, that the colour of the flowers of *Ph. multiflorus* is composed of two or more independent colours, which behave in the same way as does every other independent character of the plant. If the colour of flower A is composed of two or more independent characters $A_1 + A_2 + \ldots$ which together make the red colour then the hybrid combinations $A_1a + A_2a \ldots$ should be formed by fertilization with the different character "a" ...... It is easy to understand that from these different combinations a whole series of colours must originate ...... As the total number of combinations is 16, all colours must be spread regularily over these 16, however in different frequencies. If the expression of the colours is accomplished in this way, then the previous case, that the white flower-colour occurred only once among 31 plants, can be explained as follows: This colour is only present once in this series and therefore occurs on an average of 1 per 16 F_2-individuals or in a case dealing with three factors, only once per 64 plants" (1865, p. 34).

We now recall the previously mentioned scheme of a dihybrid segregation after hybridization of a plant AABB (purple-red flowers) and a plant aabb (white flowers), in which both of the characters A and B individually cause red colour. We have already seen that the hybrid AaBb gives rise to four kinds of pollen-grains and four kinds of egg-cells, viz. AB, Ab, aB and ab and that these reproductive cells give 16 combinations, viz., 1 AABB, 2 AABb, 2 AaBB, 4 AaBb, 1 AAbb, 1 aaBB, 2 Aabb, 2 aaBb and 1 aabb. In 15 of these 16 combinations one or more factors A or B are present and these 15 combinations are able to develop red colour, however, each combination with a different intensity. Combination 16 (aabb) on the other hand, lacks all colour factors and because of this the flower is white. Hence a ratio of 15 coloured: 1 white will be found (if we neglect the differences in intensity).

In the same way, from the trihybrid scheme discussed earlier of the cross AABBCC × aabbcc it can be derived that in the F_2 only the combination-number 64 (aabbcc) will be white, and all other 63 combinations will be more or less coloured with different intensities (from AABBCC to Aabbcc or aaBbcc or aabbCc). For this discovery also, the honour falls to MENDEL, who, as the first investigator, saw a definite constructive solution to this apparently deviated hybrid behaviour.

It was left to NILSSON-EHLE (1909) to give the final explanations of this phenomenon, which was generally accepted.

In NILSSON-EHLE's crosses between different types of wheat, which were made on a large scale, it is possible to obtain much information concerning deviated "Mendelian"-frequencies.

A wheat strain known as 0728 with dark brown glumes was crossed with different strains which had white glumes. The F_1-generation was uniform and had dark-brown glumes. The F_2-individuals were bred from 15 F_1-plants (which originated from four crosses). These showed a perfect segregation into 15 brown-glumed plants: 1 white glumed plant.

The data borrowed from NILSSON-EHLE (1909, p. 60–61) and the calculations based on these data, show this very clearly:

F_1	nF_2	observed brown : white	theoretically expected 15 : 1		D	S.E.$_{\bar{x}}$ abs. 15:1	D : S.E.$_{\bar{x}}$ abs. 15:1
1a	60	56 : 4	56·250 :	3·750	0·250	1·865	0·134
1b	113	106 : 7	105·938 :	7·062	0·062	2·574	0·024
1c	138	128 : 10	129·375 :	8·625	1·375	2·645	0·522
1d	168	158 : 10	157·500 :	10·500	0·500	3·136	0·159
2a	50	47 : 3	46·875 :	3·125	0·125	1·713	0·073
2b	91	85 : 6	85·312 :	5·688	0·312	2·304	0·135
2c	107	101 : 6	100·312 :	6·688	0·688	2·504	0·274
3a	88	81 : 7	82·500 :	5·500	1·500	2·270	0·661
3b	85	78 : 7	79·688 :	5·312	1·688	2·232	0·756
3c	38	36 : 2	35·625 :	2·375	0·375	1·498	0·250
3d	72	68 : 4	67·500 :	4·500	0·500	2·504	0·200
3e	115	111 : 4	107·812 :	7·188	3·188	2·596	1·228
3f	135	127 : 8	126·562 :	8·438	0·438	2·813	0·156
4a	95	88 : 7	89·062 :	5·938	1·062	2·362	0·450
4b	149	140 : 9	139·688 :	9·312	0·312	2·955	0·106
Total	1504	1410 : 94	1410·000 :	94·000	0·000	9·396	0·000

A better accordance with the theoretically expected F_2-segregation into 15 brown glumes : 1 white glume is not imaginable, and indeed the existence of two mutually independent genes may be considered as having been proved. These genes are individually the cause of the brown colour of the glumes and together they express this colour in a greater intensity. For the monohybrid 3 : 1 segregation we could expect a theoretical ratio of 1128 : 376, from which we can calculate $D = 282$, S.E.$\bar{x}_{abs.3:1} = 16\cdot796$ and P.E.$_{.3:1} = 16\cdot7$. Such a high P.E.-value cannot be accepted.

Another interesting case concerns the shepherd's-purse (*Capsella bursapastoris*). Here are two genes, each of which governs the typical fruit-shape. When both are absent, the fruit is not an inverted flat-triangular one, but appears ellipsoidal as in *Capsella heegeri*.

SHULL (1914) crossed both the species and continued very systematically to investigate the offspring including the F_3-generation. The F_1-individuals, without exception, have triangular fruits; among a total of 2907 F_2-plants there were 2782 *bursapastoris*- and 125 *heegeri*-types.

These observations also were not in agreement with a 3 : 1 segregation; P.E.$_{.3:1}$ was calculated as 25·775. One may suppose that we are again dealing with a segregation of 15 : 1, but here again the P.E. is higher than can be accepted (P.E.$_{.15:1}$ = 4·343). However, it seems that indications exist that this deficit in the total of *heegeri*-type plants in the F_2 can be attributed to a slight decrease in viability of this type. Offspring were bred from 18 F_2-individuals with triangular fruits. In the F_3-generation it became evident that 7 F_2-individuals gave rise only to triangular fruits in their offspring, 6 segregated in a ratio of 15 : 1 and 5 in a ratio of 3 : 1.

Figure 42 gives a picture of the composition of such an F_2-generation. If we indicate the factors which cause triangular fruitshape by C and D then one parental plant becomes CCDD (*C. bursapastoris*), the other becomes ccdd (*C. heegeri*). The F_1-plants were thus CcDd and gave rise to four forms of pollen-grains and four types of egg-cells: CD, Cd, cD, cd; thus there were 16 combinations in the F_2-generation. Among these sixteen there was one (ccdd) which was a *heegeri*-

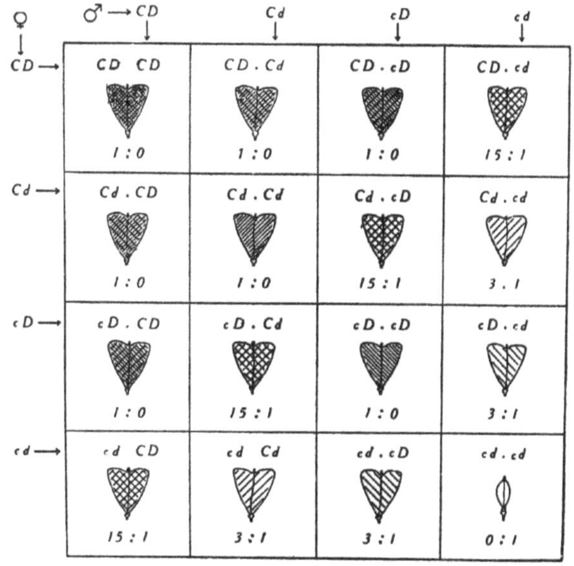

♀ \ ♂ →	CD	Cd	cD	cd
CD →	CD CD — 1:0	CD.Cd — 1:0	CD.cD — 1:0	CD.cd — 15:1
Cd →	Cd.CD — 1:0	Cd.Cd — 1:0	Cd.cD — 15:1	Cd.cd — 3.1
cD →	cD.CD — 1:0	cD.Cd — 15:1	cD.cD — 1:0	cD.cd — 3:1
cd →	cd CD — 15:1	cd Cd — 3:1	cd.cD — 3:1	cd.cd — 0:1

Fig. 42. F_2 obtained from the cross between *Capsella bursapastoris* and *C. heegeri* (after SHULL, 1914a).

plant, the other 15 had triangular fruit-shapes. Among these fifteen there must theoretically be 7 (1 CCDD, 2 CCDd, 2 CcDD, 1 CCdd and 1 ccDD) which bred true to type in the F_3 in relation to triangular fruit-shape (or in other words behave according to the ratio 1 : 0), 4 have to segregate according to 3 : 1 (2 Ccdd and 2 ccDd) and 4 have to segregate as 15 : 1 (all CcDd).

As we already saw, SHULL found, in reality, that among the 18, there were 7 true breds (which "segregated" according to a 1 : 0 ratio), 6 segregated into 3 : 1 and 5 into 15 : 1, which agrees very well with our expectations. Calculating on a basis of 18 individuals, these frequencies become 8·4, 4·8 and 4·8. The observations are thus sufficiently in balance with the theoretical expectation.

In the case of more than two equivalent factors (that is to say each of the factors causing a particular characteristic but still segregating independently), NILSSON-EHLE again found the first clear example.

Some hybridizations of wheat strains with a red colour in the grain and wheat strains with white grains gave a more or less exact 3 : 1 segregation (31 red : 9 white; 72 r. : 28 w., 32 : 6, 67 : 13, 63 : 15 in five F_2-generations). These segregations thus point to the existence of a single gene for red in which both of the original parents differed. On the other hand, other crosses of red × white wheat, in which the mother-plant had white grains, showed F_2-generations in which there was not found even a single white grained plant.

These F_2-generations were still composed of 78, 30, 49, 31, 86 and 110 individuals (NILSSON-EHLE 1909, p. 67) in the various F_2-families.

Yet another cross gave a segregation of 55 red : 1 white-grained plants. So the course must be more complicated than in the cases of the colour of the glumes of wheat or the fruit-shape of the shepheid's purse, and still again the "Mendelian"-scheme can be indicated as the basis of an explanation, viz., a scheme in which three independent genes individually cause the red colour. If we represent these 3 genes for red colour by the figures R_1, R_2, and R_3, the red-grained plant becomes $R_1R_1R_2R_2R_3R_3$ and the white-grained motherplant $r_1r_1r_2r_2r_3r_3$.

Hence the F_1-individuals heterozygous in all of these three genes have as their genotypical formula $R_1r_1R_2r_2R_3r_3$. They gave rise to 8 different types of pollen-grains and to the same 8 types of egg-cells, viz., $R_1R_2R_3$, $R_1R_2r_3$, $R_1r_2R_3$, $r_1R_2R_3$, $R_1r_2r_3$, $r_1R_2r_3$, $r_1r_2R_3$, and $r_1r_2r_3$.

The F_2-generation yields 64 combinations, comparable with the earlier discussed scheme (p. 103) for the trihybrid AaBbCc, and of these 64 combinations there are 63, which have one or more R genes (whether this is R_1, R_2 or R_3). These 63 are therefore more or less red-coloured and the 64th plant, $r_1r_1r_2r_2r_3r_3$, is white. Thus there occurs from three independent genes for colour, only one white-grained plant out of a total of 64 F_2-plants; all the other plants are more or less red.

We will now take a closer look at all these red-grained F_2-individuals and see that they differ mutually in intensity of their colour. There are a few dark-ones, a few very light and a great majority of different grades and shades of medium red. A further contemplation with relation to the theoretically obtainable F_2-individuals makes this obvious. If we arrange all the 64 F_2-combinations obtained from the F_1-plant $R_1r_1R_2r_2R_3r_3$ into groups according to the number of R genes present in the genotypical formula then it will be clear that this number can amount at the most to 6 and at the least to 0. We are able to draw the following scheme: (See page 134).

If it were possible to distinguish the grains of the individuals with 6 R genes from those with 5, 4, 3 etc., we should be able to find theoretically that the 7 groups should be represented by:

group	6 R	5 R	4 R	3 R	2 R	1 R	0 R
proportion of the total	$1/64$	$6/64$	$15/64$	$20/64$	$15/64$	$6/64$	$1/64$

or in other words in the ratio of 1 : 6 : 15 : 20 : 15 : 6 : 1 ,

At once we can establish a very important fact from this, i.e., that the distribution of 64 individuals of the F_2-generation into groups according to the number of R genes, seems to be absolutely regular and can be expressed in the same graphic representation as recorded in chapter II (the so-called *symmetric frequency distribution-curve*). In other words the binomial distribution obtained needs not always be the result of nutritional circumstances and is not a proof of the conception held by the biometric school, that the material investigated is composed of individuals which are mutually genotypically the same. We have already seen (p. 39) the possibility, that the summation of smaller frequency distributions lead to the origin of a binomial curve and we now meet, as a third possibility the presence of mutually hereditable characters, which segregate independently and act in the same way, and by so acting intensify one another.

Also, in the presence of more mutually equivalent independently

6 R	5 R	4 R	3 R	2 R	1 R	0 R
			$R_1\,R_2\,R_3$ / $r_1\,r_2\,r_3$			
			$R_1\,R_2\,r_3$ / $R_1\,r_2\,r_3$			
			$R_1\,r_2\,R_3$ / $R_1\,r_2\,r_3$			
			$r_1\,R_2\,R_3$ / $R_1\,r_2\,r_3$			
			$R_1\,R_2\,r_3$ / $r_1\,R_2\,r_3$			
		$R_1\,R_2\,R_3$ / $R_1\,r_2\,r_3$	$R_1\,r_2\,R_3$ / $r_1\,R_2\,r_3$	$R_1\,R_2\,r_3$ / $r_1\,r_2\,r_3$		
		$R_1\,R_2\,R_3$ / $r_1\,R_2\,r_3$	$r_1\,R_2\,R_3$ / $r_1\,R_2\,r_3$	$R_1\,r_2\,R_3$ / $r_1\,r_2\,r_3$		
		$R_1\,R_2\,R_3$ / $r_1\,r_2\,R_3$	$R_1\,R_2\,r_3$ / $r_1\,r_2\,R_3$	$r_1\,R_2\,R_3$ / $r_1\,r_2\,r_3$		
		$R_1\,R_2\,r_3$ / $r_1\,R_2\,R_3$	$R_1\,r_2\,r_3$ / $r_1\,R_2\,R_3$	$R_1\,r_2\,r_3$ / $R_1\,r_2\,r_3$		
		$R_1\,r_2\,R_3$ / $r_1\,R_2\,R_3$	$r_1\,R_2\,R_3$ / $r_1\,r_2\,R_3$	$r_1\,R_2\,r_3$ / $R_1\,r_2\,r_3$		
		$r_1\,R_2\,R_3$ / $r_1\,R_2\,R_3$	$R_1\,r_2\,r_3$ / $r_1\,R_2\,R_3$	$r_1\,r_2\,R_3$ / $R_1\,r_2\,r_3$		
		$R_1\,R_2\,r_3$ / $R_1\,r_2\,R_3$	$r_1\,R_2\,r_3$ / $r_1\,R_2\,R_3$	$R_1\,r_2\,r_3$ / $r_1\,R_2\,r_3$		
		$R_1\,r_2\,R_3$ / $R_1\,r_2\,R_3$	$r_1\,r_2\,R_3$ / $r_1\,R_2\,R_3$	$r_1\,R_2\,r_3$ / $r_1\,R_2\,r_3$		
		$r_1\,R_2\,R_3$ / $R_1\,r_2\,R_3$	$R_1\,r_2\,r_3$ / $R_1\,r_2\,R_3$	$r_1\,r_2\,R_3$ / $r_1\,R_2\,r_3$		
	$R_1\,R_2\,R_3$ / $r_1\,R_2\,R_3$	$R_1\,R_2\,r_3$ / $R_1\,R_2\,r_3$	$r_1\,R_2\,r_3$ / $R_1\,r_2\,R_3$	$R_1\,r_2\,r_3$ / $r_1\,r_2\,R_3$	$R_1\,r_2\,r_3$ / $r_1\,r_2\,r_3$	
	$R_1\,R_2\,R_3$ / $R_1\,r_2\,R_3$	$R_1\,r_2\,R_3$ / $R_1\,R_2\,r_3$	$r_1\,r_2\,R_3$ / $R_1\,r_2\,R_3$	$r_1\,R_2\,r_3$ / $r_1\,r_2\,R_3$	$r_1\,R_2\,r_3$ / $r_1\,r_2\,r_3$	
	$R_1\,R_2\,R_3$ / $R_1\,R_2\,r_3$	$r_1\,R_2\,R_3$ / $R_1\,R_2\,r_3$	$R_1\,r_2\,r_3$ / $R_1\,R_2\,r_3$	$r_1\,r_2\,R_3$ / $r_1\,r_2\,R_3$	$r_1\,r_2\,R_3$ / $r_1\,r_2\,r_3$	
	$R_1\,R_2\,r_3$ / $R_1\,R_2\,R_3$	$R_1\,r_2\,r_3$ / $R_1\,R_2\,R_3$	$r_1\,R_2\,r_3$ / $R_1\,R_2\,r_3$	$r_1\,r_2\,r_3$ / $r_1\,R_2\,R_3$	$r_1\,r_2\,r_3$ / $R_1\,r_2\,r_3$	
	$R_1\,r_2\,R_3$ / $R_1\,R_2\,R_3$	$r_1\,R_2\,R_3$ / $R_1\,r_2\,R_3$	$r_1\,r_2\,R_3$ / $R_1\,R_2\,r_3$	$r_1\,r_2\,r_3$ / $r_1\,R_2\,R_3$	$r_1\,r_2\,r_3$ / $r_1\,R_2\,r_3$	
$R_1\,R_2\,R_3$ / $R_1\,R_2\,R_3$	$r_1\,R_2\,R_3$ / $R_1\,R_2\,R_3$	$r_1\,r_2\,R_3$ / $R_1\,R_2\,R_3$	$r_1\,r_2\,r_3$ / $R_1\,R_2\,R_3$	$r_1\,r_2\,r_3$ / $R_1\,R_2\,r_3$	$r_1\,r_2\,r_3$ / $r_1\,R_2\,r_3$	$r_1\,r_2\,r_3$ / $r_1\,r_2\,r_3$

segregating genes (i.e., n in number), we will obtain a regular distribution: if the heterozygotes are intermediate and the AA-types can be distinguished from the Aa-types, then the distribution in the phenotypical classes will always have the character of a binomial frequency-distribution with $2n + 1$ classes; but if one gene is always dominant over the other, so that the AA- and Aa-individuals are phenotypically the same then the number of classes will amount to $n + 1$ and the frequency-distribution will be extremely skew, whereas if one given gene is already sufficient to bring out the complete phenotype, the segregation of the F_2-generation will show the ratio $2^{2n}-1 : 2$. For 1 up to and including 6 pairs of genes the segregations are summarized in the following table:

Number of gene pairs	Intermediate heterozygotes — Number of individuals with (genes)													Dominance per gene pair — Number of individuals with (genes)							Total dominance in each pair — Number of individuals with (genes)	
	12	11	10	9	8	7	6	5	4	3	2	1	0	12 or 11	10 or 9	8 or 7	6 or 5	4 or 3	2 or 1	0	1 or more	0
1											1	2	1						3	1	3	1
2								1	4	6	4	1						9	6	1	15	1
3						1	6	15	20	15	6	1					27	27	9	1	63	1
4				1	8	28	56	70	56	28	8	1				81	108	54	12	1	255	1
5			1	10	45	120	210	252	210	120	45	10	1		243	405	270	90	15	1	1023	1
6	1	12	66	220	495	792	924	792	495	220	66	12	1	729	1458	1215	540	135	18	1	4095	1

After NILSSON-EHLE a great number of other investigators performed this experiment in which two or more different and independently segregating genes had the same effect. In genetical literature one can find many examples which point to the existence of such genes: a summary of this material was given by TJEBBES (1931).

In 1911 the phenomenon of these equivalent, independently segregating genes was already being considered by LANG. He concluded that it was of such general significance that he gave it the name of "*polymery*" (1911, p. 113). "We will call this phenomenon polymery and will contrast it with poly-hybridization. Hence, we can speak of *polymery* (dimery, trimery, etc), if a definite character in the reproductive cells is caused by more than one (two, three etc.) equivalent, but independent factors, whose actions intensify each other. On the other hand we speak of *poly-hybridism* (di-hybridism, tri-hybridism etc.) if by a cross, both of the original parents differ from each other in more than one (two, three etc.) different factors". It is a pity that this definition of polymery is not strictly upheld by later investigators. It was forgotten particularly that LANG premises the equivalence of the factors and moreover requires that they intensify one another in their action.

In the present literature both these conditions are left out of the conception of *"polymeric genes"*. In the American literature the insignificant term *"multiple factors"* often occurs. MATHER (1941, 1943) gives to his conception of *"polygeny"* a still broader and vaguer significance, which is valid for all quantitative characters. Hence, a great number of segregation phenomena, which differ fundamentally are recorded as polymeric factors and moreover, the explanation "polymery" is often applied too rashly. With perfect justice TJEBBES (1931, p. 252) speaks of a "pons asinorum" and with the help of it the most complicated segregation phenomena can be explained in an elegant (but extremely shallow) manner. This objection of TJEBBES can only be met if we follow the track of PLATE (1913, p. 155), who suggested for polymery the word *homomery* which sets out clearly the equivalence of the factors. It is even better to divide polymery into *isomery* (equivalent factors) and *anisomery* (non equivalent factors) and to divide the first into *cumulative* (the wheat colour of NILSSON-EHLE) and *non-cumulative* (SHULL's *Capsella*). In the investi-

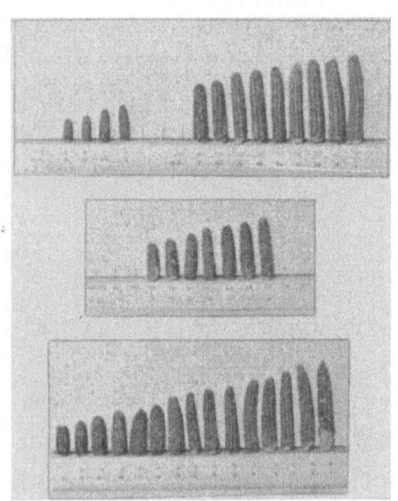

Fig. 43. A cross between corn varieties with different lengths of cobs (after EAST, borrowed from BAUR, 1930).

gations of NILSSON-EHLE it was sometimes possible to make a division between the groups which differ in the number of genes present in the individuals. In some other cases this classification is not possible, especially if the characters (which are denoted by polymeric genes) are modifiable, and by means of this capacity the classification into different F_2- groups becomes unwieldy. EAST and HAYES (1911) obtained results from a cross of two maize stocks, differing in the length of their ears (indicating the existence of polymeric factors). The starting material (fig. 43) was two corn stocks "Tom Thumb" with an ear-length of 5–8 inches, and "Black Mexican Sweet" with ears between 13–21 inches in length. Measurements of a number of ears of these stocks gave the following frequencies:

	Tom Thumb				Black Mexican Sweet								
Length of the ears in inches	5	6	7	8	13	14	15	16	17	18	19	20	21
Number of ears	4	21	24	8	3	11	12	15	26	15	10	7	2

The F_1-generation was an intermediate one and yielded 67 ears which were distributed in their length as follows:

Length of the ears in inches . .	9	10	11	12	13	14	15
Number of ears	1	12	12	14	17	7	4

The differences in length were solely due to feeding modification because the F_1-individuals were mutually the same in genotype. In the F_2-generation there occurred a greater "variability", which is evident from the following numbers:

Length of the ears in inches	7	8	9	10	11	12	13	14	15	16	17	18	19	20	21
Number of ears	2	5	17	33	33	33	27	21	13	10	11	12	1	2	1

Besides the modification in length (as a result of nutritional differences) we must also take into account some heritable factors to explain the differences in length. The increase in "variability" in the F_2-generation was the result of differing genotypical constitutions in the F_2-individuals. This greater *diversity* in the F_2 is the consequence of some polymeric genes for length, in which both of the parents differed from one another.

In numerous experiments concerning the inheritance of so-called *"quantitative characters"* (measures, weights, etc.) the existence of polymeric factors is concluded, on a similar basis. A more exact analysis of these genes is usually neglected. This neglect is often due to the technical difficulties and its complexity.

However it has become evident that in some cases the segregation obtained in relation to quantitative characters, can be discussed on the basis of very distinct and sometimes single factors. LYNCH (1921) could show, that a certain shape of short ears in mice behaves, when compared with normal ear-length, as a monohybrid recessive factor.

WRIEDT (1931) gave us nice examples of monohybrid segregations in the length of the foot and the length of the beak of pigeons and AFZAL (1930) for the ratio of lengths in *Gossypium*-crosses.

Moreover it was evident from an investigation of the measurements of stems, leaves, fruits and seeds of *Vicia Faba* (SIRKS, 1931) that these measurements are achieved on the one hand by *polymeric* (but clearly anisomeric genes) and on the other hand that every character from the polymeric combinations can be comprised of different *multiple allelomorphs*. This is shown in the measurements of the leaves in figure 44. The multiple allelomorphs G_4, G_3, G_2 and G_1 influence the growth in general and for the leaf-width there is also an "additional factor" (the pair of allelomorphs W-w), for the leaf-length the pair B-b and the multiple allelomorphs T_2-T_1-t. Hence, the factors G_{4-1} and W-w form a polymeric group for the leaf-width, the factors G_{4-1}, B-b and T_2-T_1-t form another group for leaf-length.

Hence, it is clear that research into the inheritance of quantitative characters can in some cases lead to the conclusion that one is dealing with one pair of alleles and in other cases to the supposition that we are dealing with polymeric factors. A further analysis of these polymeric factors. can lead to the establishment of definite factors, which can be partly isomeric or partly anisomeric and can moreover still be

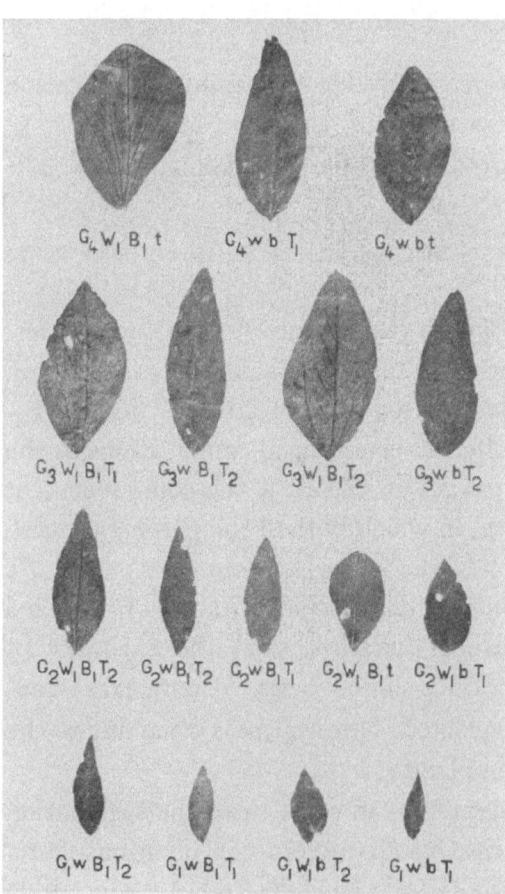

Fig. 44. Leaf-types of *Vicia Faba* with their respective genotypical formula (after SIRKS, 1931).

comprised of groups of multiple allelomorphs (SIRKS 1926 b).

It wholly depends on the condition of the material if such an analysis can be attained.

The presence of polymeric factors still gives rise to other conclusions to which NILSSON-EHLE (1915, p. 25) pointed. If one of the parents is a homozygote, in which three pairs of different polymeric factors are present (i.e., a total of six alleles), and the other parent is also a homozygote, in which the three factors are missing, then the crossing will result in a heterozygous individual (i.e., heterozygous with regard to these three factors). Of the most extreme forms in the segregating F_2-generation, one will be similar to one of the parents (i.e., with 3 pairs of allelomorphs) and the other resemble the other parent with no factors. The other individuals of the F_2-generation will be situated between these extreme forms. However, if one parent is homozygous in relation to two of these factors (e.g. $R_1R_1R_2R_2r_3r_3$) and the other parent is homozygous with respect to the third one (thus $r_1r_1r_2r_2R_3R_3$) so that the first has 4 R factors and the second only 2 R's, then the result will be an F_1-plant, in which all three factors are present in a heterozygous condition. In the F_2-generation these three factors segregate independently and hence, this F_1-individual will give rise to 8 kinds of reproductive cells, viz., the following combinations: $R_1R_2R_3$, $R_1R_2r_3$, $R_1r_2R_3$, $r_1R_2R_3$, $R_1r_2r_3$, $r_1R_2r_3$, $r_1r_2R_3$, $r_1r_2r_3$. These 8 types give rise to 64 combinations among which are now found the homozygous forms: $R_1R_1R_2R_2R_3R_3$ and $r_1r_1r_2r_2r_3r_3$. The first of these will comprise 6 factors of R and will hence be a darker red than both of the parents; $r_1r_1r_2r_2r_3r_3$ has no R-factors and because of this is white. The course of such a cross can be summarized as follows:

P_1a	$R_1R_1R_2R_2r_3r_3$	$r_1r_1r_2r_2R_3R_3$	P_1b
red 4 genes			red 2 gene s
Reproductive cells	$R_1R_2r_3$	$r_1r_2R_3$	
F_1		$R_1r_1R_2r_2R_3r_3$	
red 3 genes			

Reproductive cells	$R_1R_2R_3$	$R_1R_2r_3$	$R_1r_2R_3$	$r_1R_2R_3$	$R_1r_2r_3$	$r_1R_2r_3$	$r_1r_2R_3$	$r_1r_2r_3$
F_2 combinations	red	6 genes	5 genes	4 genes	3 genes	2 genes	1 gene	white
F_2 individuals per 64	1	6	15	20	15	6	1	

Hence, we obtain as a result of this an F_2-generation from a cross between two pure bred red-grained wheat-types, which get their red colour from different heritable factors, homozygous individuals, which are darker red than either of the parental plants. The F_2 also contains homozygous white-grained plants.

Whereas in the case of pure polymery (*isomery*, which was already mentioned) all factors are equivalent and each factor in its turn realizes the same character to a limited degree, factors were also found (in different hybridization experiments) which did not have quite the same effect, but which could still intensify one another (*anisomeric polymery*).

It was NILSSON-EHLE's investigations again, which gave us an eloquent example of this. Among his crosses of oats there were also crosses between types which had different inflorescence shapes (plume-oats with stiff and soft twigs, raceme-oats and others). The hybridiza-

tion of these plants, which individually belong to two varieties which breed true to type and which had a plume-shape inflorescence with almost stiff erect twigs (Bell-oats and Stormogul, fig. 45 upper row), yielded in the F$_2$-generation (figure 45 bottom row) types similar to the parents, and also inflorescences with plume-shape with very soft twigs which hung down, and plume-shaped ones with almost stiff erect twigs and raceme inflorescen-

Fig. 45. A cross between two plumed oat-types together with three types taken from the F$_2$-individuals (after NILSSON-EHLE, 1915).

ces in which the twigs were turned to one side (NILSSON-EHLE, 1915, p. 28).

The ratios were: 1 wide-plumed: 14 stiff-plumed: 1 raceme type. This segregation can be explained by supposing that both parents

(Bell-oats and Stormogul) owed their plume-type to different factors which were dominant over the raceme type. Plants in which both these factors are present in a homozygous condition have an extremely spread-out plume-type. The formulae for both of the parents are therefore e.g., $A_1A_1a_2a_2$ and $a_1a_1A_2A_2$; those for the F_1-plants $A_1a_1A_2a_2$ and among the F_2-individuals there thus occurred the homozygotes $A_1A_1A_2A_2$ (wide plume-type) and $a_1a_1a_2a_2$ (raceme-type). The original parents were not completely similar and hence it can be concluded, that the factors A_1 and A_2, though polymeric, were still not completely equivalent. Hence, to a certain extent we can compare this hybridization result with the previous cross between two red-grained wheat types, from which a dark-red plant and a white one originated in the F_2-generation; but the cases are not wholly similar because each of the red-grained factors have the same effect, whereas both of the plume-factors resulted in different characteristics. This example of gene co-operation forms a convenient link with another kind of phenomena, which we cannot class with that of polymery but which is nevertheless the result of co-operation between heritable characters. The word co-operation is used here in a rather wide sense since we will meet three cases in this sphere; 1st: The presence of both "co-operating" genes can lead to a different external visible characteristic, which neither of them individually is able to produce; and also, the absence of both can lead to yet another phenotype; 2nd: Both factors give rise to a different phenotype and absence of the factors has the same result as the presence of one of them. 3rd: The presence of both factors gives a result which is not different from the case when only one of them is represented. The absence of both factors gives rise to a striking result. It is possible to speak in this latter case of "negative co-operation" but this term is not very well chosen.

We will start with an example from the first group. BATESON and his co-workers SAUNDERS and PUNNETT investigated the comb-shape of poultry by means of hybridization experiments carried out on a large scale. Single combed (Leghorns, Minorcas), rose-combed (Hamburgers and Hollanders), pea-combed (Brahmas) and walnut-combed poultry were involved. If these types are produced from true breds they will always be identical to their respective parents. By systematical hybridization it became evident (see BATESON's table 1908, p. 26–27) that the single comb-shape is recessive when compared with the rose-comb and with the pea-comb. The factor for rose-comb (R) is dominant over the recessive r; this results in a segregation of 3 rose-comb (1 RR + 2 Rr) to 1 non-rose-comb (rr) in the F_2. Similarly, the pea-comb factor (P) is dominant over the recessive p and causes a segregation into 3 pea-comb (1 PP + 2Pp) to 1 non-pea-comb (pp) in the F_2.

What will the result be if the animals with pea-comb and with rose-comb are mated together? If both parents are pure breds with respect to these factors then only walnut-combed progeny will be produced. In other words. animals in which both factors R and P are present have a walnut-comb, animals with only R have a rose-comb, animals with only P, a pea-comb and animals in which both genes are absent have a single comb. On this basis the cross of a pure bred rose-comb animal (RRpp) and a pure bred pea-comb (rrPP) must take place as follows (c.f. fig. 46):

		rose-comb		pea-comb		
P_1a		RRpp	×	rrPP		P_1b
reproductive cells		Rp	×	rP		
F_1			RrPp (walnut-comb)			
reproductive cells of the F_1's	RP,		Rp,	rP,		rp

The F_2-generation is composed of:

♂	RP	Rp	rP	rp
♀ RP	RRPP nut	RRPp unt	RrPP nut	RrPp nut
Rp	RRPp nut	RRpp rose	RrPp nut	Rrpp rose
rP	RrPP nut	RrPp nut	rrPP pea	rrPp pea
rp	RrPp nut	Rrpp rose	rrPp pea	rrpp single

In short: the F_1 of the cross between pure bred rose-comb and pure bred pea-comb is always composed of walnut-combed animals and these mated with one another give rise to a segregation of 9 walnut-combed: 3 rose-combed: 3 pea-combed: 1 single combed animals in the F_2. In these crosses animals containing both factors or lacking both factors can be distinguished from each of the two parents; all the other individuals of the F_2-generation, which have only one factor resemble the parent from which this factor was inherited. The result is that the cross behaved as a normal dihybrid and the F_2-generation showed the common segregation ratio of 9 : 3 : 3 : 1.

It will be evident that a different ratio of numbers will occur in both of the other groups of examples, i.e., in the cases in which the co-operating factors express themselves only by the presence of both or by absence of both.

The second group previously listed include also crossings between individuals which differed in two factors. In this case only individuals which have both factors can be distinguished from each of the parents. The investigations of CORRENS (1912a) with *Linaria maroccana* contributed a clear example of this behaviour. Among the numerous strains of this species

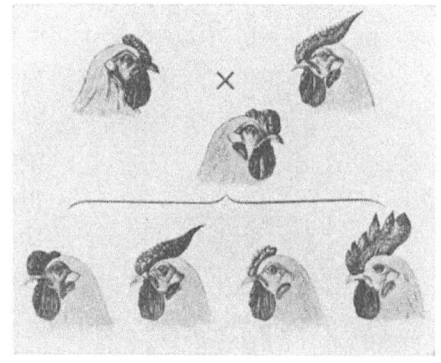

Fig. 46. A cross between pea-combed and rose-combed poultry; F_1 and F_2 (after BATESON, borrowed from BAUR and GOLDSCHMIDT).

there are white, pink and purple flowered plants. CORRENS chose a white flowering plant and a pink one for the cross; the F_1-generation was only composed of purple flowering individuals. This unexpected result led to very important conclusions. If by self-pollination, seed from these was obtained and in this way an F_2-generation was bred, it became evident that this generation was composed of purple, pink and white plants in the ratio of 9 purple : 3 pink : 4 white. What was also important, was, that the following generations (F_3, F_4 etc.) showed that among purple F_2-individuals, types occurred which in later generations behaved as true breds, so that the possibility of obtaining plants with a new phenotype had been shown to be possible by means of hybridization.

Apparently this is a complicated case, but if we take the origin of the purple and the red colour into consideration, the explanation will soon be found. The colouring matter only turns red if the cell sap, in which it is dissolved, has an acid reaction, whereas it becomes purple if this reaction is alkaline. The way in which the colour changes after hybridization becomes obvious if we suppose that the cells of the pink plants contain acid cell sap and those of the white flowers more alkaline cell sap.   Moreover, if the factor for red flowers is dominant over that of white flowers and the factor for alkaline cell sap dominant over that of acid cell sap, the F_1-generation is purple. If we choose R for red and r for non-red and A for alkaline and a for acid as symbols for these factors then the F_1-plants give rise to four types of reproductive cells (germ-cells Kc) viz., RA, Ra, rA and ra. Their 16 combinations are pictured in figure 47, in the same way as for the dihybrid segregations. As soon as R is present once or twice the plant becomes coloured (i.e., in 12 of the 16 cases) and there are 9 purple-ones (1 RRAA, 2 RRAa, 2 RrAA and 4 RrAa) to 3 pink plants (1· RRaa and 2 Rraa). In the case when r is recessive, the plant is non-coloured (i.e., 4 of the 16 cases). But these four white plants are not at all similar with respect to their cell sap. One of them is homozygous alkaline (AA), two are heterozygous (alkaline-acid, Aa) and one is homozygous acid (aa). This difference does not find expression; phenotypically all these four plants

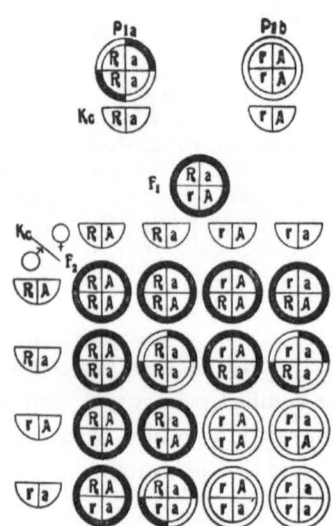

Fig. 47. A cross between *Linaria maroccana* white × pink, F_1 and F_2 (after CORRENS, 1912a).

are white flowering.

The fundamental importance in this cross of pink and white *Linaria maroccana* is this: that both the parents apparently differ in one factor but in reality differ in two pairs of allelomorphs, and that by co-operation of both of the dominant factors (colour matter and alkaline cell sap), the F_1 will have a different phenotype to that of both its parents. The absence of both the factors (or the combination of both the recessive factors) gives rise to the same phenotype as does the absence of only the colouring matter of the flower.

With regard to this, the factor for alkaline cell sap can be present in the white flowers as well as the factor for acid cell sap without expressing itself in a different phenotype. Not even after self-pollination or after hybridization of the white plants (which can differ mutually with regard to this property) will a different phenotype be obtained. The gene for the colour of the cell sap is present invisibly in the white plants. In the early days of scientific genetics this state as well as that of recessivity was called *latent*. But this conception is rather vague and was already used in many, very different meanings and is still used today. It is better to avoid this term and to speak in this case of *cryptomeric genes* (E. TSCHERMAK, 1904, p. 535).

TSCHERMAK called types *cryptomeric* if "types, which by in-breeding (self-fertilization or cross-fertilization with phenotypically identical individuals) are true breds with regard to particular characteristic properties, but which after cross-fertilization with a phenotypically different characteristic, give rise to a characteristic change of these properties", thus to new phenotypes or *"hybridization nova"*. Later, TSCHERMAK (1914, p. 186) has made the definition of the conception cryptomery more clear-cut by limiting the term to "the possession of genes, which can be active in an exceptional and newmanner, by means of combinations with other genes". Moreover it is stressed "that one cannot observe externally the presence of genes, which are able to be active in this way".

In this conception lies the very important principle that there is a possibility that heritable characters exist in animals and plants which cannot find expression, because a certain definite character with which they must operate is missing.

The previous case of CORRENS dealing with cryptomery in *Linaria maroccana* was not by any means the first case to be discovered acting in such a way, but it is one of the simplest and because of this we give it as an example in the discussion of cryptomeric properties. It is obvious that cryptomery is a very common and wide-spread phenomenon and can take the most peculiar forms.

TSCHERMAK and CORRENS had already found (1902) different examples of crosses in which one of the parents contained a hidden and hence cryptomeric gene. Since these investigations, numerous analogous cases have been discovered. The case when both parents contain mutually cryptomeric genes, which after combination in the F_2-generation are enabled to co-operate is rather more complex. Particularly striking evidence has been given by BATESON, SAUNDERS and PUNNETT (1906) concerning crosses with white flowering types of

Lathyrus odoratus. Crosses with the strain Emily Henderson, which has white flowers (but among which individuals occurred with round pollen-grains next to plants with long pollen-grains) gave the following results: self-pollination never gives anything other than a white flowering progeny with round or long pollen-grains. Mutual hybridization of Emily Henderson with round pollen always gives rise to white flowering plants similar to the flowers of the mutual hybridization of Emily Henderson with long pollen. But hybridization of two white flowering E. H., from which one has round pollen and the other long pollen, gives rise to the origin of an F_1-generation which has purple flowers. In figure 48 is pictured under 1, one parent (P_1a, an E. Henderson with long pollen), under 2 the other parent (P_1b, E.H. with round pollen) and under 3 the F_1-generation (purple with blue

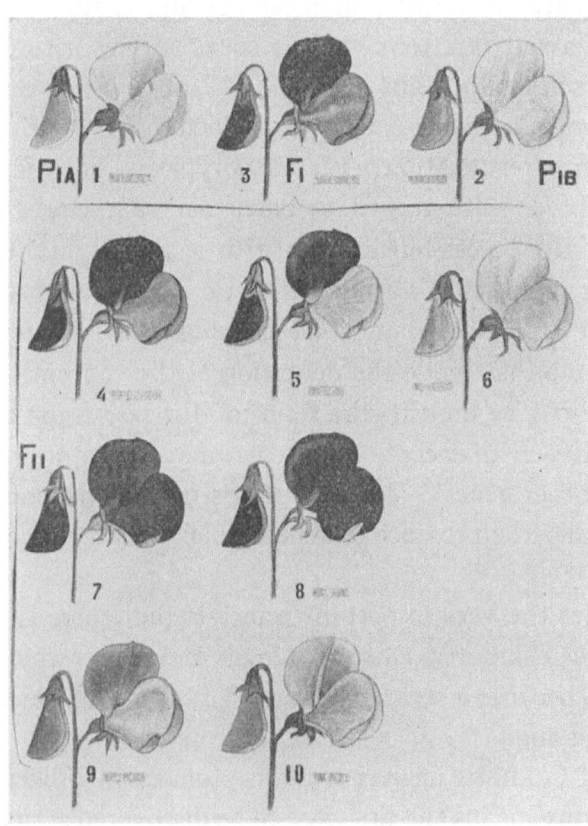

Fig. 48. A cross between white *Lathyrus* with long pollen and white *Lathyrus* with round pollen. Coloured F_1 and segregating F_2 (after BATESON, 1909).

standards). The F_2-generation of this hybrid produced no less than 7 different phenotypes (4 = purple with blue standard; 5 = red with pink standards; 6 = white; 7 = purple with purple standards; 8 = red with red standards; 9 = light-purple and 10 = light-pink). Thanks to very extensive breeding experiments, BATESON and his associates succeeded in basing this case on a simple scheme of factors (1906, p. 31–33).

A classification can be made into purple-, red- and white plants. Out of 3725 plants, which comprise the whole experiment, there were 1634 purple, 498 red and 1593 white, a ratio which does not seem at first glance to be a simple one to explain and which can still be derived from a scheme of 3 factors.

If we can accept that colour only develops in the case when two factors co-operate (called C and R by BATESON), i.e., if both factors are present, then it follows that all F_2-combinations with C and R (whether homozygous or heterozygous) are coloured and all F_2-combinations in which one or both of the factors are absent become white.

Among the 16 combinations arising from hybridization of a dihybrid there are 9 in which both factors are present and 7 in which one or both are absent. This follows at once if we repeat the F_2-scheme given earlier in this book, substituting for A and B, the symbols C and R in a hybrid scheme:

	♂ CR	Cr	cR	cr
♀ CR	1 CCRR	2 CCRr	3 CcRR	4 CcRr
Cr	5 CCRr	6 CCrr	7 CcRr	8 Ccrr
cR	9 CcRR	10 CcRr	11 ccRR	12 ccRr
cr	13 CcRr	14 Ccrr	15 ccRr	16 ccrr

It becomes clear that the combinations 1, 2, 3, 4, 5, 7, 9, 10 and 13 are coloured, because the factors C and R are present (these factors cause colour by co-operation). The other combinations 6, 8, 11, 12, 14, 15 and 16 are white (because C or R or both are absent). Thus this explains the ratio of 9 coloured to 7 white. The colour difference between purple and red depends on one factor B, which can only find its expression in colour containing plants. This factor B divides the coloured F_2-individuals into two groups: purple (BB or Bb) and red (bb) and hence in the ratio of 3 purple: 1 red. Out of the 9 coloured F_2-individuals there should be $^3/_4$, i.e., $6\,^3/_4$ purple and $^1/_4$ i.e., $2^1/_4$ red.

Thus we will obtain the ratio of $6^3/_4$ purple: $2^1/_4$ red: 7 white or rather a ratio of 27 purple: 9 red: 28 white. If we now calculate the numbers expected for a total of 3725 F_2-plants(which is shown in the scheme borrowed from BATESON, figure 49) we will have 1571 purple: 524 red: 1630 white. Actually there were experimentally 1634 purple: 498 red: 1593 white. This ratio is not in absolute accordance with the theoretical numbers, but we must bear in mind the fact that the plants were not artificially selfed but were pollinated by the visits of a bee-species *Megachile*, and this fact can explain the results i.e., too many plants with the dominant phenotype.

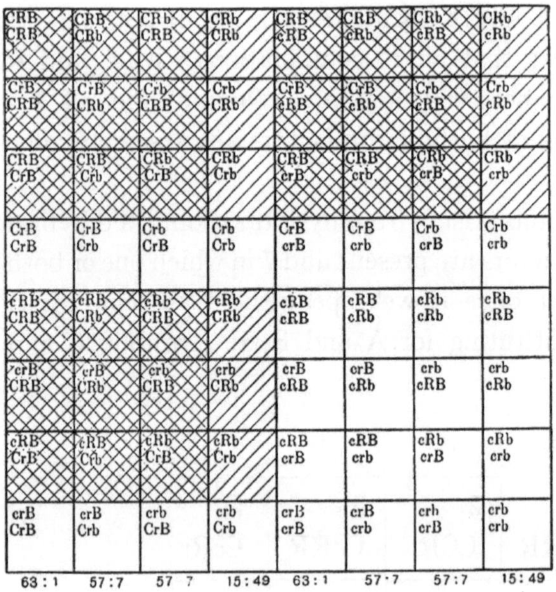

CRB/CRB	CRB/CRb	CRb/CRB	CRb/CRb	CRB/cRB	CRB/cRb	CRb/cRB	CRb/cRb
CrB/CRB	CrB/CRb	Crb/CRB	Crb/CRb	CrB/cRB	CrB/cRb	Crb/cRB	Crb/cRb
CRB/CrB	CRB/Crb	CRb/CrB	CRb/Crb	CRB/crB	CRB/crb	CRb/crB	CRb/crb
CrB/CrB	CrB/Crb	Crb/CrB	Crb/Crb	CrB/crB	CrB/crb	Crb/crB	Crb/crb
cRB/CRB	cRB/CRb	cRb/CRB	cRb/CRb	cRB/cRB	cRB/cRb	cRb/cRB	cRb/cRB
crB/CRB	crB/CRb	crb/CRB	crb/CRb	crB/cRB	crB/cRb	crb/cRB	crb/cRb
cRB/CrB	cRB/Crb	cRb/CrB	cRb/Crb	cRB/crB	cRB/crb	cRb/crB	cRb/crb
crB/CrB	crB/Crb	crb/CrB	crb/Crb	crB/crB	crB/crb	crb/crB	crb/crb
63:1	57:7	57 7	15:49	63:1	57·7	57:7	15:49

Fig. 49. Scheme belonging to the F_2 combinations with respect to the factors C, R and B as given in fig. 48 (after BATESON, 1906).

Undoubtedly it can be derived with certainty from the course of the segregations that both of the types of *Lathyrus odoratus* Emily Henderson, one with long and the other with round pollen, each have a factor which only gives a coloured phenotype by co-operation. In the white flowering plants, the factors C or R can be present cryptomerically. On the other hand the factor for blue (B) may also be present in a cryptomeric condition and can be the cause of the turning of red colour into purple. In one of the F_2-types of Emily Henderson the B-factor is present next to C or R, but this B-factor cannot find its expression because the basic combination colour (CR) is still incomplete.

Even more striking than this are the cases of varieties of white beans and of races of white mice. In white beans there can be numerous cryptomeric factors present (for brown, coffee brown, blue, purple and black colours, a factor for speckling) which cannot show their

Plate   III

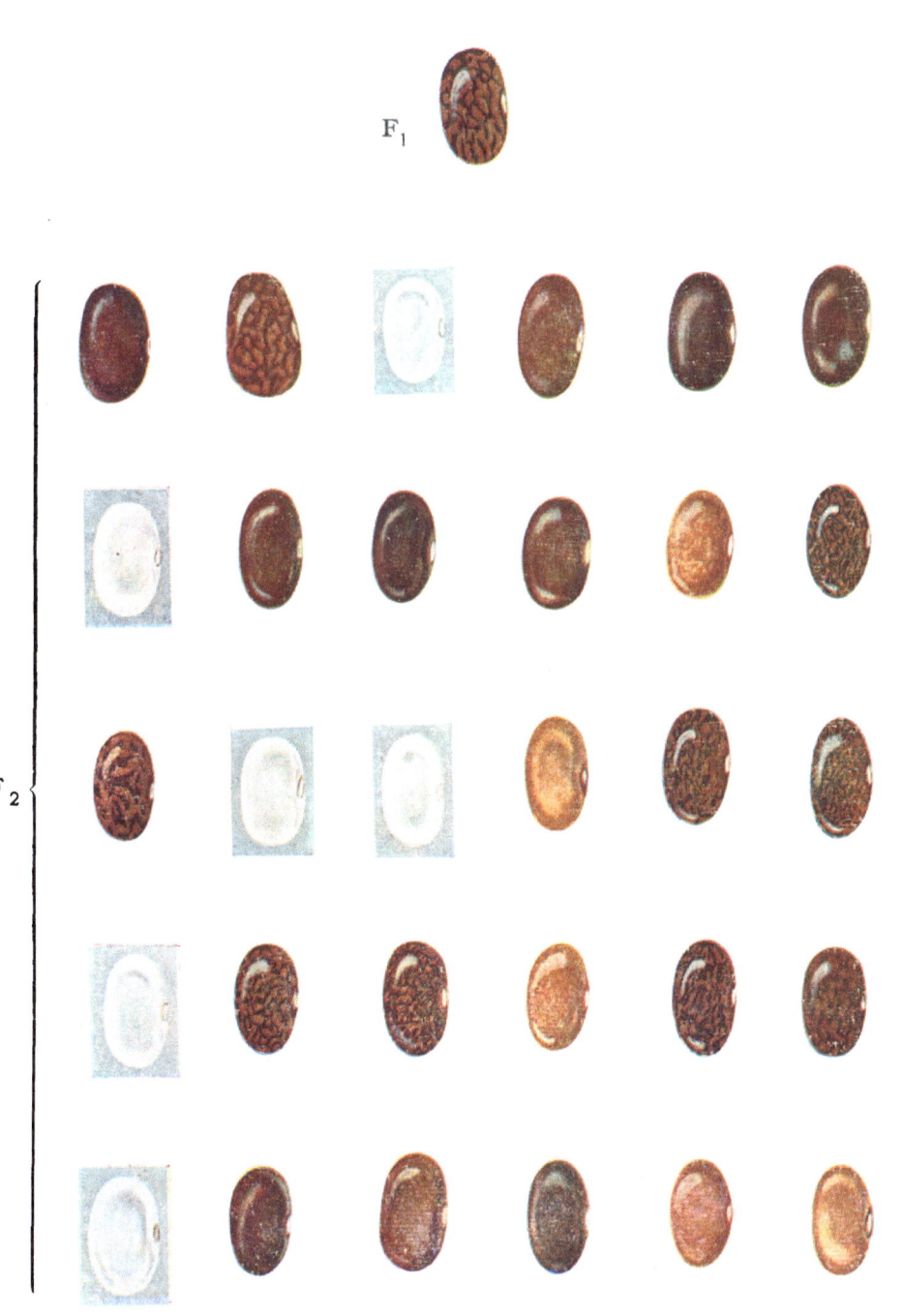

presence in the white bean, because a definite basic factor (which is sometimes called the pigment-factor) is missing in all varieties of white seeded beans. The presence of this basic factor is neither evident from selfings nor from mutual hybridization of white beans, but only when the white bean is hybridized with one or another coloured-one which possesses the pigment-factor, and because of this is coloured. This became evident from all the experiments published over this subject. Plate III will show how complicated such a segregation can be in the F_2-generation. This plate illustrates the hybridization of a brown kidney bean with a white bean. If such cryptomeric factors are active, then the phenotype will indeed be very misleading. In all white beans we will be dealing with cryptomeric factors, but the genotypical constitution of these white beans can differ to a high degree.

The same quality exists in white mice; white mice never give other than a white offspring, but pairing with wild-coloured races always gives rise to wild-coloured F_1's, and F_2's of the most unexpected combinations. Here also a number of investigators (see CUÉNOT, 1928) have tried to demonstrate the existence of 32 combinations of colour- and marking factors which can be present cryptomerically in white mice and which can only find their expression if they are paired with one or another of the coloured mice in which the pigment-factor (C) is present as the basis for all colour development.

From these examples it is possible to derive with certainty, the theorem, that there exist factors, which only in co-operation with other factors and by presence of both can cause a particular phenotype. On the other hand, absence of both of the factors has the same effect as the presence of one of them. If they are present separately, but cannot express themselves in isolation, they are called *cryptomeric factors*.

Finally an example from the last group, where the presence of both factors causes the same phenotype as one of them separately, but only the absence of both gives rise to a new phenotype. A cross between a type of oats with black glumes and one with yellow glumes gave an F_1-generation with black glumes but somewhat lighter than the black parental plant (NILSSON-EHLE, 1909, p. 44). This should lead us to expect that both of the factors for black glumes and for yellow glumes should form a pair of allelomorphs, and that in the F_2-generation a segregation into 3 black: 1 yellow should be obtained (since black is more or less covering yellow). But this was not the case i.e., the progeny of one of the black F_1-plants gave a segregation into 155 black glumed: 43 yellow glumed: 15 white glumed plants. If we group the black plants and the non-black plants, the segregation will be 155 : 58 i.e., approximately 3 : 1 (theoretically the ratio should be 159·75 : 53·25) and if we group the yellow types and the white types from the

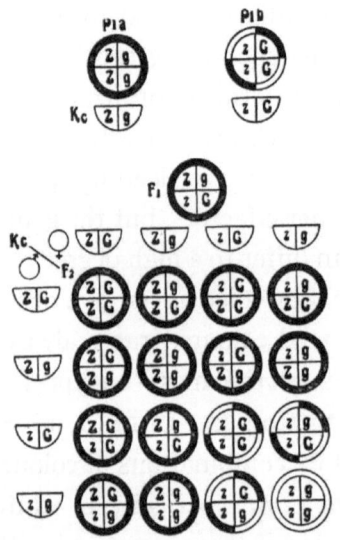

Fig. 50. A cross between black oat and yellow oat. F_1 black, F_2 segregating into 12 black: 3 yellow: 1 white (after the data from NILSSON-EHLE, 1909).

non-black plants then they also give a ratio of yellow to white of almost 3 : 1 (actually found was 43 : 15, and 43·5 : 14·5 was expected). It is possible that we deal in this cross with a dihybrid in which we distinguish black (Z) as the opposite factor to non-black (z), and yellow (G) as the opposite factor to white (g).

The reason for this is that the numbers obtained point to a segregation of 16 F_2-individuals into 12 black: 3 yellow: 1 white (actually obtained were 155 : 43 : 15; according to the theoretically 12 : 3 : 1 segregation, this should be $159^3/_4$: $39\,{}^{15}/_{16}$: $13\,{}^5/_{16}$).

Figure 50 shows the course of the segregation which we can also illustrate by a diagrammatic formula as given in this figure. Sixteen possible combinations of the reproductive cells can thus be obtained in the F_2:

1 ZZGG	2 ZZGg	3 ZzGG	4 ZzGg
5 ZZGg	6 ZZgg	7 ZzGg	8 Zzgg
9 ZzGG	10 ZzGg	11 zzGG	12 zzGg
13 ZzGg	14 Zzgg	15 zzGg	16 zzgg

As soon as one or both of the Z-factors are present, the plant is black-glumed; if all Z-factors (i.e., in the formula zz) are missing the glumes will be yellow or white. They are yellow if one or both G-factors are present; white if both of them are absent. Hence, the numbers 1, 2, 3, 4, 5, 6, 7, 8, 9, 10, 13 and 14 (that is to say 12 out of the 16 possibilities) must be black; the numbers 11, 12 and 15 are yellow and only number 16 is white. If these suppositions are correct we can derive further conclusions concerning the progeny of the various black-glumed plants. Four out of the sixteen plants (1, 2, 5 and 6) are homozygous ZZ (whether with G or g) and cannot give rise to other than an exclusively black-glumed progeny; four (4, 7, 10, 13) are similar to the F_1-generation and hence segregate again into 12 black: 3 yellow: 1 white; two (3 and 9) are heterozygously black and homozygously yellow and thus have to give rise only to 3 black (Z with G) and 1 yellow (z with G) progeny; two also are heterozygously black and homozygously white (8 and 14) and thus give rise to a segregation into 3 black (Z and g) and 1 white (z and g); of the last four, two (12 and 15) segregate into 3 yellow (z and G) and 1 white (z and g), one (11) breeds true to type yellow, and one (16) true to type white. We thus have to expect that in 16 individuals in the F_2-generation there will be:

a. 4 true breeding black,
b. 4 segregating into black, yellow and white,   ⎫
c. 2 segregating into black and yellow,   ⎬ black F_2-plants
d. 2 segregating into black and white   ⎭
e. 2 segregating into yellow and white,   ⎫
f. 1 true breeding yellow,   ⎬ yellow F_2-plants
g. 1 true breeding white   white F_2-plants

NILSSON-EHLE succeeded in winning separate selfed seed from 185 plants and in breeding an F_3-generation. Out of these 185 plants, there were 131 black, 39 yellow and 15 white. The 131 black F_2-plants were composed of 45 true-breds, 43 black-yellow-white segregating plants, 20 black-yellow segregating plants and 23 black-white segregating plants. Theoretically it should have been (for groups a, b, c and d) $43^2/_3$, $43\ ^2/_3$, $21\ ^5/_6$ and $21\ ^5/_6$ which is in very fine agreement with the numbers obtained. Among the 39 yellow F_2-plants there existed 23 which segregated into yellow-white and 16 true bred yellows (theoretically expected, groups e and f: 26 : 13) and all 15 F_2-plants with white glumes were absolutely true bred for white. This data thus forms a sufficient proof that the theoretical supposition of NILSSON-EHLE was true.

Briefly, to summarize, the results of these experiments are: — that hybridization of two plants, which seemingly differ only in one character may in reality be a dihybrid, and that in consequence of this, the double recessive plant which lacks both factors, can have a new and different phenotype.

This hybrid has another important aspect; in the F_2-generation ZZGG- and ZZgg-plants occur next to one another, the first is homozygous for the black factor and for the yellow factor; the second is homozygous black, but it lacks the yellow factor. Both of these plants remain black-glumed and are not distinguishable at sight, one from another. The gene for yellow is present in the first group but is invisible; this factor is completely covered up by the factor for black and cannot find its expression in the phenotype of the plant.

To describe the condition in which the yellow factor is found here, BATESON introduced the term *hypostatic* (1907, p. 653). By means of this the uncertain and rather vague conception *"latent'* is avoided. In contradistinction to this, the black factor which completely eclipses the hypostatic yellow factor is called an *epistatic* factor.

It must be stressed that *cryptomery* and *hypostasy* are very different conceptions, although both belong to the obsolete conception of *"latent"*. A factor is cryptomeric if its presence does not find an expression in the phenotype, and can only be observed in "another combination of genes which are arranged in a new manner". Hypostasy on the other hand is limited to those cases in which a factor is present and has an effect, but this effect is eclipsed by another factor from another pair of alleles.

Some investigators call a hypostatic gene (such as the yellow gene in the ZZGG oats) a cryptomeric-one, because hybridization of ZZGG with the white zzgg oats unfailingly gives yellow oats (zzGG and zzGg) in the F_2-generation, but such use of the term "cryptomeric" is not recommended, since it can be the cause of a misunderstanding. Indeed the conception cryptomery must be clearly distinguished from hypostasy in the way which has already been pointed out.

The phenomenon of *epistasy* can, as was already mentioned (p. 128) give rise to a false conclusion, i.e., the factors investigated form a series of *multiple alleles*. In this respect the classic results in the case of multiple allelomorphs which we often mentioned were obtained by BAUR (1911) from the hybridization experiments with *Aquilegia*-types which had different leaf-colours: green, yellowish-green (chlorina) and mottled green-chlorina (variegata).

The three possible hybrids yielded an 3 : 1 segregation:

green   × variegata: 614 green   : 185 variegata

green   × chlorina: 755 green   : 243 chlorina

variegata × chlorina: 144 variegata : 54 chlorina

In considering these ratios, this "triangular-segregation" reminds us of the existence of *multiple alleles*: G_3 green, G_2 variegata and G_1 chlorina. In this case the hybridization would be $G_3G_3 \times G_2G_2$ giving a segregation into 3 green : 1 variegata; $G_3G_3 \times G_1G_1$ gives a segregation into 3 green : 1 chlorina and $G_2G_2 \times G_1G_1$ a segregation of 3 variegata: 1 chlorina. It is a question as to whether this segregation can be explained in this way. In crosses with chlorophyll-types of *Vicia Faba* (SIRKS, 1931) the heritable behaviour of the plain green (typica), variegata and chlorina was investigated, and the following possibilities were obtained:

green × variegata : — 3 green : 1 variegata or

:— 12 green : 3 variegata : 1 chlorina

green × chlorina :— 3 green : 1 chlorina

variegata × chlorina : — 3 green : 1 chlorina

Three of these segregations are thus parallel to those of BAUR; the fourth (12 green: 3 variegata: 1 chlorina) proves only that the action of the factors is completely different. The factor A is always the cause of plain green and is *epistatic* over the *hypostatic* variegata factor. The segregation can be explained by this formula:

1) AaVV segregates into 3 green (AAVV and AaVV) : 1 variegata (aaVV);

2) AaVv segregates into 12 green (AAVV, AAVv, AAvv, AaVV, AaVv and Aavv): 3 variegata (aaVV and aaVv): 1 chlorina (aavv);

3) Aavv segregates into 3 green (AAvv and Aavv): 1 chlorina (aavv);

4) aaVv segregates into 3 variegata (aaVV and aaVv): 1 chlorina (aavv).

Relationships between genotypical factors (as we have learned to recognize in this chapter) express themselves in the following manners:

A. Instead of two genes, which together form a pair of allelomorphs, it is possible that three or more genes can form themselves into pairs (1 and 2, 1 and 3, and 2 and 3), i.e., pairs of allelomorphs which only segregate according to the scheme for the monohybrid. Such groups of factors are called "*multiple allelomorphs*" or "*multiple alleles*".

B_1. Mutually equivalent genes give rise, each in their turn, to the same phenotype; in the case of the presence of more of such factors, the phenotype can either be expressed more intensively, or to the same degree as if only one gene was present. According to LANG this is true *polymery*, or *isomery* to be more exact (colour of wheat is caused by *cumulative isomery*; fruit shape of *Capsella* by *non-cumulative isomery*).

B_2. If the genes mentioned under B_1 are unequivalent then the term "polymery" has no more value in the sense originally given to it by LANG. It is better to speak of *"anisomery"* in this latter case.

B_3. The heredity of *"quantitative characters"* (measurements, weight etc.) can be caused by one single factor or by polymeric factors (perhaps by isomeric and, certainly by anisomeric factors), also multiple alleles may play a part in this type of heredity.

C. Two mutually unequivalent factors, each give rise to a particular phenotype; when both are present an absolutely new phenotype originates; when both are absent yet another phenotype occurs (plume-shape in oats, comb-shape in poultry);

D. Out of two mutually unequivalent factors, only one gives rise by itself to a definite phenotype; the other-one is only able to co-operate with this factor. Such a co-operation is the cause of a change in phenotype. This latter factor may be present in organisms in which its partner is missing, but in this condition it is completely inactive; its presence cannot be observed and, because of this, the factor is termed *"cryptomeric"* (e.g., flower colour in *Linaria maroccana*);

E. Out of two mutually unequivalent factors, neither gives rise by itself to a definite phenotype, they are only able to co-operate with one another, and by doing so they will give rise to a definite phenotype. If separated they can occur in different organisms without being noticed and are called *"cryptomeric"* factors under these conditions (the basic factors for colours (C and R) in *Lathyrus odoratus*);

F. Two mutually unequivalent factors each gives rise to a particular phenotype; if both are present the same phenotype originates as if only one of them was present. This factor is called *epistatic*. The other factor which cannot find its expression in the phenotype is called a *hypostatic* factor. If both factors are absent, another phenotype originates (e.g., black, yellow and white colour of oat-glumes). Sometimes *multiple allelomorphism* is given as an explanation in cases in which epistasy actually occurs, however, such a conclusion is not justifiable.

CHAPTER VIII

RELATIONS BETWEEN GENES II
(The mutual dependence of factors during segregation)

Among the most important conclusions derived from MENDEL's own investigations, first place is held by the law relating to the behaviour of the different factors with respect to the ratio they bear to one another in the segregation process. As we have already seen in chapter V, MENDEL believed that the factors behave absolutely independently, that is to say, as independent units. "Hence it follows that the behaviour of every two different characteristics united in a hybrid is independent of remaining differences existing in both parental plants' (1865, p. 20).

If this supposition is correct and the independence of the factors is not limited during the segregation process then we should find that the segregation ratio 9 : 3 : 3 : 1 always occurs in the F_2-generation of a dihybrid, in the case of complete dominance of one of the factors over the other one in the same pair. At its utmost this ratio can only be altered in the sense that it appears as 9 : 3 : 4; 9 : 7 and 12 : 3 : 1 in the case of phenotypical resemblance between two or more of the groups of a 9 : 3 : 3 : 1 segregation. Similarly, the F_2-generation of a trihybrid (in the case of complete dominance of one of the factors of each of the three pairs) must be composed of 8 phenotypical groups which are represented respectively by 27 : 9 : 9 : 9 : 3 : 3 : 3 : 1 individuals.

If there are fewer phenotypically distinguishable groups then we find numbers such as 27 : 9 : 28 as in the proportion between the blue, red and white F_2-individuals in BATESON's *Lathyrus*-hybrid, described previously.

But if all phenotypical groups, which one could expect, were present, the theoretical ratios in these F_2-groups were not always obtained in later experiments.

During the neo-mendelian period this deviation between ratios expected and obtained could be observed many times. One of the most striking examples of this was found by BATESON in a hybrid of *Lathyrus odoratus* with long pollen and *L.o.* with round pollen. The course of the segregation in the F_2-generation was not only of importance because of the discovery of cryptomeric factors (which cause the expression of colour by co-operation of these factors) but also because the grouping of the F_2-plants with respect to pollen-shape brought about a peculiar complication, which was not in accordance with the primitive principle of the independence of the factors during the development of the reproductive cells. The F_2-individuals of this hybrid had long pollen and therefore, the character "long pollen" was dominant over the character "round pollen". The F_2-generation, as we have already seen on p. 148, was composed of a total of 3725 plants from which 2844 gave rise to long pollen and 881 to round pollen-grains (P.E. $=1\cdot96$). It appears, in this case, that a simple monohybrid segregation is active, which also seems to be evident from the distribution of the 1593 white flowering plants with regard to the type of pollen (1199 long: 394 round, or P.E. $= 0\cdot24$). The distribution of the purple flowering and the red flowering F_2-plants, however, with respect to pollen-shape deviated so considerably from this that there was no evidence for a 3 : 1 segregation. The 1634 purple flowering individuals were composed of 1528 with long pollen-grains and 106 with round pollen-grains (P.E. $= 17\cdot01$); out of the 498 red flowering plants there were 117 long pollen plants and 381 round pollen plants (P.E. $= 26\cdot50$). This peculiar surplus of long pollen in the purple plants and the deficit of this type among the red flowering types must have a specific cause. It was due to the perspicacity of BATESON that the significance of this phenomenon was realized, even if the explanation put out was not quite correct.

It was originally established (BATESON, SAUNDERS and PUNNETT, 1905, p. 89) that a definite relationship (coupling) between the shape of the pollen-grain and the colour existed; later they expanded this idea (BATESON, SAUNDERS and PUNNETT, 1906, p. 9) by supposing a relationship between the factor B (whose presence converts the red oclour into purple) and the factor A (A = long pollen; a = round pollen). Such a *coupling* would be active in the sense that the four types of gametes: AB, Ab, aB and ab (reproductive cells), which could be expected are not formed in equal numbers.

It was supposed that both the types AB and ab should originate seven times more frequently than Ab and aB. In this way, the ratio of the four types becomes 7 AB : 1 Ab : 1 aB : 7 ab. If this is accepted it would follow that the F_2-individuals should be classified into 177 A and B, 15 A, 15 B and 49 which are neither A nor B. This is evident, if we write down the sixteen possible combinations of a dihybrid segregation, but in this case, take into consideration the differences in numbers in which the 4 types of the reproductive cells are formed:

♂ ♀	7 AB	1 Ab	1 aB	7 ab
7 AB	1 AABB 49	2 AABb 7	3 AaBB 7	4 AaBb 49
1 Ab	5 AABb 7	6 AAbb 1	7 AaBb 1	8 Aabb 7
1 aB	9 AaBB 7	10 AaBb 1	11 aaBB 1	12 aaBb 7
7 ab	13 AaBb 49	14 Aabb- 7	15 aaBb 7	16 aabb 49

The combinations 1, 2, 3, 4, 5, 7, 9, 10 and 13 all contain A and B and are represented by $(49 + 7 + 7 + 49 + 7 + 1 + 7 + 1 + 49) = 177$ individuals out of 256; the combinations 6, 8 and 14 have the A factor only and together yield $(1 + 7 + 7) = 15$ individuals. The combinations 11, 12 and 15, which have no A but only B, give $(1 + 7 + 7) = 15$ individuals, whereas the combination 16 which has neither A nor B gives 49 individuals out of a total of 256 F_2-individuals. Calculated on the basis of a segregation scheme of 177 : 15 : 15 : 49 BATESON's F_2-generation, concerning 2132 coloured plants should be composed of 1474 purple plants with long pollen (AB), 125 red with long pollen (Ab), 125 purple with round pollen (aB) and 408 red with round pollen (ab). Experimentally, BATESON obtained the numbers 1528, 106, 117 and 381 for these four groups which is in good accordance with the theory for such a complicated case.

According to this supposition the ratio of numbers of the segregation long pollen: round pollen should also differ according to which of the respective F_2-individuals belonged to the homozygotes for the factor B (i.e., BB), to the heterozygotes for this factor (Bb) or to the homozygous recessive plants (bb). In the first group (figure 49, columns 1 and 5) we only find the combinations (7 AB + 1 aB) with (7 AB + 1 aB) hence there occur 49 AABB, 14 AaBB and 1 aaBB or 63 long pollen plants to 1 round pollen plant (63 : 1). In the second group (figure 49, columns 2, 3, 6 and 7) the combination of 7 AB with (1 Ab + 7 ab), of 1 Ab with (7 AB + 1 aB), of 1 aB with (1 Ab + 7 ab) and of 7 ab with (7 AB and 1 aB). These combinations give rise successively to 7 AABb + 49 AaBb, 7 AABb + 1 AaBb, 1 AaBb + 7 aaBb, 49 AaBb + 7 aaBb; that is, a total of 14 AABb : 100 AaBb : 14 aaBb, or 114 with the factor A : 14 with the factor a, or 114 plants with long pollen : 14 plants with round pollen (114 : 14 = 57 : 7). The third group (figure 49, columns 4 and 8) represent all combinations (1 Ab + 7 ab) with (1 Ab + 7 ab), from this it follows that there are present 1 AAbb : 14 Aabb : 49 aabb, or 15 long pollen individuals : 49 plants with round pollen. These ratios 3 : 1, 57 : 7 and 15 : 49 are indicated in figure 49 under the respective columns.

The fundamental importance of this investigation lies in the fact that from this, it becomes evident why the types of reproductive cells which can be expected, do not originate in equal numbers. Due to a coupling between the factor B, for purple colour of the flower, and the factor A for long pollen, on the one hand, and the gene b (red colour of the flower) and a (round pollen) on the other hand, there come into being far more reproductive cells formed with the factors A and B (or with the factors a and b) than reproductive cells with the combinations Ab and with aB. In the case investigated the ratio of these reproductive cells was approximately 7 AB : 1 Ab : 1 aB : 7 ab.

Other experiments dealing with this subject quickly followed. The great importance of these exceptions to one of MENDEL's fundamental rules was the motive for BATESON's further work in this direction, and so, there then appeared more of such cases from his crosses with

Lathyrus. These cases showed, that other ratios of gametes could occur. Hence BATESON obtained a wholly different coupling ratio, when he investigated the relationship between the colour of the flower and the shape of the standard in *Lathyrus*. Sufficient data had already shown him that the erect shape of the standard of the corolla is dominant as a simple factor over the bent standard (hooded). Likewise it was well-known to him that the blue colour was dominant over the red colour. In a cross between a white *Lathyrus* with erect standard (Emily Henderson) and a white *Lathyrus* with hooded standard (Blanche Burpee) a number of purple-flowering F_1-plants were obtained, all phenotypically like the stock Purple Invincible. The F_2-generation segregated naturally into coloured (figure 51, 4–9) and uncoloured (10–11), blue (4, 5, 7, 8), and red (6, 9), erect standards (4, 6, 7, 9, 10) and hooded (5, 8, 11) standards.

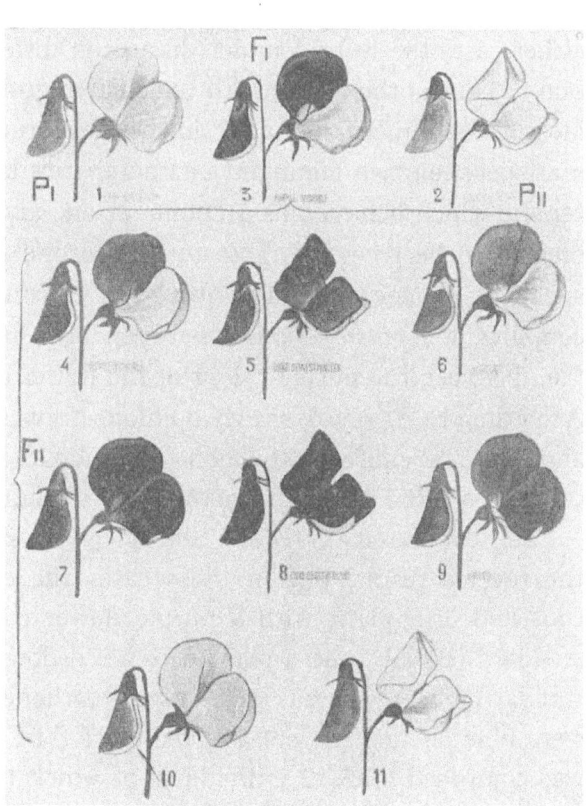

Fig. 51. Cross between white *Lathyrus* with erect standard and white *Lathyrus* with hooded standard. Coloured F_1 and segregating F_2 (after BATESON, 1909).

But the striking and peculiar fact was that considering the colour and standards shape only, among 4000 F_2-plants no single red flower occurred which had a hooded standard. All plants with red flowers had erect standards. A number of families composed of progenies of F_1-plants, with purple colour and erect standard, included 3761 coloured individuals, from which

902 were purple with hooded standards, 1920 with erect standards and 939 red with erect standards. The ratio of these three groups was almost that of a 1 : 2 : 1 — segregation (a ratio of 940·25 : 1880·50 : 940·25 was expected theoretically). It seemed as though the hybrid behaved as a monohybrid, and due to this, both of the characters (i.e., the blue colour and the hooded shape of the standard) were caused by only one single factor. However, this was not in accordance with the fact that one of the characters (blue colour) was dominant, the other one recessive (hooded standard). In either case, the hybrid under discussion differed from the preceding one : 1) one of the theoretically calculated groups in the F_2 was always lacking experimentally and 2) in these experiments a coupling did not exist between two dominant characteristics but between one characteristic which behaved as a dominant-one and a characteristic which seemed to be recessive. Due to this second difference one may not speak of proper coupling; quite the contrary in fact: a negative coupling, a *repulsion*, exists between both of the dominant factors (i.e., between the purple colour of the flower and the erect standard). According to BATESON's early opinion this was a complete *"repulsion"* and was the cause of the non-formation of reproductive cells containing both the dominant factors next to one another.

Later communications of PUNNETT (1913, p. 89 and 98) have given the reverse result, but in these cases the original parental plants consisted of a plant with a purple flower colour and with an erect standard (BBEE) and a plant with a red flower colour with a hooded standard (bbee), whereas in BATESON's earlier experiments, the parents were blue hooded (BBee) and red-erect (bbEE). The F_2 of PUNNETT was composed of 2712 individuals of which there were 2036 BE: 12 Be: 10 bE: 654 be. These numbers have the closest relationship with the theoretical ratio due to a coupling between the factors B and E of the magnitude 127 : 1 : 1 : 127. The ratio calculated was 2023 BE : 11 Be : 11 bE : 667 be, which is in good accordance with the ratio obtained.

Hence, a new principle had been discovered, that is to say: the repulsion of factors as the opposite of coupling.

The point concerning the completeness of the repulsion is not supported by BATESON later on. In 1911 he and PUNNETT (1911, p. 297) wrote: "In most cases this could not be tested in practice owing to the very large number of plants required. Thus the coupling between erect standard and blue is on the 127 : 1 : 1 : 127 system and if the repulsion were of similar intensity we should expect only one hooded red in 65·536 plants".

The same factors can act in one case as coupled and in the other case as a repulsion. This depends on the constitution of the original parents. In order to distinguish this difference between coupling and repulsion clearly, it is better to write the genotypical formula for an F_1-individual, not in the primitive form BbEe, but in the following manner which is preferable: BE. be for coupling (coupling phase) and Be. bE for repulsion (repulsion phase).

By means of this notation the constitution of the parental genotypes is made clear.

It was accepted, that both phenomena are fundamentally similar; nowadays one term is usually used for this dependence of factors upon one another (linkage). Coupling occurs in the case when one of the parents of the hybrids possesses both of the dominant factors, the other-one possesses both recessive factors. Repulsion occurs when one parent possesses a dominant and a recessive factor and the other parent the factors which are allelomorphic with these, i.e., a recessive factor next to a dominant-one.

The fundamental points in which the investigations of BATESON and his co-workers yielded new views, were as follows:

Hybridization of a type having two dominant factors and a type with two recessive factors does not always give rise to an F_2-segregation according to the 9 : 3 : 3 : 1 scheme. This scheme can only be followed if the factor-pairs are mutually independent. Linkages (coupling or repulsion) can occur in different grades, which can be expressed by the ratio of the numbers of the four types of gametes or reproductive cells expected. By independent segregation, these four types are formed in equal numbers (1 : 1 : 1 : 1); for coupling, this ratio can be expressed as n AB : 1 Ab : 1 aB : n ab, for repulsion as 1 AB : n Ab : n aB : 1 ab.

In 1912 a paper of MORGAN and LYNCH was published in which the discovery of a coupling between two factors in a fly-species (the banana fly, *Drosophila melanogaster*) was communicated. This was the first in a series of very important contributions from the school of the American *Drosophila*-investigator T. H. MORGAN. In this investigation they opened up a hitherto almost unknown but fruitful subject matter, with the linkage phenomenon as the object of study.

Some of the F_1-flies bred by MORGAN were mutually paired, others were back-crossed with flies recessive in both linked factors. Both of these series of experiments yielded respectively F_2- and F_1R-generations. In the publication of MORGAN and LYNCH (1912) the cross is discussed (see figure 52) of a *Drosophila* with a black body (black, bb) and deformed wings (vestigial, vv) and a *Drosophila* with grey body (BB) and normal wings (VV), i.e. the combination bbvv × BBVV. The F_1 was grey and had normal wings (BbVv) from which the dominance of both of these characteristics could be derived. From backcrossing the females with a bbvv-male (black, vestigial wings), in the F_1R-generation the ratio of the gametes formed by the F_1-flies could be obtained: the F_1R groups correspond with the gametes of the F_1-generation. It was discovered that this F_1-generation gives rise to all the four types of gametes (BV, Bv, bV and bv), but that these gametes were not formed in equal numbers. Instead of half, only 17% of the total number of gametes were new *recombinants* (Bv and bV together). MORGAN named these new recombinants "*crossovers*". Hence a linkage exists between the genes B and V on the one hand, and between b and v on the other hand. One half of the crossover gametes (i.e., 8·5%) were Bv, and the other half bV-gametes.

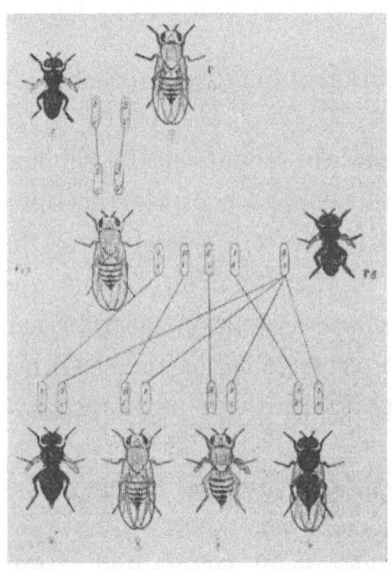

Fig. 52. Cross between two different *Drosophila*-types which differ in two pairs of factors (BB and VV) and the backcross with the double recessive fly from which a 17% crossover-frequency was obtained (after MORGAN, 1909).

A wealth of facts concerning the *Drosophila* investigations are published in the News Letter D (rosophila) I (nformation) S(ervice) devoted to this material, and which contains the most recent data (see also MULLER and HERSKOWITZ' bibliographies of 1939 and 1953).

MORGAN's school, and numerous later investigators did much to forward this field of interest. The following list (which was borrowed from MORGAN and BRIDGES, 1916, p. 83) gives some idea of crossover percentages between some *Drosophila*-genes obtained. Each factor is indicated by a word which characterizes its distinct character. E.g., "Yellow" means "yellow body"; "white" and "vermillion" have reference to the eye-colour; "miniature" and "rudimentary" indicate two peculiarly deformed wings and "bar" is used for a distinct property of the eye.

Genes	Total numbers of individuals	Crossovers,	Percentages
Yellow and white	46·546	498	1·07
Yellow and vermillion	10·603	3·644	33·4
Yellow and miniature	18·797	6·440	34·3
Yellow and rudimentary . .	2·563	1·100	42·9
Yellow and bar	191	88	46·1
White and vermillion	15·257	4·910	32·1
White and miniature	41·034	13·513	32·8
White and rudimentary . . .	5·847	2·461	42·1
White and bar	5·151	2·267	44·0
Vermillion and miniature . .	5·329	212	4·0
Vermillion and rudimentary .	1·554	376	24·1
Vermillion and bar	7·514	1·895	25·2
Miniature and rudimentary .	12·567	2·236	17·8
Miniature and bar	3·112	636	20·4
Rudimentary and bar	195	7	4·4

A worthy parallel to *Drosophila* was found in maize (EYSTER, 1934; EMERSON, BEADLE and FRASER, 1935) and in peas (WINGE, 1936). A great number of genes were tested in twos for their mutual ratio of gametes, and typical examples of linkage came to the fore. Maize has the advantage that a whole generation of descendants is gathered on one single cob which makes it very demonstrative material. Figure 53 gives a picture of this; between both of the genes C (C coloured aleurone, c uncoloured aleurone) and Sh (Sh non-shrunken endosperm, sh shrunken endosperm) there exists, in the left hand figure a coupling, in the right hand-one a repulsion. The experimental numbers in the back-crosses were (HUTCHISON, 1922):

a) $\dfrac{\text{C Sh}}{\text{c sh}} \times \dfrac{\text{c sh}}{\text{c sh}}$ b) $\dfrac{\text{C sh}}{\text{c Sh}} \times \dfrac{\text{c sh}}{\text{c sh}}$

	coloured	coloured	white	white
	non-shrunken	shrunken	non-shrunken	shrunken
a)	4032	149	152	4035
b)	638	21379	21096	672

From the series a) a crossover-percentage of 3·6 can be calculated, from the series b) one of 3·9%.

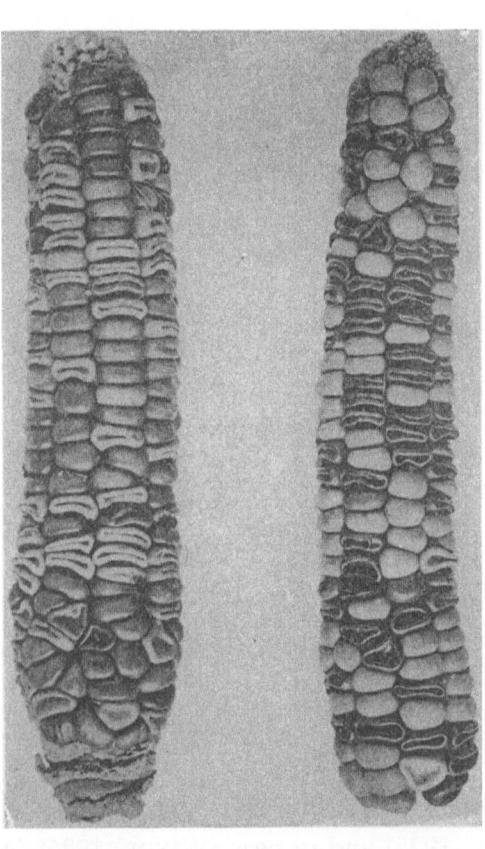

The investigation with relation to genetical coupling between genes, which was already performed about thirty years ago on a large scale with different species of plants and animals has also made it very clear that linkage is not at all an exception and that each value between 0 and 50 (whole numbers or fractions) can be a crossing-over percentage. If in a definite F_2- or F_1R-generation no individuals are observed which owe their origin to the crossover-gametes, then the crossover-percentage will be zero (0) and thus the linkage can

Fig. 53. Two cobs of F_1-R-generations of corn kernels. Left, the cob, originating from the backcross $\dfrac{\text{C Sh}}{\text{c sh}} \times \dfrac{\text{c sh}}{\text{c sh}}$ with predominantly coloured-non shrunken and white-shrunken combinations. Right, the cob of the backcross $\dfrac{\text{C sh}}{\text{c Sh}} \times \dfrac{\text{c sh}}{\text{c sh}}$ with predominantly coloured-shrunken and white-non shrunken kernel types (after HUTCHISON, 1922).

be considered absolute. If a crossover-percentage is fixed at 50% then all four types of gametes are represented in equal numbers and hence the result is the same as in an independent segregation of alleles.

Hitherto in all cases of linkage discussed attention was drawn only to the tie which occurs between two seemingly independent genes, but it soon became evident from investigations, that more than two genes can be mutually linked.

Although first examples of linkage between more than two factors were noticed in plants (*Primula sinensis*, peas, *Matthiola*), the best species from the point of view of investigation is again *Drosophila melanogaster* in which more than 1000 genes have been investigated by MORGAN and his co-workers up to the present day.

The first fact which appeared was, that a linkage between a factor A and a factor B, when accompanied by a linkage B and a third factor C, entailed a linkage between A and C; a conclusion which can also be derived from the table dealing with *Drosophila*-factors on p. 163. Since Yellow seems to be linked with White and this latter in its turn is linked with Miniature there also exists a linkage between Yellow and Miniature. If we consider for a moment the crossover-percentages experimentally obtained of these cases of linkage, it is evident that the highest percentage is almost the sum of both the others: the percentages are for example for Yellow-White 1·07, for White-Miniature 32·8 and for Yellow-Miniature 34·3. This regularity, in spite of some exceptions, turned out to be of great theoretical importance (see Chapter XI).

The first exception is, that if the highest crossing-over-percentage from any two of the three factors involved is less than 15%, then as a rule, the crossing-over phenomenon will be observed only in one of the remaining two linkages. Suppose that in a trihybrid F_1-individual AaBbCc both of the factors A and C show a crossing-over of 14·5%, whereas A–B give 10%, then a crossing-over between B and C will not usually be observed. Theoretically one would expect nearly 4·5% crossing-over but nevertheless the coupling has expressed itself as an absolute-one. This phenomenon, that the occurrence of an exchange between two factors out of a triplet, which are mutually linked, prevents another exchange between two of these three linked factors, occurs only if the highest percentage of crossing-over is rather small (i.e., less than 15%). STURTEVANT (1913) called this *"interference"*.

A second exception to the rule that two crossingover-percentages can be added together to find the third, occurs when these percentages are found to be rather high; as a rule the highest percentage observed in this case is smaller than the sum of both the others. We will borrow some numbers from the table given on p. 163: For Yellow-Vermillion, a percentage of 33·4 was obtained, for Vermillion-Miniature 4·0, but the percentage Yellow-Miniature was not equal to the sum of both numbers observed (i.e., 37·4). A smaller number, 34·3, was obtained. This exception is due to the occurrence of *"double crossing-over"*. In chapter XI we will return to the significance and explanation of this phenomenon.

The whole crossing-over investigation dealing with some hundreds of factors in *Drosophila melanogaster* again led MORGAN to conclusions of theoretical importance, i.e., that all these factors are distributed amongst four groups. Within each group the factors are more or less mutually strongly linked. One group contains almost 150 factors, a second group almost 100, a third almost 80 and a fourth only nine factors. A number of heritable characteristics observed in *Drosophila melanogaster* occur in a series of multiple alleles; the mutual degree of linkage between the different members of such a series and the others factors of the same linkage-group, was found to be constant.

Other *Drosophila*-species have also been investigated (e.g., *D. virilis*, *D. obscura* and *D. willistoni*) and have given the result, that all their factors could be classified into 6, 5 or 3 groups respectively. As an analogy in plants we can refer to *Zea Mays, Pisum sativum, Hordeum vulgare, Pharbitis Nil, Antirrhinum majus, Vicia Faba, Oryza sativa* and others, but the analysis in these cases has not been carried out to the same extent as in *Drosophila*. However, all the plants quoted here gave examples of a great number of factors which belonged to a limited number of linkage-groups. It may be especially stressed here that the continuous increase of factors which seem to belong to one and the same linkage group has made higher demands in exactness in determining the percentage of crossovers, since this determination carries with it great theoretical consequences to which we will return later in more detail (chapter XI).

It is not surprising that a very comprehensive literature exists over this subject (MATHER, 1938a) when one considers the different angles from which an attempt was made to find as reliable a method as possible for these calculations. In the first place the most obvious procedure appeared to be that of calculating the ratio of gametes from a mathematical treatment of the composition of the F_2-generation. For a summary of the different methods of calculation we will refer to the studies of WELLENSIEK (1927) and ALAM (1929). It may be sufficient to point out the great advantages which the so-called *product-ratio method* of FISHER (and BALKAMUND, 1925, 1928) has, due to its great clarity of result and simplicity of application (for tables see IMMER, 1930).

If the four types of gametes in a dihybrid (AB, Ab, aB, ab) are respectively represented by w, x, y and z, then in the case of "coupling" the parental gametes will be $w + z$ and the number of cross-overs amount to $x + y$. In the case of "repulsion" there are $x + y$ parental gametes and $w + z$ cross-overs. This is to say that if, in the latter case,

the cross-over percentage is p (or rather o·p) then $w = z = \dfrac{p}{2}$ gametes

while $x = y = \dfrac{1-p}{2}$.

The gameto-types, formed in the case of repulsion (Be.bE), occur in the following frequencies:

gameto-type	AB (w)	Ab (x)	aB (y)	ab (z)
frequency	$\dfrac{p}{2}$	$\dfrac{1-p}{2}$	$\dfrac{1-p}{2}$	$\dfrac{p}{2}$

Although strictly speaking the gameto-ratio is valid for repulsion only, the same discussion and the same notation can be used when dealing with the phenomenon of "coupling", since in this case, a value for p will be obtained in which the difference between p and 1 indicates the theoretical crossover-percentage. Such a ratio of gametes will now lead to a definite ratio of phenotypical F_2-groups, and if each individual factor (A or B) shows a theoretical segregation of 3 : 1 we will obtain:

phenotypical group	AB	Ab	aB	ab
frequency	$\dfrac{2+p^2}{4}$	$\dfrac{1-p^2}{4}$	$\dfrac{1-p^2}{4}$	$\dfrac{p^2}{4}$

This calculation can easily be derived by using the scheme on p. 100. The product-ratio method tries to calculate the value of p by deriving

the ratio $P = \dfrac{\text{number of AB} \times \text{number of ab}}{\text{number of Ab} \times \text{number of aB}}$ from the frequencies observed and tries to apply a connection between this value of P and the theoretical p. This connection can be calculated as follows:

$$P = \frac{\text{number of AB} \times \text{number of ab}}{\text{number of Ab} \times \text{number of aB}} = \left(\frac{2 + p^2}{4} \times \frac{p^2}{4}\right):$$

$$:\left(\frac{1 - p^2}{4} \times \frac{1 - p^2}{4}\right) = \frac{2p^2 + p^4}{1 - 2p^2 + p^4}$$

From this it follows conversely that $p^2 = \dfrac{(P + 1) - \sqrt{3P + 1}}{(P - 1)}$

Hence for every value of P observed which goes with p, we are able to fix the percentage of crossing-overs. It is obvious that this calculation is only valid providing that each of the pairs of factors A-a and B-b individually fulfil the theoretical ratio of 3 : 1 and that linkage occurs in the gametes of both sexes. But if the factors are transferred in a linkage-phase in one of the sexes only (i.e., the factors behave independently in the opposite mating type) or if one of the alleles of a pair has a more complicated character (e.g., polymeric), then completely different formulae will be necessary.

We are indebted to ALAM for a very clear presentation of these formulae which are given on p. 169. All these formulae are suited to direct application and their results are sufficiently reliable for our purpose.

The second method, as we already saw on p. 162, which was soon taken up by MORGAN is a direct method of finding the ratio of the gametes. It is evident that in the case of backcrossing of the F_1-individuals (AaBb) with the double recessive (aabb) a much simpler formula can be used, which will lead to the same result.:

$$p = \frac{\text{number of Ab} + \text{number of aB}}{n} \quad \text{for "coupling"}$$

$$p = \frac{\text{number of AB} + \text{number of ab}}{n} \quad \text{for "repulsion"}$$

It is obvious that in the calculation of crossover-percentages the method of backcrossing with a double recessive individual is preferable. It must be taken into consideration however, that possible differences in the viability or developmental capacity between the different types of gametes or individuals can occur.

In exceptional cases a third method can also be useful in fixing the proportion of the gamete-types. There are some plants and animals

Calculation of the value of p² from the value found for p, for different phenotypical ratios.

No.	Phenotypical ratio	Phenotypical classes in %				Value of P expressed in p	Value of p² derived from P
1	1 : 1, backcross	$\dfrac{p}{2}$	$\dfrac{1-p}{2}$	$\dfrac{1-p}{2}$	$\dfrac{p}{2}$	$\dfrac{p^2}{1-2p+p^2}$	$\left(\dfrac{P-\sqrt{P}}{P-1}\right)^2$
2	3 : 1 and 3 : 1, coupling in one sex	$\dfrac{2+p}{4}$	$\dfrac{1-p}{4}$	$\dfrac{1-p}{4}$	$\dfrac{p}{4}$	$\dfrac{2p+p^2}{1-2p+p^2}$	$\left(\dfrac{(P+1)-\sqrt{3P+1}}{(P-1)}\right)^2$
3	3 : 1 and 3 : 1, coupling in both sexes	$\dfrac{2+p^2}{4}$	$\dfrac{1-p^2}{4}$	$\dfrac{1-p^2}{4}$	$\dfrac{p^2}{4}$	$\dfrac{2p^2+p^4}{1-2p^2+p^4}$	$\dfrac{(P+1)-\sqrt{3P+1}}{(P-1)}$
4	9 : 7 and 3 : 1, cooperating genes for one character	$\dfrac{6+3p^2}{16}$	$\dfrac{3-3p^2}{16}$	$\dfrac{6-3p^2}{16}$	$\dfrac{1+3p^2}{16}$	$\dfrac{2+7p^2+3p^4}{6-9p^2+3p^4}$	$\dfrac{(9P+7)-\sqrt{(3P+5)^2+192P}}{6(P-1)}$
5	9 : 7 and 9 : 7, cooperating genes for both characters	$\dfrac{18+9p^2}{16}$	$\dfrac{18-9p^2}{16}$	$\dfrac{18-9p^2}{16}$	$\dfrac{10+9p^2}{16}$	$\dfrac{20+28p^2+9p^4}{36-36p^2+9p^4}$	$\dfrac{2((9P+7)-2\sqrt{63P+1})}{9(P-1)}$
6	15 : 1 and 3 : 1, dimeric genes for one character	$\dfrac{11+p^4}{16}$	$\dfrac{4-p^2}{16}$	$\dfrac{1-p^2}{16}$	$\dfrac{p^2}{16}$	$\dfrac{11p^2+p^4}{4-5p^2+p^4}$	$\dfrac{(5P+11)-\sqrt{(3P+11)^2+60P}}{2(P-1)}$
7	15 : 1 and 15 : 1, dimeric genes for both characters	$\dfrac{56+p^2}{64}$	$\dfrac{4-p^2}{64}$	$\dfrac{4-p^2}{64}$	$\dfrac{p^2-}{64}$	$\dfrac{56p^2+p^4}{16-8p^2+p^4}$	$\dfrac{4((P+7)-\sqrt{7(P+7)+7P})}{(P-1)}$
8	15 : 1 and 9 : 7, dimeric genes for one and cooperating for the other character	$\dfrac{33+3p^2}{64}$	$\dfrac{3-3p^2}{64}$	$\dfrac{27-3p^2}{64}$	$\dfrac{1+3p^2}{64}$	$\dfrac{33+102p^2+9p^4}{81-90p^2+8p^4}$	$\dfrac{(15P+17)-4\sqrt{(3P+4)^2+15P}}{3(P-1)}$
9	63 : 1 and 3 : 1, trimeric genes for one character	$\dfrac{47p^2}{64}$	$\dfrac{16-p^2}{64}$	$\dfrac{1-p^2}{64}$	$\dfrac{p^2}{64}$	$\dfrac{47p^2+p^4}{16-17p^2+p^4}$	$\dfrac{(17P+47)-\sqrt{(17P+47)^2-64(P^2+P)}}{2(P-1)}$
10	27 : 37 and 3 : 1, three cooperating genes for one character	$\dfrac{18+9p^2}{64}$	$\dfrac{9-9p^2}{64}$	$\dfrac{30-9p^2}{64}$	$\dfrac{7+9p^2}{64}$	$\dfrac{14+25p^2+9p^4}{30-39p^2+9p^4}$	$\dfrac{(39P+25)-\sqrt{(21P+11)^2+3072P}}{18(P-1)}$

from which these cells (originating after the reduction division) can give rise to an individual without fertilization; in this case we will speak of "personified gametes". This occurs in lower plants which have a true life-cycle of generations of individuals with a double, and individuals with a reduced single chromosome number, such as algae, fungi and mosses. This also occurs occasionally in some of the *Hymenoptera* (in some insect species the males originate from unfertilized egg-cells, for instance bees, ichneumon flies, etc). and also in some groups of grasshoppers (*Tettigidae*). However, although such cases are interesting from the point of view of linkage relationships, they are of no vital significance.

Apart from this mathematical basis of certainty, the experimental reliability of the investigation in the determining of the frequency of cross-overs is very important. Different experiments concerning the same genetical factor have led many times to different results. In many cases this was the cause of some investigators becoming sceptic over the fundamental significance of the crossingover frequency. But on the other hand these different results were also the cause of a closer inquiry into the underlying process and so it became evident that the process of crossing-over depends largely on environmental influences as well as on heritable-ones. We refer for details concerning the documentation to STERN's monography (1933, p. 230–271).

In the first place it has been found from investigations that the percentages obtained for coupling and repulsion are not always equal. For instance in the maize cross carried out by DEMEREC (1926), concerning the factor pairs non-shrunken and shrunken endosperms and green and white seedlings, the crossingover-percentage in the coupling-phase was 13·40% and in the repulsion-phase 31·25%. IMAI (1926) found that the leaf-characteristics of *Pharbitis nil* in the cross dealing with the factor pairs smooth-irregular surface, and plain green-variegated gave 16·38% cross-overs by coupling and 33·90 by repulsion. ELOFF (1932) in his experiments dealing with the influence of centrifuging upon *Drosophila* obtained in the coupling-phase, a percentage of 13·86, and in the repulsion-phase one of 21·44. It is still unknown how far the coupling and the repulsion respectively are responsible for these differences, or how far environmental influences affect both forms of linkage.

ELOFF's investigation points to the latter, though the control experiments gave in cases of repulsion a higher number of cross-overs than by coupling (20·44 and 19·59 to 14·4 and 16·37%). It seems to be an important fact that repulsion always yields higher numbers of crossovers than coupling; we will return to this point later-on.

Within the environmental influences, which as we have said appear to have an important effect on the percentage of crossovers, temperature, X-rays, centrifuging and U.V-light all have a definite influence. On the other hand the crossingover-intensity is still dependent on the individual inherent factors i.e., sex, age and other specific conditions.

We are indebted to PLOUGH (1917) for the first investigation into the influence of the temperature on the crossingover-frequency. He crossed males which had three recessive characteristics, which according to MORGAN, belonged to the same linkage-group, with females of the wild type. The F_1-females originating from this cross were therefore heterozygous in these three factors (B, Pr and C). Each of these F_1-females were paired at a certain temperature with a male, who was recessive in all the three factors and the progenies were bred under the same environmental conditions. Ten different temperatures were used. At 5°C the eggs did not develop; at 35°C the females were sterile. Other temperatures used were 9°, 13°, 17·5°, 22°, 27°, 29°, 31° and 32°C. The crossing-over values for these temperatures between the two factors involved were respectively 13·6, 17·5, 8·2, 6·0, 6·2, 8·7, 18·2 and 15·4. The graphical representation on the following page will make this clear (figure 54).

From the investigation of PLOUGH it was also evident that not all groups of factors were equally sensitive to the influence of temperature, and that also, within the same linkage-group differences exist in sensitivity to such influences. On the other hand, according to STURTEVANT (1919) these temperature influences can be eliminated by the presence of a definite heritable factor CIIL. It is already an ascertained fact that the temperature, during a definite sensible period in the course of the development of the ovula, can influence the crossingover process in some linkage-groups or parts of groups.

The results of irradiation with X-rays or with radium are more irregular: the investigations of MAVOR and SVENSON (1923–1924) and of MULLER (1925–1926) have shown, that the same couplings of factors (B, Pr and C) investigated by PLOUGH, are sensitive to X-rays to almost the same extent, though the influence of the latter produced

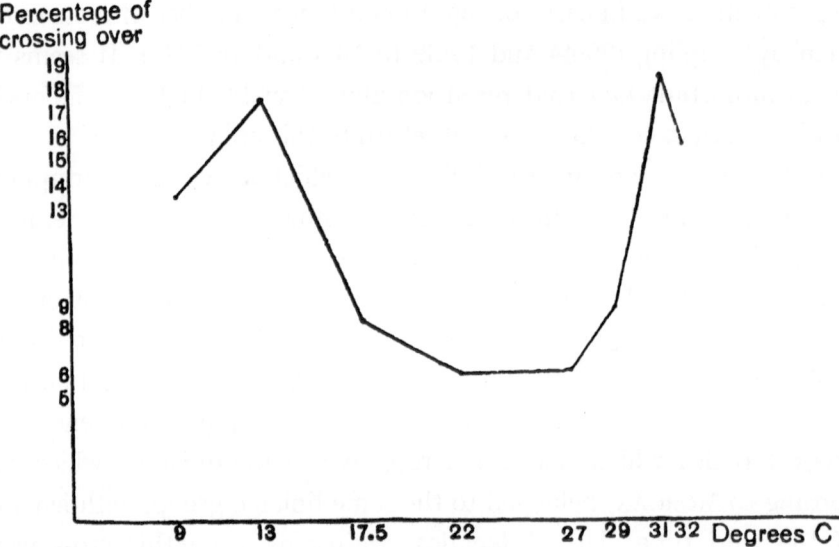

Fig. 54. Graph of the influence of temperature upon the crossover-percentage; the horizontal axis represents temperature, and the vertical the percentage (after PLOUGH, 1917).

longer after-effects than that of temperature. The contrary occurred with the same irradiation and the same intensity in another group of factors, and caused a decrease in the crossover-percentage.

As to the matter of the influence of centrifuging and of ultraviolet light we are indebted to ELOFF (1932). He found that centrifuging also increases the crossover-percentage in the coupling-phase as well as in the repulsion-phase, while ultra-violet light has an increasing effect in the case of repulsion and a decreasing effect in the case of coupling.

Summarizing, these experiments prove clearly that the crossover-percentage depends to a large extent on external environment and that the environmental influences affect different groups of factors or parts of groups to a differing degree.

These external circumstances as well as the internal conditions play a significant rôle in the crossing-over process. In the first place the influence of sex: in some cases it appears that the crossover-frequencies in the male and in the female reproductive cells from individuals with the same genotypical constitution are not always equal. A botanical example of this is *Primula sinensis* (GREGORY, DE WINTON and BATESON, 1923) in which the crossover-percentages in the pollen are nearly always higher than in the ovula, as also in *Zea Mays* (COLLINS and KEMPTON, 1926), about which the investigators could not give a final conclusion, even though important differences were obtained: in crossingover-percentages sometimes the ovula had the advantage and sometimes the pollen. In animals apparently the behaviour of the gametes appeared to be more clear; in a number of cases thoroughly investigated, it became evident that the crossover-percentage in the egg cells was rather higher than in the spermatozoa: this is valid for the grasshopper *Paratettix texanus*, for the Crustacea *Gammarus chevreuxi*, as well as for mice and rats. The best expression of this difference is found in *Drosophila*, where the male shows a very sporadic but undeniable crossing-over (see RECK, 1936). On the other hand, in birds, the frequency of crossing-over seems to be much higher in the male and in the silkworm (*Bombyx mori*) it was noticed that crossing-over occurs only in males.

Apart from the sex, the crossover-frequency depends also on the age of the individual. BRIDGES (1929) and PLOUGH (1917, 1921) pointed out very clearly in *Drosophila*, that the crossingover-frequencies show a rhythmic periodicity connected with the age of the mother animal. BRIDGES found the following numbers:

Age of the mother in days	1-3	3-5	5-7	7-9	9-11	11-13	13-15	15-27
Number of descendants	732	1692	1427	1286	1363	1305	1181	986
Average number of coupling breakage per descendant	0·717	0·655	0·594	0·504	0·463	0·459	0·483	0·509

17-19	19-21	21-23	23-25	25-27	27-29	29-31	31-33	33-35	35-37
926	683	871	859	619	566	309	263	166	204
0·515	0·514	0·464	0·443	0·401	0·486	0·524	0·502	0·500	0·575

BRIDGES calculated the total number of cases in his crosses in which two of the seven factors (ru, h, D, st, p. ss and e), for which the mother-individuals were heterozygous, occurred separately in the progeny. Naturally this can either take place in the same individual with respect to one pair of factors or with respect to more pairs so that the mean number of linkage breakages could be established per descendant. In figure 55 the distribution of these numbers is shown in which the rhythmic behaviour of the crossingover-frequencies during aging can be seen.

Further, there are examples in some animals (cocks, *Gammarus*, *Habrobracon*) of a remarkable decrease in the crossover-percentage

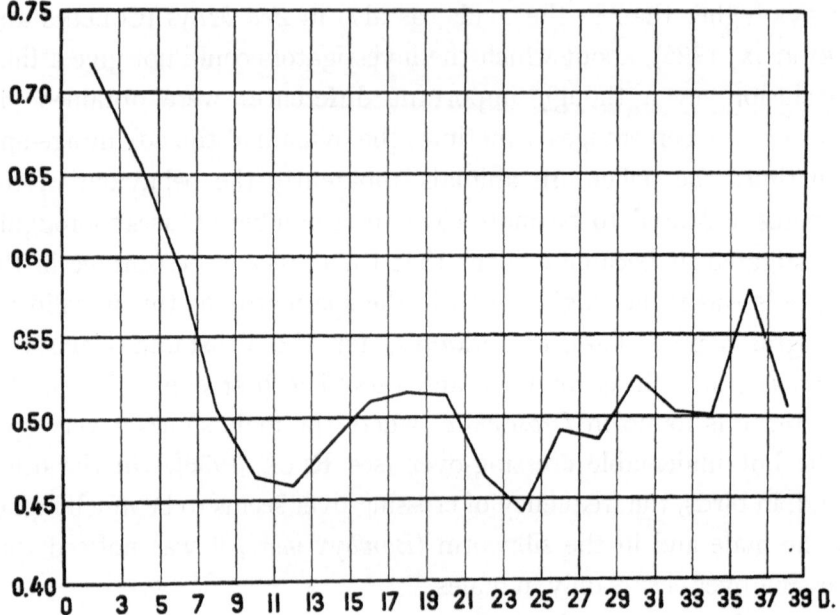

Fig. 55. Average number of coupling breakages per descendant under influence of the age of the mother (after BRIDGES, 1929).

upon increase of the age. No reliable cases of this influence of age are known in plants.

The crossover percentage also depends on definite genes. A great number of such genes are known in *Drosophila*. These genes can increase the crossover percentage or decrease it. Their action is limited to the linkage group to which they belong, or to a completely different linkage group; sometimes they are only recognizable as *"crossover-modifiers"*, or perhaps they possess another genotypical significance. They can be intensified by selection, or possibly they are indifferent to this. Their action is less strong in a homozygous state than in a heter-ozygous one. Such genes, which influence the crossover percentage are found in *Gammarus*, the silk-worm (*Bombyx*), in maize, in *Pisum*, and in *Phaseolus*.

The fact that crosses between different types involving the same factors can yield different exchange percentages (KOLLER, 1931), in some cases can be attributed to some definite factor. There are also other possibilities (such as ring-formation between chromosomes in peas) which will be discussed further in Chapter XII. As an extreme case of such a difference between hybrids we recall the phenomenon in peas, where two factor pairs A-a for coloured and white flower, and Gp-gp for green or yellow pods in some hybrids segregate absolutely independently whilst in other hybrids, a very strong coupling with one or two percent crossing-over occurs (HAMMARLUND, 1923, 1928). In hybrids between *Zea Mays* and *Euchlaena mexicana* there is an absolute   absence of any exchange of four factors although these factors show a normal exchange in the hybrids between maize-varieties (BEADLE, 1932).

Summarizing the above it becomes clear that in the establishment (standardization) of a reliable percentage of exchanges, equally high demands must also be made with regard to experimental reliability, as well as to the mathematical calculations; and that the following must be taken into account:

1) the phenomenon of interference (prevention of exchange between definite factors by means of an exchange of other factors linked with them);

2) the fact that in repulsion, another frequency of exchange is often obtained than that which would be found in the case of coupling;

3) the role played by external environment (temperature, irradiation, centrifuging, ultraviolet light) during the exchange process;

4) the possibility that the percentage of exchanges can vary between individuals of different sex;

5) a contingent influence wielded by the age of the individual;

6) the existence of definite factors, which can intensify or hinder the actual exchange-process;

7) possible differences between ordinary hybrids of varieties and hybrids between biological races, sub-species and species.

It appears that definite groups of genotypical factors exist in every organism which are more or less linked with the group, but which, as groups, form almost independent wholes. These wholes mutually segregate independently. This phenomenon is called linkage. For reliable summaries of the literature concerning this process of coupling we are indebted to STERN (1933) and LUDWIG (1938). JUST (1934) gives a summary of the significance of this process in man. A mathematical contemplation of the theory is given by OWEN (1950).

We will refer later on to the theoretical consequences (Chapter XI). First we will discuss a series of experiments, which originally had a completely different approach but which came to the same result, that is to say, the bringing to light of the existence of important groups of linked factors.

The genius and intuition of HUGO DE VRIES led him before 1900 to the conception that the question of origin of species in nature could only be solved by experimental research in this field. His *Oenothera* studies now form the foundation of a monumental whole, to which so many younger botanists, who have followed him, have lent their strength. After many years of hard struggles over all sorts of points, which from the theoretical aspect were of great significance, a great number of problems have been solved. The *Oenothera* experiments deal with a great number of different subjects; one of them deals with the multiformity of the F_1 generation which can be established in many hybrids.

Hybrids between different species of *Oenothera*, and especially those between *O. Lamarckiana* on the one hand and *O. biennis*, *O. muricata* and *O. suaveolens* on the other, showed in every experiment that by self-pollination the species gives offspring which are almost uniform, and that the F_1-generation were still composed of mutually different types.

The first follower of DE VRIES, MACDOUGAL (1905, p. 17) records that he could distinguish four different types among the F_1-individuals in the crosses *O. Lamarckiana* ♀ × *O. biennis* ♂ and *O. Lamarckiana* ♀ × *O. muricata* ♂ which already showed differences at a very early stage of their development; all individuals could easily be classified into four types (i.e. quadruplets) and intermediate forms were completely lacking.

Following this discovery came a publication of DE VRIES (1907), who discussed the reciprocal cross *O. biennis* ♀ × *O. Lamarckiana* ♂ and he stated that this cross did not yield a uniform F_1-generation in the sense that 50% of the hybrids showed a mutual resemblance and the other 50% showed another mutual resemblance.

The first type was called *laeta* by DE VRIES, the other *velutina*. The difference between the types is elucidated by the figures 56 and 57, in which rosettes and fully grown plants are depicted (figure 56 on the left, *laeta*, on the right *velutina*; Figure 57, on the right *laeta*, on the left, *velutina*). The difference between the types is based chiefly on the properties of leaf-shape and the hairing. In 1908, DE VRIES obtained such "twin" *laeta* and *velutina* next to one another from each of the

50%'s of the F_1-individuals, not only when he made hybrids with *O. biennis* as mother, but also when he used *O. muricata*.

This discovery was of extreme importance from different points of view; not only because through this the "Doppelnatur" of *Oenothera Lamarckiana* (HONING, 1909, 1911) was recognized, but especially because it provided the impulse to carry out a very detailed investigation with regard to the occurrence of such twins in *Oenothera* crosses. This investigation was undertaken

Fig. 56. Rosettes of the twin plants *laeta* and *velutina* (after GATES, 1915b).

by DE VRIES and a number of others (see summaries by LEHMANN, 1922 and RENNER, 1929, p. 82–90). It was RENNER's work which yielded the best results. In the main the results were as follows: if *Oenothera Lamarckiana* is used as a father plant in hybridization with

Fig. 57. Fullgrown twin plants of *laeta* and *velutina* (after LOTSY, 1917).

12

O. biennis or with *O. muricata*, then there originate in the first and second case two groups, *biennilaeta* and *biennivelutina*. In the second case there originate two other groups, *murilaeta* and *murivelutina*.

The individuals of the first cross are in accordance with a number of characters which originate from *biennis*, and those of the second cross also exhibit a number of characters, contributed by *muricata*. However the *biennilaeta* and the *murilaeta* plants resemble one another in very striking characters. On the other hand a resemblance exists between *biennivelutina* and *murivelutina* individuals. On the grounds of this, RENNER now believed that *O. Lamarckiana* transferred a number of characters in two great complexes to its offspring. He calls these complexes "*gaudens*" and "*velans*". If 50% of the pollen formed by *O. Lamarckiana* contains the *gaudens*-complex and the other 50% contains the *velans*-complex, then the result of the hybridization with *biennis* (in which the ovula are similar, and which transfer the character specific for this species, called "*albicans*") is that the F_1 generation is composed of 50% plants with *albicans* and *gaudens* complexes, (the *biennilaeta*-ones) and 50% plants with the *albicans*-and *velans*-complexes (the *biennivelutina* ones). Likewise, the hybridization of *muricata* which transfer in their ovula a character-complex called "*rigens*", gives rise to *rigens-gaudens* plants (*murilaeta*) and to *rigens-velans*-plants (*murivelutina*) (Figure 58 gives a schematical view of this).

The cross of *Oenothera suaveolens* as mother with *O. Lamarckiana*

Fig. 58. Left: Origin of the twin plants in the cross *O. biennis* ♀ × *O. Lamarckiana* ♂.   Right: Origin of the twin plants in the cross *O. muricata* ♀ × *O. Lamarckiana* ♂.

as father is more complex. *O. suaveolens* gives rise to two types of ovula (the "*albicans*"-type, which also originates from *biennis* and the "*flavens*"-type). Because of this, four groups must originate in the F_1-generation in the following combinations: *albicans-gaudens* (*biennilaeta*), *albicans-velans* (*biennivelutina*), *flavens-gaudens* (*suavilaeta*) and *flavens-velans* (*suavi-velutina*). Figure 59 represents this process

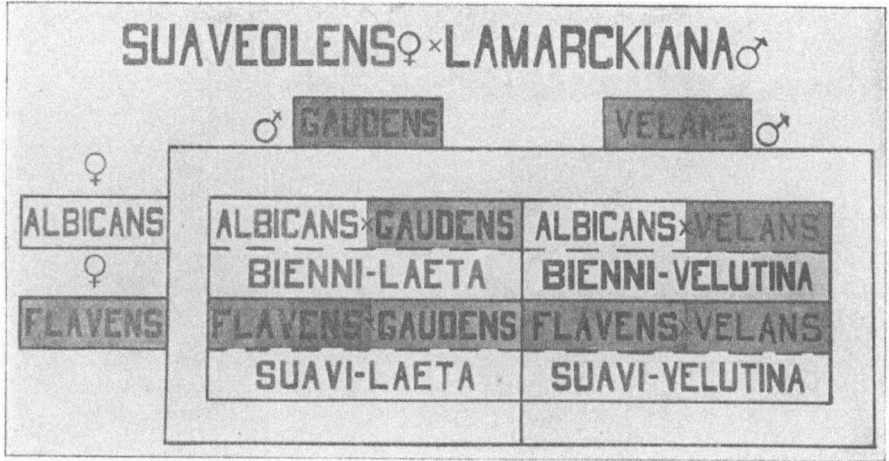

Fig. 59. Origin of the quadruplets in the cross *O. suaveolens* ♀ × *O. Lamarckiana* ♂.

schematically. The three groups, *biennilaeta*, *murilaeta* and *suavilaeta* are thus all characterized by a complex of characters as a consequence of the presence of the *gaudens*-complex of factors.

According to RENNER, this *gaudens*-complex in the *laeta*-types is the cause of the following characters: high growth, angular tough stems, broad light green leaves, erect flat bracts, a little rough hairing, thin fruits, anthers poor in pollen-grains, pale yellow corollas, absence of red colour. On the other hand, the *velans*-complex transfers to all *velutina*-types: low growth, round brittle stems, narrow dark green leaves, bent gutter-shaped bracts, dense soft hairing, thick flower buds, thick fruits, anthers rich in pollen-grains, dark yellow corolla, red colour of the calyx and fruits, and red colour at the foot of the stronger hairs on the stem. This is the reason that RENNER (1917) proposed the term "*complex heterozygote*" for *Oenothera Lamarckiana* to cover its behaviour, viz., the keeping together into two great complexes of a number of apparently incoherent genes. This term seems to be very well chosen. In other connections we will have the opportunity to return many times to the complex heterozygosity of some of the *Oeno-thera* species. We will only mention in passing, that pollen-grains with different complexes can often be distinguished microscopically (so called *analysis of tetrads*, RENNER, 1919a and b). Such pollen-grains are drawn in figure 60. On the left is the *velans*-type (big with thin walls and spool shaped starch) and on the right, the *curvans*-type (smaller, thick walls, spherical starch).

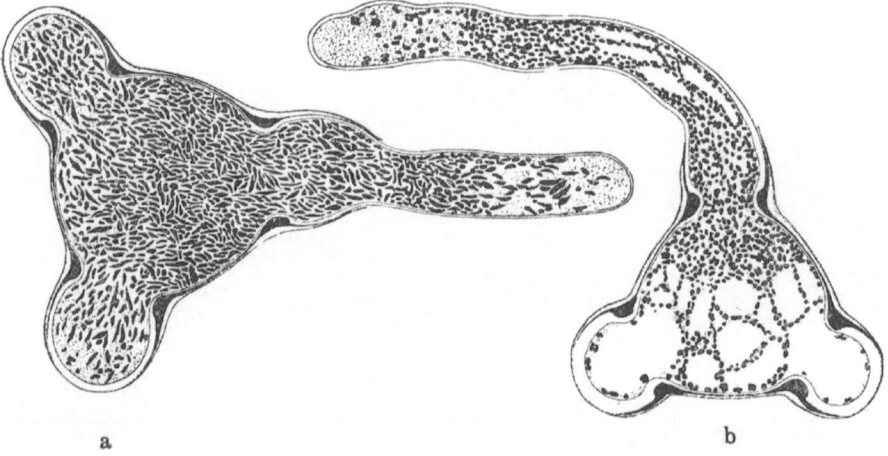

a   b

Fig. 60. Pollen-grains of different *Oenothera*-complexes (after RENNER, 1919a).

The process of linkage is closely connected with two others which are often very hard to distinguish from this: viz. *pleiotropy* and *multiple allelomorphism*.

The conception *pleiotropy* was introduced by PLATE (1910, p. 597) for the phenomena, of a single factor expressing simultaneously different characteristics in different organs. A great number of examples of this could be referred to here, e.g., blue-flowering individuals of *Lupinus angustifolius* always have a darker seed-coat whilst the white-flowered ones have a lighter coloured one. Broad beans (*Vicia Faba major*) with a black spot in the keels of the flowers always have brown colouring seeds; the seeds of plants with absolutely white flowers remain white; broad beans with long stems also have large leaves and big fruits.

In such cases it is plausible to suppose that one or more pleiotropic factors exists and a detailed genetical analysis has proved this many times. But besides this there are examples in which the connection between both of the characters, which could be attributed to the same pleiotropic gene, is very indistinct. NILSSON-EHLE (1911 c.p. 15) believed that the formation of the ligula on the leaf-foot and of the racemose inflorescence in oats could be considered as an expression of a single pleiotropic factor (L_1), but the question still remains as to whether pleiotropy really does exist here. If two separate factors are present; one for the formation of the ligula and one for the shape of the inflorescence, and if these factors are strongly linked, then it will be extremely difficult (and many times impossible) to decide if the cooperation of both of the characters is based upon one pleiotropic factor or if intensive linkage of both of the factors is the cause of frequent occurrence of both of the characters appearing together. NILSSON-EHLE originally considered awns, hairs and the occurrence of a ring on the spikelet of some oat types to be the result of a single pleiotropic factor, but later he considered them to be the consequences of a complex of factors. If definite hybridizations could lead to the conclusion that different characters could be expressed by only a single factor, whereas other hybridizations show that separable factors were responsible for this, it will have been proved that we have ruled out pleiotropy. Therefore the view of DOBZHANSKY (1928, p. 387) "that every gene acts on all parts of the body and that each part develops under the action of the totality of the genes" is an exaggeration of the conception of "pleiotropy". Difficulties arise in the same way because of the possibility of mistaking a very strong linkage for *multiple allelomorphism* (see STERN, 1930b). If two or more factors show a very strong degree of linkage, then these factors will normally remain together in a dominant or a recessive condition in the same gamete. Such dubious cases occur in *Drosophila*, *Lebistes* and *Paratettix*. The smallest of the four linkage groups in *Drosophila* is composed (as we have already seen) of only a few factors, and some are very strongly linked, so that for a long time they were considered to be multiple alleles. However it was pointed out by MORGAN and his co-workers and repeatedly established by MULLER that a slight breakage in the linkage is possible, so that it is preferable to speak of a very strong linkage rather than of a multiple allelomorphism. The same question occurs in a group of 9 genes in *Lebistes* (WINGE, 1927b), which are sometimes interpreted to be multiple alleles, and sometimes as being very intensively linked factors. Finally, in a group of 23 factors in *Paratettix texanus* (NABOURS, 1929), which were originally recorded as being one series of multiple alleles, it was later found (NABOURS, 1930) that a very slender breakage of the linkage occurred.

Pleiotropy and multiple allelomorphism thus undoubtedly occurred, but one must always bear in mind the fact that an absolute or very intensive linkage can present the same appearence as if one single factor had a pleiotropic behaviour: or that the factors which are in reality coupled, seem to form a series of multiple alleles.

These possibilities must be considered with particular care, when the phenotypical characters expressed by the factors involved are of a very divergent nature. Contrary to the more or less naive way, which was to the forefront in 1900 just after the rise of genetics, viz., that all heritable characters should be independent of each other, and that an organism was nearly comparable to a box of bricks; it was put forward that an organism keeps its heritable characters together in a very well defined structure, and that a tie exists between the genes which strongly influences their behaviour during the segregation processes. The recognition of this was only made possible by the work of a great many investigators.

CHAPTER IX

THE MUTUAL SIMILARITY OF THE F₁-INDIVIDUALS

With regard to the exactitude of the rule concerning the uniformity of the F_1-individuals, it seems that there existed only a slight difference in view-point between investigators prior to 1900. For example, GAERTNER (1849, p. 237) believed that "there originated from seeds produced by a cross between two species, which breed true to type, plants of one definite type, which resemble one another completely. There sometimes also occurs (however seldom) among the seedlings, from one and the same hybridization, (that is to say, from seeds from one and the same fruit) alongside a great number of hybrid plants of wholly the same normal hybrid type, some which show very aberrant forms and shapes. We will call these aberrations "exceptional types", since they occur only in very small numbers, indeed often only as one specimen". NAUDIN (1865, p. 146–147) gave these statements: "I have always found that the hybrids which I have obtained myself and whose origin was well known to me, had a great mutual uniformity with other individuals of the first generation which also originate from the same cross, regardless of the number of individuals" and "To put it into a few words, the hybrids of one and the same cross mutually resemble one another in the first generation, precisely or nearly so, as do individuals originating from one and the same pure species". MENDEL found in his hybridization of peas a very clear mutual resemblance in the F_1-individuals: "In *Pisum*, the hybrids obtained directly from hybridization of two types, in all cases have the same external appearance, but their offspring are variable and indeed, change according to definite rules" (1869, p. 31).

FOCKE (1881, p. 469) rightly put the state of affairs as follows in very carefully chosen words: "All individuals originating from the hybridization of two pure species or races are, if they are bred or cultivated under similar circumstances, mostly mutually similar, or at least do not differ more than specimens can do from one and the same species".

In spite of these views of GAERTNER, NAUDIN, FOCKE, and MENDEL, (which are of nearly the same tenor), and their considerable experience in this respect, the validity of their views is questioned today in genetical literature. The few exceptions which the investigator noticed are therefore placed in the forefront, and on the basis of this the validity of the laws of "uniformity of the F_1-generation" is disputed.

As an argument, those investigations are used which cannot be explained immediately by the simple mendelian theories, such as crosses between plant species (bulbs, shrubs, fruit-trees and other woody plants) which are vegetatively multiplied and which then as clones seem to show an apparent constancy in their characters notwithstanding their heterozygosity. But, apart from this, heterogeneous F_1-generations are also found after hybridization between those organisms which can only be multiplied sexually. For instance, the investigations of COLLINS and KEMPTON (1911) led to very striking results in their hybridization of maize types between starchy and waxy endosperm. The F_1 seeds contained starch so that the factor for this was dominant; out of a total of 22339 F_2 seeds, 5179 were waxy and 17160 were starchy. This points to a normal monohybrid segregation, but there were too few waxy seeds. An investigation into 45 F_2 families (each ear representing one family) revealed a great difference in segregation numbers in the F_2 of these families. A large enough number of these families showed a deviation from the numbers expected that one could no longer simply accept an explanation of a monohybrid segregation.

If we take as our limit for the probability, a deviation of 3 times the standard error, then there were 16 out of 45 families which were not in accordance with the segregation, which was considered to be a monohybrid one: 12 of these gave a deficit of waxy endosperms, 4 a deficit of starchy endosperm seeds, as can be seen from the following figures borrowed from COLLINS and KEMPTON: (p. 185).

These apparently inexplicable deviations from the segregation ratio expected were considered by COLLINS and KEMPTON to be a result of individual differences, which should exist between the F_1-individuals concerned, which are the parents of the F_2 seeds. From

Deficit of waxy seeds					Deficit of starchy seeds				
Family	Starchy	Waxy	Expected	P.E. abs.	Family	Starchy	Waxy	Expected	P.E. abs.
115	573	104	508 : 169	5·8	116	249	125	279 : 93	3·7
117	634	174	606 : 202	2·3	153	406	175	436 : 145	2·9
120	384	75	344 : 115	4·3	199	317	132	336 : 113	2·1
201	439	106	409 : 136	3·1	200	133	65	148 : 50	2·5
202	443	112	416 : 139	2·6					
204	441	112	415 : 138	2·5					
213	441	103	408 : 136	3·3					
222	531	104	477 : 158	4·9					
229	366	92	344 : 114	2·4					
233	554	119	505 : 168	4·3					
234	514	134	486 : 162	2·5					
237	297	75	279 : 93	2·1					

this it should also follow that the F_1-generations should not be composed of homogeneous types, but of mutually different types.

In hybridization experiments with animals, numerous non-uniform F_1-generations were also obtained, especially in fowl. In 1910 (a and b), DAVENPORT compiled a great deal of data about this. We will borrow the following examples from his material: Animals with a single comb were paired with polish fowls which had a double comb. The F_1-generation gave combs which had a Y-shape but which were very variable with regard to the single base and the double part. There were cases in which the middle split was complete, in which case the comb appeared double, and there were also animals where up to $^9/_{10}$ths of the whole comb-length was single. In breeding the F_2-generation it was evident that these F_1-animals differ mutually not only phenotypically but also genotypically. If all F_1-individuals are now classified into 10 classes where 0 indicates a pure double (two row) comb, and 10 indicates the animals in which the comb is single over from 1–10% of its entire length, 20 indicates those animals in which the comb is single over 11–20% of the entire length, and so on up to 90 in which 81–90% of the comb-lengths are joined. Hence the F_2-animals were spread over the following three simple groups: single comb, Y-comb and double comb:

middle part of the comb as a % age of total comb-length in F_1	0	10	20	30	40	50	60	70	80	90	less than 50%	more than 50%
(single comb . . .	29	25	30	25	24	29	33	35	40	51	25·5	32·6
F_2 (Y-comb	46	50	43	51	53	48	46	47	38	34	49·8	45·2
(double comb . . .	25	25	27	24	23	23	21	18	22	15	24·7	21·2
number of animals											900	1108

The F_1-animals which had complete double comb (group 0) gave an F_2 which segregated into 29% simple, 46% Y-combs and 25% double combs; those F_1-animals in which the simple part amounted to 90% of the total comb-length, showed a much higher percentage (51%) of single combs in the F_2 and accordingly, a lower number of Y-combs (34%) and double combs (15%). Therefore it seems that a difference in genotype exists among those F_1-animals.

Other races of fowl show a similar behaviour with respect to other characters. For instance, Houdans and Dorkings have 5 toes each instead of the normal number of 4 on each foot. This phenomenon is known as *hyperdactylism* (literally: over-fingered). However characteristic the property may be in these two races, it is indeed peculiar that sometimes pure-breeding strains of these races yield 4 or 5% four toed animals (both feet) and sometimes 4 on one foot and 5 on the other. The possession of 4 toes on each foot is indicated by 4–4; 4 on the one and 5 on the other by 4–5. If now we pair such a hyperdactylic animal with 5 toes with normal races, then the F_1-generation is not at all homogeneous. DAVENPORT (1910b) found for instance, that in such F_1-generations 27·3% of the animals were 4–4; 19% were 4–5 and 53·6% were 5–5. It thus seems that the factor for hyperdactylism is dominant, but that, according to DAVENPORT this *dominance is incomplete*. If an absolute dominance existed, then the number of hyperdactylic animals in the F_1-generation must be 100%. The F_2-generation bred from this was composed of 47·4% normal and 52·6% hyperdactylic animals (instead of a ratio of 25 : 75).

Another race of hyperdactylic fowls, — silk fowls —, has this character to such a high degree that every foot has 6 toes (i.e. 6–6). If now such fowls are mated with normal 4–4 animals then the F_1 is composed of 96% hyperdactylic and only 4% normal animals. Here it appears that the dominance of the factor for hyperdactylism should be stronger than in Houdans and Dorkings which have 5 toes, this dominance is still incomplete.

This phenomenon was also observed in mice and could also be ascribed to an *"incomplete dominance"*. RABAUD (1919, p. 76) gave the striking result that hybridization of wild-coloured mice from a strain of known origin with albinos takes a different course dependent on the ancestry of the albinos. If the albinos had no single grey-yellow mouse among their ancestors the F_1 was exclusively wild-coloured. If

however albinos originate from grey-yellow mice, then the F_1-gener-
ation was not homogeneous any more than was hybridization of wild-
coloured mice with grey-yellow mice themselves. In both of the latter
cases, the F_1-generation was composed of wild-coloured and yellow
animals, almost in the ratio of 1 : 1. These pairings always gave the
same result, and for this reason, RABAUD accepted that the wild-
colour factor from the known strain was incompletely dominant.
Great numbers of similar cases were obtained in other mice crosses,
especially those in which albinos were involved.

The question now arises as to whether it is possible to find an acceptable explanation for the various investigations discussed here
concerning the apparent exceptions, which would show the incorrectness of the uniformity rule. Indeed the hybridization results have
led us to consider these exceptions very sceptically and not to proceed
at once to an acceptance of "incomplete" dominance, which expresses
itself in the one individual and not in the other, or to any other
emergency solution.

It is obvious in the case of heterogeneity of the F_1, after hybridization
of the two clones of vegetatively multiplied plants, to make the
supposition that these types of bulbs, perennial plants or woody plants
are already heterozygous to a large degree in themselves, so that the
"latent" recessive factors can express themselves: Self-pollination of
such plants contributes to the conclusive argument for this assumption
because in this case the offspring is not at all uniform. Moreover there
still remain other forms of "latent" characteristics,

a) *cryptomeric* factors which can be present in plants and still not
come to expression because they are not able by themselves to wield
any phenotypical effect, since the required partner for this is missing;

b) *hypostatic factors* which can be covered up by other non-allelo-
morphic epistatic genes and

c) *pure latent factors*, those which do not express themselves either
in the growing or in the fully grown individual, but which intervene
temporarily during the life-cycle before the possibility for observation
takes place.

Cryptomeric factors can exist homozygously or heterozygously,
but nothing can be ascertained about them from self-pollination or
from pairing with identical genotypical individuals. Sometimes
cryptomeric factors cannot even be traced by crossing genotypically

different individuals. If we once again take the example of the *Linaria* investigation of CORRENS (see p. 143), then we will find that the hybridization of a pink-flowering *Linaria maroccana* was not only interesting because a purple flowering F_1-generation originated from this cross, but also, and especially, because there occurred next to one another in the F_2-generation, white flowering plants which were either homozygous with respect to the factor for alkaline cell sap or with respect to acid cell sap, or heterozygous for alkaline-acid cell sap. These plants of the third group (heterozygous alkaline-acid) are especially important because they give rise (after hybridization with pink flowers) to an F_1-generation which is not uniform, but which is composed of 50% purple and 50% pink flowering individuals. This is due to the fact that 50% of their gametes had the factor for alkaline cell sap and the other 50% for acid cell sap, whereas one parent (the pink one) was a real true-bred and the other (the white), seemed to be a true-bred, but was not. Consequently, the F_1-generation was not uniform, but divided into two types.

Among albino mice there also occur those which are quite different from a genotypical view-point. This is due to the absence of the basic factor for colour which leads to animals being white even though a great number of other factors for colour can be present in homozygous or in heterozygous states. If these cryptomeric factores are heterozygous, the F_1 of a cross with coloured animals will not be uniform.

If a factor is *epistatic* over another *hypostatic* factor (as the factor for black colour of oat glumes is epistatic over the yellow factor) then we can never ascertain whether this factor for yellow colour is homozygously or heterozygously present unless specific crosses led to the removal of the black factor and enabled the factor yellow to express itself. Plants with the formula ZZGg (see p. 150) always yield black plants when self-pollinated or hybridized with black ones. Nevertheless they are heterozygous with respect to the factor for yellow. If such ZZGg is crossed with zzgg-individuals then an F_1-generation originates which is phenotypically uniform (all plants having black glumes), but which are still genotypically heterogenous. The F_1 contains ZzGg-types as well as Zzgg-plants; the first segregate in the F_2 into the ratio of 12 black: 3 yellow: 1 white, the latter into 3 black: 1 white.

Perhaps the irregular results of COLLINS and KEMPTON with regard to their maize hybridization (p. 185) can be considered to be the result of the presence of latent *"certation"* factors (see p. 115 and Chapter XX) which influence the rate of growth of pollen tubes possessing varying heritable tendencies. Whatever the case may be BRINK and MACGILLIVRAY (1924, 1926) found important data in this direction. As these factors can be present to different degrees in different individuals, it is possible to explain the differences between the F_1-individuals by this latent *"certation"* factor.

A third cause of dissimilarity in the F_1-generation, after hybridization between so-called "pure" breeding types may lie in the presence of *polymeric genes*. In the discussion concerning this we have seen (p. 129) that hybridization of an individual homozygous with respect to two polymeric factors ($R_1R_1R_2R_2$ for instance) with an individual recessive with respect to both factors ($r_1r_1r_2r_2$) gave an F_1-generation with the formula $R_1r_1R_2r_2$. From this an F_2 was produced which gave only 1 recessive individual next to 15 individuals, which possessed one or more of these polymeric factors. Backcrossing of the F_1, viz., $R_1r_1R_2r_2$ with the recessive $r_1r_1r_2r_2$ gave a higher number of recessives among the F_2-individuals viz, 1 to 3 individuals which appeared to possess a character that was more or less dominant. Even more striking than this is the case when 3 polymeric factors are present, then there originate by self-pollination in the F_2-generation 1 recessive next to 63 which have one or more of the polymeric factors. In the case of the backcrossing with an individual recessive in all factors, the F_2 is composed of 1 recessive to 7 more or less dominant ones.

This may have the consequence that the heterozygous nature of the polymeric factors does not easily express itself either by the pairing of similar individuals or by self-fertilization. But the nature of these factors appears in crosses with individuals which are recessive in all characteristics. In relation to this we will take DAVENPORTS mating of Houdans (which always possess *hyperdactylism*) and normal fowl with 4 toes as an example. According to DAVENPORT's statement there often occur among "pure"-breeding Houdans, 4–5% individuals with 4 toes; when mated with normal fowl, DAVENPORT found 27·3%. This can be explained completely supposing that hyperdactylism is caused in these races by two polymeric factors, which can be called H_1 and H_2. Animals with the genotypical formula $H_1h_1H_2h_2$ give, upon mutual mating of each of the 16 descendants only one with $h_1h_1h_2h_2$ which has a normal foot, that is nearly 6%; hybridization of an $H_1h_1H_2h_2$-animal with a normal $h_1h_1h_2h_2$ gives 1 $h_1h_1h_2h_2$ once in every 4 individuals (that is to say 25%). The explanation is in a good accordance with the data given by DAVENPORT, so we may be able to rule out *"incomplete dominance"* of the factor for hyperdactylism. But still it is not quite certain what the explanation for this irregularity (called hyperdactylism) is. In all experiments with rather small numbers of offspring, pure breeding of the race can apparently be simulated through polymeric factors, if the factors are in a heterozygous state.

A fourth cause of differences between the individuals of an F_1-generation was found before (p. 176) in the complex heterozygous nature of *Oenothera Lamarckiana*. This plant species gives, by self-pollination, if we disregard some aberrant forms which occur, nothing other than individuals of the type *Oenothera Lamarckiana*, but still there originate after hybridization of these plants with other *Oenothera*-species, an F_1-generation which is not homogeneous, but which segregates into two or more very easily distinguishable groups. This is due to the fact that *O. Lamarckiana* always forms two types of gametes called gaudens and velans and that these types each contain a number of factors which are mutually linked. An F_1-generation of which *O. Lamarckiana* is one of the parents, will always be dimorphic. Just why this complex-heterozygosity of the plants does not reveal itself on self-pollination, and just why the presumed homozygotes: gaudens-gaudens and velans-velans do not originate, is in itself a problem, but this does not alter the fact that *Oenothera Lamarckiana* always gives rise to two types of gametes. This problem will be discussed more fully later on. The causes of dissimilarity in the F_2-generation given under 2, 3, and 4 thus led to the theorem that an organism can be apparently pure-bred and still form gametes which differ in mutual heritable tendencies. We will define the term *"breeding true to type"* as being *the homozygous nature in relation to all heritable characters present in that organism*. The great difficulty in demonstrating this purity of breed is often the cause of confused results. Plants and animals can absolutely resemble one another externally and, after self-fertilization or planned pairing they can appear to be thoroughbred but in reality this may not be the case. Every investigator who wants to prove the dissimilarity of the F_1-generation must ask himself seriously the question: was the material with which I started the investigations completely homozygous in all its factors? In a large number of publications not enough attention is paid to this, many research workers speak about "pure" and yet take no trouble to prove that the material is of genetical purity. As long as these proofs are lacking geneticists who investigate exact science are hardly in a position to doubt the uniformity rule which was defined by LANG (1905) for the first time, and which still holds good.

Besides these cases of apparent true to type breeding, which can be the cause of mutual dissimilarity among the F_1-individuals, there is still another source of baseless doubt about the exactitude of the uniformity rule. Every genetical character, in its phenotypical expression, depends on the external environment, and this simple fact is often forgotten. The environment for different members of an F_1-generation can sometimes differ.

The investigations of HOGE have been particularly instructive. HOGE (1915) worked with MORGAN on selection investigations in *Drosophila melanogaster*, the banana fly mentioned earlier, and she discovered an aberrant form in her material, which attracted attention through its duplication of feet or through duplication of some of the segments of the feet. Hybridization of such an individual with a normal fly yielded different results; sometimes the F_1 was entirely normal and the F_2 was composed of a great number of normal and a small number of abnormal individuals; another time normal and abnormal animals occurred next to one another in the F_1 as well as in the F_2. The cause of this peculiar behaviour was found to lie in the environment. At high temperatures heterozygotes had a normal phenotype, at low temperatures, an abnormal phenotype. Even homozygously abnormal animals could have the common, normal external appearance when bred at high temperatures. If the experiments were undertaken at high temperatures then it was possible to consider the normal animals as the dominant-ones (factor for normal phenotype N, for abnormal phenotype n). Nn will be normal at high temperatures, though in the cold the F_1 will be abnormal and hence the factor for this (n) could be considered as dominant.

Important investigations into this were carried out by TIMOFÉEFF-RESSOVSKY (1927, 1931, 1934c). In collaboration with VOGT (1926) and REINIG (1928) he investigated the dependence of various genes on the temperature during the development of *Drosophila funebris*; especially the incompleteness of the vein-system of the wings. A gene v_{ti} is, in its extreme effect, the cause of the disappearance of the transverse vein between the second and the third long-veins. Apart from this extreme type, a number of normal individuals and intermediate forms always occurred. Hence, these investigators distinguished three concepts as follows: 1) *penetration* of the factor which indicates the percentage individuals which show the abnormal characteristic, 2) *expressivity* of the factor, by which is understood the average degree of expression of the factor in the progeny concerned and 3) *specificity* of the factor, which indicates the manner in which the expression of the factor is taking place. It is of great significance that by means of selection, strains of *Drosophila funebris* could be obtained which are absolutely similar for the extreme forms (normal type, and those absolutely lacking in the transverse-vein) but from which the inter-mediate forms have a different specificity for the factor. Due to this a different variation series occurred, for which VOGT (1926) suggested the term *"eunomy"*. In figure 61 the expressivity of three such eunomies is drawn beside a normal wing; the start and finish are thus mutually completely similar but this final result is obtained in three different ways.

It was found that several strains also differed in their sensitivity to

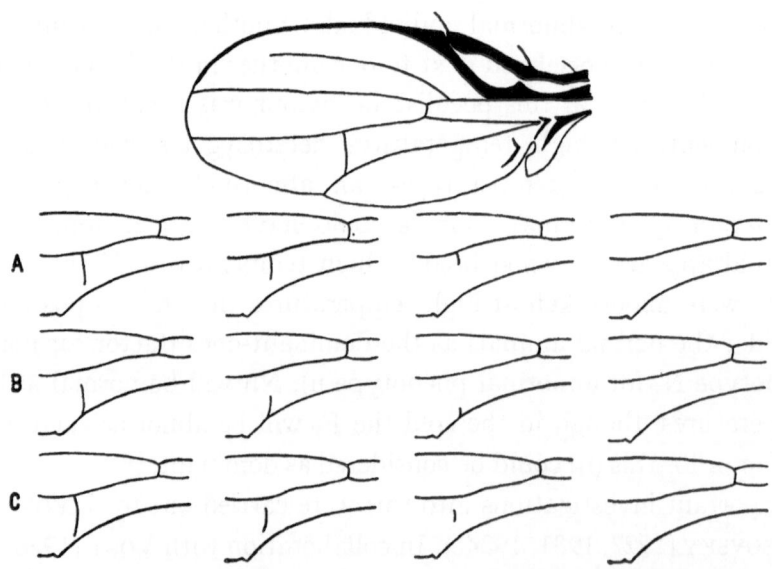

Fig. 61. Expressivity of three eunomies (A, B and C) in *Drosophila funebris* (vena transversa interrupta) (after TIMOFÉEFF-RESSOVSKY, 1931).

temperature. Investigating six lines of the eunomy (a) for this temperature sensitivity, the following were found as penetration-percentages:
57, 66, 71, 79, 99 and 100; for five lines of eunomy (b): 56, 62, 74, 97
and 100, and for three lines of eunomy c: 57, 91 and 100.

It becomes clear that the development of a definite character can be
influenced to a high degree by environment. On the other hand, the
processes of development can be dependent on differences in heritable
characters even in cases in which each of these processes of development leads to a uniform final result. If one cannot control the environment of the F_1-individuals absolutely then as a consequence of
this, one can see that not all the F_1-individuals are phenotypically
similar, although their genotype can still be identical.

The environment appears to exert an influence on the phenotype
not only after fertilization, but from data available it also appears that
the *environment of the gametes* (from which the F_1-generation will
originate later on) is an influence on the dominant and recessive genes,
which will express themselves in this F_1-generation.

Genetical literature is very rich in studies concerning hybrids
between different species of sea-urchins; the species *Sphaerechinus
granularis* and *Echinus microtuberculatus* are used many times for
this purpose because of their very typical shape in the larval stage
(fig. 62). A larva cultivated from this cross was in most cases almost

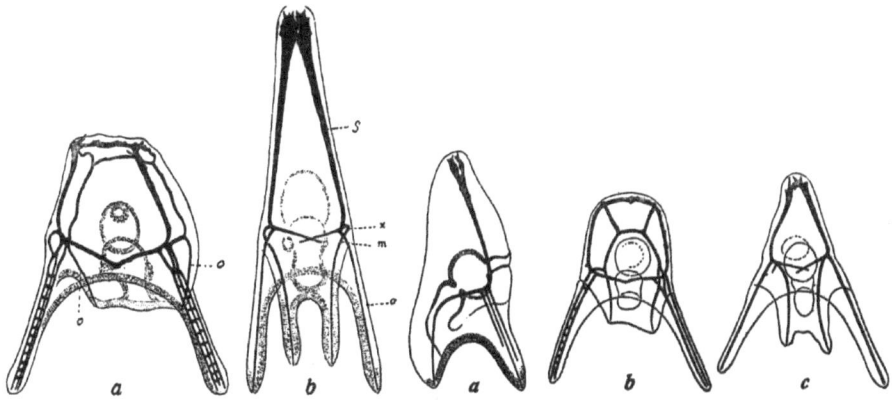

Fig. 62. Larvae of the sea-urchin
species: *Sphaerechinus granularis* (a)
and *Echinus microtuberculatus* (b).
(after Boveri, 1895).

Fig. 63. Larvae of the hybrids between
the species represented in fig. 62. a)
intermediate type, b) metroclinous type,
c) patroclinous type (after Boveri, 1895
and Herbst, 1907).

intermediate in shape between both its parents (figure 63a) but in some experiments, the hybrid bore more resemblance to the mother (*Sphaerechinus*) and because of this the hybrid was called *metroclinous* (figure 63b), whereas in other experiments the hybrid resembled more the father and hence was called *patroclinous* (figure 63c). This difference in external appearance between the larvae also occurred if they were descended from the same pair of parents in the reciprocal cross. Lack of evidence that the parents are true breds can be held responsible for the differences among the F_1-individuals arising from the same pair of parents, but thanks to the work of a great number of investigators (see HERBST, 1926) another cause for these differences has been brought to light. This was formulated very clearly by O. KOEHLER (1916, p. 282–283) in two of his theorems: viz., "Every gamete in the reproductive organs of the parental animal passes through a periodic fluctuation in its ability to transfer heritable characters. This ability increases from a low startingpoint to a maximum value and then falls again slowly during ageing of the animal. The individuals which originate from the fertilized eggcell are therefore the result of two counteracting forces of heredity, one is present in the egg and the other in the spermatozoon. The "variability" of children from the same pair of parents is due to this i.e., that the different gametes of the father-animal, as well as those of the mother-animal, differ in age (that is to say in stages of development)", and: "The value (valence) of the genotypical characters in every gamete depends on its degree of maturity. Hence the difference in maturity-degree of both conjugating gametes at the moment of fertilization is the cause of the variability expressed by the total of all the genotypical factors".

We have now discovered a very important principle, that is to say, the potency of the genotypical characters (present in the reproductive cell) is not always the same but varies. This variation depends on the age of the reproductive cell.

Besides causes which lie in the genotypical heterozygosity of the parents (which may or may not be observable), environment (active, outside the individual) and age differences of the parents or between the reproductive cells (both acting inside the individual) may give rise to an F_1-generation which is not always uniform. However all these exceptions will not alter the general validity of the theorem (origi-

THE MUTUAL SIMILARITY OF THE F_1-INDIVIDUALS

nating before 1900) and which can be put in these words: If both parents are homozygous (true breds) with regard to all the heritable characters present, then the F_1-generation (which will originate from this cross) is genotypically equivalent and completely uniform, provided that this generation grows up under similar environments.

CHAPTER X

THE MUTUAL SIMILARITY OF RECIPROCAL CROSSES
(heterogamy, sex-linked inheritance, maternal inheritance)

In his pioneering hybridization experiments with two species of tobacco (*Nicotiana paniculata* and *N. rustica*) KOELREUTER (1763, Fortsetzung, p. 24) concluded that it was a matter of indifference which of these species was taken as the mother and which as the father. In both cases, the same result was obtained; the F_1-generation of *N. paniculata* ♀ × *N. rustica* ♂ was not distinguishable from that of the reverse or reciprocal cross *N. rustica* ♀ × *N. paniculata* ♂. Although he found deviations from this rule of the similarity of reciprocal crosses in some other hybridization experiments, however, these were not sufficient to prove the rule invalid. His successors could only corroborate his findings, that in general no difference worth mentioning occurred between reciprocal crosses. Hence FOCKE, in his meritorious 'Pflanzenmischlinge' (1881, p. 479) comes to the conclusion 'that in general, in true species of the vegetable kingdom, the shape-determining ability of the male and the female element are absolutely equal at fertilization'. Nevertheless this rule does not have so firm a foundation as the previous one. There are many exceptions to FOCKE's rule which later turned out to have a fundamental significance (PELLEW, 1929). We find important examples in which both of the reciprocal hybridization products are dissimilar. Quoting again from rich *Oenothera*-investigations we find (besides the strong linkage of genes in groups and dissimilarity with in F_1-generations) another striking phenomenon in many of the species of *Oenothera* investigated. This is the phenomenon of *heterogamy* (DE VRIES, 1911, p. 99). In 1903 for instance DE VRIES had already concluded (1903, p. 471) that the hybrids *Oenothera biennis* ♀ × *O. muricata* ♂ showed many more '*muricata*' characteristics than the reciprocal cross *O. muricata* ♀ × *O. biennis* ♂ which tends to resemble the *biennis*-plants (figure 64).

In 1911 he stated this result more precisely as follows: 'In the pollen of *O. biennis* L. and *O. muricata* L. other typical characters are transferred to the descendants, than those from the egg-cells of the same plants. The 'pollen-picture' (complex of characters transferred by the pollen) agrees, in outline with the visible properties of the species, but the 'egg cell-picture' is another story'.

With this, DE VRIES had discovered the extremely important principle which

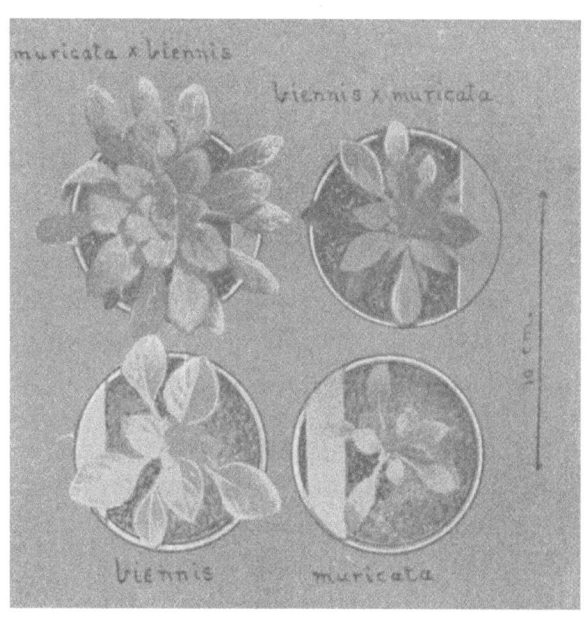

Fig. 64. *Oenothera biennis* and *O. muricata* with both reciprocal hybrids (after DAVIS, 1914).

could give us the explanation of the differences between these reciprocal hybrids. It turned out that this principle played an important part in the genetics of the species *Oenothera*, as became evident from DE VRIES' later work (1913) and that of RENNER (1917, 1918 a-c, 1925). From this point of view there is an important difference between the *Oenothera*-species which are now thoroughly investigated: *O. Lamarckiana* gives rise (as we already saw) to egg-cells and pollen-grains with the so-called *velans*-complex of factors and similarly to egg-cells and pollen-grains with the *gaudens*-complex. Thus in both kinds of gametes, both of the complexes occur. Therefore the species is called *isogamous*. On the other hand *O. muricata* gives rise to egg-cells with a *rigens*-complex of factors only (e.g., very erect stems) and to pollen-grains with the *curvans*-complex only, which causes a very sharply bent stem-top. This species is thus *heterogamous* in the highest sense.

Between *O. Lamarckiana* and *O. muricata* there lie two other European species: (1) *O. biennis* which according to the material of RENNER (see p. 178) makes two types of egg-cells (*albicans*-complex and *rubens*-complex) and one type of pollen-grain (*rubens*-complex) and (2) *O. suaveolens*, which also gives rise to two forms of egg-cells (*albicans*-complex and *flavens*-complex) and one type of pollen-grain (*flavens*-complex). Hence, both latter species are described as *semi-heterogamous*. Summarizing, RENNER (1918a) gives the following view concerning these four species:

O. Lamarckiana	= *velans* ♀, ♂	*gaudens* ♀, ♂.	isogamous
O. biennis	= *albicans* ♀.	*rubens* ♀, ♂.	semi-heterogamous
O. suaveolens	= *albicans* ♀.	*flavens* ♀, ♂.	semi-heterogamous
O. muricata	= *rigens* ♀.	*curvans* ♂.	heterogamous

The consequence of this heterogamy is that the reciprocal hybridization products differ as soon as one of the parents is heterogamous. The *biennis* ♀ × *muricata* ♂ cross, for instance, will give rise to a dimorphic F_1-generation composed of the types *albicans-curvans* and *rubens-curvans* but the reciprocal *muricata* ♀ × *biennis* ♂ cross gives rise to a simple F_1: *rigens-rubens*-plants only. However complicated these phenomena are, it is still obvious that DE VRIES considers the Dutch *O. biennis* as absolutely heterogamous and found in his material only one type of egg-cell. RENNER on the other hand found in his *O. biennis* which was collected in the neighbourhood of Munich, apart from a great majority of *albicans*-egg-cells, a much smaller number of *rubens*-egg-cells also.

The cause of the heterogamy in the *Oenothera*-species is not at all clear; this phenomenon is even attended with a considerable sterility in the reproductive cells (that is to say in the egg-cells as well as in the pollen-grains).

It would be tempting to ascribe heterogamy of *O. muricata* to the following (as some investigators have done) that the egg-cells with the *curvans*-complex and the pollen-grains with the *rigens*-complex are eliminated through the presence of a definite, lethal acting factor (see chapter XX), but there are many objections to this obvious, but rather too simple explanation. One objection is that an isogamous species such as *O. Lamarckiana* still has considerable sterility of its gametes (see RENNER, 1925, 1940).

Up till now this clear-cut form of heterogamy was only found in the *Oenothera*-species but many other phenomena were noted in other organisms which can be classed under heterogamy. The occurrence of double-flowering plants in the stock-gilly flower (*Matthiola incana*), which was so thoroughly investigated by Miss SAUNDERS (summary 1928), is a very nice example of this. She established several important facts: viz., that in this plant two types of single flowers occurred which are undistinguishable by eye. One type after self-pollination, or after mutual hybridization of individuals belonging to this type, never gives anything other than single flowers in the progeny. Specimens of the other type (by self-pollination as well as by mutual pairing) always gave a definite number of double flowering descendants, and these occurred very nearly in the ratio 15 single: 17 double (47% : 53%); the latter are always absolutely sterile. These single flowering individuals are never true breds; they always segregate into single and double and they are called 'ever-sporting'. The differences between these two types can be very easily shown by the following schematical pedigree:

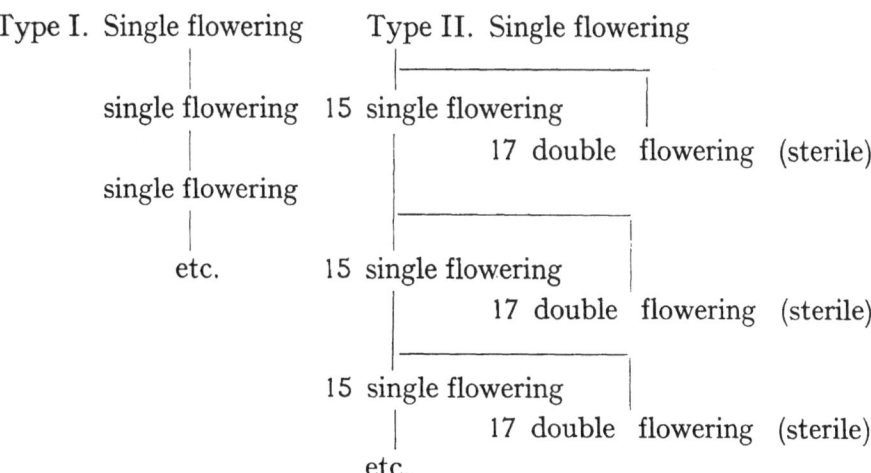

Type I. Single flowering

single flowering

single flowering

etc.

Type II. Single flowering

15 single flowering

17 double flowering (sterile)

15 single flowering

17 double flowering (sterile)

15 single flowering

17 double flowering (sterile)

etc.

When Miss SAUNDERS discovered this and had sufficient material of both types at her disposal, she carried out both reciprocal crosses as follows: I ♀ × II ♂ and II ♀ × I ♂ with the result that in both cases only single flowering individuals were obtained.

Both reciprocal crossing-products were apparently similar but in reality this is not so. The crossing II ♀ × I ♂ gave an F_1-generation of single flowering individuals which were partly composed of true breeding single flowering plants, and for the remainder segregating individuals, whereas the reciprocal cross I ♀ × II ♂ gave an F_1 of single flowering plants, which were all 'eversporting'. Miss SAUNDERS indicates this as follows:

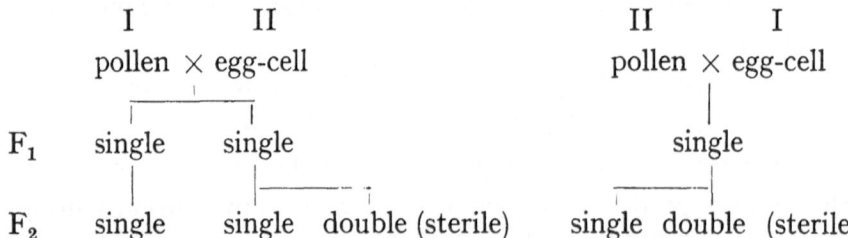

That a difference between both reciprocal F_1-generations really exists is only evident in the F_2-generation; phenotypically the F_1-generations are similar, however, they are genotypically different. Since there is no reason to believe the egg-cells and pollen-grains of type I (pure single flowering plants) are different, it is obvious that this difference in reciprocal crosses is due to a peculiarity in the reproductive cells of type II, viz., that the individuals of this type only give rise to one kind of pollen-grains, but on the other hand to more than one type of egg-cells. The data of Miss SAUNDERS indicates that the single flower is a result of the presence of two linked genes X and Y; the presence of only one of these genes (or absence of both) causes double flowers. Plants of type I have the formula XXYY and give rise to XY pollen-grains and XY egg-cells; plants of type II, with constitution XxYy, form pollen-grains xy and 4 types of egg-cells i.e., XY, Xy, xY and xy, however, with the limitation that between X and Y a linkage exists with a crossingover-percentage of 6%, so that the 4 types of egg-cells occur in the ratio 15 XY : 1 Xy : 1 xY : 15 xy.

If we accept this view concerning the constitution of factors then from the results obtained it follows that: Type I ♀ (XY) crossed with type II ♂ (xy) gives only XxYy-individuals which all segregate in the F_2 into single and double plants. Type II ♀ (15 XY : 1 Xy : 1 xY : 15 xy) crossed with type I ♂ (XY) gives homozygously single flowering (XXYY) plants and also heterozygously single flowering plants in the F_1-generation (XXYy, XxYy and XxYY). These latter single flowering plants segregate in the F_2-generation.

The principal consequence of this investigation is that individuals of the type II ('eversporting') of *Matthiola* crossed with those of type I can give different reciprocal F_1-generations, and that this difference can be ascribed to a difference in heritable tendency between the pollen-grains and the egg-cells of type II; the pollen-grains are all similar and possess both of the recessive factors x and y; the egg-cells belong to 4 types and occur in the ratio 15 XY : 1 Xy : 1 xY : 15 xy. Thus, since the pollen-grains have the same heritable tendency as part of the egg-cells, *Matthiola* can be classified under *semi-heterogamous* plants by virtue of its double flowers. SAUNDERS does not give any further explanation of the absence of pollen-grains possessing the factor combination expected, i.e., XY, Xy and xY. Because of this an attempt was made (following the paths of FROST, 1915) to ascribe the heritable behaviour of these 'eversporting' plants to a single factor S and by assuming that the pollen-grains with the dominant factor are destroyed by a 'lethal' gene. In this way every self-pollination of the 'eversporting' plants should result in a back-cross of Ss (egg-cells) × ss (pollen-grains). This explanation was put forward as follows by WINGE (1932b): the factor S for single flowers (s for double flowers) is linked with a factor V which causes viability of the pollen (pollen with v is inviable). Concerning these two factors the following fomulae can be derived:

$$\text{pure single } \frac{SV}{SV}, \text{ eversporting single } \frac{Sv}{sV}, \text{ double } \frac{sV}{sV} \text{ or } \frac{sV}{sv}.$$

The gameto-types Sv and sV are non-crossover gametes formed by the 'eversporting' plants, Sv's do not survive to pollen maturity. SV and sv are the cross-over gametes of the 'eversporting' plants. Once again the latter (sv) does not occur in the pollen in a viable form. The progeny of an 'eversporting' plant is therefore composed of single (SsVv) and double (ssVV) plants originating from non-crossover gametes and also of products from cross-over × non-crossover gametes (SSVv, SsVV and ssVv). The products of cross-over gametes when mutually hybridized give the rare combination SSVV and the common eversporting' SsVv. In the progeny of a selfed 'eversporting' plant WINGE found that a 3 : 1 segregation (139 : 47) occurred in the F_3-generation.

In the fourth generation two progenies were true breds and two segregated again according to 3 : 1 (1075 : 344). This indicated that the F_2-plant, from which the third generation descended has as its formula SsVV with a segregation into 1 SSVV : 2 SsVV : 1 ssVV. The too high a number of double flowers, which always occurs among the descendants of an 'eversporting'-individual, is explained by the supposition, that the heterozygous Vv-individuals should be weaker than the VV-homozygotes.

Very careful investigations of KUHN (1937) have produced evidence that the conception of heterogamy in *Matthiola* is correct; the lethal factor prevents the function of the pollen type involved, either by prevention of germination or through an interruption of the growth of the pollen-tube. That this lethal gene is linked with the S-gene was evident from experiments carried out by KAPPERT (1937) who showed zygotes with a non-viable factor constitution. The work of WESTERGAARD (1936) and of KUHN (1938) proved that the cytological explanation for this lethal gene, originally published by PHILIP and HUSKINS (1931), was incorrect.

METZ (1927) observed a more or less analogous case of heterogamy as a consequence of elimination of a gameto-type in the male heterozygote, in his crosses with regard to a wing aberration (truncate, recessive factor t) in *Sciara coprophila*. The result of the reciprocal F_1-generations, their F_2-offspring and back-crosses between the reciprocal F_1-types and their recessive tt-parents, was a follows:

	Parents	Offspring	Parents	Offspring
F_1	tt ♀ × TT ♂ = all tT		TT ♀ × tt ♂ = all Tt	
F_2	tT ♀ × tT ♂ = 50% tt + 50% tT		Tt ♀ × Tt ♂ = 50% Tt + 50% TT	
F_1 × P_1 tT ♀ × tt ♂ = 50% tt + 50% tT			Tt ♀ × tt ♂ = 50% Tt + 50% tt	
P_1 × F_1 tt ♀ × tT ♂ = all tt			tt ♀ × Tt ♂ = all Tt	

Briefly, the course of heredity of the truncate wings is based on the fact that in the tT-males (originating from truncated mothers and normal fathers) the T-factor for normal wings is eliminated, whereas in the reciprocal Tt-♂♂ just the reverse occurs, that is, the t-factor is not found in the gametes. In contradiction to this elimination of gameto-types in *Matthiola*, METZ succeeded in finding a cytological cause for this behaviour which will be discussed in chapter XII.

These examples of heterogamy, which were observed in heterozygotes, can be classified together with those cases dealing with the influence of sex on the crossingover-frequency (see p. 172). As an explanation of the phenomena in *Matthiola* (p. 199–201) it was put forward that a continuous back-crossing of a heterozygote (egg-cells) with a recessive homozygote (pollen-grains) took place. This explanation of 'eversporting' types is very closely related to another group of explanations of the causes underlying the difference in reciprocal crosses, i.e., underlying the important phenomena of linkage between a factor and the factor for sex, known by the term 'sex-linkage'.

The first investigator who worked with this subject was DONCASTER (1908); he took his material from the currant geometer *Abraxas grossulariata*, in which a light coloured form occurred, *lacticolor*, which up till then was only found in females (figure 65). The first

Fig. 65. *Abraxas grossulariata* and its variety *lacticolor*
(after DONCASTER and RAYNOR, 1906).

cross made by him was between a *grossulariata* ♀ and a *lacticolor* ♂; from these crosses 45 *grossulariata* ♂ and 50 *grossulariata* ♀ were obtained (fig. 66.1). A part of these F_1-individuals were mutually paired (fig. 66.2) and yielded 14 *grossulariata* ♂ and 4 *grossulariata* ♀ and 7 *lacticolor* ♀; *lacticolor* ♂ was absolutely missing. F_1-males were back-crossed with *lacticolor* ♀ (figure 66.3) and 63 *grossulariata* ♂, 62 *grossulariata* ♀, 65 *lacticolor* ♂ and 70 *lacticolor* ♀ originated from these crosses. For the first time *lacticolor* males occurred which were up to this point unknown and which could now be used for back-crossing with *grossulariata* F_1-females (figure 66.4). This cross gave a segregation into *grossulariata* and *lacticolor*, but with the condition that all *grossulariata*-individuals (130) were males and all *lacticolor* animals (145) were females. And in his final experiment, DONCASTER obtained nothing but *lacticolor* (♀ and ♂) from *lacticolor* ♀ × *lacticolor* ♂.

The crosses 3 and 4 (in figure 66) in which the first (i.e., *grossulariata* ♂ × *lacticolor* ♀) yielded both forms in both sexes are very important. The reciprocal cross of *lacticolor* males with *grossulariata* females yielded only *grossulariata* males and *lacticolor* females.

This result was surprising and demands an explanation. Shortly before this SMITH (1906) and CORRENS (1907) had already considered that the determination of sex, that is to say the cause of difference between male and female organism, was based on a heritable factor. Hence one of the 2 sexes should be homozygous with regard to this

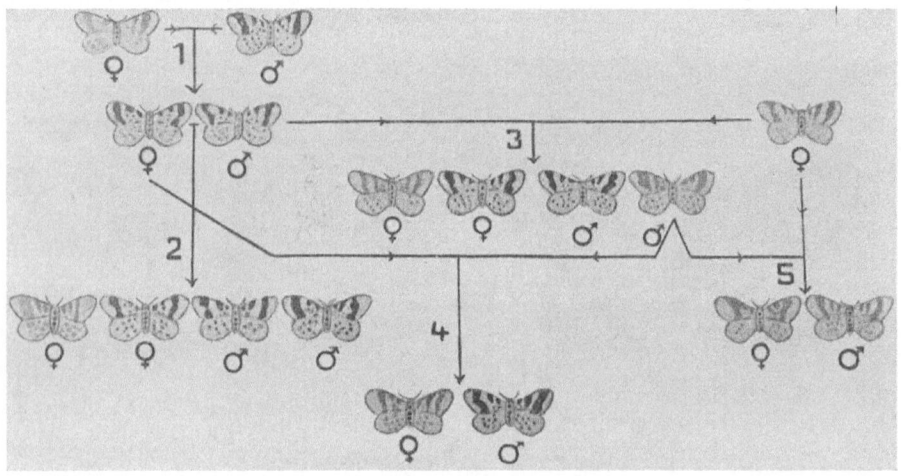

Fig. 66. DONCASTER's scheme of the several crosses between *Abraxas grossulariata* and its variety *lacticolor*.

factor and the other sex heterozygous. This principle had already been thought of by MENDEL (1905, p. 241). We will discuss later on in detail the importance of the conclusions which can be derived from this (Chapter XVIII) and its fundamental basis. It is sufficient for the moment to consider only the theorem. It is a tempting explanation which solves for ever ancient problems. The theorem is that: the one sex is heterozygous (XY) and thus forms two types of gametes — one is X and the other Y. The other sex is homozygous (XX) and gives only one type of gametes (all X) and their pairing gives rise to equal numbers of XX- and XY-individuals and hence equal numbers of ♀ and ♂ individuals.

DONCASTER took this principle to be the explanation for the peculiar course of his *Abraxas*-crosses. The facts led him (viz., DONCASTER's

crosses 3 and 4) to believe that in *Abraxas* the female could be considered heterozygous and the male as homozygous for the sex factor
He thought that the explanation was composed of the following three
parts: (1) that Mendelian factors cause the sex-determination and that
the factor for female is supposed to be dominant; (2) that female
individuals are heterozygous with respect to this factor and have
a formula Ff (the females (Ff) give rise to egg-cells containing F from
which females originate, and f which will give rise to males). The males
are homozygously recessive ff and give rise to only one type of sperm-
cells (f); (3) since a repulsion exists between the factor F and the factor G
for *grossulariata*, which is dominant over g (*lacticolor*), all F-egg-cells
possess g, and all f-egg-cells possess G. With the help of these three
suppositions it is easy to explain the result of *Abraxas*-crosses and
the following scheme indicates the course supposed:

		Results	
1. Lacticolor ♀ (Ffgg) × grossulariata ♂ (ffGG)		50% FfGg	50% ffGg
Gametes ♀ Fg and fg, gametes ♂ fG		gross. ♀ F₁	gross. ♂ F₁
2. F₁ grossulariata ♀ (Fg.fG) × F₁ grossulariata ♂ (ffGg)		25% FfGg	25% Ffgg
Gametes ♀ Fg and fg, gametes ♂ fG and fg		gross. ♀	lactic. ♀
		25% ffGG	25% ffGg
		grossulariata ♂	
3. Lacticolor ♀ (Ffgg) × F₁ grossulariata ♂ (ffGg)		25% FfGg	25% Ffgg
Gametes ♀ Fg and fg, gametes ♂ fG en fg		gross. ♀	lactic. ♀
		25% ffGg	25% ffgg
		gross. ♂	lactic. ♂
4. F₁ grossulariata ♀ (Fg.fG) × lacticolor ♂ (ffgg)		50% Ffgg	50% ffGg
Gametes ♀ Fg and fG, gametes ♂ fg		lactic. ♀	gross. ♂
5. Lacticolor ♀ (Ffgg) × × lacticolor ♂ (ffgg)		50% Ffgg	50% ffgg
Gametes ♀ Fg and fg, gametes ♂ fg		lactic. ♀	lactic. ♂

The results of Doncaster are in sufficient accordance with the
theoretical calculations, that we are forced to see a strong support
in the data for his contemplation over the problem.

An important amplification of these results was obtained through
the possibility that the *lacticolor*-males which were finally obtained,
could now be crossed with wild *grossulariata*-females. From this cross
it was evident, that these matings yielded a progeny from which the
male were always *grossulariata* and all females *lacticolor*, that is to say,
the same result was obtained as in cross 4.

The wild *grossulariata* females used in this experiment originated from different parts of England, partly from areas where *lacticolor* was never found. These experiments still prove that the wildtype females were genotypically similar to the F_1-*grossulariata* ♀ and thus they had Fg. fG as their formula; that is to say, they were heterozygous with regard to the factor for *grossulariata*-pattern but this heterozygous nature of females of the geometer does not find an expression, because in nature they always pair with males which are homozygous ffGG.

Hence, the cross *lacticolor* ♀ × *grossulariata* ♂ gives all (♀ and ♂) *grossulariata*, whereas *grossulariata* ♀ × *lacticolor* ♂ gives all ♀ *lacticolor*, all ♂ *grossulariata*. The result of this last cross, in which a characteristic is passed from the mother to the sons and from the father to the daughters, is known as 'criss-cross inheritance'. From these crosses it is evident that differences in reciprocal hybridization products do occur.

Naturally this did not remain the only case of typical sex-linked heredity for long, that is to say, linkage with the factor for sex. There are plenty of examples among several different species of animals but we will restrict ourselves to some which are very strikking. The hybridization experiments, by which the existence of sex-linked factors was proved for fowls, are completely comparable with its course in *Abraxas*. We will mention only two cases: barred Plymouth Rocks, when compared with non-barred black Cornish Indian Game, Langshans and Minorcas, and the silver-pencilled Assendelver fowl when compared with the gold-pencilled form of the same race.

By reason of the relatively little data available, SPILLMAN (1908, 1909) had already formulated the hypothesis that the factors for barred feathers and for sex in fowls were linked as follows: the hen is always heterozygous with respect to both factors, whereas the cock is always homozygous concerning the sexfactor, and homozygous or heterozygous with respect to the factor for barred feathers. This supposition was wholly corroborated by experiments concerning this question, especially by those of PEARL and SURFACE (1910 a and b).

They made a number of crosses between the dark race Cornish Indian Game (C.I.G.) and the Barred Plymouth Rock (B.P.R.). A scheme of the course of both reciprocal crosses is represented in figure 67; the F_1-generation of the C.I.G. hen × B.P.R. cock was composed of 96 black hens, 95 barred cocks (*criss-cross inheritance*);

the F_2 gave black and barred cocks and dark and barred hens. The reciprocal cross B.P.R. cock × C.I.G. hen, gave a homogeneous barred F_1-generation and yielded 68 ♀ and 70 ♂. The F_2 was composed of barred cocks only and hens of both colours. The result of these crosses

Fig. 67. Reciprocal crosses between Cornish Indian Game (black) and Barred Plymouth Rock (after the data of PEARL and SURFACE, from MORGAN, 1913).

is similar here to that in *Abraxas*; the whole question is explained if the following formulae are supposed for both races:

B.P.R. ♂ = BfBf, with gametes Bf.
B.P.R. ♀ = BfbF, with gametes Bf and bF.
C.I.G. ♂ = bfbf, with gametes bf.
C.I.G. ♀ = bfbF, with gametes bf and bF.

The factor for barred feathers is represented by B (b for black); Ff are the genes for the female sex, ff is the formula for male sex. The factors B and F show mutual complete repulsion so that they never will occur in the same egg-cell. The crosses indicated in figure 67 can also be drawn as a pedigree from which the validity of the formulae is obvious:

Parents	CIG ♂ × BPR ♀					CIG ♀ × BPR ♂		
Formula	bf.bf × Bf.bF					bf.bf × Bf.Bf		
Gametes	bf	Bf and bF				bf and bF	Bf	
F_1	Bf. bf bF.bf					Bf.bF	Bf.bf	
	barred ♂ dark ♀					barred ♀	barred ♂	
F_2	Bf.bF	Bf.bf	bf.bf	bF.bf	Bf.Bf	Bf.bF	Bf.bf	bF.bf
	barred ♀	barred ♂	dark ♂	dark♀	barred ♂	barred ♀	barred ♂	dark ♀

The barred feathers of some races of fowl are thus attributed to a typical sex-linked factor, which is present heterozygously in the female and homozygously in the male.

The same situation occurs in silver-pencilled and gold-pencilled animals of the Assendelver Fowls. The data compiled by HAGEDOORN was given by him to BAUR (1911; 1930) who analysed it.

HAGEDOORN's material (1921, p. 157) contained crosses of silver-pencilled hens with gold-pencilled cocks and the reciprocal cross: gold-pencilled hens with silver-pencilled cocks. The first cross yielded a total of 492 animals of which 249 were cocks (all silver-pencilled) and 243 were hens (all gold-pencilled); the reciprocal cross gave 650 individuals in the ratio of 162 silver-pencilled cocks to 163 silver-pencilled hens and 165 gold-pencilled cocks to 160 gold-pencilled hens. Plate IV gives a figure showing both these reciprocal crosses. If we represent the factor for the female sex by F (Ff = ♀, ff = ♂) and that for silver-pencilled by Z (gold by z), and if we suppose that both factors are in the repulsion phase, then we can reproduce both mentioned crosses as follows:

| Parents | Silver-pencilled × Gold-pencilled | | Gold-pencilled × Silver-pencilled | |
	hen	cock	hen	cock
Formula	Zf.zF	zf.zf	zF.zf	Zf.zf
Gametes	Zf and zF	zf	zF and zf	Zf and zf
F₁ {	Zf.zf	zF.zf	Zf.zF zF.zf	Zf.zf zf.zf
	s.p.♂	g.p.♀	s.p.♀ g.p.♀	s.p.♂ g.p.♂

These formulae were corroborated by mutual mating of gold-pencilled animals and silver-pencilled animals: gold-pencilled cocks (zzff) with gold-pencilled hens (zzFf) gave 53 gold-pencilled cocks and 51 gold-pencilled hens. Silver-pencilled cocks (Zzff) with silver-pencilled hens (Zf.zF) gave 11 silver-pencilled (ZZff and Zzff) cocks, 4 silver-pencilled hens (Zf.zF) and 5 gold-pencilled hens (zzFf). These results are in a good accordance with the theory.

Contrary to this heterozygosity of the ♀ we now examine a series of investigations concerning other groups of animals, where sex-linked genes also occur and in which the male is considered to be heterozygous.

This started with experiments of MORGAN (1910 a and b) concerning the inheritance of the red eye colour in *Drosophila melanogaster*. Usually both sexes have red eyes; but when one day a white eyed ♂ appeared in the cultures, it was a very nice opportunity to investigate the heredity of this phenomenon, which was gratefully accepted

Plate IV

162

165

163

160

Reciprocal crosses of Assendelver poultry

by MORGAN. On pairing of this male with red-eyed females a nearly complete red-eyed progeny was obtained (1234 ♂ and ♀, only 3♂ flies with white eyes appeared). These heterozygous red eye animals, when mutually paired, yielded 2459 females with red eyes, 1001 males with red eyes and 782 males with white eyes. Heterozygous red-eyed females paired with white-eyed males gave rise to flies with red and white eyes in both sexes. White-eyed ♀♀ crossed with white-eyed males yielded only animals with white eyes. From this we ascertain the

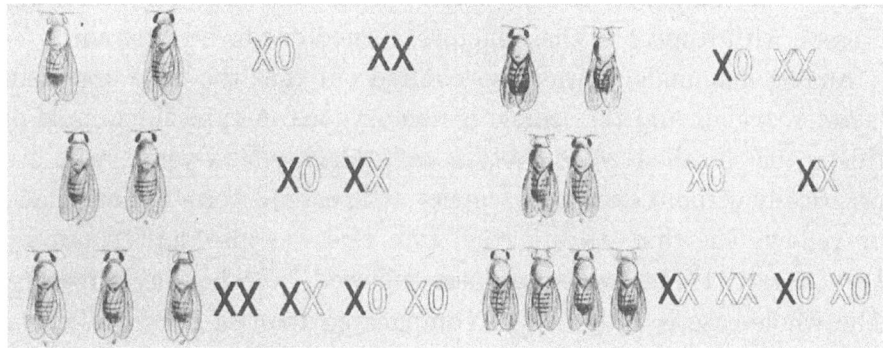

Fig. 68. Reciprocal crosses between white-eyed ♂ and red-eyed ♀ and between red-eyed ♂ with white-eyed ♀ together with F_1's and F_2's (after MORGAN, 1913).

fact that white-eye was recessive. Finally white-eyed females were mated with red-eyed males; this cross gave rise to white males and red-eyed females only (*criss-cross heredity*). In figure 68 both reciprocal crosses are represented with their F_1's and F_2's. In this figure XO always represents a male, and XX a female; the black X stands for a factor for red eyes: the outlined white X for a factor for white eyes, which is in accordance with what is now the general concept concerning the determination of sex (see chapter XVIII). However, if we want to give a schematic explanation of these *Drosophila* experiments in the manner we have adopted upto now, we can represent the factor for red eyes by R, that for white eyes by r, and a female by FF and a male by Ff.

Further, a complete coupling exists between the factors R and F. Both crosses are represented in figure 68 and are as follows:

Parents	White-eyed ♂	×	Red-eyed ♀		Red-eyed ♂	×	White-eyed ♀
Formula	rF.rf		RF.RF		RF.rf		rF.rF
Gametes	rF and rf		RF		RF and rf		rF
F_1 {	RF.rf		RF.rF		rF.rf		RF.rF
	r.e.♂		r.e.♀		w.e.♂		r.e.♀

14

Gametes of F[1]	RF and rf		RF and rF		rF and rf		RF and rF	
F[2] {	RF.RF r.e.♀	RF.rF r.e.♀	RF.rf r.e.♂	rF.rf w.e.♂	RF.rF r.e.♀	rF.rF w.e.♀	RF.rf r.e.♂	rF.rf w.e.♂

The discovery of the red factor, as being sex-linked, was not MOR-GAN's only investigation in this field. A great number of others followed, and in a paper, which MORGAN, BRIDGES and STURTEVANT published in 1925, they could summarize about 150 factors which were all sex-linked, and could state the fact that, with respect to all these factors the male was always heterozygous and that the reciprocal crosses, with respect to these factors, turned out to be different.

Among mammals, it was also pointed out that the male organism is heterozygous and the female is homozygous. A typical example of this is the so-called 'three coloured cat'', black-yellow-white, who are practically without exception females, whereas the tom-cats are black or yellow. The first investigation into this was also carried out by DONCASTER (1914a) who was later followed by other investigators. The whole case is rather more complicated than in *Drosophila*, not only because it appears that two factors are involved viz., Y (yellow colour) and B (black colour), but also because a relatively high percentage of unexpected combinations occur which is evident from the following summary of the results of DONCASTER:

Parents		Offspring					
		Males			Females		
		Orange	Black	Threec.	Orange	Black	Threec.
1. Black	♀ × orange ♂	—	46	**1**	—	**13**	48
2. Orange . . .	♀ × black ♂	20	—	—	—	—	16
3. Three colours .	♀ × orange ♂	54	38	**1**	47	**5**	43
4. Three colours	♀ × black ♂	35	29	**1**	—	12	21
5. Orange . . .	♀ × orange ♂	48	—	—	40	—	**3**

The question depends on whether DONCASTER's data is absolutely reliable; since he did not obtain it from his own experiments, but by compiling data given to him by several other people, among which mistakes could very easily occur. Without any doubt the females mentioned in cross 5 were not three coloured or did not descend from an absolutely orange pair of animals. The limitation of black colour and the spreading of orange differ so much among three coloured animals that it is quite possible that a three coloured animal is described as an orange one. If the bold type numbers in the table (which are all relatively low) were not there a very satisfactory explanation could be given by proposing that two factors are involved: B whose presence causes black and Y, the factor for yellow colour. These factors together cause the characteristic of the three coloured cat and can both be linked with the sex-factor F (FF = female, Ff = male).

Between Y and B a repulsion always exists, hence the black females are, for instance, of the genotypical constitution yyBBFF (see cross 1) with yBF as the type of egg-cells; orange tom-cats are YbF. ybf with both of these types as sperm-cells and their hybridization gives yBF. YbF (three coloured cats) and yBF.ybf (black tom-cats). In this way all the four crosses can be explained. We refer to BAMBER (1927) for a possible explanation of the unexpected products (see bold type numbers in table).

Analogously 'sex-linked' factors are also found in man, in which the most typical example is that of green-red colour-blindness.

It has been known for a long time that colour-blindness generally occurs more in men than in women. Nowadays four different types of colour-blindness are distinguishable which form two series of multiple alleles, viz., deuteranomaly (weak green blindness) and deuteranopy (green blindness) next protanomaly (weak red blindness) and protanopy (red blindness) and both series of alleles are sex-linked. Through studies of pedigrees, to which this investigation is natu-

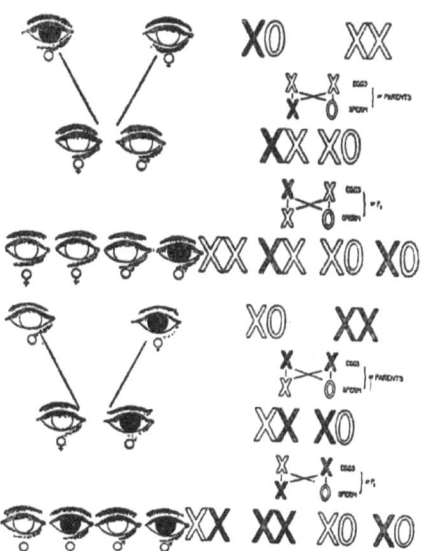

Fig. 69. Scheme of heredity of colour-blindness (after MORGAN, 1916).

rally limited (see WAARDENBURG, 1932), sufficient data has been compiled to describe the heredity of the colour-blindness. The explanation is as follows: Males are either colour-blind or normal; women can be colourblind, normal or apparently normal. ('carrier'). This means that a normal man or woman never transfers the aberration to any of the children; an apparently normal woman, who is thus not colour-blind herself can do this. Figure 69 explains this heredity; the upper half of this figure shows that the marriage of a colour-blind man with a normal woman gives only normal children from which the girls are apparently normal; the marriage of such an apparently normal woman with a normal man gives 50% of the girls normal and 50% apparently normal; 50% of the boys are normal and 50% are colour-blind. The lower part of the picture indicates the result of a marriage of a normal man with a colour-blind woman: all girls are apparently normal, all boys colour-blind (*criss-cross* inheritance).

The children from a couple in which the man is colour-blind and the woman apparently colour-blind are as follows: 50% of the girls are apparently normal and 50% are colour-blind; 50% of the boys are normal and 50% colour-blind. Finally from a marriage between a colour-blind man and a colour-blind woman there originate only colour-blind children (boys and girls).

The formula is obtained by representing the gene for sex by F and that for normal eyes by N. Further we have to accept that a coupling takes place between F and N.

Hence, the following scheme can be given:

normal man . . . .   NFnf   Gametes NF and nf.
colour-blind man . . . .   nFnf   Gametes nF and nf.
normal woman . . . .   NFNF Gametes NF.
apparently normal woman 'carrier'   NFnF Gametes NF and nF.
colour-blind woman   . . . . . .   nFnF Gametes nF.

From this the six possible marriages and their consequences can be derived.

Summarizing the principal items:

Female heterozygosity.
Abraxas.

1) *lacticolor* ♀ × *grossulariata* ♂ = all *grossulariata* or 1 : 1 : 1 : 1.
2) *grossulariata* ♀ × *lacticolor* ♂ = *lacticolor* ♀ + *grossulariata* ♂
   Fowls.
1) dark ♀ × barred feathers ♂ = all barred feathers or 1 : 1 : 1 : 1.
2) barred feathers ♀ × black ♂ = dark ♀ + barred feathers ♂.
1) gold-pencilled ♀ × silver-pencilled ♂ = all silver-pencilled or
   1 : 1 : 1 : 1.
2) silver-pencilled ♀ × gold-pencilled ♂ = gold-pencilled ♀ + silver-pencilled ♂.

Male heterozygosity.
Drosophila

1) white eye ♀ × red eye ♂ = white eye ♂ + red eye ♀
2) red eye ♀ × white eye ♂ = all red eyes or 1 : 1 : 1 : 1.

Man

1) colour-blind ♀ × normal ♂ = colour-blind ♂ + normal ♀
2) normal ♀ × colour-blind ♂ = all normal or 1 : 1 : 1 : 1.

In the case of female heterozygosity, the matings under 2) (in which the dominant character originates from the mother, and the recessive from the father) produce recessive female and dominant male descendants.

In the male heterozygosity the crosses sub 1): recessive female × dominant male, show an analogous 'criss-cross heredity'. From these crosses originate only dominant female and recessive male individuals.

In none of these crosses a transfer of the characteristic involved from mother to daughter or father to son occurred. Nevertheless, examples of such a *one-sided heredity* or *Y-heredity* are known. From the work of SCHMIDT (1920) and WINGE (1927b) it became evident that such a heredity occurs in the fish species *Lebistes reticulatus*. All females of this species are mutually phenotypically similar but recessive factors can still occur in these animals and the behaviour of these characters is comparable to the white eye colour in *Drosophila*. Therefore, four types of females were obtained which can only be distinguished by their male progeny, viz., the types X_o, X_{co}, X_{li} and X_{ti} (neutral, coccineus, lineatus and tigrinus). Apart from this, nine factors for colour find expression in the male but only one of these can occur in any individual and never more than one at a time. These nine factors (maculatus, iridescens, oculatus, ferrugineus, sanguineus, aureus, armatus, pauper and variabilis) are thought to be Y-factors for reasons we will come to understand better in chapter XI. They are indicated by the following symbols: Y_{ma}, Y_{ir}, Y_{oc}, Y_{fe}, Y_{sa}, Y_{au}, Y_{ar}, Y_{pa} and Y_{va}. These 13 colour factors, which all cause stripes, are represented in figure 70.

By systematic crossing between females (of the Y_o-type) and males bearing the different colours (i.e., any of the 9), and also by systematic crossing of the F_2-generations, and by several back-crosses, WINGE could ascertain that the X-factors behave in the same way as 'sex-

linked' factors do in the case of male heterozygosity; a female with
the X-factor produces sons with this X-factor only but the daughters
are dependent on the male for te second X-factor, (necessary
for the genotypical XX constitution), transferred by the male.
In the body-cells the female always has two X factors as alleles, in her
reproductive cells she has only one X factor; the male, on the other
hand, always possesses one X factor in its somatic cells and a Y factor

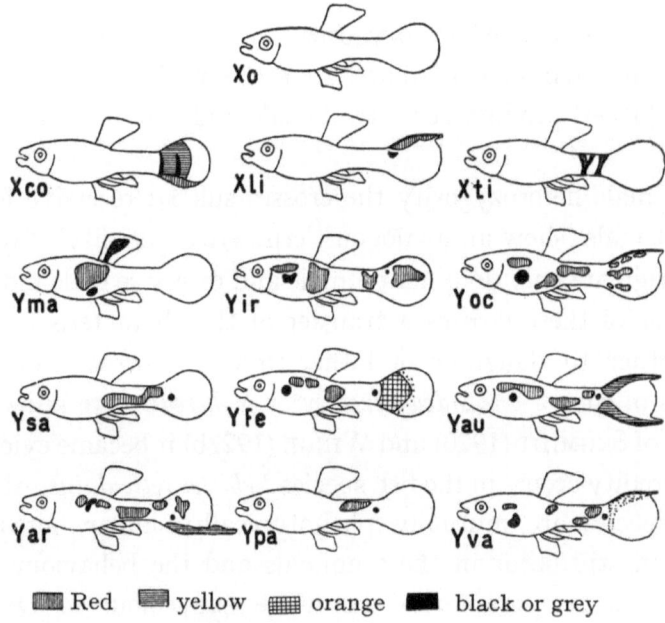

Fig. 70. Patterns of *Lebistes reticulatus*; X- and Y-heredity (after WINGE, 1927b).

as its allele. Hence, either X or Y factors are found in the male reproductive cells. The following example will explain the course of such
a cross:

	Parents	Offspring
F_1	$X_0X_0 \times X_{tl}Y_{au}$	= all ♀ X_0X_{tl}; all ♂ X_0Y_{au}
$F_1 \times F_1$	$X_0X_{tl} \times X_0Y_{au}$	= $50\%♀X_0X_0 + 50\%♀X_0X_{tl} + 50\%♂X_0Y_{au} + 50\%♂X_{tl}Y_{au}$
Parent ♀ × F_1 ♂	$X_0X_0 \times X_0Y_{au}$	= all ♀ X_0X_0, all ♂ X_0Y_{au}
F_1 ♀ × Parent ♂	$X_0X_{tl} \times X_{tl}Y_{au}$	= $50\%♀X_0X_{tl} + 50\%♀X_{tl}X_{tl} + 50\%♂Y_0X_{au} + 50\%♂X_{tl}Y_{au}$

The X_{tl}-factor originating from the father, is transferred to all
the daughters and is transferred to the grandsons only in the F_2
arising from the heterozygous X_0X_{tl} F_1-female. On the other hand the

Y factors are passed on directly from the father to the male progeny; a male of the Y_{au}-type produces only Y_{au}-males and a father with the factor Y_{fe} only Y_{fe}-sons. These Y factors thus do not exhibit the criss-cross inheritance but all the same they are still 'sex-linked' to a high degree (that is to say, linked with a factor which causes sex). The explanation of this is obvious: the X-factors are coupled with a factor which causes the female sex; the Y-factor is linked with the gene through which the male sex finds its expression. The general validity of this explanation will be discussed in detail in chapter XVIII.

Fig. 71. Plants of *Melandrium album*: right, broad leaved; left, narrow leaved (after SHULL, 1914b).

Two forms of sex-linked inheritance can be easily distinguished:

1) *criss-cross inheritance* that is to say, passing on of definite characteristics from mother to sons and from father to daughters.

2) *one-sided inheritance* in the male line, that is to say, the case in which the characteristics are transferred directly from father to all the sons.

It is clear that all the cases discussed concerning sex-linked heredity were borrowed from the animal kingdom. In contrast to animals (where the division into two sexes is the usual rule) the vegetable kingdom is very poor in dioecious species, and due to the difficulties involved in using dioecy for the investigation into heredity, not many experiments have been carried out with the existing species. By a fortunate chance BAUR (1912b) found the first typical example of *sex-linked inheritance in plants* in an unusual aberrant male plant of

Melandrium album, with narrow leaves as contrasted with the normal broad-leaved form (fig. 71). This plant was used for a cross with a normal leaved female plant; the F_1-generation had normal broad leaves. Out of 300 F_2-individuals in the seedling stage almost 75% were broad-leaved, 25% narrow leaved. Out of these, 52 narrow leaved seedlings were kept along with 3 broad-leaved plants used for comparison, and to BAUR's great surprise all narrow-leaved were ♂ ♂ and the 3 broad-leaved plants were ♀.

His experiments were continued by SHULL (1914b) who concluded that *Melandrium album* is homozygous as a female plant with regard to the factor for sex (FF); the male plant is Ff. Further, the factor F in a coupling phase with factor B, is responsible for the character broad-leaved: BB- and Bb-forms are broad-leaved, bb-plants have narrow leaves. Four types of plants had now been found by BAUR and SHULL:

BF.BF = homozygously broad-leaved female.
BF.bF = heterozygously broad-leaved female.
BF.bf = heterozygously broad-leaved male.
bF.bf = homozygously narrow-leaved male.

Normal plants of the species are BF.BF (♀) and BF.bf (♂); the unexpected aberrant forms were narrow-leaved males, that is bFbf (♂); the F_1-plants were partly BF.bF (broad ♀) and partly BF.bf (broad ♂); the F_2 was composed of BF.BF (broad ♀), BF.bf (broad ♂), bF.BF (broad ♀) and bFbf (narrow ♂). According to this theory it is small wonder that BAUR found that all selected narrow-leaved F_2-plants were male.

It is not always easy to fix with certainty whether the heritable tendency for a definite character is sex-linked as a factor, or not.

There are also numerous characters which find expression only in one of two sexes, and hence they are only observable in the successive generation in that particular sex. From this, one could suppose that they are 'sex-linked'. This immediately leads us to think of secondary sex marks which occur in all forms in animals and dioeceous plants. The research into the inheritance of such secondary sex characteristics is not easy; a genetical investigation into the subject of these secondary sex characteristics is not always possible. The question as to whether the characteristic is primarily caused by sex-linked factors or influenced secondarily by the sex, which has already been determined by other factors, is not always flexible. The cases of 'Y-chromosome inheritance' in the male line of *Lebistes* already mentioned, certainly belong to the first group; the second possibility can be investigated only by means of genetical experiments if one of the races used for the crosses bears the secondary sex characteristics in none of its sexes and the opposite mating type (belonging to another race) expresses this sex-characteristic in both sexes.

The first example of such an inheritance was investigated by Wood (1909) who made crosses between two races of sheep, with the idea

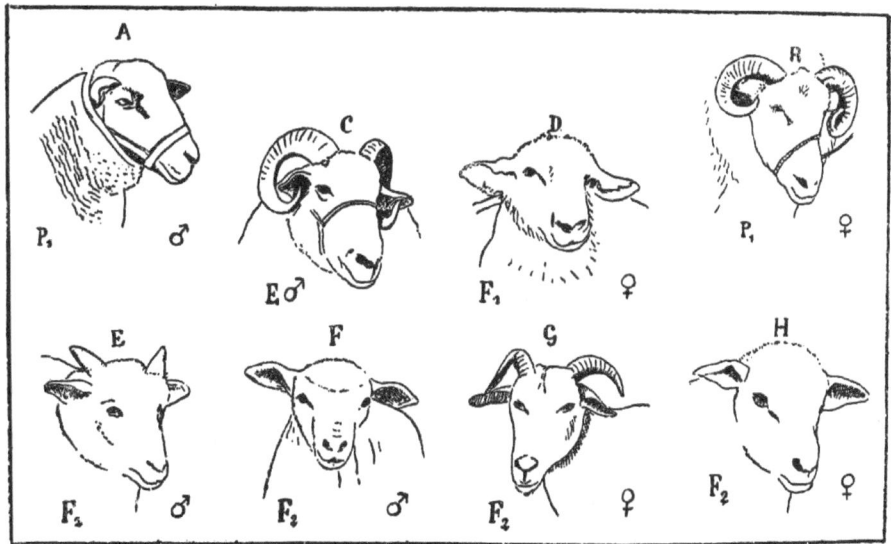

Fig. 72. Scheme of the sheep-crosses between Dorset and Suffolk of Wood (after Lang, 1914).

of studying the inheritance of its horns. The races used by him were: Dorset in which both ♀ and ♂ were horned, and Suffolk in which both sexes were hornless. Both reciprocal F_1-generations (fig.72) were similar; all rams (38) were horned: the ewes (35) hornless. An F_2-generation was composed of 33 animals: 21 rams (7 with large horns, 7 with round knobs, 3 with loose skin knots and 4 without horns) and 12 ewes of which 8 were hornless, 1 with round knobs and 3 with large horns. Different back-crosses were also achieved. Briefly, from these investigations (later completed by Arkel and Davenport, 1912), it can be derived that the factor for horned is dominant in the male and recessive in the female. In accordance with this, in the F_2, 75% of all rams should be horned and 25% hornless and for the ewes just the reverse (i.e., 25% horned and 75% hornless).

Though the data is not sufficient to prove this, it is even difficult in such a simple example to consider this inheritance to be due to a 'sex-linked' factor.

With perfect justice MORGAN (1914 b) distinguishes under the heading *sex-limited inheritance* the above mentioned phenomena found by WOOD. Sex-linked factors (as discussed earlier in this chapter) are always linked with the factor responsible for the expression of sex; sex-limited factors are influenced in their action by the sex of the individual in which they are present. Sometimes they are reinforced and sometimes opposed by the sex of the individual (see FÖYN, 1932, MONTALENTI, 1939).

Besides these causes of differences between reciprocal crosses in which either both parents play a rôle, or only the father has an influence, there are also other cases in which only the elements brought in from the maternal side are of significance (CORRENS, 1937). The strong dependence of the maternal reproductive cell and the embryo on the mother body is of primary importance. The principal difference between egg-cell (which apart from providing a nucleus usually provides also a considerable quantity of plasm to the young individual) and male gamete (in which the nucleus is perhaps in exceptional cases attended with a small quantity of plasm but in most cases functions only as nucleus) undoubtedly will support the idea that in some cases a greater significance is attached to the egg-cell than to the male gamete.

In the introduction (p. 3) it has already been mentioned that *pseudo-heredity* must be distinguished from true inheritance and that all strange organisms or all strange influences present in the reproductive cells, which affect the embryo or the gametes, must be excluded from true inheritance.

Some of these cases can be ascribed to the influences of feeding which take place after fertilization and originate from the mother organism during the period of embryonic life. For instance, it is generally considered that a definite difference exists between mules and hinnies. The former's father is a donkey and the mother a horse; the latter's father is a horse and the mother a she-ass.

How far this view is exact, has not yet been fixed with certainty; among the connoisseurs there are some, who deny absolutely that a sharp difference exists between the reciprocal crosses, and believe that this general view is due to the result of incomplete observation. However, in the case that it is true — that is to say, that mules and hinnies do differ mutually in important points, it is not necessary that this difference has a genetical basis because, during the embryonic life, the animals were for a long time submitted to different nourishment so that the differences could be ascribed to this.

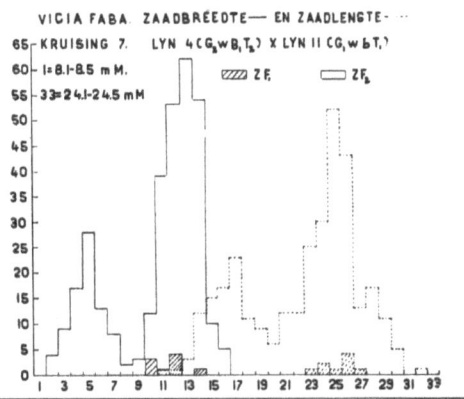

In the vegetable kingdom we can refer to the significance of the (pure maternal) seed-coat for the development in size of the enclosed embryo. Several investigations concerning heredity of seed sizes have shown that these are based on definite genes and that the seed size depends as well on the gene combination (which is brought into the embryo by egg-cell + pollen-grain) as well as on heritable factors in-

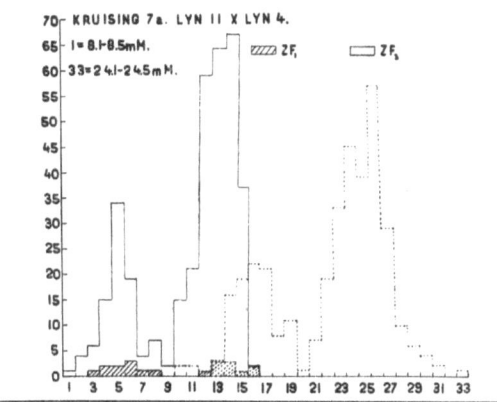

Fig. 73. Heredity of seed width (——) and seed length (...) in reciprocal crosses of *Vicia Faba*: top strain 4 × strain 11; bottom strain 11 × strain 4. The small frequency curves at the bottom of each drawing represent the F_1 seedgeneration, the large curves the F_2-seedgenerations (after SIRKS, 1931).

fluencing the seed-coat from a maternal side. The result of the fertilization of a small seeded plant, as mother, with pollen of a large seeded stock has as its consequence, a small seed, since the embryo cannot reach the size to which its factor constitution should enable it, because the small seed-coat (originating from the mother) acts as a brake. In the reciprocal cross, with the big seeded stock as mother, the hindrance fails to appear so that these seeds with the same factor combination in the embryo, can become bigger than the reciprocal ones (SIRKS, 1931, see figure 73).

In the F_2-generation of the seeds (originating by self-pollination from the F_1-plants) this difference between the reciprocal crosses does not occur because the seed-coats are now of the same factor combination (i.e., the F_1-individuals). It was possible to show that the basic factor for colour in beans (*Phaseolus*) is linked with such a factor for growth so that the white colour of the seed-coats is of greater or lesser significance for the size of the seed (SIRKS, 1925).

For symbiontic organisms which can be transferred to the progeny by reproduction, we will refer to the book of BUCHNER (1921).

It is obvious that among the gametes the egg-cells are especially susceptible to the external influences and the consequences can produce after-effects over a long period. This was pointed out by the investigations of SITOWSKI (1910) who fed the larvae of different moth-species (*Tineola*, *Endrosis* and others) with wool which was soaked in a dye-stuff in fat (Sudan-red, Brilliant-blue, and others). These larvae take the dye-stuff into their fat-cells and are intensively dyed by it. With the further development of the larvae to imago it remains; it is also present in the egg-cells, and the consequence of this latter is, that the progeny of the dyed mothers remain coloured even when fed with undyed wool. On the other hand, progenies of a father treated in this way are undyed. The plasm of the egg-cells which had absorbed the dye-stuff is responsible for this behaviour; the sperm did not bear plasm. In the generation which follows, the colour disappears more and more; the quantity of dye applied was limited, and new dye-stuff was not added after the first treatment of the mothers.

Related to these feeding influences are all those cases in which the egg-plasm has taken up material from outside (immunization experiments) or is changed by action of other organisms; we will discuss this later on in chapter XIX.

Contrary to these strange organisms or materials which are brought in from outside, and which do not form an inherent part of the normal reproductive cell, are the *plastids* which occur in plant cells such as: chlorophyll grains which lie scattered in the egg-cell plasm. Variegated plants (green-white or green-yellow) are very important in the investigations of this. Since BAUR's investigations (1909b) into the variegated varieties of *Pelargonium*, a very extensive series of investigations has been carried out which have shown that variegation in leaves has several different causes (see DE HAAN 1933, RENNER 1934, CORRENS 1937). These investigations (which will be discussed in chapter XVI) have proved that the plastids in this case can be absolutely independent of the plasm and of the nucleus, that is to say, they behave independently.

The combination of an egg-cell with green plastids and a pollen-grain originating from a green plant, will yield a normal green plant; the combination of an egg-cell in which the plastids are colourless, with a pollen-grain from a flower of a white plant, gives rise to an early-dying white seedling. The cross of a green egg-cell with a white pollen-grain or the reciprocal cross of a white egg-cell with a green pollen-grain gives rise to a green-white variegated seedling. In itself it is not necessary that the nature of the plastids causes reciprocal differences. These differences can be the consequence of plastids passing from pollen-grains to the egg-cell, as seems to have been proved for a number of plant-species. However, if the pollen-grain does not

Fig. 74. *Mirabilis jalapa*: left normal, right albomaculata (after CORRENS, 1909a).

transfer plastids (or only a very few plastids) into the egg-cell, then it is possible that reciprocal differences can indeed originate from this independent character of the plastids.

Certainly this will be the case if the plasm also influences the colour of the plastids, because then it is possible that the influence of the egg-cell plasm is dominating, and differences will exist between the reciprocal crosses. We can take a very nice example of this from the excellent investigations of CORRENS (1909 a and b) concerning *Mirabilis Jalapa*. These investigations have reference to the peculiar variegation which is shown by some individuals of the species, called the *albomaculata* form. The leaves are not plain green but spotted green and white (figure 74) in which the green colour sometimes dominates and another time the white colour. Such plants sometimes form completely green branches in which the white colour has absolutely disappeared and even (but more rare) whole white branches occur which are quite lacking in chlorophyll. This interesting phenomenon was investigated more closely by CORRENS. In the first instance he kept the flowers of such branches under self-pollination. The result was, that all (88) descendants originating from green branches were green, and in the next generation they did not yield other than green plants; from flowers of variegated branches, three sorts of descendants were obtained: green, variegated and white; of which the latter died as seedling (as a consequence of the lack of chlorophyll), and finally from some of the rare flowers on the white branches seed was obtained, that only gave rise to white, early dying, seedlings. More interesting still are the crosses made between the normal plants and flowers of the white branches of the *albomaculata* form: two plain green plants

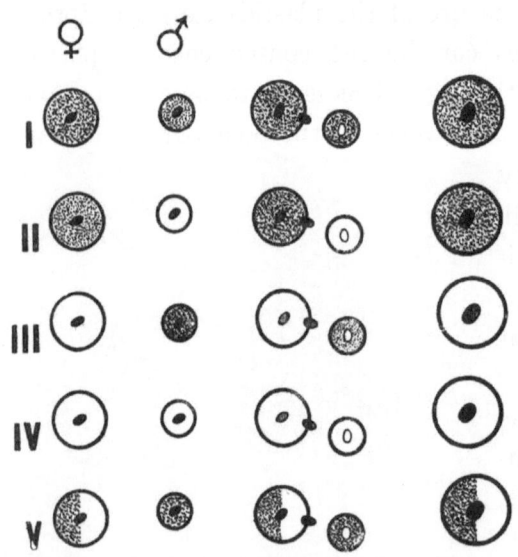

were used as mothers and they gave, after pollination with pollen of white branches, only (29) green descendants; the reciprocal cross 'white' flower as mother, with pollen of a green plant gave only 22 seedlings of which 17 were white and soon died and the other 5 (which were very white variegated plants) also succumbed at an early stage as a result of their weak health.

Fig. 75. Scheme of self-pollination and crosses in *Mirabilis jalapa* normal and *albomaculata*.

● healthy plasm   ○ sick plasm

(after CORRENS, 1909b).

By means of these experiments, a case of differences between both the products of reciprocal crosses was shown with certainty but the tug of war turned to the problem of giving a good explanation of these remarkable deviations from the prescribed rule. The explanation here was relatively obvious, and with aid of a scheme given by CORRENS himself (1909 b) is easy to understand (figure 75). If the view held by many investigators is correct that is, that in fertilization, the nucleus of the pollen-tube only penetrates into the egg-cell, whereas this egg-cell contributes the surrounding plasm as well as a nucleus, then the cause of the lack of chlorophyll can be found to lie in the disorder of this plasm which establishes an abnormal predisposition for chlorophyll-grains. Case (1) thus represents the self-pollination of a 'green' flower; (2) a hybridization of a 'green' flower as mother with the pollen of a 'white' flower; (3) the reciprocal cross: 'white' flower ♀ and 'green' flower ♂; (4) the self-pollination of a 'white' flower and (5) indicates how, by means of the plasm of the egg-cell, a variegated plant can originate. The explanation of the occurrence of these variegated seedlings is still a difficulty because it also presupposes a 'variegated' nature for the egg-cells. CORRENS believes that one is not allowed to suppose a mixture of green and white plastids in the egg-cell because, such mixed cells are not present in the vegetative parts; the variegation of the egg-cells could be ascribed to a labile state of the plasm.

In these *Mirabilis* investigations the plasm alone is responsible for the nature of the descendants, and this conclusion leads us to the very important question as to how the different parts of the cell: nucleus and plasm, play a rôle in the processes of inheritance.

In the following chapters we will consider the localization of the factors and the rôle of the plasm.

CHAPTER XI

LOCALIZATION OF THE FACTORS

With the admission of the existence of very definite genotypical factors (which would be responsible for the expression of several observable characteristics in plants and animals), the question follows at what point are these factors localized within the organism.

The premendelian investigators, HERTWIG (1884), STRASBURGER (1884) and more especially WEISMANN (1885) had drawn up the hypothesis (as we already saw in chapter III) that the *cell-nucleus*, and especially the *chromatic substance* present in the nucleus, should be the 'bearer of the heritable characteristics'. Their point of view was very hypothetical, but based on several observations, especially with regard to the precise manner in which this chromatic substance, concentrated in the *chromosomes* is divided among both daughter-nuclei originating from one nucleus.

We now have to discuss the view of modern geneticists concerning this hypothesis and the answer that later investigation has given to this question.

At once we meet great difficulties: up to now the problem has not been easy to approach from a direct karyological view point but experiment is demanded nowadays as a necessary proof. Hence, conclusions based solely on karyological grounds are excluded. Every contemplation concerning the question, which now interests us, has to be content with an indication of the probabilities which can, by accumulation, lead to practical certainty.

The first step, which could lead to a conclusion, was carried out by the well known *merogony* investigations of BOVERI with eggs of sea-urchins (1895). On shaking the eggs in a tube they fell into small parts and the nucleus disappeared. Hence, by all appearances, fragments of eggs were obtained which possessed no nucleus but which could be fertilized by a spermatozoon of another species and gave rise to larvae with distinct paternal properties. In his last publication (1918) BOVERI indicated some sources of errors which weakened the force of the argument in his earlier statements. Even if the work of BOVERI is not valid today, we are still indebted to him for his example which stimulated other investigators to choose hybridization and fertilization in sea-urchins as subjects for investigation. HERBST

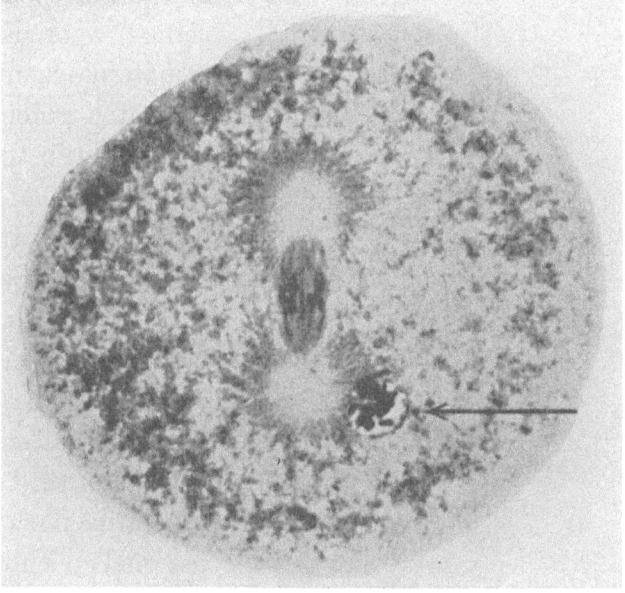

Fig. 76. Division of a non-fertilized egg nucleus before fusion with the sperm nucleus (←) has taken place (after BĚLǍR, 1920).

(1909) succeeded in giving qualified proof of the localization of the heritable factors in cell-nuclei. It happens, that the egg-cell nucleus of the sea-urchin sometimes divides before the fusion with the sperm-nucleus (figure 76). The delayed fertilization then concerns only one of these division-nuclei and the consequence is, that the young organism partly is composed of cells which possess a maternal nucleus and also partly a fertilization-nucleus. Because of this, parts of the larval body can show the maternal characteristics and next to these can be other organs (raised from the cells containing the fertilization-nucleus) possessing hybrid characteristics. Such a larva is shown in figure 77 with the maternal and hybrid parts of the body indicated. This investigation of HERBST clearly showed that the sperm-nucleus brings in factors at fertilization.

Also on the botanical side, early evidence had already been produced by CORRENS (1899) to show that, the nucleus acts as a factor carrier. In his Maize crosses (published before the rediscovery of MENDEL's work) CORRENS pointed out that in the results of reciprocal crosses (in so far as the endosperm is concerned) a preponderance of maternal characteristics was often undeniable. He could explain this on the grounds of the fact, just discovered at that time, that the endosperm originates as a fusion-product of three nuclei: two of which are polar nuclei of the maternal embryo-sac and the other, the second generative nucleus of the paternal pollen-tube. The strong influence of the mother is, according to CORRENS, due to a 'surplus of maternal nucleic substance'.

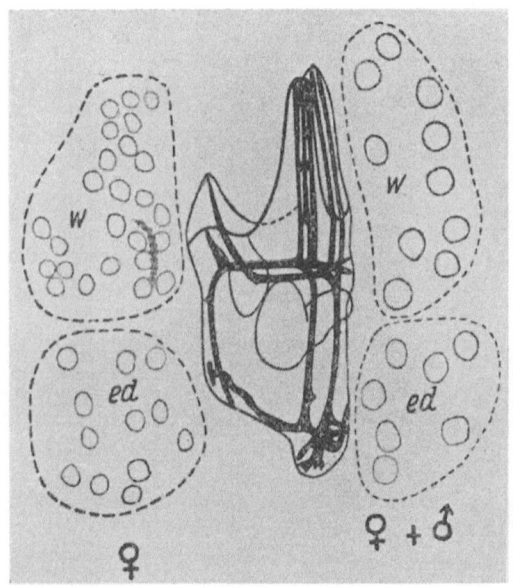

Fig. 77. A sea-urchin larva shows a partly maternal and partly hybrid phenotype. Larva originated from the egg-cell of fig. 76 (after HERBST, 1926).

Thus by these results of CORRENS and HERBST, the evidence can be considered to have been produced that the cell-nucleus transfers heritable characters, but still it has not been shown that the supposition of WEISMANN and other investigators, concerning the rôle of the 'chromatic substance', was right.

Before discussing the relationship between this 'chromatic substance' or chromosomes and the heritable factors, we must recapitulate the life-cycle of these chromosomes.

It is evident, that we can do this only very superficially on the basis of illustration; for a more detailed study we refer mainly to the books of BELAR (1928), GEITLER (1934), GUILLIERMOND and others (1933), DE ROBERTIS and others (1953), SCHRADER (1953), SHARP (1943), TISCHLER (1951) and WILSON (1937). Figure 78 gives a representation of the *somatic cell-division* or *'mitosis'* in *Allium cepa*. In the resting nucleus (*interphase*, a) the chromatic substance arranges itself into threads which become more and more visible (*prophase*, b, c). The *prometaphase* (d, e) shows the release and arrangement of the chro-

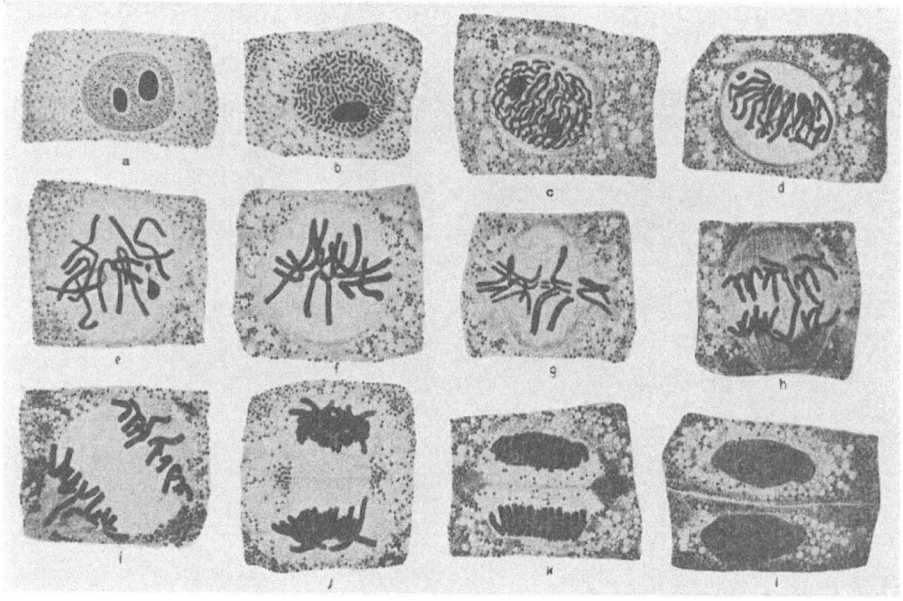

Fig. 78. Course of a somatic cell division (after BĚLǍR, 1928).

mosomes along the equator (*equatorial plate*) which is completed in the *metaphase* (f); there next occurs a longitudinal cleavage of the chromosomes, after which the pieces move apart from one another, each going towards its nearest pole, of which there are two, at opposite ends of the cell (*anaphase*, g-i). The next process is the contraction of the chromosomes and their vacuolization and the completion of both daughter nuclei (*telophase*, j-l).

There is still much difference of opinion and doubt concerning the *mechanism of this nuclear division*. This finds a clear expression in the books mentioned (especially that of SCHRADER, 1953). The recent publications of ÖSTERGREN (1949, 1950) also seem to be of importance.

He considers the whole process of the division of the nucleus from a physico-chemical stand-point.

Both divisions of the nucleus, which are together called *'meiosis'* and which give rise to four *gametes* from one single somatic cell, are absolutely different. Figures 79–81 represent a number of succes-

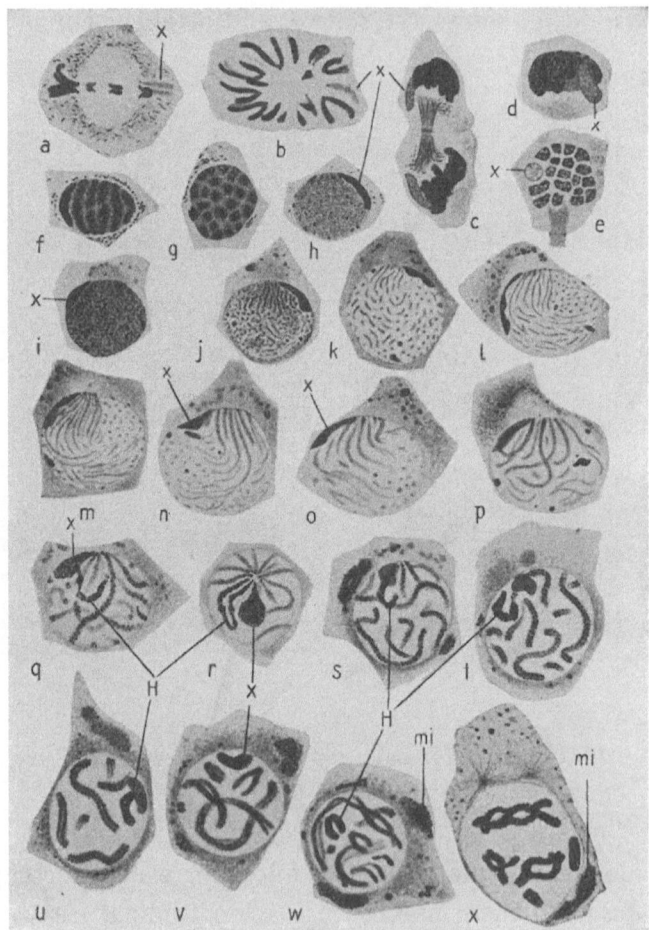

Fig. 79. Formation of gametes in *Stenobothrus lineatus*, 1
(after BĚLĂR, 1928).

sive stages of division which take place in the formation of gametes in the grass-hopper species *Stenobothrus lineatus*.

In the somatic cell of male animals of this species, 17 chromosomes are found, 16 are so-called *autosomes* and one is an *X-chromosome*. We will return later on to the significance of this distinction (chapter

XVIII). This number of chromosomes was exactly ascertained from the metaphase of the karyogenesis of a spermatogonium (figure 79 b) from which four spermatozoa originate later on. The following figures (c–h) represent the successive *interkinesis stages* (resting or metabolic stage) after which, in i. j and k (*leptotene stage*), the chromosomes once

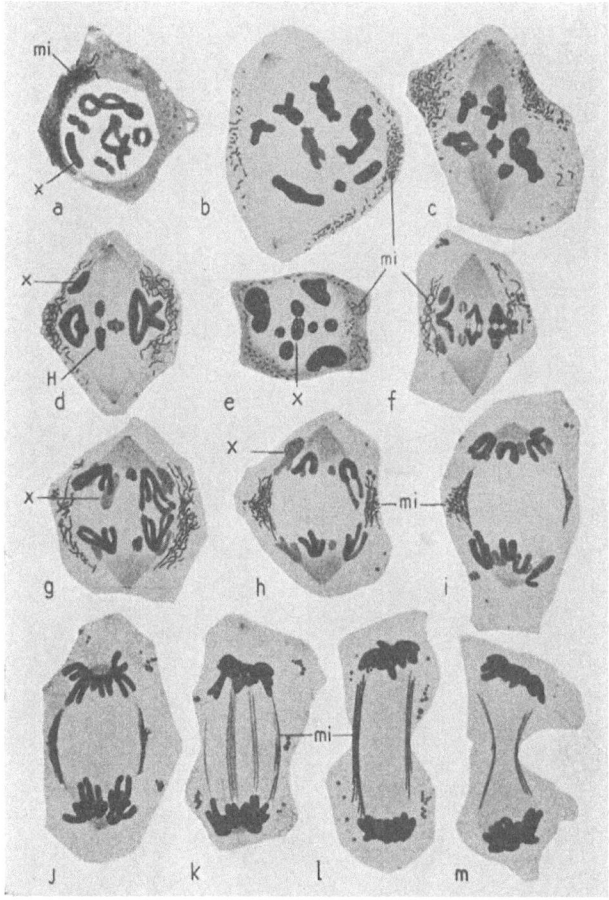

Fig. 80. Formation of gametes in *Stenobothrus lineatus,* 2
(after BĚLǍR, 1928).

again become more easily visible. In l and m the important process of the chromosome-pairing (*conjugation*) occurs: the chromosomes pair up into eight pairs (*gemini*) which are formed from the so-called *homologous autosomes*, whereas the *X-chromosome* remains unpaired. This pairing does not, as a rule, take place synchronously in all the chromosomes so that both paired and unpaired threads are found (*amphitene-stage*, n–o). Next, the pairs of chromosomes become shorter and thicker (*pachytene-stage*, p-t) after which, longitudinal splitting of the chromosomes occurs so that every pair of gemini is composed of four pairs of threads exactly alongside one another;

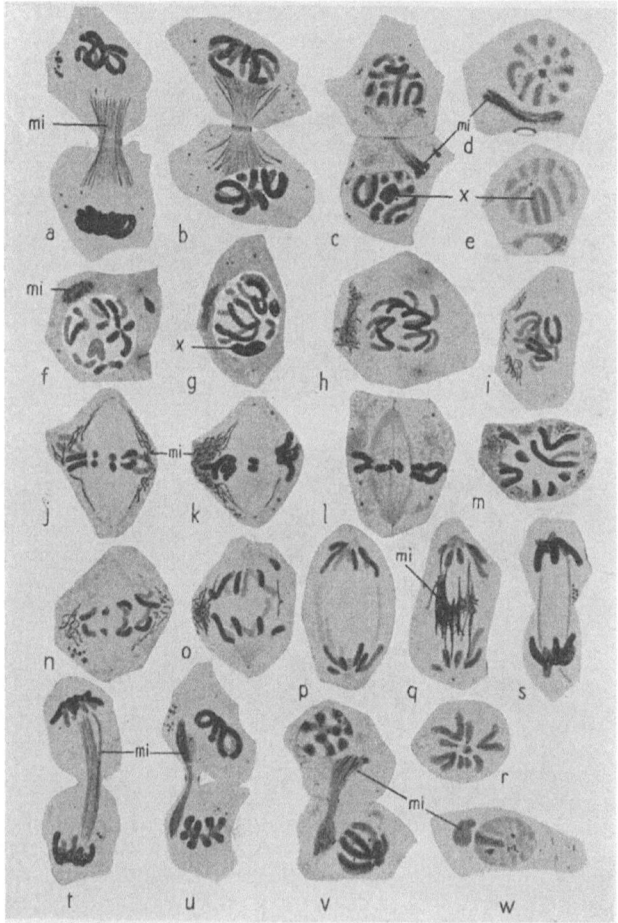

Fig. 81. Formation of gametes in *Stenobothrus lineatus*, ♂
(after Bĕlăr, 1928).

these are called *chromatids* (*chromatid tetrads* can be easily observed in the optic section, figure 79, u). Chromatids wind themselves round the longitudinal axis (*strepsitene-stage*, v.w,) and proceed to the so-called *diakinesis* (figure 79, x; figure 80, a).

After this the tetrads are arranged in the equatorial plane (*meta-*

phase, fig. 80, b–d from the side; e from the side of the pole); in the *anaphase* (f-j) both homologous chromosomes of a pair of gemini begin to pull apart and go to the opposite poles (the chromatids of one chromosome remain together) and form in the *telophase* (k-m), on reaching the poles, the daughter-nuclei; hence, the number of chromosomes is reduced (*reduction division*). The X-chromosome in the meantime has gone to one of the poles so that one daughter-nucleus contains eight chromosomes besides the X-chromosome, the other one contains only the eight autosomes. Next to this the first cell-division (figure 81, a–e) follows and in the mean time the nucleus again passes through a more or less resting-stage (*interkinesis*).

The following stages are similar to those of a somatic division (f–h *prophase*, i–m *metaphase*, n–r *anaphase*, and s–v *telophase*) in which the chromatids of each chromosome separate and become chromosomes (also the X-chromosome passes through a longitudinal splitting) so that finally, four spermatids (later spermatozoa) originate from each spermatogonium (16 *autosomes* + X-*chromosome*): two with 8 *autosomes* and two with 8 *autosomes* + 1 X-*chromosome*.

The so-called '*reduction division*' has the result that from one body-cell, with the double (*somatic* or *diploid*) number of chromosomes, four gametes are formed with the *single* or *haploid* number.

It is evident that this remarkably precise course of the distribution of the chromosomes during the formation of the reproductive cells soon leads to a comparison of the experimental data with the results of karyological investigations, and it was this method which enabled us to throw light upon the question of localization of the factors. The first attempts were made by CORRENS (1902 c), SUTTON (1903) and HUGO DE VRIES (1903) who sought for a relationship between both these methods of investigation. These were soon followed by BOVERI (1904), who gave the prophetic statement:

'If a hybridization is concerned with numerous characters then by continued mating it will be evident, that the number of combinations, in which these characteristics can occur together, is greater than the number of combination-possibilities of the chromosomes present. This will then lead to the conclusion that the characters located in one chromosome distribute themselves over both of the daughter-cells during the reduction division, which indicates that an exchange of parts of homologous chromosomes takes place' (1904, p. 118).

These first attempts of a consideration of chromosomes as staples of the heritable characters, was followed by MORGAN (1910 c) who was, in the beginning, rather sceptical about the possibility of obtaining evidence for it: 'Since the number of chromosomes is relatively small and the characters of the individual are very numerous, it follows on the theory that many characters must be contained in the same chromosome. Consequently many characters must Mendelize together. Do the facts conform to this requisite of the hypothesis? It seems to me that they do not. A few characters, it is true, seem to go together, but their number is small, and it is by no means evident that their combination is due to a common chromosome. It is true that in no one species do we know much concerning the behaviour of many characters, but so far as we know them there is no evidence that they Mendelize in groups commensurate with the number of chromosomes ... But even admitting this possible way of eluding the objection, the other point raised above concerning the absence of groupings of characters in Mendelian inheritance, seems a fatal objection to the chromosome theory, so long as that theory attempts to locate each character in a special chromosome' (1910 c, p. 467–468). And later: 'Our general conclusion is, therefore, that the essential process in the formation of the two kinds of gametes of hybrids in respect to each pair of contrasted characters, is a reaction or response in the cells, and is not due to a material segregation of the two kinds of materials contributed by the germ cells of the two parents' (1910 c, p. 479).

But very soon after this MORGAN and his students (especially BRIDGES, MULLER and STURTEVANT) had data at their disposal which was of the highest importance: this contained all that had become known concerning *linkage* and factors which are linked to the factor responsible for the sex (*sex-linked*). MORGAN used the banana fly, *Drosophila melanogaster* for his material which was a successful object for two reasons: 1) the great number of heritable aberrations which originated in the cultures, and 2) the very small number of chromosomes found in the nuclei (8 in the body-cells and 4 in the gametes).

Moreover, these chromosomes are characteristically built: in the body-cells of the female (figure 82, left) there are two pairs of more or less bent chromosomes, one pair of small spherical ones, and one pair of oblong. In those of the male, the first three pairs are similar to those of the female, but the fourth pair is composed of two unequal chromosomes: one of these is similar to those present in the female and one is longer and rather bent (fig. 82, right).

The first of this pair is usually called the *X-chromosome*, the second-one the *Y-chromosome*. This was established by numerous investigations. In relation to the view growing in that time over the determination of sex (to which we will return in chapter XVIII) it was derived from this chromosome constitution that the factor for sex (with regard to which the male was heterozygous and the female homozygous) should lie in this latter pair of chromosomes. Hence, this pair is called

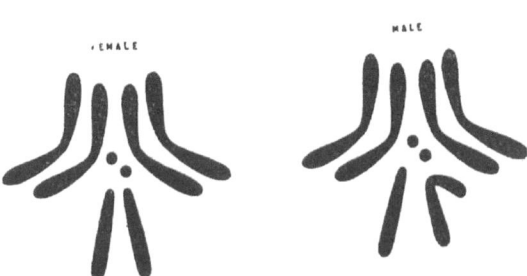

Fig. 82. Schematic representation of the chromosomes in the body cells of *Drosophila melanogaster* (left ♀, right ♂) (after MORGAN, 1916).

'sex-chromosomes', the others are called 'autosomes'. One chromosome out of each pair is present in the gametes. Thus, in all egg-cells, the same set of chromosomes occurs. Fifty percent of the sperm-cells however have the same set of chromosomes, the remaining 50% have a set with one Y-chromosome.

Over and above this, it was obvious that a second fact entered in: all or nearly all the 500 factors studied in *Drosophila melanogaster* each of which causes a definite aberration, belong to four groups, in the sense that the factors of one and the same group are mutually linked to a lower or higher degree. The same tie (*linkage*) between factors of different groups was never observed. Moreover, all the factors which belong to MORGAN's group I were sex-linked that is to say linked to the gene for sex. This group is by far the greatest and included in 1916 nearly 150 of the factors investigated; groups II and III are smaller and were at that time respectively composed of 100 and 80 factors. Group IV is a very small one in which only nine factors are known.

This parallel between the four chromosomes in the gametes and the four groups of mutually linked factors was the occasion for MORGAN and his co-workers, to make a rather daring assumption: group I of the factors is, he believes, wholly situated in the chromosome which we have just now held responsible for the sex; groups II and III lie in both bent and dumb-bell-shaped chromosomes and the small group IV lies in the small spherical chromosome. Completely independent segregation of factors should mean that both factors are situated in different chromosomes; a more or less strong linkage between two factors indicates that these factors are situated in the same chromosome.

Starting from this principle or hypothesis, MORGAN tried to illustrate

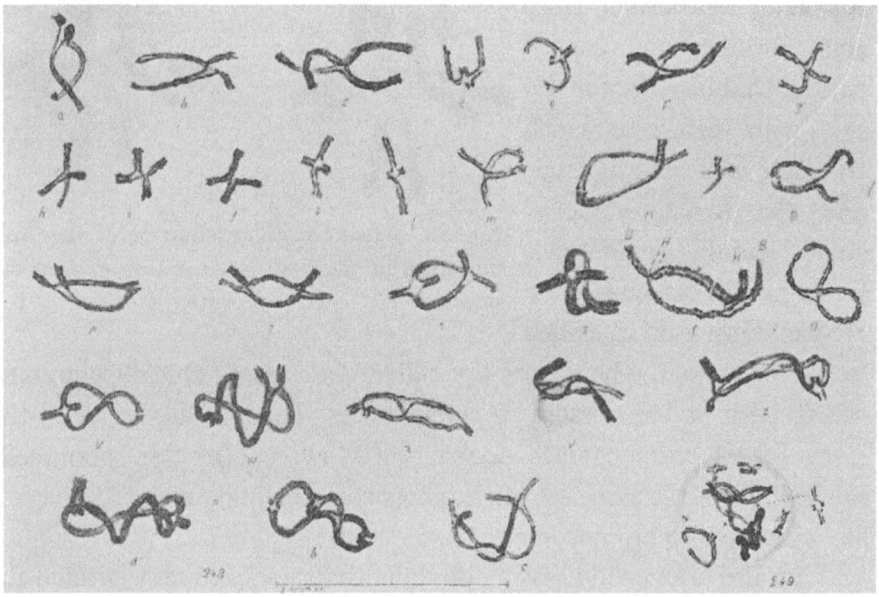

Fig. 83. Coiling of homologous chromatids (chiasmatype)
(after JANSSENS, 1924).

still better the position of the heritable factors (genes) in the chromosome.

For this he made use of the phenomenon of *linkage* between two factors and especially of the different degrees of linkage intensity. In the discussion of the linkage phenomenon we saw that the *crossover-percentage* between different factors differs greatly and that, between complete linkage and completely independent segregation all kinds of transitions are found.

When determining the standard crossover-frequency one must take account of the disturbing influences mentioned in chapter VIII, but by doing so one can calculate a definite crossover-percentage for every two alleles which are mutually linked. This is evident from the table on page 163.

What is the cause of these differences in linkage intensity? To this question also, MORGAN gives us an answer. If we consider with STURTEVANT (1913) that the genes in the chromosomes are linearly arranged, then we can accept that the lesser or greater strength of the linkage between two genes depends on the distance between these two genes.

If we try to find a connection between experimental results and observations on chromosomes, then we may find in the latter the cause of the breakage of a linkage. It does not seem that both homologous chromosomes which pair during the formation of the gametes always remain next to one another, but there is a stage (*strepsitene*) in which from each of the two chromatid pairs one chromatid (originating from one of the homologous chromosomes) coils itself around one chromatid of the homologous chromosome. The other two chromatids remain separate. JANSSENS (1909, 1924; see fig. 83) observed this for the first time and gave the name '*chiasmatype stage*' to this phenomenon; karyological research has finally recognized this as a very common phenomenon.

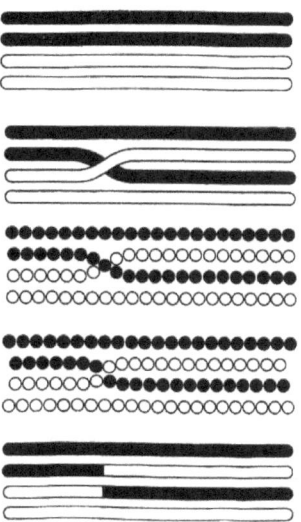

Fig. 84. Scheme of crossing-over of chromatids (modified after MORGAN, 1916).

If it is correct that on the crossing of the coiled chromatids (figure 84 represents this schematically) such a close contact is made that they fuse, then it is evident that finally, a part of the one chromatid sticks to the rest of the second chromatid and that both remaining parts of these chromatids also fuse.

This should result in the origin of two new chromatids each of which is a consequence of an exchange between parts of the old chromatids. This is indicated in the diagram by representing one original chromatid as black and the other as white. In the new chromatids, one is partly black and partly white. In the other, the corresponding parts are just the reverse.

If now a number of dominant factors are located and arranged linearly, and similarly the recessive alleles are located in the white one, then the new chromatids will each contain a part of the recessive

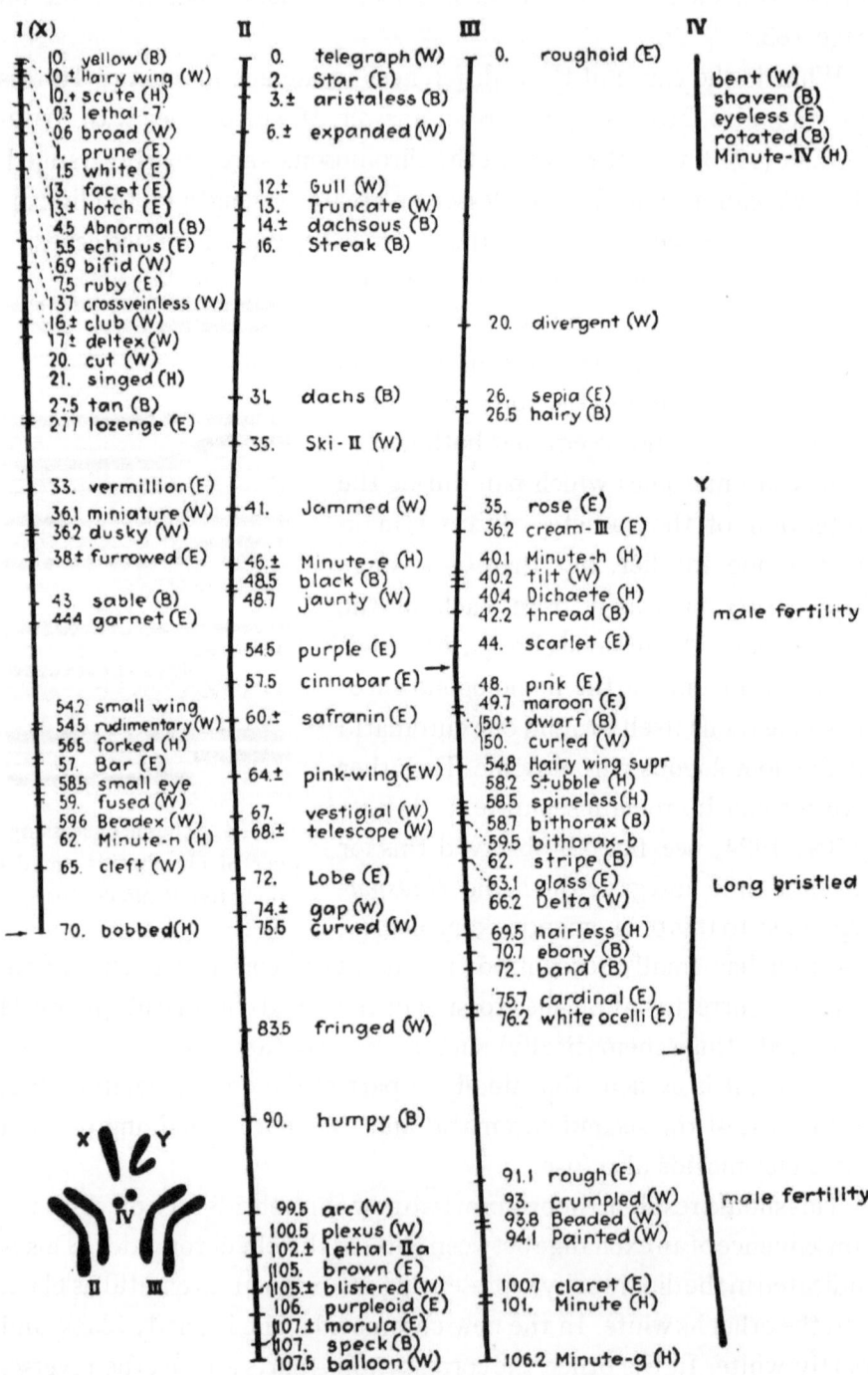

I (X)

0.	yellow (B)
0±	Hairy wing (W)
0.+	scute (H)
0.3	lethal-7
.06	broad (W)
1.	prune (E)
1.5	white (E)
3.	facet (E)
3.±	Notch (E)
4.5	Abnormal (B)
5.5	echinus (E)
6.9	bifid (W)
7.5	ruby (E)
13.7	crossveinless (W)
16.±	club (W)
17±	deltex (W)
20.	cut (W)
21.	singed (H)
27.5	tan (B)
27.7	lozenge (E)
33.	vermillion (E)
36.1	miniature (W)
36.2	dusky (W)
38.±	furrowed (E)
43	sable (B)
44.4	garnet (E)
54.2	small wing
54.5	rudimentary (W)
56.5	forked (H)
57.	Bar (E)
58.5	small eye
59.	fused (W)
59.6	Beadex (W)
62.	Minute-n (H)
65.	cleft (W)
70.	bobbed (H)

II

0.	telegraph (W)
2.	Star (E)
3.±	aristaless (B)
6.±	expanded (W)
12.±	Gull (W)
13.	Truncate (W)
14.±	dachsous (B)
16.	Streak (B)
31.	dachs (B)
35.	Ski-II (W)
41.	Jammed (W)
46.±	Minute-e (H)
48.5	black (B)
48.7	jaunty (W)
54.5	purple (E)
57.5	cinnabar (E)
60.±	safranin (E)
64.±	pink-wing (EW)
67.	vestigial (W)
68.±	telescope (W)
72.	Lobe (E)
74.±	gap (W)
75.5	curved (W)
83.5	fringed (W)
90.	humpy (B)
99.5	arc (W)
100.5	plexus (W)
102.±	lethal-IIa
105.	brown (E)
105.±	blistered (W)
106.	purpleoid (E)
107.±	morula (E)
107.	speck (B)
107.5	balloon (W)

III

0.	roughoid (E)
20.	divergent (W)
26.	sepia (E)
26.5	hairy (B)
35.	rose (E)
36.2	cream-III (E)
40.1	Minute-h (H)
40.2	tilt (W)
40.4	Dichaete (H)
42.2	thread (B)
44.	scarlet (E)
48.	pink (E)
49.7	maroon (E)
50.±	dwarf (B)
50.	curled (W)
54.8	Hairy wing supr
58.2	Stubble (H)
58.5	spineless (H)
58.7	bithorax (B)
59.5	bithorax-b
62.	stripe (B)
63.1	glass (E)
66.2	Delta (W)
69.5	hairless (H)
70.7	ebony (B)
72.	band (B)
75.7	cardinal (E)
76.2	white ocelli (E)
91.1	rough (E)
93.	crumpled (W)
93.8	Beaded (W)
94.1	Painted (W)
100.7	claret (E)
101.	Minute (H)
106.2	Minute-g (H)

IV

bent (W)
shaven (B)
eyeless (E)
rotated (B)
Minute-IV (H)

Y

male fertility

Long bristled

male fertility

Fig. 85. Map of the four chromosomes of *Drosophila melanogaster*, showing the position of the genes (after SHARP 1943, from MORGAN, STURTEVANT and BRIDGES, 1928, and from STERN, 1929a).

factors and a part of the dominant factors so that with this 'chiasma-type' an exchange of factors has taken place. Both pairs of chromatids are now distributed over two daughter-cells (that is, every daughter-cell obtains one pair of chromatids). The original linkage between the factors before exchange of chromatids (the factors were located in the same chromatid) may break during the process of chromatid-distribution over the two daughter-cells because after this process of distri-

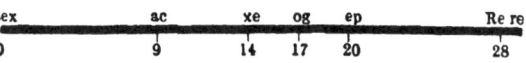

Fig. 86. Preliminary map of the sex-chromosome in man (after HALDANE, 1936).

bution, the factors can be located in different chromatids. It is obvious, that a linkage will break more easily when the factors are situated further from one another along the chromatid length.

Again MORGAN takes a rather daring step; he supposed the percentages in which definite cross-overs occur to be measures of the distances between both genes involved. If the cross-over percentage of the factors W and M for instance amounted to 32.8 and of W and Br to 44.0 he concludes from this that gene W is not so far from M as it is from B.

By taking all the factors in pairs and by equating their cross-over percentages with the distance between these genes, he succeeded in allocating a place in chromosome no. I for all factors which are sex-linked. It is supposed that these factors are located as genes in this chromosome and one is able to draw an exact map for this chromosome with the arrangement of the genes and the places which they occupy (MORGAN called such a place a 'locus'). In this way MORGAN, together with his co-workers, proceeds for the three other pairs of chromosomes and in consequence of this he came to know the cartography of the four chromosomes of Drosophila. In figure 85 the maps of the four chromosomes are represented with a statement of the number of genes involved.

For years now we have been trying such 'chromosome maps' for other organisms (even for man who has 24 chromosome pairs). HALDANE (1936) succeeded in mapping the position in the chromosome of some sex-linked factors (achromatopsy, ac; xeroderma pigmentosum, xe; disease of Oguchi, og; epidermolysis bullosa dystrophica, ep; respectively dominant and recessive retinitis pigmentosa, Re and re). See figure 86. More details are now given by GATES (1954).

A neccessary consequence of this hypothesis should be that from the known distances of two factors a and b, and that of factors a and c, the distance between b and c could be calculated; this distance should be the difference or the sum of both values mentioned. That is, (a–b) — or + (b–c) = (a–c). Hence, in a given case we can check the cross-over percentage of the factors W and M as 32.8 and that of W and Br as 44.0 and in this way, that of M and Br should be 11.2 (i.e., 44.0 — 32.8). In reality this is not the case, and the value for M and Br obtained (20.4) is much too high. Or in other words, W–M (32.8) + M–Br (20.4) should be equal to W–Br (44.0) and in reality it is much higher But through a very ingenious auxiliary hypothesis, MORGAN gets out

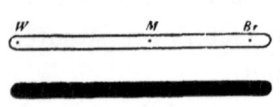

of this difficulty: it is quite possible that both chromatids are coiled around one another, have two fusion points and that differently arranged chromatids originate.

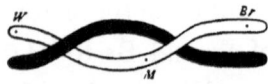

Each of these chromatids consists of the middle part of one of the original chromatids together with the ends of the other original chromatid.

Through this process of 'double crossing-over' it is possible that two genes in the same chromosome will finally remain together in spite of the fact that the distance between them is rather great. Figure 87 shows this

Fig. 87. Scheme of double crossing-over (after MORGAN, 1916).

schematically: the genes mentioned i.e., W, M and Br are indicated with regard to their localization. And so, it could be explained why the cross-over percentage W–Br was found to be too small; because, if such a double cross-over takes place, then the linkage between W and M as well as between M and Br will be discontinued but not the linkage between W and Br.

The sum of the distances between genes a and b and between b and c can thus be greater than the distances calculated for a and c from the cross-over percentage and in many cases a higher sum is obtained.

It is a pity that the term 'locus' (as defined by MORGAN) to mean 'mapping' location of a definite factor (character) was not used by all scientists in the strict sense of its definition. There is nowadays a tendency in American literature to define a locus as the location of factors (no matter how many) in the chromosome which are responsible for a definite phenotypical characteristic. By means of this a locus acquires broader significance. The phenomenon of *pseudoalleles* (an apparent *multiple allelic* series) can be shown to be fractionable by crossing-over, whereas the adjacent loci interact in position effects, suggesting a close relationship in developmental action of the adjacent genes (SRB and OWEN, 1953) and also of *step-allelism* (see chapter XIV) can often be due to a group of very closely linked factors which are as a group often called a locus.

In *Neurospora crassa* an absence of the enzyme tryptophandesmolase is believed to be due to several strongly linked genes (BONNER, 1951; WEIJER, 1954; YANOFSKY, 1952). Since the oneness of the gene became a matter of discussion the concept of the locus as the location of a gene in the chromosome changed into the location of genetical units (McCLINTOCK, 1951), which subdivide the gene and which are separable by chromosomal breakage or crossing-over.

This brings us into a position to define what we will understand by the following:

1st. A *factor* or *character*: A factor or character is the Mendelian concept of a heritable unit, a product of interaction between gene and environment which shows this interaction in the phenotype of the organism.

2nd. A *gene*: A gene is that part of the actual chromosome presumed to be responsible for the expression of a definite phenotype by interaction of the environment and not subdivisible by chromosomal breakage or crossing-over; a gene is presumed to be a unit of crossing-over, a unit of mutation and a unit of biochemical reaction.

3rd. A *locus*: A locus is the spot on the chromosome map where a Mendelian factor or character is presumed to be located as a gene.

The difference between *factor* and *gene* can be stated as follows: a gene is presumed to be the chromosomal unit partly responsible for a heritable phenotype — a factor is responsible for a definite phenotype under the influence of a certain environment.

Modern geneticists like to speak over 'a gene for Red' or 'a gene for shrunken endosperm', but we must not forget that such an equation of gene-factor is still one of the desires of modern genetics. There is evidence to show that such an identification between gene and factor is really the case (we will refer to this later in the book) but nevertheless it is necessary for a generally intelligible terminology to stick to the original definition of the gene-factor concept, since final proof of such an identification does not yet exist for most of the cases investigated. There are however situations in which it is difficult to make a distinction between gene action and factor action.

In the field of gene-factor equations, the boundary between these two concepts was precisely defined in early days of genetics. However, since then investigators have come to realize that this boundary must be considered as only a vague one as is the case in many fields of science when progress is made in the explanation of phenomenon through searching for a mutual causal basis.

We will not give further details here of MORGAN's localization theory; for this we can refer to summaries such as those of MORGAN, BRIDGES and STURTEVANT (1925) and of STERN (1928). However, it should be pointed out here, that the whole study represents a brilliant attempt to find the solution to a very important problem of inheritance. The whole study is composed of three hypotheses and we are permitted to consider these as having been proved.

Firstly, the equality existing between the number of chromosomes in the nuclei of the gametes, and the number of groups of linked factors; secondly, the supposition that the genes in the chromosomes are arranged linearly and the hypothesis based upon this, that the distance between the different genes is connected with the frequency of exchanges observed between them, and thirdly the existence of a chiasmatype stage in the sense of a real fusion of parts of homologous chromatids as a basis for the explanation of the exchange of genes.

The first hypothesis is not founded on investigations with *Drosophila* only; in all plant and animal studies an absolute justification of this equality was shown. We found that maize was also a good object (see RHOADES and MCCLINTOCK, 1935; LONGLEY, 1941).

The second hypothesis is again composed of two parts: the supposition of the linear arrangement of the genes in the chromosomes and the equality of the frequency of the gene exchanges and the distance between the genes in the chromosome. It is clear that the latter part is based upon the first part. It has taken many years for this view of the linear arrangement to become generally accepted. It was first necessary for STERN (1926, 1927) to obtain more certainty than MORGAN could give us. STERN with his comparative karyological-genetic experiments could prove, that the so-called translocation (see p. 280) of a part of the Y-chromosome (in which some definite gene was supposed to lie) to the X-chromosome, did indeed cause a linkage between these Y-factors and the X-factors in the progeny. By this it became certain that the Y factor in question was situated at the place (locus) on the Y-chromosome derived by means of calculation. But still this is not a final proof; for this, a detailed research into the structure of the chromosome was necessary.

In the infancy of cytological research PFITZNER (1882) described how, within the chromosomes, during a definite stage (especially between *leptotene* and *pachytene*), large or smaller thickenings occurred along their threadlike structure, like knots in a cord, and how these thickenings were submitted to the splitting process of the chromosomes into two chromatids.

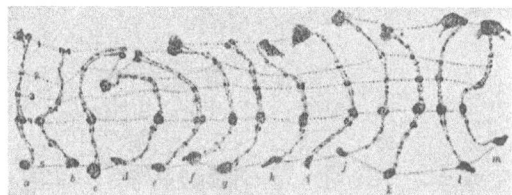

These observations were also stated by VAN BENEDEN (1883) soon after, who accentuated the complete similarity in structure of both chromatids which originated from one chromosome. In this stage

Fig. 88. A pair of chromatids taken from the pachytene stage of 13 different individuals of *Phrynotettix magnus*, a grasshopper. (The most striking chromomeres are joined by dotted lines) (after WENRICH, 1916).

these thickenings seem to be a regular phenomena and because of this WILSON (1896) dubbed them '*chromomeres*'. They were so regular that the chromatids of corresponding chromosomes which could be recognized in different cells of the same organism or of its congeners, show accordance in arrangement of the chromomeres (figure 88).

This structure of the *chromomeres* attracted more and more attention and it was asked whether a possible connection existed between the linear arrangement of the genes in the chromosome (as supposed by MORGAN) and the observable linear structure of the chromosome itself during the reduction division. Numerous coy attempts were undertaken to obtain some evidence in this direction until BELLING (1928, p. 316) stated rather daringly that: 'chromomeres are genes, doubtless with more or less of an envelope'.

But this statement remained a hypothesis: the chromosomes are

16

only visible for a short time in the process of reduction division and because of this even the most refined karyological section cannot give cogency. Nevertheless, assistance came from an absolutely unexpected source.

Old observations of BALBIANI (1881), later repeated by VAN HERWERDEN (1910, 1911) a.o., had shown that in the group of insects to which *Drosophila* belongs, different species have extremely large chromosomes in the nuclei of their *salivary gland-cells* (i.e., *salivary chromosomes*) and that these chromosomes are composed of thin slices which are easily visible microscopically.

These observations were not used for a long time and only were considered as data, until attention was turned to these results in 1933 from two sides (HEITZ and BAUER, 1933; PAINTER, 1933) and their significance in the testing of MORGAN's localization theory was realized (see HEITZ, 1935; METZ and LAWRENCE, 1937).

Through their work it was ascertained that this structure in the cell-nuclei of the salivary gland did indeed represent a set of gigantic chromosomes. The size ratio between normal chromosomes in the body cells and these 'salivary chromosomes' is from 1 : 50 to 1 : 100, therefore, they are very suitable for microscopical research.

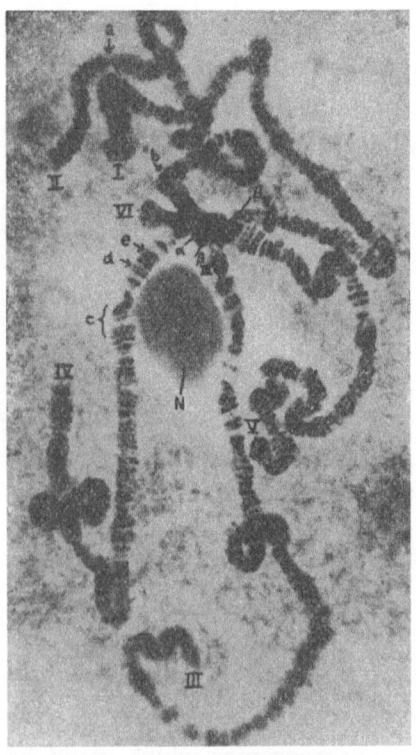

Fig. 89. The six chromosomes in a salivary gland of *Drosophila virilis* (after HEITZ, 1935).

In a species such as *Drosophila virilis*, with a haploid number of six chromosomes (figure 89), these are all easily visible. These six chromosomes indicated by Roman numerals are oriented like spokes of a wheel. This number is constant and always in agreement with the haploid number which is typical of the species; also, the differentiation into thin slides is characteristic for each chromosome. In different cells, the same arrangement of thick and thin, broad and narrow slides are found in corresponding chromosomes. According to the very probable view of BRIDGES (1935) the large salivary gland chromosomes originate because both homologous chromosomes arrange themselves alongside one another and then undergo a series of longitudinal divisions. But now the splitting parts do not separate; they remain alongside one another and, after a number of successive divisions, each chromosome is composed of a bundle of numerous parts parallel to one another. The consequence of this is, that the corresponding chromomeres of these divided chromosomes are located on the same place and together they form, according to their size, broken or closed rings, or slides of different sizes (figure 90). The precise structure of chromosomes can be readily observed in these special nuclei and the study of them opens several possibilities of importance for testing MORGAN's localization theory. The quintessence of this investigation is that the behaviour of genes (which

Fig. 90. Part of a salivary gland chromosome of *Chironomus Thummii* (after BAUER, 1935).

can be supposed, on grounds of linkage studies) sometimes deviates from calculations, and that in this case, corresponding irregularities in the structure of the chromosomes could be ascertained by means of microscopical research of the salivary gland chromosomes.

We will discuss these irregularities in the chromosome configuration in the next chapter but the main point can be established here: that is, comparison of a chromosome map which was drawn on the grounds of crossing-over data, with the map derived from microscopical investigation, has in many cases led to a complete agreement (fig. 91). This agreement was at first accepted with reserve but later on accepted unconditionally as more evidence became available.

From the comparison it was apparent that the idea that the distances which could be determined by the cross-over-frequency could not be considered as absolutely exact. This was the opinion of STURTEVANT (1913) and many other investigators. They took 1% cross-over as the distance unit but even the most convinced followers of the school of MORGAN went back on this and now consider the cross-over percentages

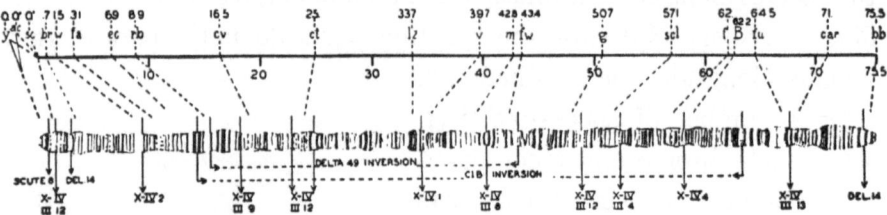

Fig. 91. A comparison between a genetical and cytological chromosome map (after PAINTER, 1934a).

as a 'function' of the distance, or as STERN (1931, p. 554) puts it even: 'that the gene maps do indeed represent the exact arrangement but not the exact distance between the genes'. This also found its explanation: a chromosome is composed of parts of different composition; so-called *euchromatic parts* which comprise active genes, and *hetero-chromatic* parts in which probably non- or nearly non-active genes are located. The difference between these two views will be discussed later; originally this difference presented a difficulty in acceptance of the localization theory but this was later ruled out. The final proof of the localization theory is now almost complete and we are indebted for this to the study of H. J. MULLER (1929). It was possible to destroy a small part of the chromosome by X-rays after which the change in the structure of the salivary gland chromosome could be observed, whereas the genes which should have been present in this chromosome part were proved to be missing by culture experiments.

The third supposition, which comprises MORGAN's localization theory, is that of exchange of factors during the chiasmatype stage.

For a long time there was a fight over this but finally nearly all had to confess that the original observations made by JANSSENS generally occur, and their theoretical interpretation, as was pointed out above, is indeed justified. We will come back to this later on in chapter XIII.

If we may consider this question of localization of the genes as solved, then there still remains another question as to how far these genes are responsible for the differences which are characteristic for the species, genus and families. All factors searched till now by genetical means have, with a single exception, reference to detailed characteristics. This is even the case when we speak of *'developmental factors'* (see chapter XXI) which deeply affect the normal formation of the organism. As well as these detailed characteristics, there exists a complex of characteristics by which one species is distinguished from another. The chapter dealing with 'Constant hybrids and interspecific hybrids' will show, that in this sphere of research only a few positive results have yet been obtained. Contrary to the optimistic enthusiasm of some investigators (MULLER, 1929), who believed that nearly all systematical differences, even those between the highest and the lowest organized organisms, can be based on the genes located in the chromosomes, there are other views, according to which the plasm and its components and composition should be of great significance. Owing to an almost complete lack of precise data it is not possible to draw a conclusion about this, undoubtedly very important question.

CHAPTER XII

The recognition of the significance of the chromosomes as 'house rooms' of the genes led to the establishment of co-operation between genetics and karyology and also to a strong development of cell-nucleus research. In former days cytology was a more or less separate part of histology and was left to its own devices; in the last 30 years this sphere of research came under the influence of genetics and by that was more and more concerned with the study of questions in other fields of science, such as taxonomy.

All individuals of a normal species have in their somatic cell nuclei, a definite, usually even, number of chromosomes which form pairs of homologous chromosomes which either resemble one another or are sometimes morphologically distinguishable. In nature we find a variety of chromosomes as great as the variety of nature itself. This is clearly shown in figure 92. The cells originating from the reduction division comprise one half of this somatic or diploid number, that is to say, the gametes obtain one chromosome from each homologous pair and therefore, they are called 'haploid'. Even in some groups of bacteria, structures resembling chromosomes are found (DUBOS, 1946; KNAYSI, 1946; DELAPORTE, 1950).

It can be seen in the tables of chromosome numbers that for each species (or in any case, for each variety) there is a characteristic fixed number of chromosomes. These tables are given by MARCHAL (1920), GAISER (1926–1933), TISCHLER (1927–1938, 1950) and DARLINGTON and JANAKI AMMAL (1945) who have compiled them for plants, whereas HARVEY (1916–1920), BRESZLAU and HARNISCH (1927), OGUMA and KAKINO (1932), McCLUNG (1940), MATTHEY (1949) and MAKINO (1950) did the same for animals. This constancy in the somatic number (which was considered absolute for a long time) is no longer so certain;

more and more data was obtained which indicated that in 'polyploid' organisms (see p. 249 onwards) these numbers can differ in different

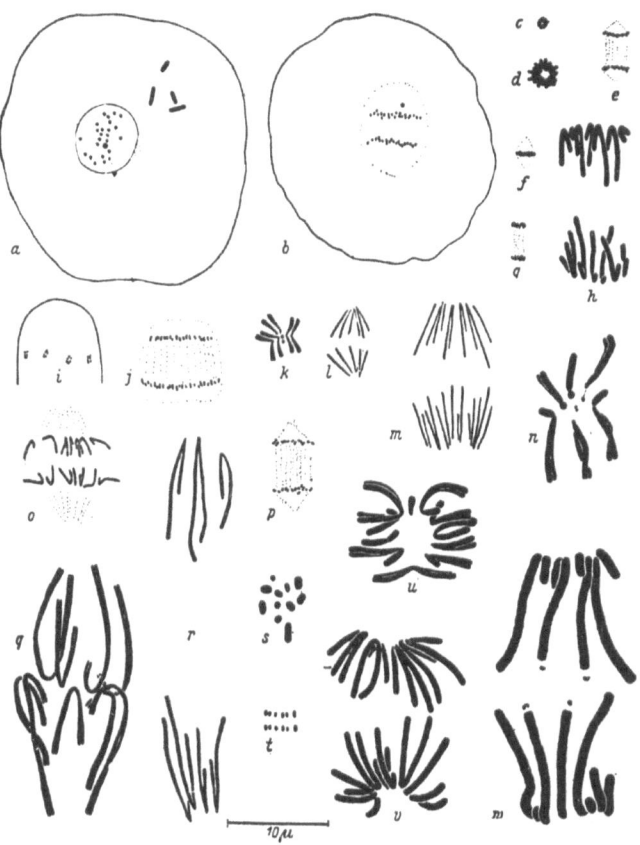

Fig. 92. Chromosome sets of: a, b. *Hartmannella Klitzkei* (amoeba); c, d. *Cocconeis placentula* (diatom); e. *Vaucheria* (green alga); f. *Chrysomonad*; g. *Sphaeroplea annulina* (green alga); h. *Pitophora kewensis* (green alga); i. *Amanita muscaria* (fungus); j. *Sparmannia africana* (angiosperm); k, l. *Drosophila melanogaster* (fly); m. *Salmo salvelinus* (fish); n. *Crepis capillaris* (angiosperm); o. *Sauromatum guttatum* (angiosperm); p. *Pygaera pigra* (butterfly); q, r. *Oedogonium* (green alga); s. *Gerris lateralis* (whirling beetle); t. *Macrotylus quadripunctatus* (flower-bug); u. *Gomphocerus rufus* (orthoptera); v. *Amblystoma tigrinum* (amphibian); w. *Aloe strigata* (angiosperm) (after GEITLER, 1938).

parts of the body and in different tissues (HÅKANSSON, 1950; KOOPMANS, 1951; VAARAMA, 1949).

It is true that this constancy of the haploid number is no longer valid for all varieties which belong to one taxonomical species.

BLACKBURN (1928) ascertained, for instance, the existence of three varieties within the 'species' *Silene ciliata* which cannot be distinguished morphologically. However, remarkable differences exist in their haploid chromosome numbers (12, 24 and 96). The point of view can still be upheld that we should consider these three 'varieties' as separate 'species'.

Moreover the chromosomes which form the chromosome set are often so different from one another that they are recognizable and a 'chro-

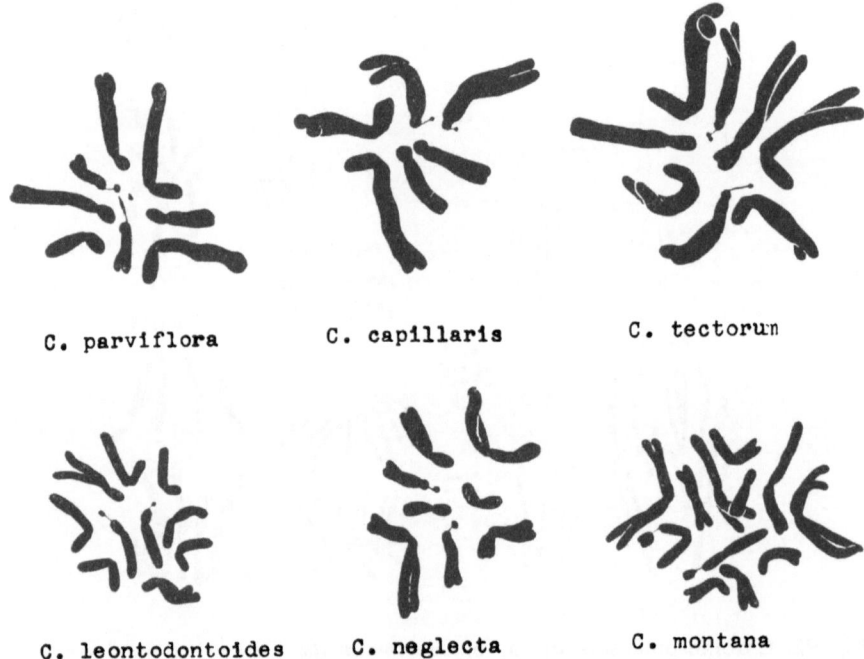

C. parviflora   C. capillaris   C. tectorum

C. leontodontoides   C. neglecta   C. montana

Fig. 93. Chromosome portraits of six species of *Crepis* (*C. capillaris* 2n = 6, *C. parviflora* 2n = 8, *C. tectorum* 2n = 8, *C. neglecta* 2n = 8, *C. leontodontoides* 2n = 10, *C. montana* 2n = 12) (after BABCOCK and NAVASHIN, 1930).

mosome picture' can be drawn which is typical for each species; figure 93 shows this for six species of the genus *Crepis*.

All species from one and the same genus show a remarkable regularity in their chromosome pictures (chromosome configuration).

In 1917 WINGE recognized in principle the significance of this regularity, and since then its general validity has been proven by several investigators, so that there is now no doubt about it.

We can refer for some examples of this phenomena to the genus *Rosa* (TÄCKHOLM, 1922; somatic numbers 14, 21, 28, 35, 42 and 56; see figure 94), *Chrysanthemum* (TAHARA, 1921 and SHIMOTOMAI, 1938: 18, 36, 54, 72 and 90), *Senecio* (AFZELIUS, 1924: 10, 20, 40, 50, 60 and 180), *Papaver* (LJUNGDAHL, 1922: 14, 28, 42, 70, but also 22 and 44) and *Nymphaea* (LANGLET and SÖDERBERG, 1927: 28, 56, 84, 112 and 224). We can calculate a basic number of chromosomes for the following: *Chrysanthemum* 9, *Nymphaea* 14, *Papaver* 7 and 11, *Rosa* 7 and for *Senecio* 5. This basic number (which is more or less typical for the whole genus) is called by LANGLET (1927a) the '*monoploid chromosome number*' and indicated by the figure '*p*'. The variety with the lowest number is somatic diploid (double); the higher numbers are respective triploid, tetraploid, pentaploid, hexaploid etc. (generally called polyploid). It seems that there is a great regularity concerning the chromosome number which is present in the somatic cells

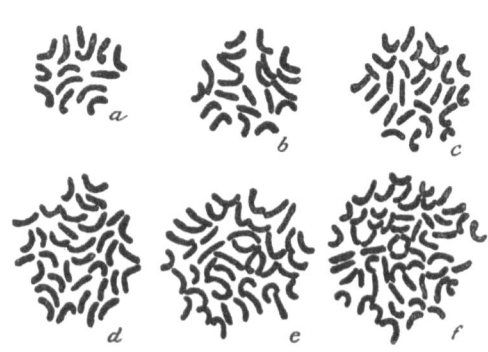

Fig. 94. Chromosome portraits of six species of *Rosa*. a. *R. webbiana* 2n = 14, b. *R. chinensis* 2n = 21, c. Rose 'Konrad Ferdinand Meyer' 2n = 28, d. *R. tomentosa cuspidatoides* 2n = 35, e. *R. nutkana* 2n = 42, f. octoploid hybrid 2n = 56 (after TÄCKHOLM, 1922).

and this number is generally an even one. Some exceptions are found in the above: among the *Rosa*-species there occur some which have uneven somatic numbers (the triploid and pentaploid numbers 21 and 35) whereas the genus *Papaver* seems to have two basic numbers (7 and 11). Notwithstanding this, the whole creates the impression that corresponding numbers are found in related species. Therefore, it seems possible to base a hypothesis concerning the taxonomical relationship of these species on the basis of these chromosome numbers: SENN (1938) attempted this for the whole family of the Leguminosae. In 1951 LÖVE gave a general discussion of these problems.

In animals the occurence of polyploid 'races' is rather rare (see DARLINGTON, 1953; MATTHEY, 1953; SACHS, 1952; VANDEL, 1937; WHITE, 1954); only in few groups were cases of polyploidy observed and then practically only, if, apart from a normal bisexual reproduction, a parthenogenetic reproduction also took place (SEILER, 1942; SUOMALAINEN, 1950).

This regularity in the number of chromosomes in related species that is to say, species in which this number is an exact multiple of the basic number, is called *euploidy* by TÄCKHOLM (1922), and has attracted the attention of scientists as to the question whether, and how, a true bred species can give rise to a progeny in which the whole chromosome complex in the gametes (called *'genome'* by WINKLER (1920, p. 165)) is doubled under the influence of one or another cause. Such a duplication process was observed in the vegetative as well as in the generative cells and was in many cases caused experimentally.

One of the first investigations to obtain an artificial accumulation of chromosomes was carried out by GERASSIMOW (1902, 1904) who studied algae (*Spirogyra* and *Zygnema*) and avoided the separation of the chromatin-mass of a nucleus previously divided by means of refrigeration. Despite this the dividing cell wall was formed in the normal place. The final result was that cells occurred next to one another which had no nuclei, and cells which contained either double size nuclei or two nuclei of normal size. From such cells with double nuclei he could cultivate cell-threads which were composed of polychromatic cells only. Chemicals such as ether and chloroform caused such phenomena as these. GERASSIMOW limited his studies to morphological observations concerning the correlation between cell size and chromotine-mass, vigour and division capacity of cells etc.

Since that time GERASSIMOW's work has remained in the background until the same method was used by VAN WISSELINGH (summary in 1919) with algae and by F. WETTSTEIN (1923–1924; 1924–1928; 1925) with mosses. To both methods (originating from GERASSIMOW, refrigeration and chemicals), centrifuging was added by VAN WISSELINGH.

We are especially indebted to WETTSTEIN for the beautiful investigations in this field of research. The vegetative divisions of the protonema of musci seem to be very susceptible to narcotic chemicals as well as to refrigeration, and by means of this WETTSTEIN succeeded in influencing the mitosis so that he could cultivate diploid cells from a haploid protonema and from this again, wholly diploid protonemata. The breeding of tetraploids from these diploids was a step further but a limit was reached in this direction.

Another experimental method by which plants can be obtained with double chromosome number in the vegetative tissues, is that of regeneration. The well-known investigations of the brothers MARCHAL (1907–1911) with the musci genus *Amblystegium* had shown that parts of the sporogonium cultivated on differing media, divided after some days and gave rise to a protonema which is diploid (like the sporogonium) and develops into a moss plant which is thus also diploid and gives rise to diploid gametes. Fertilization of these diploid gametes is possible, under certain circumstances, not only between diploid gametes but also between a diploid and a haploid gamete. In the first case there originate tetraploid sporogonia, in the second case, triploid sporogonia. Since corresponding polyploid protonemata can originate from these polyploid sporogonia by regeneration, WETTSTEIN succeeded in obtaining a consecutive series of haploid, diploid, triploid and tetraploid protonemata of one and the same species.

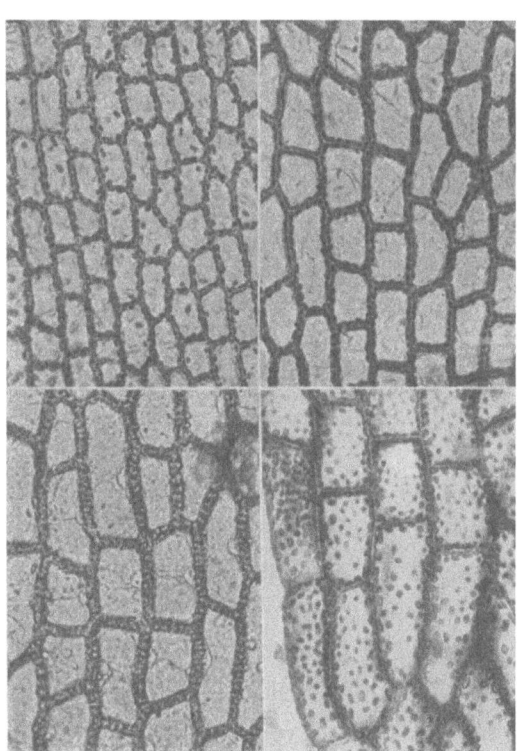

These plants thus have respectively one times, two times, three times and four times the same set of chromosomes; in WINKLER's nomenclature they are thus

Fig. 95. Parenchym cells of haploid, diploid, triploid and tetraploid moss plants (after WETTSTEIN, 1927).

monogenomic, digenomic, trigenomic and *tetragenomic*. A limit seems to be reached with these tetraploid or tetragenomic protonemata because protonemata originating from these moss plants are not able to give rise to viable gametes. These forms are very important from a morphological point of view because the four types show great differences in their organ sizes. This is represented in figures 95 and 96 in which drawings are given of parenchymatic cells and of leaf-sections in which the differences in size are evident.

Closely related to this is the following example borrowed from higher plants. In 1895 DE VRIES found the variety *O.L. gigas* (see DE VRIES, 1901, p. 157) in his cultures of *Oenothera Lamarckiana*. GATES (1909) established the somatic chromosome number *O.L. gigas* to be 28, whereas *Oenothera Lamarckiana* itself has 14. Since this discovery many *Oenothera* species of the gigas-forms were found (see

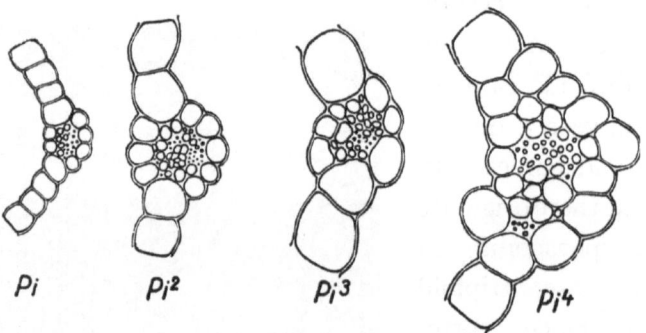

Fig. 96. Sections of leaf-ribs of haploid, diploid, triploid and tetraploid moss plants (after WETTSTEIN, 1928).

GATES, 1928). The investigations of WINKLER (1907–1916) followed: a duplication of the chromosome number was obtained in the callus tissue which developed vegetatively in grafts of two different *Solanum* species upon one another. Tetraploid plants originated from *Solanum lycopersicum* (tomato) and from *Solanum nigrum* (black nightshade). The finer differentiation of tissue, which characterizes the Phanerogamae compared to Musci, makes it sometimes more difficult to trace which tissue in the *Solanum* plants has given rise to these tetraploid forms. In each plant (especially in the cambium) there are some cells which already have the tetraploid chromosome number (see LESLEY, 1925) and one can imagine that the tetraploid plants arose from these cells by means of regeneration. But, if this is the case the question as to how these cells originated, still remains. WINKLER considered it obvious that on grafting, that is to say, on injuring these cells, a vegetative fusion of two cells should take place. It does seem that WINKLER succeeded (1938) in growing plants from cells by the application of this method. These cells bear chromosomes originating from the tomato as well as from the black nightshade by means of incomplete fusion of the nuclei; he calls such a *'vegetative hybrid'* a *'burdo'*.

The occurence of tetraploid (tetragenomic) cells in somatic tissue was observed in several plants; DE LITARDIÈRE (1923, 1925) studied this phenomenon in *Spinacia oleracea* and in *Cannabis sativa*; LANGLET (1927a and b) also tested the same plants and two *Th..lictrum*-species. According to their results, the origin of such tetraploid cells can be the result of two processes: either, fusion of two nuclei (this occurs in *Solanum* and *Thalictrum*-species) or by means of a double longitudinal division of the chromosomes during the prophase (*Spinacia, Cannabis*). Since that time numerous other cases of polyploid body-cells were observed not only in plants but also in animals.

How far this phenomenon is comparable with that of the salivary gland cells of *Drosophila* is not certain. GEITLER (1953) has given a summary of these phenomena.

Research into the effect of chemicals fell into the background as did also the result of the discovery of regeneration, however, recently study into chemical influences upon the cell was again taken up, especially as a result of the discovery that colchicine (the alkaloid of the meadow saffron, *Colchicum autumnale*) has a great influence upon the division of the nucleus. The development of this colchicine method is very interesting: originating from medical science (where it is applied in cancer research) it attracted botanists (e.g. O. J. EIGSTI and D. F. JONES) after the discovery was made that colchicine had a stimulating effect as well as an interrupting effect upon the nuclear division. This became evident from the work of DUSTIN (1936) and his co-workers.

As a result of these investigations, the first papers dealing with this strange effect of colchicine were soon published (BLAKESLEE, 1937; BLAKESLEE and AVERY, 1937; NEBEL and RUTTLE, 1938), and were followed by almost an endless series of studies in this sphere: a summary of five years colchicine research was given in 1942 by KRYTHE and WELLENSIEK. The list of literature comprised about 400 titles. And since that year the number of colchicine studies has grown enormously; EIGSTI and DUSTIN Jr. are continuing this bibliography (1947, 1949).

The effect of colchicine upon the division of the nucleus is of primary importance for cytology because the mechanism of the nuclear division is disturbed and this phenomena brings all sorts of interesting questions to light (see LEVAN and OESTERGREN, 1943). From a genetical point of view this phenomenon also has significance because the process of the normal somatic division of the nucleus is stopped at the metaphase stage. This process was called 'Stathmokinesis' by DUSTIN (1938) that is to say, the division of the chromosomes takes place normally but the division products remain together with the result

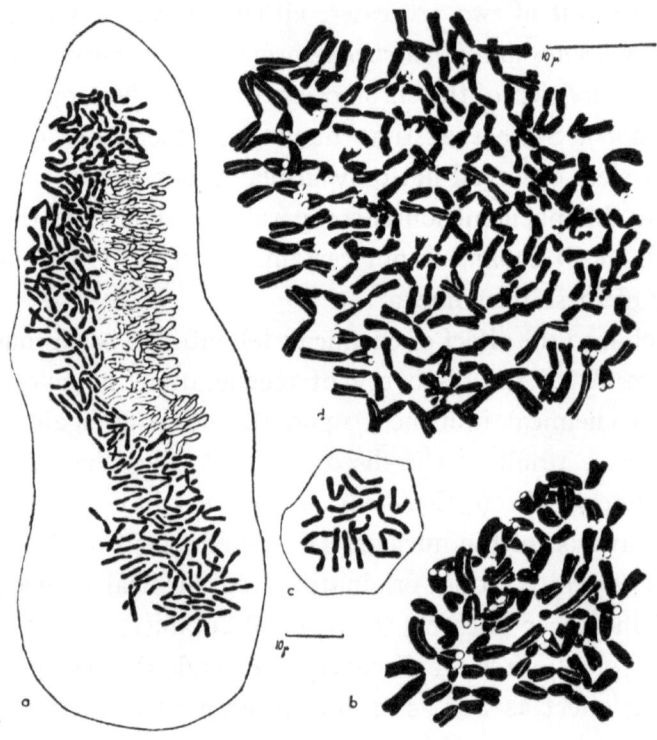

Fig. 97. Nuclei of *Allium Cepa* which are treated with colchicine; a) nucleus with 500 chromosomes; b) part in detail; c) normal diploid nucleus with 16 chromosomes; d) nucleus with approximately 128 chromosomes (after LEVAN, 1938).

that from a diploid body-cell a tetraploid one will originate which can be the origin of wholly tetraploid tissue.

Sometimes the division of the chromosomes can repeat without intermediate division of the nucleus. In this way, from one single cell of *Allium cepa* with 16 chromosomes (figure 97c) a cell can originate with 128 chromosomes (figure 97d) or even one with about 500 chromosomes (figure 97a).

The generative cells are also sensitive under given circumstances. The formation of generative cells with deviating chromosome numbers can be caused by treatment with chemicals, or by cultivation at low temperatures and so on. If the colchicine treatment is applied during reduction division then four normal haploid gametes do not originate from one single diploid body-cell, but two gametes with the diploid number of chromosomes or even with four times or a tetraploid number of chromosomes. It is of great significance that by application of these methods viable gametes can originate (which are capable of fertilization) with somatic chromosome numbers which thus give rise to polyploid descendants by fertilization (see summary HEILBORN, 1930).

In all the cases mentioned, the chromosome configuration of one single species is doubled or multiplied. Therefore, the offspring can be called *autopolyploids*. This name was introduced by KIHARA and ONO (1926), and is usually nowadays abbreviated to *autoploids*. In all probability, this autoploidy

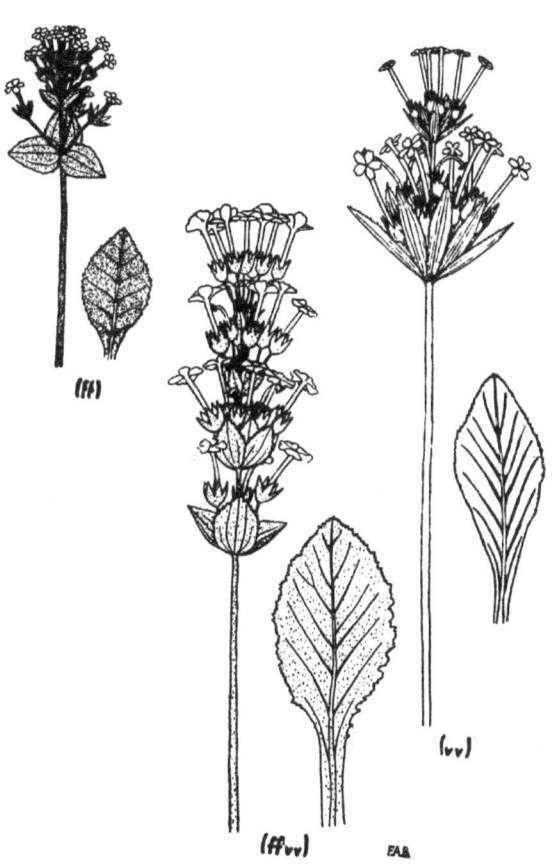

Fig. 98. Left, *Primula floribunda* (ff); middle, *P. kewensis* (ffvv); right, *P. verticillata* (vv) (after ANDERSON, 1936).

plays a substantial part in the process of evolution: MÜNTZING's monographic study (1935) has compiled much evidence for this point of view.

Finally, we must refer to the interspecific crosses or species hybrids as an important source of accumulation of chromosomes, especially to those interspecific crosses in which both parents are characterized by the same chromosome number. A first contribution concerning the remarkable form *Primula kewensis* was made by DIGBY (1912) and later elaborated by NEWTON and PELLEW (1929). This species occurred in 1900 in Kew among plants of *P. floribunda* and was for good reasons considered as a hybrid between *P. floribunda* as mother and *P. verticillata* as father (see figure 98).

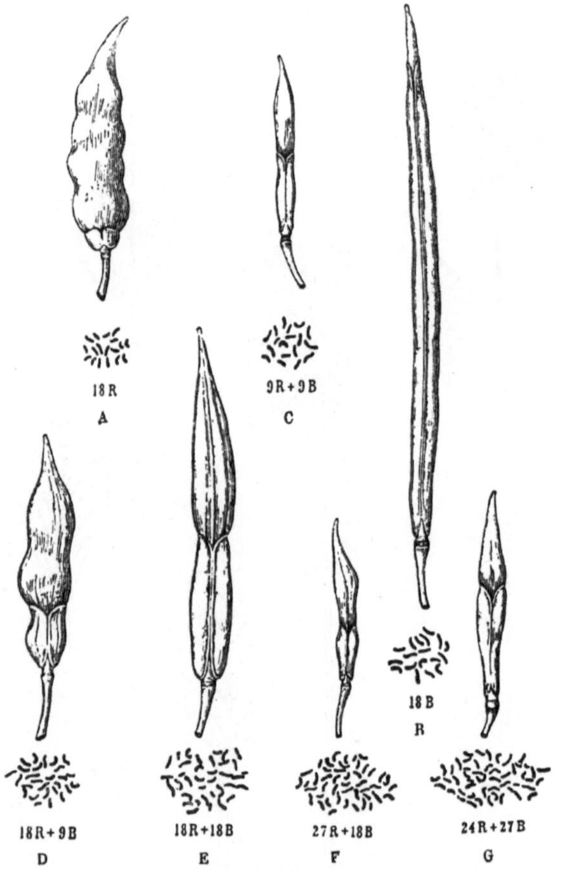

Both parents had n = 9 chromosomes and the hybrid 2n = 18; for some years the plant behaved as if completely sterile. Then in 1905 one of the inflorescences turned out to be fertile after self-pollination and yielded a number of almost identical descendants which formed the starting point of the species *P. kewensis* which is now cultivated frequently in our gardens. However, this progeny differed in chromosome number; generally *P. kewensis* forms gametes with n = 18 and somatic cells with 36 chromosomes but in rare cases gametes with other numbers of chromosomes were also obtained.

Fig. 99. Fruit shapes of polyploids from the cross *Raphanus sativus* (A) with *Brassica oleracea* (B); the chromosomes are indicated by the letters R and B (after KARPECHENKO, 1928).

The tetraploid *P. kewensis* is the first known example of a form which originated by hybridization of two species and which has as its chromosome number, the sum of the haploid numbers of both parental plants. This investigation was followed by many others. We will restrict ourselves in our examples. We can mention the cross between

Raphanus sativus (n = 9) × *Brassica oleracea* (n = 9) which, in an investigation of KARPECHENKO (1928) yielded a normal diploid F_1 with 2n = 18. Self-pollination of these F_1-plants gave only a few descendants however, which all had the somatic number of 36 chromosomes, and hence could be considered as *tetraploids*.

On the other hand descendants were obtained by back-crossing with *Raphanus*; these bore 27 chromosomes in their somatic cells and were therefore *triploid*. In later progenies *pentaploids* also occurred (somatic number 45) and also some more or less *hexaploids* (*hypohexaploids* with 51–53 chromosomes instead of 54 chromosomes expected). Even by hybridization of a triploid (27) and a hexaploid (51), KARPECHENKO obtained an individual with 78 chromosomes in the somatic cells. Figure 99 gives a representation of the fruit shapes of some of these types.

The above mentioned cases of accumulation of chromosomes led to the origin of *euploid* series in the sense defined by TÅCKHOLM (1922). The numbers of chromosomes in the original species and in the progenies are in the ratio 1 : 2 : 3 : 4 : 5 etc., without any intermediate form. An important difference can still be found between the former polyploids originating from one parental individual and the latter forms due to hybridization; the first comprises, in one or another multiple form, the same configuration of chromosomes (*genome*), which was present doubly in the original somatic cells and are therefore called *autopolyploids* by KIHARA and ONO (1926) as mentioned earlier. If the hybridization products comprise two different genomes which are both double, that is to say, the tetraploids contain two of each set, they are called *allopolyploids* (K. and O., 1926). Originally the term *amphidiploid* was used but nowadays the abbreviated terms *alloploids* and *amphiploids* instead of allopolyploids have come more into the vogue.

Another source of irregular chromosome configuration were hybridization products of two different species which do not have the same number of chromosomes but each possess a multiple of the same basic number. We will take as an example the description of the cross between *Papaver nudicaule* (n = 7) and *P. striatocarpum* (n = 35, LJUNGDAHL, 1922, 1924). She obtained a fertile form with 42 chromosomes. The hybrids of *Nicotiana glutinosa* (n = 12) and *N. tabacum* (n = 24) made by CLAUSEN and GOODSPEED (1925), which were almost sterile in the F_1-generation, yielded one single fertile plant which gave rise to a uniform F_2-generation completely resembling the F_1-generation. We will also mention the investigation of TAHARA and SHIMOTOMAI (1927) concerning the cross of *Chrysanthemum marginatum* (n = 45) and *C. lavandulaefolium* (n = 9) which yielded an F_1-generation with 72 chromosomes in the somatic cells instead of the expected number of 54, and finally BREMER's studies (1922–1931, 1949) over sugar-cane and its hybrids in which he found, that if *Saccharum officinarum* (n = 40) was crossed with *S. spontaneum* (n = 56) he would obtain an F_1 with 2n = 136 (56 + 2 × 40) and F_2-individuals which, although mutually somewhat different, still resembled in type the F_1-plants. Their somatic number was about 136 (sometimes rather lower).

From these experiments it is evident that a simple duplication of the whole chromosome configuration does not always occur. It is true in CLAUSEN's example concerning *Nicotiana*, a hexaploid (2n = 72), originated from the cross of a diploid (2n = 24) with a tetraploid (2n = 48). It thus seems that in this case all the chromosomes were doubled but in TAHARA's *Chrysanthemum* cross it seems that duplication takes places for 9 chromosomes of *C. lavandulaefolium* and also for 9 of *C. marginatum* only. In BREMER's *Saccharum* studies there occurred duplication of the *officinarum* chromosomes, whereas in the *Papaver*-cross of LJUNGDAHL no duplication of the number of chromosomes took place.

Finally we may mention as also having importance, the classical investigation of FEDERLEY (1913) concerning species of butterflies viz., *Pygaera anachoreta* and *P. curtula*. The first species has in its generative cells 30, the second 29 chromosomes. The F_1-generation has in its somatic cells the sum of these numbers, that is 59. The backcross of F_1 ♂ × *anachoreta* females gave rise to butterflies with 89 chromosomes in their somatic cells next to some with 87 or 88 chromosomes.

Therefore, the conception of polyploidy is that of a collective name for the common phenomenon of accumulation of chromosomes; a good recent review is given by STEBBINS (1947).

It is obvious, that soon after the time that this phenomenon became known the cytological background of this process would become an object of research and interest. Many hypotheses were given (see

HEILBORN, 1934). The phenomenon that polyploidy occurs less frequently in animals, came especially to the front and for this many explanations were given.

We have already shown that (according to the investigations of LITARDIÈRE and LANGLET) a duplication in the number of chromosomes in the somatic cells can be the result of two phenomena: either fusion of two nuclei, or double longitudinal splitting of the chromosomes during the prophase of the nuclear division. These two are the processes by means of which autotetraploids can occur in somatic tissue. Such an explanation of a double longitudinal splitting of chromosomes seems to be acceptable in the case of *Primula kewensis* mentioned earlier, because, whereas all F_1-plants except one were sterile, the tetraploid progeny originated from one single individual, so that in this case a somatic duplication of the chromosomes may be presumed.

Also, CLAUSEN and GOODSPEED believe that for their *Nicotiana* cross, a somatic duplication of all chromosomes in the young embryo (from which the fertile F_1-plant originated) should have taken place, however, for such a statement there is as yet, not enough evidence.

In BREMER's example of the noble sugar cane *Saccharum officinarum* (used as the female plant) in crosses with wild *S. spontaneum* the doubling of the chromosomes of the mother plant is due to *endomitosis*. Investigation into the female gametophyte of sugar cane showed that without exception, a normal chromosome reduction occurred in the first meiotic division. After the second meiotic division the megaspores come to lie in a row, from which the innermost (which is called chalazal megaspore cell) develops into an embryo-sac. A splitting of all the chromosomes inside the nucleus of the young embryo-sac cell was found. Increase of chromosome number in these species crosses of sugar cane is therefore due to an *endomitotic process* which takes place in the female gametophyte. Such a process may be involved in other species crosses as well. WINGE (1932a) considered a number of other similar cases, also to be the result of somatic doubling of chromosomes.

Contrary to this, cases are also known in which a definite disturbance of the reduction division occurs. A primary condition for a normal course of the reduction division is no doubt the grouping into pairs of the homologous paternal and maternal chromosomes into gemini and it is obvious that the mutual affinity between chromosomes which originate from different species will not always be so strong as in the case between the homologous chromosomes of one species. Nevertheless, pairing of chromosomes originating from different species is not an exception: in LJUNGDAHL's *Papaver* crosses it was evident that in the reduction division of the F_1-plants the 7 *nudicaule*-chromo-

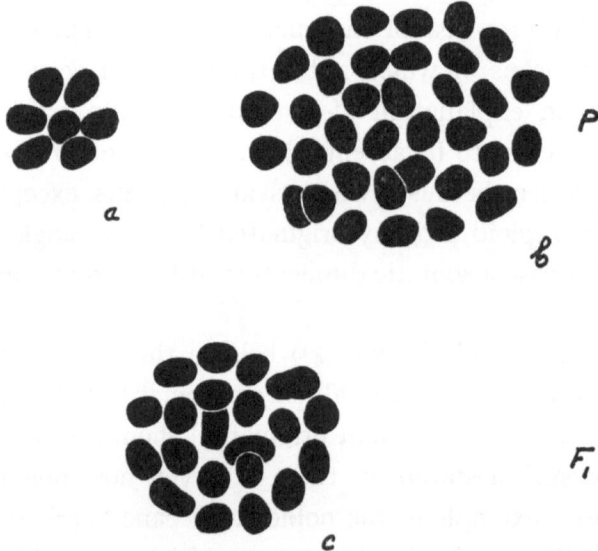

Fig. 100. Chromosomes of *Papaver nudicaule* (a), *P. striatocarpum* (b) and their hybrid (c) (after LJUNGDAHL, 1924).

somes paired with 7 partners of *striatocarpum*-chromosomes while the remaining 28 *striatocarpum*-chromosomes were grouped into 14 gemini. The result is that 21 gemini were formed and all gametes received 21 chromosomes. If now this F_1-hybrid (n = 21) is back-crossed with *nudicaule* (n = 7) there will originate from this cross, plants with a haploid number of 14 chromosomes. Figures 100 and 101 give a representation of this.

The fusion of chromosomes which originate from the same species was called *autosyndesis* by LJUNGDAHL (1922). If gemini are formed by two chromosomes from different species then we are dealing with *allosyndesis*.

Another extreme is found in the *Raphanobrassica* crosses of KAR-PECHENKO and the *Pygaera* cross of FEDERLEY. In the reduction division of the F_1-individuals of *Raphanobrassica* no conjugation into gemini takes place. This is caused by the absence of affinity of the chromosomes. The final result is that the chromosomes are distributed irregularly over both daughter cells originating from the first division, so that gametes with from 1 up to and including 17 chromosomes are formed. These are not viable but there is also the possibility that the first division is nullified that is to say, the nucleus does not pass

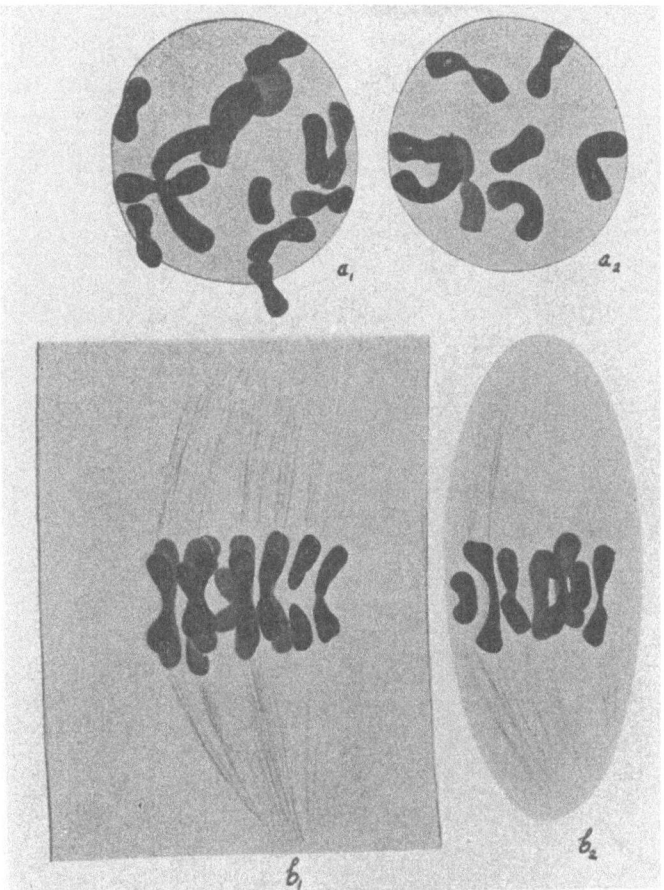

Fig. 101. Backcross of the hybrid *Papaver nudicaule* × *P. striatocarpum* (n = 21) with *P. nudicaule* (n = 7). (after LJUNGDAHL, 1924).

through the stage of the first division but comes directly into the stage of the second division. This nucleus divides into two nuclei after the 18 chromosomes split longitudinally. This process gives rise to viable gametes. Figure 102 gives a schematic representation of the possibilities (only 6 chromosomes out of the 18 are drawn). The gametes with 18 chromosomes are suitable for fertilization, and can in this way be the origin of tetraploids.

Between both these extreme types in which either all chromosomes form gemini by conjugation, or the conjugation is completely absent, there are still several other possibilities which can lead to the distur-

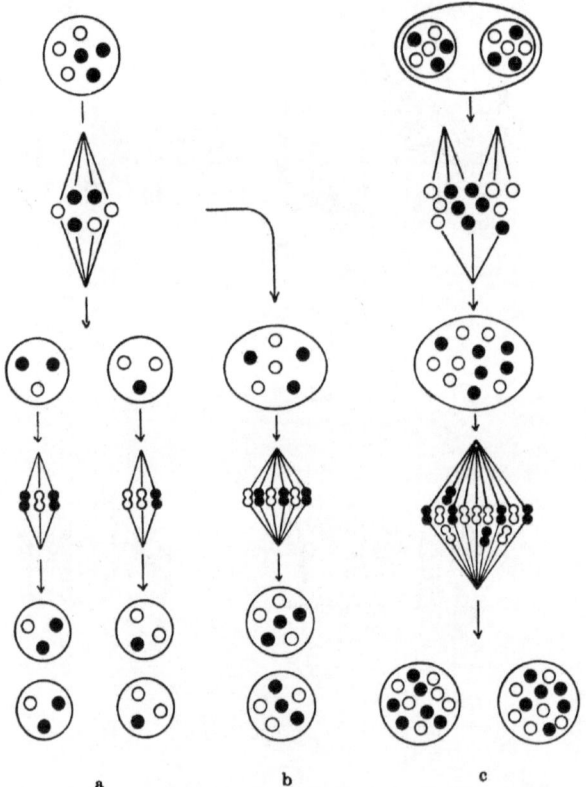

Fig. 102. Scheme of the formation of gametes in the F$_1$ individuals of *Raphanobrassica* (see text) (after KARPECHENKO, 1927).

bance of the normal reduction division. In the case when pairing is observed between chromosomes the further course of the first division can be broken off and a so-called *restitution nucleus* (ROSENBERG, 1927) can originate. After this the restitution nucleus undergoes a normal somatic division and gives rise to a 'dyade' of two cells, which have each the somatic chromosome number. This process is drawn in figure 103 as it occurs in *Hieracium boreale*.

Finally, another possibility seems to occur for the formation of diploid gametes: there are indications that the cell division which precedes the reduction division does not always come wholly to an end, so that the gamete mother-cell contains two nuclei which later fuse into one nucleus which has in that case the double (somatic) number of chromosomes. This phenomenon, which can also be classified under

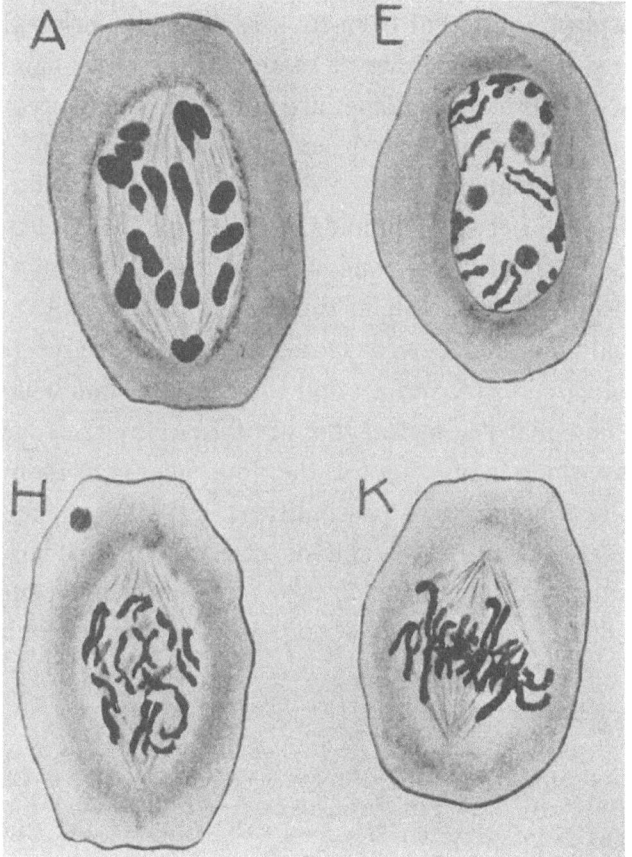

Fig. 103. Four stages in the formation of a restitution nucleus in *Hieracium boreale* (after ROSENBERG, 1927).

the same heading as the formation of the 'restitution nuclei' was called as 'syndiploidy' by STRASBURGER (1907). KARPECHENKO holds that such a syndiploidy is responsible for the occurrence of tetraploid gametes in his F_1-individuals which therefore should undergo the process of restitution nuclei formation twice (see figure 102c).

These polyploids are also important for their genetical behaviour apart from their karyological importance. It is obvious that duplication of the number of chromosomes in somatic cells and in gametes causes a change in the heritable behaviour of these individuals. Hence, it cannot be a matter of indifference which way these polyploids are accomplished. In the simplest case of autotetraploids (in which the polyploidy is a consequence of a single duplication of the same genome and hence, all gametes have a double genome) the genetical constitution will be completely different from the case when the original genome is present three times, five times or more. On the other hand, allopolyploids in which different genomes are united in different combinations give an even more complicated composition. We will limit ourselves to the simplest tetraploids.

The difference between diploids and tetraploids is clear from the fact that heterozygosity for one single factor (A) is only possible in one way in diploids (Aa) but in three ways (AAAa, AAaa and Aaaa) in tetraploids. Diploid heterozygotes as we know, give rise only to two sorts of gametes (A and a) and the first question which we must answer is: how many gameto-types are formed by their heterozygous tetraploids (which have two of the four genes in their gametes). Theoretically there are two possibilities: 1) the four genes which are present somatically in the tetraploids, can be combined independently from one another into pairs or, 2) they may not be independent of one another and the combinations which can originate are limited.

The first possibility has thus the consequence that the three heterozygotes mentioned can give rise to the following gametes:

$A_1A_2A_3a$: A_1A_2, A_1A_3, A_1a, A_2A_3, A_2a and A_3a, thus 1 AA : 1 Aa.
$A_1A_2a_1a_2$: A_1A_2, A_1a_1, A_1a_2, A_2a_1, A_2a_2 and a_1a_2, thus 1 AA : 4 Aa : 1 aa.
$A_1a_1a_2a_3$: A_1a_1, A_1a_2, A_1a_3, a_1a_2, a_1a_3 and a_2a_3, thus 1 Aa : 1 aa.

The zygotic progeny, after self-fertilization of these three forms of heterozygotes, can thus be expected to occur in the following ratios:

$A_1A_2A_3a$	Gametes	AA	Aa
	AA	AAAA	AAAa
	Aa	AAAa	AAaa

In the case of dominance of A over a — apparent true breeding for A.

$A_1A_2a_1a_2$	Gametes	1 AA	4 Aa	1 aa
	1 AA	1 AAAA	4 AAAa	1 AAaa
	4 Aa	4 AAAa	16 AAaa	4 Aaaa
	1 aa	1 AAaa	4 Aaaa	1 aaaa

In the case of dominance of A over a — a segregation into 35 A : 1a.

$A_1a_1a_2a_3$	Gametes	Aa	aa
	Aa	AAaa	Aaaa
	aa	Aaaa	aaaa

In the case of dominance of A over a — a segregation into 3 A : 1 a.

The theoretical progeny of these heterozygotes from mutual crossing or from back-crossing with the complete dominant AAAA, and with the completely recessive aaaa can easily be calculated.

We will summarize this as follows:

Parents	AAAA	AAAa	AAaa	Aaaa	aaaa
AAAA	∞ A : 0 a	∞ A : 0 a	∞ A : 0 a	∞ A : 0 a	∞ A : 0 a
AAAa	∞ A : 0 a	∞ A : 0 a	∞ A : 0 a	∞ A : 0 a	∞ A : 0 a
AAaa	∞ A : 0 a	∞ A : 0 a	35 A : 1 a	11 A : 1 a	5 A : 1 a
Aaaa	∞ A : 0 a	∞ A : 0 a	11 A : 1 a	3 A : 1 a	1 A : 1 a
aaaa	∞ A : 0 a	∞ A : 0 a	5 A : 1 a	1 A : 1 a	0 A : ∞ a

The first example in this sphere was given to us by GREGORY (1914) who investigated the progeny of a tetraploid *Primula sinensis*. After this study, a rather extended series of experimental investigations followed concerning the segregation ratios of separate pairs of allelomorphic genes and studies of linked genes which occur in the offspring of such autotetraploids (see LINDSTROM, 1936). The mathematical basis for such segregations was elaborated by HALDANE (1930), and by BARTLETT and HALDANE (1934).

It is obvious that the progeny of more complicated polyploids, such as higher autopolyploids and especially alloploids, can show a very complicated constitution and can then differ even more from the norm of the diploids as is the case in autotetraploids. We will refer to the work of SANSOME and PHILP (1932). The most important difference between *autotetraploids* and *allotetraploids* is this: in the former, the original chromosome configuration is doubled by means of splitting of the chromosomes and the four products of the splitting of a pair of homologous chromosomes can be mutually combined into two times two homologous chromosomes; this will give at the upmost, complications in the case of heterozygosity of the original diploid organism. However, in the allotetraploid, none, or only in exceptions, will homologous chromosomes be present in the original diploid chromosome configuration. In the allotetraploids it is possible that either both chromatids of one original chromosome act as homologues and will therefore enter into a conjugation during the reduction division (so-called *autosyndesis*), or a chromosome of one parent is homologous with a chromosome of the other parent and their chromatids can undergo conjugation during the reduction division (*allosyndesis*). The one parent for example delivers the chromosome A and its split halves A_1 and A_2, the other parent B with B_1 and B_2. In the case of autosyndesis conjugation only takes place between A_1 and A_2, and between B_1 and B_2. In allosyndesis A_1 can also conjugate with B_1 and B_2 and likewise A_2 with B_1 and B_2. Autosyndesis of all chromosomes gives rise to mutually identic gametes and hence to a progeny which breeds true to type. On the other hand allosyndesis of one or more chromosomes gives mutually different gametes and a segregating progeny.

These autotetraploids are also important from the physiological point of view: it now becomes possible to compare the rôle of a tetraploid chromosome configuration with that of a diploid one (see GYÖRFI, 1941; PIRSCHLE, 1942). From the extensive literature, which is already published in this field, we will refer to the investigation of GREIS (1940) who was able to compare the morphological and physiological properties of tetraploid forms of barley, obtained by treatment of colchicine, with those of the original races. For instance, it was evident that many important differences could be established in the stomata (figure 104). For the diploids an average of 23.6 μ was obtained and a standard error of 1.80 μ. These values were 29.8 μ and 1.47 μ for the tetraploids, that is to say an important difference in average existed. In chapter XXI we will discuss this important fact more extensively.

Apart from these cases of the origin of polyploids by means of accumulation of complete genomes there are now also examples of the occurrence of *haploids* which thus have the reduced number of chromosomes of their parents in their somatic cells. Whereas in lower plants and in some groups of animals (see ANKEL, 1927–1929; SCHRADER, 1931) the occurrence of haploids is a rule and gives rise to a more or less regular life-cycle, in higher plants; on the other hand, the question of the viability of haploid individuals remained undecided for a long time and is still very dubious in other groups of animals. For instance in *Drosophila melanogaster* which is the best species for investigation from a genetical and cytological point of view, a haploid individual was never observed; BRIDGES (1925, 1930) indeed found female animals who showed recessive male characters in parts of their body and which had accordingly, a haploid chromosome series in

these parts. Therefore, these flies form a mosaic of diploid and haploid cells. However, haploid individuals are found among the Phanerogames. The first example in *Datura* was published by BLAKESLEE, BELLING, FARNHAM and BERGNER in 1922. Other examples were obtained in the genus *Brassica, Crepis, Matthiola, Oenothera, Oryza, Solanum* and *Triticum* (see the general reviews of GATES and GOODWIN, 1930; KUHN, 1930; IVANOW, 1938 and of KOSTOFF, 1941).

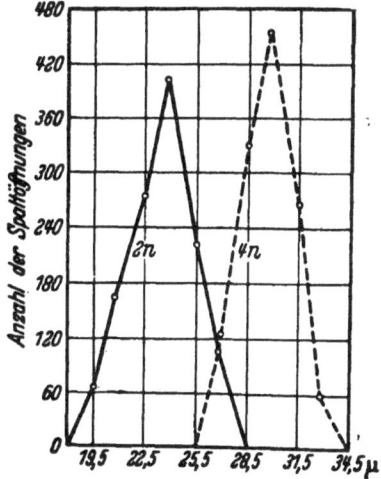

Fig. 104. Sizes of stomata in a diploid (2n) and a tetraploid (4n) barley (after GREIS, 1940).

In *Datura* cultures which were submitted to influences of low temperature, BLAKESLEE and his co-workers found two individuals among a great number of other abnormal plants which possess the haploid number of chromosomes (12) in their somatic cells. Both these specimens were sterile to a high degree. 80% of their pollen or more was wholly shrunken, and they gave rise to only a few descendants. The cell division from which the gametes originated, showed no gemini but in most cases a reduction of the chromosomes into groups of 3 + 9, 4 + 8, 5 + 7, 6 + 6 could be detected in the first reduction division. These groups gave rise, by means of longitudinal splitting in the second division, to four pollen grains with incomplete chromosome numbers. However, in some cases (10–29%) no 'reduction' took place in the first division, and a 'restitution nucleus' was formed which in its turn gave rise to a 'dyade' each with 12 chromosomes. The gametes originating from this process were normal and viable and gave rise to wholly diploid individuals on mutual fertilization.

Apart from the application of low temperatures, high temperatures and irradiation with X-rays were also used as methods for stimulating the occurrence of haploids. Nevertheless, most of the haploids in Phanerogames originated from inter-specific crosses or from crosses between varieties.

Besides the fact that they are viable and therefore form extremely good material for comparative-karyological investigation (in *Datura* haploids, diploids, triploids and tetraploids are now known) the haploids are also important for other reasons.

Firstly it was previously believed, that a diploid descendant, originating from a haploid parental plant (by means of fusion of two identical genomes), should be homozygous for all genes. In other diploids this certainty is never absolute so that a diploid progeny of haploids should provide excellent material for the investigation into gene mutations (i.e., spontaneous changes in the genic constitution which, in this case, cannot be explained as due to the heterozygous nature of the individual involved).

Indeed BLAKESLEE, MORRISON and AVERY (1927) and SATINA, BLAKESLEE and AVERY (1937) established a number of factor mutations in such diploid material of *Datura* (originating from haploids), and since then it has become evident that we have to be very careful in drawing conclusion. The experiences of KOSTOFF (1941, 1943) have shown that there are two possibilities: 1) That, by means of somatic duplication of the number of chromosomes the haploid form can obtain two identical sets of chromosomes and can give rise to identical gametes during the reduction division with the consequence that 'ideal homozygous' diploid descendants occur. 2) It is also possible that there are chromosomes in this same set of haploid chromosomes which conjugate autosyndetically so that the gametes need not be homozygous. In chapter XIX we will discuss this problem in more detail.

The second reason is that among haploids, cases of somatic duplication of the nucleus were often observed, which gave rise to diploid organs (stem, leaf and flower). Such plants give excellent material (as do the autotetraploids) for comparative investigation of growth-ability and growth-phenomena which wholly depends upon the quantity of chromosomes of the nuclei. Figure 105 gives a picture of the flowers of one and the same individual of *Crepis capillaris* (HOLLINGS-HEAD, 1930): a pure haploid flower, a diploid and a mixed haploid-diploid one. The establishment of the chromosome configuration in the haploid and in the diploid tissues is made easier in this case because

Fig. 105. Haploid, diploid and diplo-haploid forms of *Crepis capillaris* (after HOLLINGSHEAD, 1930).

of the divergent shape of the chromosomes. It is obvious from figure 106 that the diploid chromosome configuration has all the haploid chromosomes twice.

Fig. 106. Haploid and diploid chromosome sets of *Crepis capillaris* (after HOLLINGSHEAD, 1930).

Finally there are two cases among the haploids which are of extreme significance: CLAUSEN and LAMMERTS (1929) crossed a *Nicotiana digluta* (n = 36) with red flowers as mother, with pollen from a white *N. tabacum* (n = 24) and in the F_1-generation obtained (apart from a great number of plants with all sorts of chromosome aberrations) a plant with white flowers and a somatic chromosome number of 24. It is a pity that this plant was sterile; restitution nuclei were not observed and all reduction divisions led to an irregular distribution of the 24 chromosomes so that viable gametes did not originate. The only explanation which is possible here is that the generative nucleus from the pollen-tube can hold its own in the egg-plasm whereas at the same time the whole egg-nucleus is eliminated. Hence, this is an example of the phenomenon called by WILSON (1937, p. 464) — 'androgenesis' (in contradistinction to 'merogony' of DELAGE (1899) that is to say fertilization of an egg-fragment, lacking in a nucleus, by the sperm-nucleus). A case of androgenesis was also described by KOSTOFF (1929): a haploid descendant of the cross *Nicotiana tabacum* (n = 24) as mother and *N. langsdorffii* (n = 9) as father has a somatic chromosome number of 9, and is from a morphological point of view a reduced edition of the *langsdorffii*-father (see KUHN, 1930).

In later years a great number of cases dealing with multiples of chromosome numbers and also of halving of the somatic number became known and altogether these compose a whole series of euploid forms.

WETTSTEIN divided these *euploids* into *orthoploids* (even numbers as multiple) and *anorthoploids* (odd numbers as multiple).

Mention has already been made many times to chromosome irregularities which do not lead to pure multiples. These aberrations very often give rise to non-viable gametes and therefore form a great source of '*gamete lethality*'. But still a number of examples of viable individuals became known which do not have an absolute multiple of the chromosome number. In contrast with the above mentioned euploid forms, these irregular individuals are characterized as '*aneuploids*' or '*unbalanced polyploids*'.

The classic example of aneuploidy is again borrowed from the genus *Oenothera* which is a very important genus from a genetical point of view. Among *Oenothera Lamarckiana* DE VRIES (1887) discovered the form *Oenothera lata* (1901, p. 168). It was established by Miss LUTZ (1912) that this form had 15 chromosomes instead of the 14 somatic chromosomes characteristic of *O. Lamarckiana*. Since then, this fact has been found in a number of other *Oenothera* forms (see GATES, 1928). Before the time of the discovery of the occurrence of forms with an extra chromosome, *Oenothera* investigators had already found karyologically (firstly by GATES, 1908) that the reduction division in *Oenothera Lamarckiana* can sometimes have a deviating course when one of the chromosomes does not resort in the first division to its allotted pole but accompanies its homologous partner to the same pole so that four pollen-grains originate, two with 6, two with 8 chromosomes.

This phenomenon of both homologous chromosomes not being separated has since then become known by the name 'non-disjunction' (BRIDGES, 1913).

In the beginning, the occurrence of non-disjunction was only concluded on grounds of genetical composition of particular progenies; later BRIDGES (1916) could completely corroborate the exactness of this assumption in his karyological investigations.

Non-disjunction can apparently take place in completely normal diploids but it plays a specially significant rôle in the formation of gametes in triploids because in this instance, the whole genome is present in triplicate and at reduction division there will now be three homologous chromosomes to divide between two poles. Therefore, two can go to one pole and the other to the opposite pole during the first division. Thus, the consequence of non-disjunction is the origin of gametes which do not possess an exact n- or 2n-number of chromosomes but one or more extra chromosomes that is to say: n + 1, n + 2, 2n + 1 etc. The zygotes originating from the fertilization of such gametes then will have 2n + 1, 2n + 2, 3n + 1 etc., as chromosome numbers.

Such combinations became known in a great number from a series of publications of BLAKESLEE and his co-workers (see BLAKESLEE, 1928; 1931; BLAKESLEE and AVERY, 1938). They succeeded in finding a number of aberrant types in the species *Datura Stramonium*, and even a number of twelve plants which all had somatically 25 (2n + 1) chromosomes (see figure 107). As this original number of twelve aberrant types agreed with the haploid number of 12 chromosomes, BLAKESLEE supposed that in each of the aberrant forms, a particular chromosome is present in triplicate instead of duplicate. These triplets of a homologous chromosomes were called 'trisomics' by him, the other chromosomes are then, as in normal diploid plants, present as 'disomics'.

In the case that the same homologous chromosome is present in quadruplicate we will speak then of 'tetrasomics'. If there is more than one homologous pair of chromosomes present in a higher number we will refer to 'double trisomics' (2n + 1 + 1). Later it became known that more aberrant forms could be distinguished, all with a chromosome number of 25. BLAKESLEE supposed that it is possible to classify these as '*secondary trisomics*' within a definite type of trisomics. Therefore 'Sugarloaf' and 'Polycarpic' should be secondaries to the primary form 'Rolled' (fig. 108) and owe their characteristics to a halving of

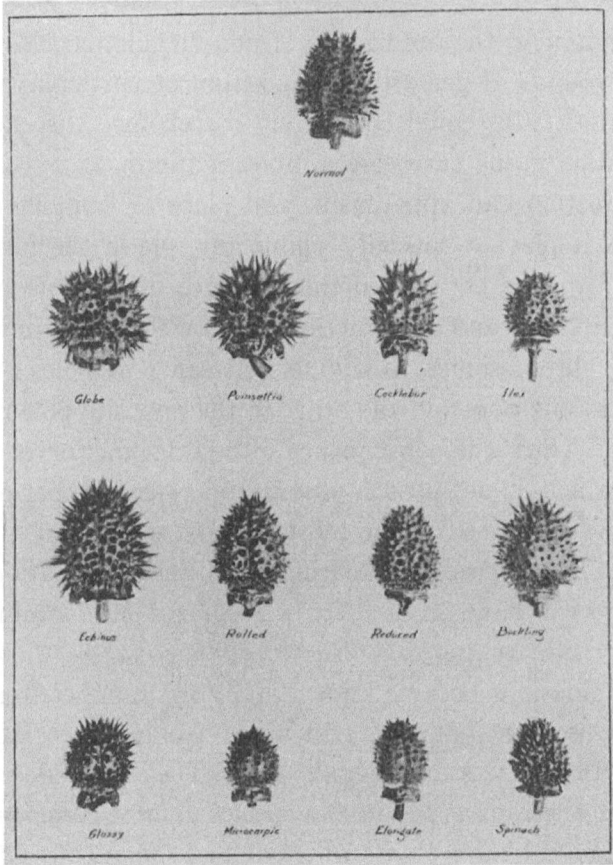

Fig. 107. Normal fruit of a diploid *Datura stramonium* and the 12 forms with 2n + 1 chromosomes (after BLAKESLEE, 1931).

the chromosome concerned; that is to say, it is presumed that each of the secondaries should possess a duplication of the same half as its extra chromosome. Hence, their formula can be indicated by 2n + 2/2. If one of the chromosomes is trisomically present it will affect the genotypical constitution of the progeny. These consequences of the aberrant chromosomic constitution are analogous to the inheritance in autotetraploids discussed above. There is no need to discuss this here in detail; we will refer to the relating investigation of BLAKESLEE and FARNHAM (1923).

Whereas on the one hand gametes and zygotes could be formed by non-disjunction with one or more extra-chromosomes, gametes will also appear from this process which lack a chromosome and hence

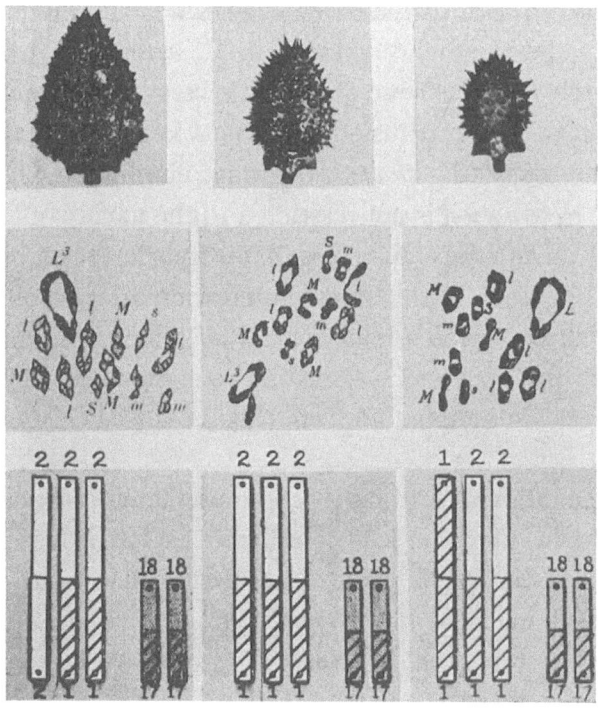

Fig. 108. The primary *Datura* 'Rolled'*a* (middle) and both the secondaries 'Sugarloaf' (left) and 'Polycarpic' (right); their chromosome constitution and a schematical representation of the chromosomes involved (after BLAKESLEE, 1931).

possess n-1 chromosomes. However, such gametes and the zygotes originating from these (2n-1) are much less viable and only occur as exception. The best example of this view is the so-called Haplo-IV-form of *Drosophila melanogaster* found by BRIDGES (1921); in its somatic chromosome configuration chromosome IV is only present once. It was observed karyologically that spermatozoa without chromosome IV can be formed as a rare exception, and are suited for fertilization. Therefore, in the offspring the factors located in this chromosome are transferred by the mother only.

Such a chromosome elimination not only occurs as an opposite process of non-disjunction but is also a frequent phenomenon in hybrids which originated from parents with different chromosome numbers. In CLAUSEN's crosses between *Viola arvensis* (n = 17) and *V. tricolor* (n = 13) he observed that in 60% of the total number of gametes, one of the unpaired chromosomes was eliminated and that in the progeny, the factor W present in *V. arvensis* (which fades the violet and the yellow colour of these flowers) is also lacking in 60% of the gametes. The conclusion is obvious i.e., that in this cross the same chromosome of *V. arvensis* is always eliminated. This process of *chromosome elimination* is not restricted to the nature of the cross that is to say, to an inter-specific cross. After hybridization of two strains of *V. arvensis*, each with an absolutely normal karyological course and with a chromosome number of n = 17, many chromosome eliminations were observed which came best to expression in one individual with n = 16 which originated from this cross (see CLAUSEN, 1926).

In all the above mentioned chromosome aberrations, chromosomes could be more or less considered as individualities. Modern karyological research has made some alterations in our conceptions; using different material *fusion of two or more chromosomes* into one, and the reverse case, that is to say, *fragmentation of a single chromosome* into different parts was observed.

A nice example of temporary fusion of four chromosomes into one complex chromosome was obtained by SEILER and HANIEL (1921) in *Lymantria monacha*. The mature gametes have 28 chromosomes, among which one is strikingly tall. Hence, the zygote has 56. Soon after fertilization (for instance in blastoderm nuclei plates) 62 chromosomes were clearly seen. It has been found that in the formation of gametes in the male individuals the number has already been reduced from n = 31 to n = 28 in the first of both reduction divisions. In future egg-cells this reduction takes place by means of fusion of four small chromosomes into one large chromosome between the first and the second reduction division. Figure 109 explains this course for the chromosome number involved.

This possibility of *chromosome fusion* is used by many investigators as an aid in the explanation of chromosome numbers occurring in several races of species which do not form an exact multiple of a basic chromosome number. We saw earlier that FEDERLEY (1913) found chromosome numbers of n = 30 and n = 29 in the butterfly species *Pygaera anachoreta*. SEILER (1925) studied several local races among the species *Phragmatobia fuliginosa* of which one has 29 haploid chromosomes in both sexes, another has n = 28, whereas a third race has only gametes with n = 28 in the males, and two types of gametes respectively n = 28 and n = 29 in the females. A comparison

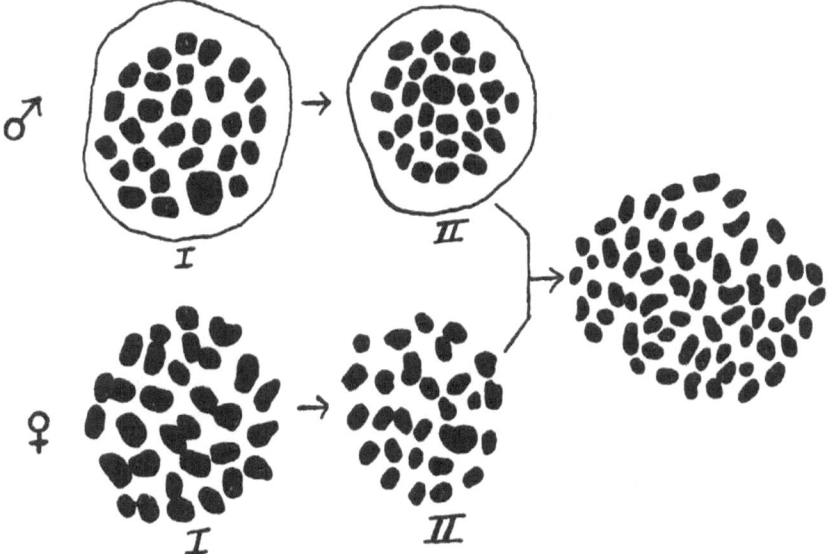

Fig. 109. Chromosome constitutions of gametes and zygotes in
Lymantria monacha (after SEILER and HANIEL, 1921).

of the somatic chromosomes in the latter race shows that two striking-ly long chromosomes are present in male individuals (2n = 56), whereas in females (2n = 57) one equally long and one other rather shorter chromosome are found (figure 110). The supposition here is obvious, i.e., that in this case one of the long chromosomes in the individuals with n = 28 takes the place of the short plus one of the small chromosomes of the type with n = 29. Such a view was held by HEILBORN (1924, 1925, 1932) in his *Carex* investigations: this very polymorphous genus does not follow the rule that the species should have the multiple of the same basic number of chromosomes. Not less than 22 haploid numbers are found in the different *Carex* species: 9, 15, 16, 19, 24, 25, 26, 27, 28, 29, 31, 32, 33, 34, 35, 36, 37, 38, 40, 41, 42 and 56.

However, it was evident from accurate length measurements of chromosomes that these deviations are not so great as appears; in the species *Carex pilulifera, C. panicea* and *C. ericetorum* e.g. three lengths of chromosomes are observed (A long, B intermediate, and C short) and the chromosome configurations of the three species are composed respectively of 3A + 4B + 2C, 3A + 2B + 11C and 1B + 14C. It is striking that the A chromosomes are always equally

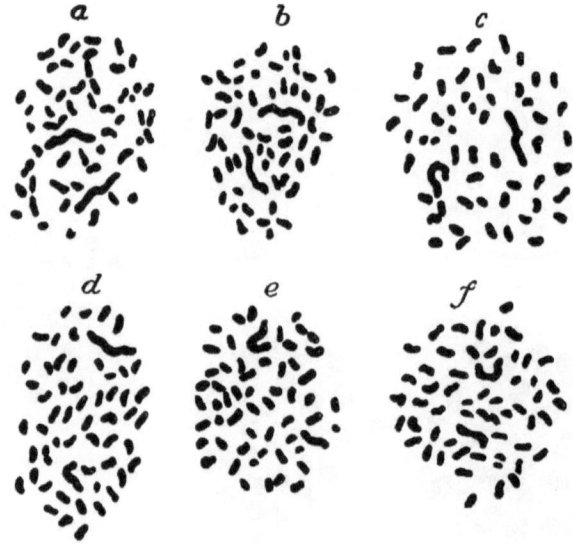

Fig. 110. Chromosome constitutions of gametes and zygotes in *Phragmatobia fuliginosa*; a–c male, d–f female (after SEILER, 1925).

long as B + C together, and that in all the three species observed the same ratio of length A : B : C was observed. This justifies the supposition that chromosome fusion (or fragmentation) has played a rôle in the origin of these species.

Fragmentation has yet another consequence: the fragment can be added as an extra to a normally diploid chromosome configuration and can then remain separate or it can be transferred to another place in the same chromosome, or even to another chromosome. This phenomenon, which was observed by BRIDGES (1919) in cultures of *Drosophila* which were treated with X-rays, he termed *'duplication'* (figure 111C). It is not easy to establish with certainty the existence of this phenomenon because karyological proof is often lacking, and hence the occurrence of duplication is only derived from genetical data. A botanical example of the *Drosophila* case is found in the investigations of LESLEY (1930) with tomatoes.

The reverse process was described by BRIDGES (1917) as '*deficiency*', that is to say, the loss or inactivity of a whole, particular and measurable section closely linked genes (figure 111B).

In a cross of a *Drosophila* male with the factors w (white eye colour) and B (bar eyes) located in the X-chromosome with normal females (W, red eye colour and b, round eyes) it could be expected that all daughters should have inherited the recessive factor for white eye-colour, and the dominant one for bar eyes from the father; from the mother, the factors for dominant red-eye-colour and the recessive round eye shape, so that they should be heterozygous with regard to both factors (an analogous case is discussed on p. 209 as

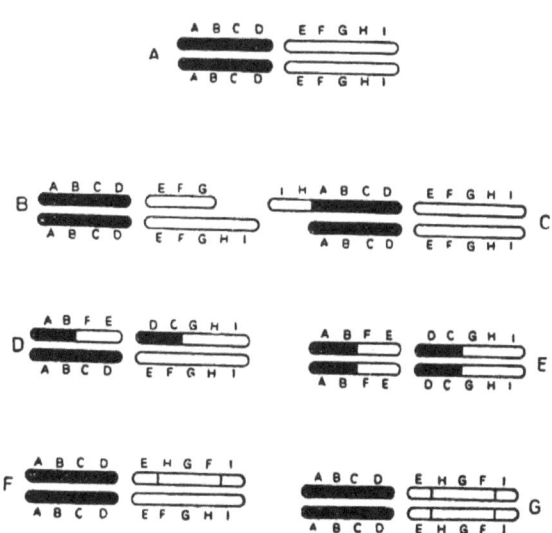

Fig. 111. Schematic representation of chromosome aberrations. A, normal diploid; B, deficiency; C, duplication; D, heterozygous translocation; E, homozygous translocation; F, heterozygous inversion; G. homozygous inversion (after DOBZHANSKY, 1937).

sex-linked heredity). This expectation was indeed corroborated in all female descendants with exception of one individual which was heterozygous with respect to eye-colour, however it had the recessive round (b-type)- shaped eye.

By further crossing with this aberrant female, BRIDGES could establish that in this animal the B-factor was eliminated (or at least inactive). A similar result was also observed with regard to the recessive factor f (forked bristles) which is situated close to B. Therefore, BRIDGES concluded that the area of the X-chromosome, in which the genes B and f are located, was changed into an inactive state which he called *'deficiency'*. Since this investigation numerous other cases of this same phenomenon have become known. MOHR (1919, 1929) observed the interesting fact that in heterozygotes which have a chromosome with the genes w, fa and A next to a chromosome in which this part (which should contain these three factors) is deficient, the action of these factors comes to a stronger expression than if they are present in a homozygous state (that is to say, double). This phenomenon is called *exaggeration*. This latter phemenon is

still insufficiently explained, but the occurrence of deficiencies is an ascertained fact. Karyological investigation could show especially a clear shortening of the chromosome involved. A nice example of such a striking cytological proof is taken from the study of McCLINTOCK (1931) on maize (Figure 112).

Fig. 112. Chromosome deficiency in Maize (after McCLINTOCK, 1931).

A special case of deficiency i.e., *'deletion'* can be distinguished if the group of missing genes was situated in the middle area of the chromosome. On the other hand a 'deficiency' is the wider conception of the loss of a factor, or a group of factors, from any part of the chromosome (that is to say, one must in this case also include the ends of the chromosome). PAINTER and MULLER (1929) have also shown karyologically the occurrence of 'deletions': in figure 113 the chromosome configurations of three 'deleted' flies are given and next to it the shortened X-chromosome is also mapped. The shortened X-chromosome contains only the two black ends from the original chromosomes (the genes are indicated by complete lines), whereas the middle white part of the X-chromosome (where genes are indicated by dotted lines) has disappeared.

An extremely nice analysis of such aberrations is given by MULLER and PROKOFYEVA (1935): At the actual top of the X-chromosome of *Drosophila* lie two genes y (yellow, yellow body colour) and sc (scute, bristles on the body). Of these factors several multiple allelomorphs are known. In the material involved the genes y^{3p} and sc^4, sc^8 and sc^{19}

occurred. Genetical investigation suggested that the gene sc[19] was present as a deletion and the other scute genes as deficiencies. It was evident from the structure of salivary gland chromosomes in all these types that chromomere No. 2 was lacking in different degrees; in sc[19] chromomere No. 1 was present, in other types it was missing. Figure 114 gives a clear view of the structure of the salivary gland chromosomes, the scheme added makes the location of the genes clear (figure 114g). From this it is evident that the supposition of BELLING (see p. 241) i.e., that in each chromomere only one gene was located, is

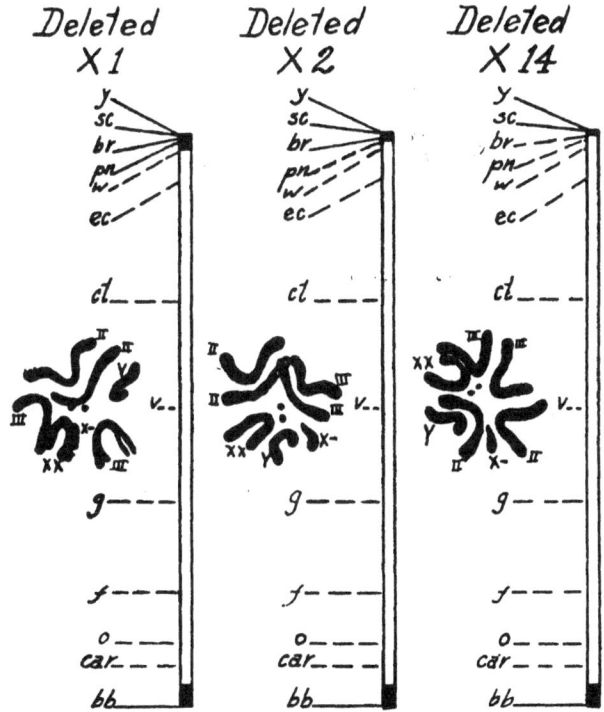

Fig. 113. Chromosome constitution and scheme of the chromosomes involved in three 'deleted' forms of *Drosophila*. In the deleted chromosomes (X-) black parts only are preserved (after PAINTER and MULLER, 1929).

inaccurate because certainly more genes are present in chromomere No 2.

In duplications and deficiencies the chromosome part involved is either doubled or changed into an inactive state. However, cases are also known in which a part of a chromosome changed place within the chromosome configuration, without being duplicated (as in the case of a 'duplication'), or became inactive (as in the case of a 'deficiency').

The first example of transplanting a chromosome segment upon another non-homologous chromo-some was described on genetical ground by BRIDGES (*translocation*, 1923a, figure 111 D and E), but it was not possible for him to prove karyologically his hypo-thesis. Soon after this STERN (1926b, 1929b) was more successful and succeeded in pointing out by parallel genetical and cytological investigations that a part of the Y-chromosome in *Drosophila* can break away and can attach itself to the X-chromosome. In this manner females were obtained with abnormal X-chromosomes, for example XY', composed of a complete X-chromosome and the arm of the bent Y-chromosome (Y'). Figure 115 gives a repre-

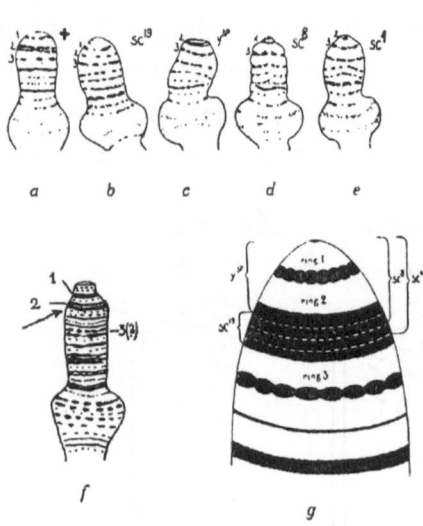

Fig. 114. Deficiencies in *Drosophila melanogaster*; a. and f. normal structure of sex chromosomes; b. deletion of gene sc[19]; c-e deficiencies of y³ᵖ, sc⁸ and sc⁴ (after MULLER and PROKOFYEVA, 1935).

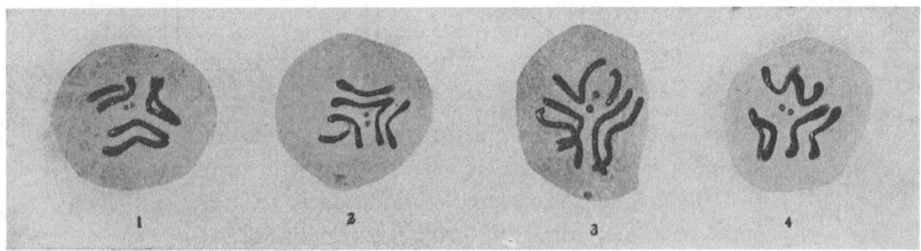

Fig. 115. Oögonia of *Drosophila*; 1–3 with XXY' structure, 4 with XY'XY' (after STERN, 1929).

sentation of three oögonia in which one X-chromosome is changed in this way and of one oögonium in which both X-chromosomes are enlarged with a Y'-part. Since then the process of translocation was karyologically observed in *Drosophila* by BRIDGES (MORGAN, BRIDGES, STURTEVANT, 1928) and by DOBZHANSKY (1929) and investigations in plants also produced evidence for this (*Crepis tectorum*, NAVASHIN, 1931; *Zea Mays*, McCLINTOCK, 1930, 1931).

Finally another possibility exists of displacement of chromosome parts. STURTEVANT (1921, 1926) could ascribe genetical results in his investigation into a change in chromosome constitution in the following sense: An end of a middle part of a chromosome could be broken away and after turning through 180° could again be attached (*inversion*, figure 111 F and G). In chromosome III of *Drosophila* the location of a number of genes is suggested to occur in this order: ru – st – sr – $+^1$ – e – ro – ca – $+^2$, whereas after inversion, the group ($+^1$ – $+^2$) would be the other way round so that the order of succession is ru – st – sr – $+^2$ – ca – ro – e – $+^1$. The existence of such inversions is now proved karyologically by investigation into the salivary gland chromosome.

It is obvious that such chromosome aberrations should act as a disturbance to the normal course of a reduction division if the organism is heterozygous with regard to the aberration, that is to say, that a pair of homologous chromosomes is in this case composed of a normal one (original) and an aberrant form. Such heterozygotes are known as '*structural hybrids*'. In a heterozygous deficiency a part of the chromosome is paired during the conjugation in the reduction division and another part is unpaired, whereas the total of genes in both homologous chromosomes differ. In a heterozygous duplication the same question arises; in a heterozygous translocation and a heterozygous inversion the total of genes present in both homologous sets are the same, but in the former there occur two different pairs of chromosomes, and in the latter only one (see figure 111). Such heterozygous inversions show still another interesting particularity which can also be favourably observed in salivary gland chromosomes. It is evident that in cases dealing with a heterozygous inversion in the middle area of the chromosome, the inversion is not longer homologous in its location of genes with the gene order of its conjugation partner. On the other end the terminal parts, which are not aberrant, behave normally. Therefore, in the middle part no gene homology occurs.

Both mutual chromosome parts try to take a normal location by means of mutual attraction of homologous genes, with the consequence that the salivary gland chromosome shows a loop in the middle. How such a loop is accomplished is evident from the additional scheme in figure 116. Such a change in chromosome constitution will naturally lead to very important consequences from a genetical point of view.

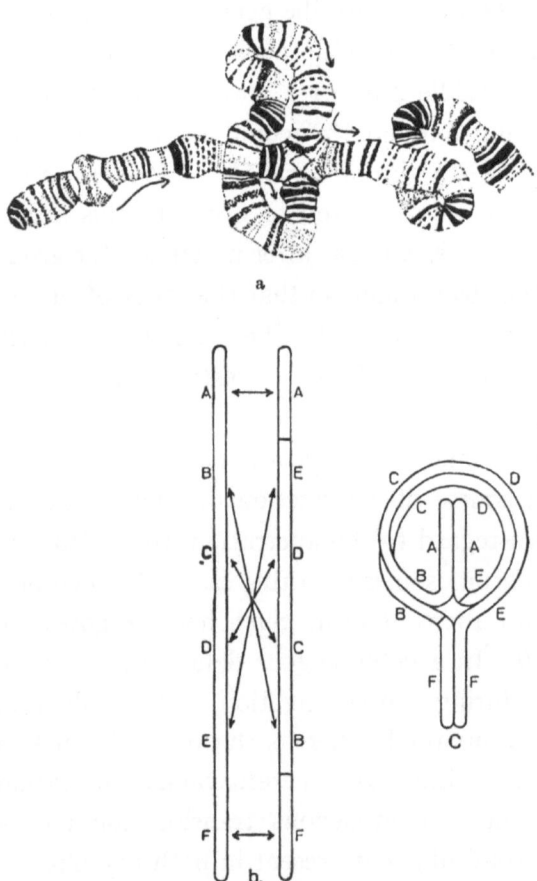

Fig. 116. Pairing of a heterozygous inversion in *Drosophila melanogaster* (after PAINTER, 1934a) with a schematical explanation of the formation of a looped chromosome (after BAUER and TIMOFEEFF-RESSOVSKY, 1943).

It is obvious that the phenomenon of linkage between genes in these chromosomes undergoes a radical change in the sense that independent factors of two non-homologous chromosomes now show linkage (e.g., in translocations) and the cross-over percentages of already linked genes, may change.

Temporary fusions of homologous and non-homologous chromosomes have an important significance in the course of hybrid segregation. We are indebted for this to CLELAND (1923, 1925, 1928, 1936) and later investigators whose karyological studies, concerning the mutual relationship between chromosomes by means of ring-formation, have explained so very much concerning the complicated phenomenon of heredity in *Oenothera*-species (see GATES and FORD, 1938). A number of species of this genus have a haploid chromosome number of 7, but differ in strength of mutual independence. For instance, in the now almost classic species *Oenothera Lamarckiana* the pairing is only observed in one pair of chromosomes during the beginning of the reduction division; the remaining 12 form a ring in which homologous chromosome parts always join each other at one end and the other end is attached to a non-homologous part of the chromosome. According to the modern view each chromosome in such a ring should thus be composed of two halves of which each half meets its homologue in its neighbouring chromosome. The composition of such a ring can be indicated as follows (a. b) – (b. c) – (c. d) – (d. e) – (e. f) – (f. a) (see CLELAND and BLAKESLEE, 1930; EMERSON and STURTEVANT, 1931). Such a process of ring formation is of course possible in the most diverse combinations: some species (*O. deserens*, *O. latifrons*, *O. Hookeri*) usually have 7 pairs of conjugated chromosomes, other species form a ring of 4 and 5 pairs (*O. franciscana*), a ring of 6 and 4 pairs (*O. rubrinervis*), a ring of 8 and 3 pairs (*O. rubricalyx*), a ring of 12 and 1 pair (*O. Lamarckiana*), two rings of 6 and 8 (*O. biennis*) or one single ring of 14 chromosomes (*O. muricata*). Figure 117 gives a picture of some of these species.

Theoretically such a process of *chromosome attachment* hinders the process of crossing-over and CLELAND and OEHLKERS (1930) came to the conclusion (after detailed compara- tive genetical-karyological investi- gation) that by increasing the num-

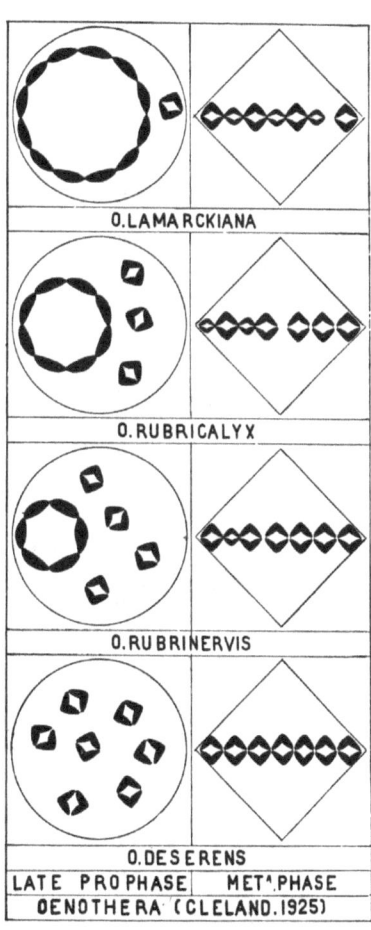

Fig. 117. Ring formation in *Oeno- thera*-species (after CLELAND, 1925).

ber of independent pairs of chromosomes the ability for exchange of factors between the complexes (that is to say, the segregation ability of the chromosome combination) was also increased. In crosses in which a ring of 14 chromosomes originates, no exchange of genes takes place; in crosses with a small ring and many pairs of chro- mosomes, exchange was observed for a number of factors.

Such a ring formation has also been found since that time in a number of other plant species; especially local races of *Godetia Whitneyi* (related to *Oenothera*) are rich in ring-shaped chromosomes (HÅKANSSON, 1942, figure 118). Moreover they were found in the hy-

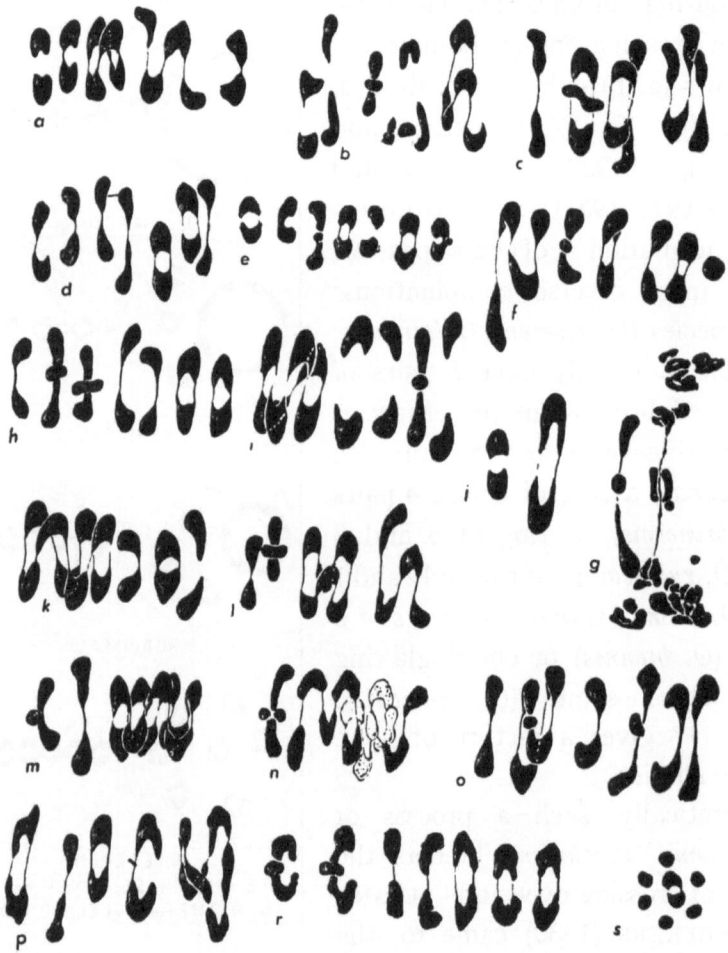

Fig. 118. Ring-shaped chromosomes in a number of strains of *Godetia Whitneyi* (after HÅKANSSON, 1942).

brids between *Zea Mays* and *Euchlaena mexicana* (ARNASON, 1936) and in the genus *Pisum* (HÅKANSSON, 1929; HAMMARLUND and HÅKANSSON, 1930; RICHARDSON SANSOME, 1929; PELLEW and RICHARDSON SANSOME, 1931).

Such chromosome rings can be the consequence of a preceding heterozygous translocation: groups of four chromosomes can arise (two normal and two translocated chromosomes) in which the homologous parts pair together, so that the whole forms a cross (figure 119a).

From this cross a ring of four chromosomes (figure 119c) can be

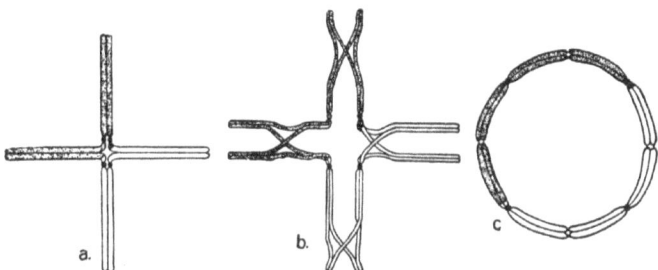

Fig. 119. Origin of a chromosome ring from translocated chromosomes (after BAUER and TIMOFEEFF-RESSOVSKY, 1943).

formed after the splitting of the chromosomes into chromatids (fig. 119b).

The occurrence of chromosome aberrations in the genus *Drosophila* has opened up the possibility of making comparisons in chromosome patterns between different species of this genus. Hence, we can make a comparison between the chromosomes of *Drosophila pseudoobscura* and *D. miranda*. It could be established that parts of the homologous chromosomes of both species are identical in genic composition, other parts show inversions and translocations etc. (see figure 120).

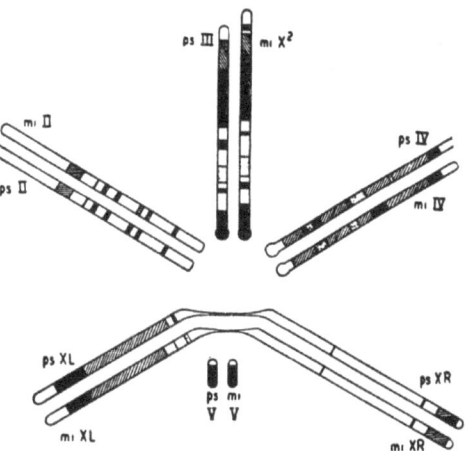

Fig. 120. Comparison of the location of genes in the chromosomes of *Drosophila pseudoobscura* and *D. miranda* (white represents identical parts; striped represents inversions; dotted represents translocations and black is unknown (after DOBZHANSKY, 1937).

Finally we will refer to a more deviated behaviour of the chromosomes on its own, borrowed from the studies of METZ dealing with *Sciara* (see also p. 202).

Analogous to the earlier investigations of BOVERI, METZ and his co-workers found an important difference in chromosome configuration between the common somatic cells and those cells of the body which later give rise to the origin of gametes (the so-called 'germ-line'). The egg-cells have three autosomes (see p. 228), a 'limited' chromosome (of which the course of the cycle is abnormal), and an X-chromosome. The spermatozoa have three such autosomes, two 'limited' chromosomes, and two X-chromosomes. All zygotes have thus three pairs of autosomes (a), three 'limited' chromosomes (l) and three X-chromosomes. During the development of the germ-line in the male, successively, one of the 'limited' and one paternal X-chromosome are elimi-

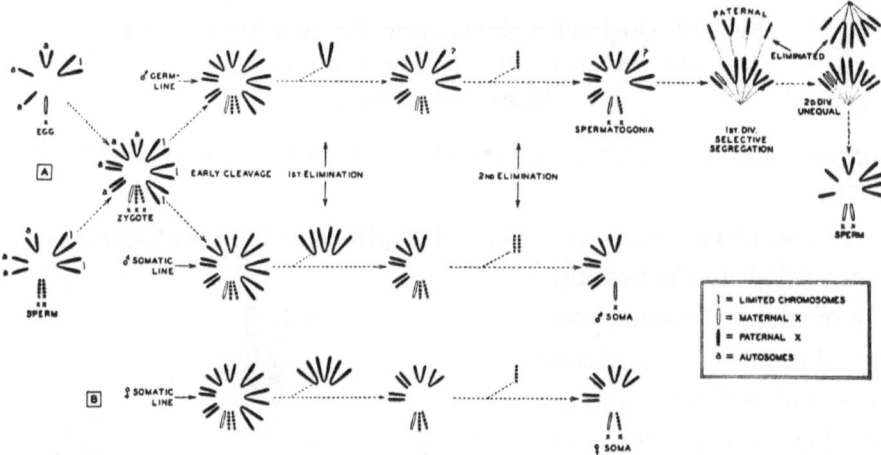

Fig. 121. Chromosome elimination during the development of *Sciara* (after METZ, 1938).

nated. One of each pair of autosomes and the second paternal X-chromosome disappear in the first of the two divisions which together form the reduction division. During the second division, the maternal X-chromosome is doubled and the other split-halves from the remaining chromosomes are again eliminated. The final result is that of the four possible spermatozoa only one originates which has the previously given chromosome configuration (see figure 121, ♂ germ-line). In the course of the division of the cells of the male (which are predestined as body cells) firstly the three 'limited' chromosomes disappear, and later the two paternal X-chromosomes, so that the body cells only obtain the three pairs of autosomes together with the maternal X-chromosome.

The bodily development of the female is parallel to that of the male with this difference, that besides the three 'limited' chromosomes, only one of the paternal X-chromosomes is eliminated so that all body cells now contain three pairs of autosomes and two X-chromosomes; the development of the egg-cells is normal. Though this is an undoubtedly interesting example, it is still very hard to give an explanation at this moment of what has up to now been a more or less isolated case of chromosome aberrations.

Summarizing, the following phenomena may be mentioned as aberrations of the normal haploid-diploid change of phase in plants and animals:

Autopolyploidy: duplication of the same genome in multiples of the chromosome number;

Allopolyploidy: fusion of different genomes into multiples of the basic chromosome number;

Euploidy: progeny with the single or with the multiple of the haploid chromosome number of the parents ((*anorthoploidy* (odd multiple): *haploidy, triploidy, pentaploidy*; *orthoploidy* (even multiples): *diploidy, tetraploidy, hexaploidy*));

Aneuploidy: progeny in which one or more chromosomes are present in a triplicate or manifold form, instead of a duplicate form as in diploids (*trisomics, tetrasomics*);

Elimination: one or more of the chromosomes are eliminated during reduction division;

Duplication: a part of a chromosome which is doubled remains to be a part of the chromosome configuration;

Deficiency: a part of a chromosome is inactive or is eliminated;

Deletion: a chromosome region in the middle of the chromosome becomes inactive due to a deficiency;

Translocation: this aberration occurs when a part of a chromosome is broken off and later becomes attached to another chromosome;

Fusion: chromosomes are preliminarily or permanently united to one complex chromosome;

Fragmentation: a chromosome falls into two or more pieces;

Ring formation: a number of chromosomes, homologous or non-homologous, mutually form a ring in which the halves of both homologues are always next to one another.

Each of these aberrations of the normal chromosome constitution bears a genetical consequence; in some cases the occurrence of the aberration was only previously concluded on genetical grounds; the investigation of the salivary gland chromosomes has borne out this supposition in *Drosophila*. In other organisms it is much harder to produce karyological evidence. Because of this it is recommended that one must be very careful in the interpretation of genetical results which arise from one of these forms of chromosome aberrations without having the karyological proof to go with it.

CHAPTER XIII

THE MECHANISM OF THE EXCHANGE OF GENES

The acceptance of the hypothesis of JANSSENS, concerning the chiasmatype as a basis for the mechanism of the exchange of linked genes (p. 235) has for a long time been considered as a weak point in MORGAN's theory. MORGAN expressed himself very carefully in this respect: 'While it must be admitted, then, that the cytological evidence of crossing-over has not been demonstrated, and from the nature of the conditions it will be extremely difficult to actually prove; nevertheless, it has been shown in a number of cases that the chromosomes are brought into a position where such an interchange might readily be supposed to take place' (MORGAN, 1928, p. 44). This care was often too little practised by geneticists who were followers of MORGAN's point of view; too often this critical point in the genetical literature is ignored and the exactness of the chiasmatype is accepted as a matter of course.

Undoubtedly the representation of the exchange mechanism (as was given by MORGAN) is very suggestive through its simplicity and its apparently very plausible nature, but a scientific problem may not be solved entirely by suggestion. Previously very little was known of the cytological nature of these phenomena. One of the most clever among the younger karyologists (BĚLAŘ, 1928, p. 357) arrived at the following careful statement after a critical contemplation of the data on hand: 'considering these facts we must conclude that for the time being, the chiasmatype hypothesis is highly plausible as opposed to being improbable, however, it remains a completely unproven supposition'. BĚLAŘ was not the only karyologist to hold this careful point of view; we can also mention the names of other famous investigators here, such as WILSON, McCLUNG and SEILER. WILSON stated in 1928 (p. 959): 'In any case it is clear that the original basis on which JANSSENS founded his theory is still unsufficiently based'.

19

Likewise McClung said (1927, p. 359–360): 'Because of the need to find a structural reorganisation, as exposed by the 'crossing-over' of characters in genetical analysis, Janssens' seductive suggestion required careful study. It has had its day in court and been found wanting ... The 'chiasms' are optical effects and not structural conditions'. And Seiler (1926, p. 282): 'No observations can be found, which carry any conviction for the chiasmatype hypothesis'.

With this state of affairs existing, which does not give much satisfaction with regard to the karyological proof for the chiasmatype as an explanation of the exchange of genes, it was no wonder, that this point has been the subject of much recent cyto-genetical research. A brilliant review of this is given by Mather (1938b).

The first thing necessary in order to reach a better insight into the mechanism of the crossing-over process of the genes, was a test for the possible exchange of recognizable parts between both homologous chromosomes. Therefore, favourable material was obtained in the so-called heteromorphous chromosomes. These are chromosomes which have a different shape at one end to that at the other. These chromosomes were found in a number of organisms; they had a pin's head-like appendix at one end connected with the main part of the chromosome by a very thin thread. This appendix was called a 'satellite' or 'trabant'.

The epoch-making publication in this direction was by Creighton and McClintock (1931). In investigations mentioned earlier (p. 281) over translocations in maize, a type was obtained in which chromosome No. 9 (recognizable by a satellite on the one end) has lost a part of its other end, however, this shortened chromosome was enlarged to about its normal length again by means of the translocation of a segment of chromosome No. 8. Through systematic crossing a plant could be obtained which had a knob (K) in one chromosome No. 9 and also the dominant factor C (aleuron colour), the recessive factor wx (waxy endosperm), and a part of chromosome 8 (I, interchange, obtained by translocation).

In the homologous chromosome 9 there was no knob (k), the recessive c, the dominant Wx and no translocation (i). This plant was crossed with a normal one (in both homologous chromosomes 9 no knob (k), c, and no translocation (i), however, heterozygous for waxy endosperm: $\frac{Wx}{wx}$).

We can express this cross in a formula as follows:

$$\frac{k}{K} \left| \frac{c}{C} \right| \frac{Wx}{wx} \left| \frac{i}{I} \right. \quad \times \quad \frac{k}{k} \left| \frac{c}{c} \right| \frac{Wx}{wx} \left| \frac{i}{i} \right.$$

$$\quad 1 \quad 2 \quad 3 \qquad\qquad 1 \quad 2 \quad 3$$

It was possible to distinguish the homozygotes WxWx from the Wxwx-plants in the F_1-individuals by staining the pollen. Therefore the full-grown plants, originating from Wx-seeds, could be classified as homozygous, and heterozygous plants respectively.

The composition of the F_1-generation is evident from the following table:

Class   I. C-wx-seeds.   3 individuals.   K-C-wx-I
Class II. c-wx-seeds.   2 individuals.   k-c-wx-I
Class III. C-Wx-seeds.   1 individual.   K-C-WxWx-i (5 others not completely investigated).
Class IV. c-Wx-seeds.   a) 2 individuals.   k-c-WxWx-I
                        b) 5 individuals.   k-c-WxWx-i
                        c) 2 individuals.   k-c-Wxwx-I
                        d) 5 individuals.   k-c-Wxwx-i

The three plants of class I are evidently not exchange-products; those of class II must arise by means of exchange between c and Wx in region 2. One out of the 6 individuals of class III was investigated sufficiently. However, as it turned out, this plant was homozygous for the Wx-factor and had a knob but no translocated part and hence, it was obvious that this plant must have its origin in the combination of an egg cell K-C-Wx-i with the pollen k-c-Wx--i. Therefore, crossing-over has taken place in the mother between C and Wx. The fourth class with c-Wx-seeds comprised four types: the five individuals of type b), and the 5 of type d) are obviously products of non-exchanged gametes, whereas those of type a) originated from an exchange between Wx and I (region 3) and those of type c) by an exchange between c and Wx (region 2) or between Wx and I (region 3). In this way it seems that evidence has been produced for a mutual exchange of segments between homologous chromosomes. However, another interpretation is also possible (WINKLER's 'konversion') which we will discuss later.

Soon after this investigation an analogous study was published by STERN (1931) which led to the same results. He started off with female individuals of *Drosophila melanogaster* in which the chromosome constitution differs from the normal. They showed fragmentation of the X-chromosome into two smaller parts (x^p and x^d) or translocation of the Y'-arm of the Y-chromosome to the X-chromosome. Therefore, because of their heteromorphy, the chromosomes could easily be recognized. So STERN could obtain female flies which showed both aberrations at the same time. One X-chromosome was normal, but the

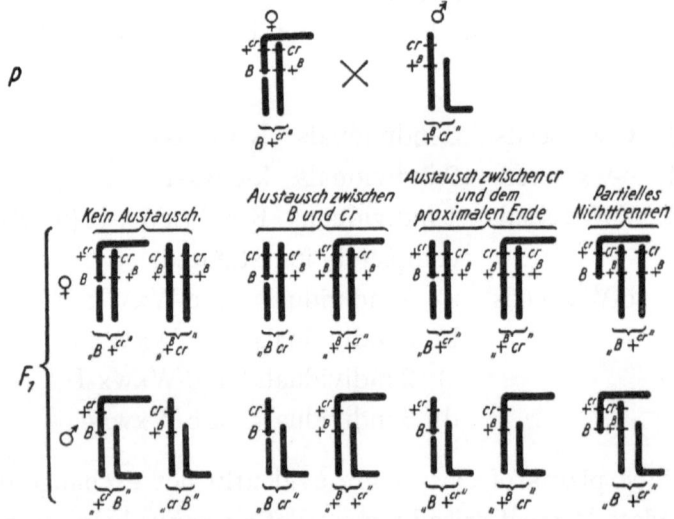

Fig. 122. *Drosophila* X.x^pY'.x^d (female) × XY (male) and their descendants (after STERN, 1931). Explanation of original german text: 'Kein Austausch' = 'no crossing-over'; 'Austausch zwischen B und cr' = 'crossing-over between B and cr'; 'Austausch zwischen cr und dem proximalen Ende' = 'crossing-over between cr and the proximal end'; 'Partielles Nichttrennen' = 'partial nondisjunction'.

other X-chromosome was fragmented and moreover the Y'-arm was attached at the end of the x^p-part. These females were heterozygous for two factors: 'cr' (carnation, a type of eye-colour obtained by MULLER after X-ray irradiation, and recessive to the normal $+^{cr}$; see MULLER, 1930b) next to the normal ' $+$ cr' and 'B' (the dominant gene for bar-eyes) next to the recessive normal '$+^{B}$' (round shaped eyes).

These females were crossed with male individuals with both recessive genes cr and $+ B$. The constitution of the mother was thus $\dfrac{B. + ^{cr}.}{+ ^{B}. \text{cr}}$

and that of the father $+ ^B$.cr. Figure 122 gives a schematic representation of the different expectations which can be developed from this cross. By non-exchange between the genes cr and B, two types of females could originate (as a consequence of the heterozygous nature of the mothers), i.e., B. $+ ^{cr}$ and $+ ^B$ cr. The first type must have both fragmented parts x^p with the Y'-arm and x^d next to the normal X-chromosome originating from the father. The other type must receive the normal X-chromosome from both parents. But if exchange occurred between $+ ^{cr}$ and B on the one hand and between cr and $+ ^B$ on the other hand, then the result must be that two other female types appeared, i.e., the combinations B.cr and $+ ^B$. $+ ^{cr}$. As both these genes could be supposed to be located in the proximal part of the X-chromosome with 'cr' close to the end and B further away, then their mutual exchange should also take place in this proximal part as a result of chromosome crossing-over.

The Y'-arm with the proximal end of x^p (containing cr) may thus have changed place with the corresponding part of the X-chromosome and by means of such a mechanism B.cr egg-cells could arise which after fertilization by $+ ^B$.cr spermatozoa caused the phenotype B.cr in the daughters. These egg-cells must therefore have the proximal part without the Y'-arm (x^p) next to the loose distal part of the X-chromosome (x^d). Hence, the reciprocal cross-over products must have the $+ ^{cr}$ gene in the X-chromosome with Y'-arm (egg-cells $+ ^{cr}$ $+ ^B$; sperm-cells cr. $+ ^B$; hence, zygotes $+ ^{cr}$. $+ ^B$; females have the normal X-chromosome together with an X-chromosome with Y'-arm).

An exchange between $+ ^{cr}$ and the proximal end can also be expected theoretically in which only the arm of the proximal x^p-part should move into the complete X-chromosome, whereas the $+ ^{cr}$.B proximal parts should get back their normal shape. These flies should be phenotypically equal to the non-exchange products but the cytological structure of the females is in this case (x^p.x^d + X) and (XY' + X). Finally an egg-cell can arise by non-disjunction in which both x^pY' and X are present. Such egg-cells therefore contain the genes cr. $+ ^{cr}$, B and $+ ^B$ and after fertilization with cr. $+ ^B$ males give rise to $+ ^{cr}$. B – zygotes. However, it was not possible to study this because it would take a life time to investigate the 8000 individuals karyologically. Therefore, a sample was taken by STERN from each phenotypical group. His comparative cyto-genetical results can be summarized in the following table:

Phenotype	Number	Chromosomeconstitution (besides X)				Non-dis-junction
		Non-exchange		Exchange B-cr		
		$x^pY'.x^d$	X	x^px^d	XY'	$x^pY'X$
B. $+^{cr}$	♀ 1978	23	—	1	—	2
	♂ 2023	1	—	—	—	—
$+^B$.cr	♀ 2118	—	26	—	—	—
	♂ 2039	—	—	—	—	—
B.cr	♀ 30	—	—	26	—	—
	♂ 31	—	—	19	—	—
$+^B. +^{cr}$	♀ 7	—	—	—	6	—
	♂ 5	—	—	—	3	—

All 107 flies karyologically investigated correspond in their chromosome configuration to the expectation. A remarkable parallel exists between the phenotype and the karyological structure, which is considered by STERN to be absolute proof of his hypothesis that an exchange of genes is based on true crossing-over of the chromosomes, that is to say, on an exchange of parts between the homologous chromosomes. Nevertheless, in this undoubtedly excellent investigation some weaknesses can be pointed out: i.e., the phenotypical group B. $+^{cr}$ should theoretically be rather larger than the group $+^B$.cr because the first comprises also the flies which originated from non-disjunction; both groups B.cr and $+^B. +^{cr}$ should be equal in numbers, however, as can be seen they deviate quite a lot from this 1 : 1 ratio (61 : 12).

A further objection can be made with regard to the fact that the investigation was carried out with flies in which translocation and fragmentation was caused by artificial methods (X-ray), so that it is perhaps possible that in the remaining physical structure of the chromosomes something was altered, and due to this the exchange of chromosome parts observed may not be equated with a normal cytological process.

Moreover, WINKLER in his discussion with STERN (concerning the interpretation) gives a completely different view, but we will refer to this later. Undoubtedly the investigations of McClintock and Creighton and of Stern may be considered as important contributions towards the question of the karyological crossing-over, and by their work, the interpretation of the chiasmata as breakage places, has gained in probability.

In order to obtain a better understanding of this theoretical matter (viz., how an exchange of chromosome parts is caused), a further discussion into the *fine-structure of the chromosomes* is first necessary.

Chromosome morphology has in the last twenty years made remarkable progress; the suggestive force of the salivary gland chromosome investigations and the refinement of karyological technique have led to the possibility of framing a clear, general picture of the structure of the chromosomes (see Heitz, 1935; Kaufmann, 1936; Geitler, 1938; Nebel, 1939; Lorbeer, 1941, and see figure 123).

The life-cycle of the chromosome during the somatic nuclear division (mitosis) and during the reduction divisions (meiosis) was superficially sketched earlier (p. 227ff). These processes are, from a physico-chemical view especially, based on a very complicated mechanism. Even the process of splitting of the chromosomes must be of a very complicated nature, because it became known that the chromosome shows a very definite longitudinal differentiation. The first 'structure-element', which is striking in all chromosomes, is a constriction arising at the place where both the split-halves become separated for the first time during the division process, and from which point they are pulled to both poles by the so-called 'spindle fibers'. This *primary constriction* (figure 123, p.i.), also known as 'kinetic constriction', 'centromere', or 'kinetochore', can lie near the centre (*median or sub-median attachment*) or almost at one of the ends of the chromosome (*subterminal*), but in such a position it is hard to observe. The functional necessity of this centromere is proved in the following manner: chromosomes without centromeres can arise by translocation of a part of the chromosome and these centromere-less chromosomes do not survive at all, whereas on the other hand, chromosomes with two centromeres are torn into parts during the splitting of chromosomes. Genetically this centromere is important, because it obstructs the exchange of genes in its neighbourhood.

Next to these primary constrictions, in a number of (but not in all) chromosomes one or two 'secondary constrictions' occur (figure

123, s.i.). The secondary constrictions are most obvious if they are situated sub-terminally and the chromosome is composed of only one thin thread at that place through which an appendix ('*satellite*', figure 123, s) is connected with the chromosome proper.

The function seems to be a purely physiological one; the secondary

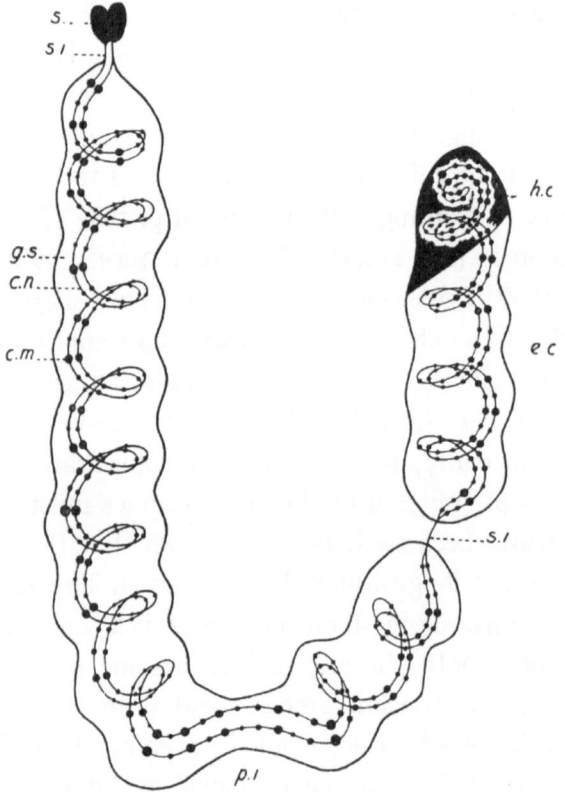

Fig. 123. Scheme of the structure of the chromosome (changed after HEITZ, 1935). Explanation: s = satellite; s.i. = secondary constriction; g.s. = enveloping substance; c.n. = chromonema; c.m. = chromomere; p.i. = primary constriction; e.c. = euchromatin; h.c. = heterochromatin.

constrictions do not play a genetical rôle. A third 'structure-element' of the chromosome is obtained in the differentiation into *heterochromatin* and *euchromatin*. We have already shown the difference between these elements earlier in this book (p. 244), karyologically the difference is that, during interkinesis (or rest stage between two successive somatic divisions) the heterochromatic parts are able to absorb stains and it is therefore possible that they can be made visible by staining, whereas the euchromatic parts lack this ability and are thus temporarily unobservable. Apart from stain absorption, both parts of the chromosomes still show a number of interesting details which are of a purely karyological nature and are therefore not very important from our point of view. Genetically speaking, the fundamental difference lies in the fact that the genes which are located in the euchromatic part are active, whereas those in the heterochromatic part remain inert or passive. This explains the distinction of *genetically active regions* and *genetically inert regions* made by MULLER in 1932 (MULLER and PAINTER, 1932). These inert regions were identified by HEITZ (1933) as the heterochromatic part of the chromosome. He was able to show an accordance between the invisible longitudinal differentiation of the chromosomes as genecomplexes and the visible differentiation in eu- and heterochromatin. This comparison led to a complete parallelism. Hence, the passive nature of the strong heterochromatic Y-chromosome became more comprehensible. Up to 1926 this chromosome was considered to be empty. Since then STERN (1926c) pointed out that this part still contains some genes.

Apart from these three structures which differentiate the chromosome longitudinally (the primary and secondary constriction and the difference between euchromatin and heterochromatin), we will consider a fourth and very important structure element: *chromonema* (figure 123, c.n.), a spirally central thread which is embedded in a basic substance (figure 123, g.s.). We are able to observe this very easily during the meta- and anaphase of the first reduction division because in this state they are coiled rather loosely as a spiral and are visible. On the other hand, in the metaphase of a somatic division they are very closely and compactly wound, and hard to recognize as spirals. WADA (1933) showed that spirals also exist in this stage. He pulled out the spiral in the living nucleus by microdissection. Properly this chromonema is composed of two spirals which are attached to each other and coiled in opposite direction. At the position of the centromere the coiling direction always reverses.

By means of a closer investigation it became clear that the chromonema coils are double: two parallel strands are visible which lie next to one another so that they can be divided laterally from one another. A very extensive literature exists over the spiral structure of the chromonema (see GEITLER, 1938 and KUWADA, 1939), but this is principally of a karyological nature and not so important from a genetical point of view.

Nevertheless, this chromonema is not a homogeneous thread; it has a regular and visible segmentation in its grains which we already met earlier in the salivary gland chromosomes as *'chromomeres'* which were connected by fibrils. As a consequence of the contraction of the coils during meta- and anaphase the differentiation becomes invisible in this stage. Chromomeres can only be observed well in the prophase of the first of the two reduction divisions and then only in the stages leptotene, amphitene and pachytene; in the strepsitene they are already invisible.

These chromomeres are of especial genetical importance. As was discussed above (p. 241) we can consider them to be the places where the genes are located, i.e., the *'genophores'*. Besides this the interest of the geneticist has been attracted by the chemical nature of the chromomeres. We will therefore come back to this point in chapter XIV.

It is obvious that, owing to such a complex structure of the chromosome, the question of the karyological crossing-over (that is to say of the exchange of parts of homologous chromosomes) will not be easy to solve. Nevertheless, in some matters an explanation can be given whereas in others, different views clash with one another.

Two questions come immediately to the foreground: 1) at what moment of the reduction division does this exchange of chromosome parts take place? 2) what are the processes which can be considered as prime causes of this exchange?

To the first question a conclusive answer can be given, at least with help of genetical arguments. In 1916 BRIDGES made it clear that it appears probable that the exchange of segments does not take place between homologous chromosomes (which are still undivided) but between one of two chromatids originating from one chromosome and one chromatid originating from the homologous chromosome. The argument of BRIDGES was based upon the origin of so-called *disomic gametes* i.e., gametes which possess an extra chromosome as well as the normal haploid set of chromosomes. These he obtained by the process of *non-disjunction*.

In a dihybrid in which the one parental chromosome contains the genes A and B, and the other contains a and b the zygote will be ABab. These chromosomes AB and ab should be only formed again by a lack of crossing-over. Hence, in this case disomic gametes have both these chromosomes (AB + ab).

From crossing-over between unsplit chromosomes the types Ab and and aB arise from one mother-cell; a disomic gamete can consequently only contain the combination Ab + aB. However, in the case that an exchange takes place between two chromatids (one from each pair)

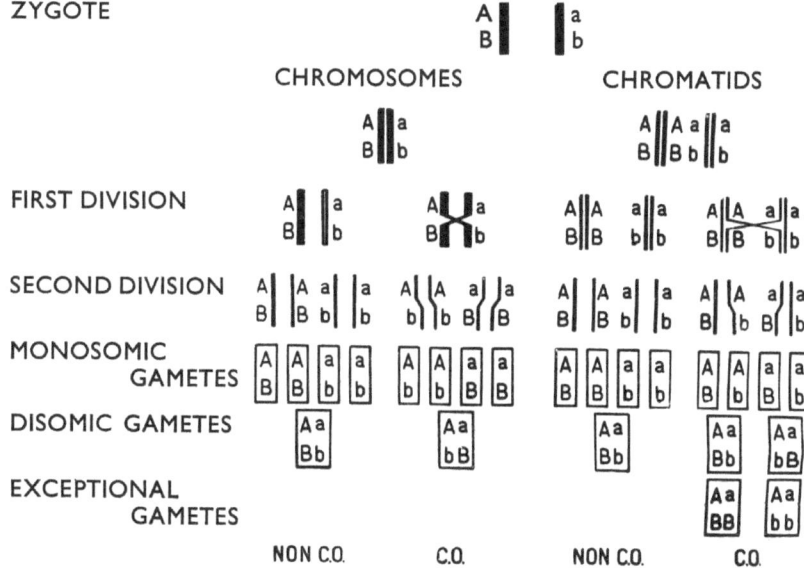

Fig. 124. Proof of the chromatid exchange (after BRIDGES, 1916).

four different chromatids originate (AB, Ab, aB and ab) from one mothercell so that other combinations are also possible (e.g. AB + aB and Ab + ab) for which otherwise no possibility of origin exists. These types did indeed occur in BRIDGES' investigations and a genetical proof was thus produced.

This proof, which is represented schematically in figure 124, was followed by many others (a.o., the elegant method of the tetradanalysis in a representative of Hepaticae *Sphaerocarpus Donnellii* of KNAPP, 1936). We can therefore consider that exchange of parts can take place only between chromatids of different chromosomes.

The karyological proof is still very hard to obtain due to the difficulty of a precise interpretation of the microscopical slides, however, the evidence obtained from good microscopical sections, points to a confirmation.

But how can we picture this exchange of chromatid segments? The simplest view concerning the karyological basis of the occurrence of chiasmata was given by SEILER (1922, 1924). As we have already seen, we are much indebted to him for the brilliant investigations into chromosome linkage in *Lymantria monacha* and *Phragmatobia fuliginosa*. His studies in *Solenobia pineti* led to the conclusion that the so-called 'combination-chromosomes' may very well be related with a factor exchange. A variety of this species with n =30 seems to have one chromosome which in another variety (n = 31) is replaced by two chromosomes, and in a third variety (n = 32) by three chromosomes. In a cross between a variety with 30 chromosomes and a variety with 32 chromosomes, this large chromosome acts as an homologue of the three other chromosomes in the other variety and it seems that stages occur in which these chromosomes fuse to one single combination-chromosome, or even the reverse, that the combination-chromosome breaks into three parts. It is obvious that factors which seem to be linked in the combination-chromosome become independent if they are distributed over three separate chromosomes. The intensity of exchange should therefore depend on the binding forces which hold the combination-chromosome together as a whole. SEILER (1924, p. 684) expresses this as follows: 'The binding forces between the separate parts of a chromosome, at least in a short chromosome, are sufficient to maintain the cohesion under all circumstances. Nevertheless, they may be also less strong so that they are only just sufficient in particular stages of the chromosome life-cycle. Therefore, every tension or pressure or chemical influence etc. can easily cause a breakage of the whole system in one or another place ... As the same forces bring about the cohesion at corresponding places in the homologous chromosomes, it is possible to consider that the same cause which leads to a breakage of the cohesion between two particular factors, can also give rise to a breakage at the same place in the homologous chromosome'.

According to SEILER's hypothesis, the process of chiasmatype should not be held responsible for the exchange of factors in the heterozygotes, but for a fragmentation of the homologous chromosomes and subsequent recombination of these segments.

It cannot be denied that such an attempt to explain the cross-over mechanism (considering the complicated structure of the chromosome) seems to be rather elementary, and can therefore not be entertained. However, this explanation is closely related to a theory which had already been put forward on genetical grounds before SEILER by MORGAN (WILSON and MORGAN, 1920) and which is nowadays still held by SAX (1930, 1936) in a slightly moderated form. These investigators considered the chiasmata as true 'cross-overs'. During the splitting of the chromosomes into chromatids (i.e., between the pachytene- and strepsitene stage), one of the chromatids of each homologous chromosome fuses. Due to the coiling of both homologous chromosomes, both chromatids break and become attached again, but this time reciprocally. Figure 125 gives a schematic representation: in a, in the pachyte-

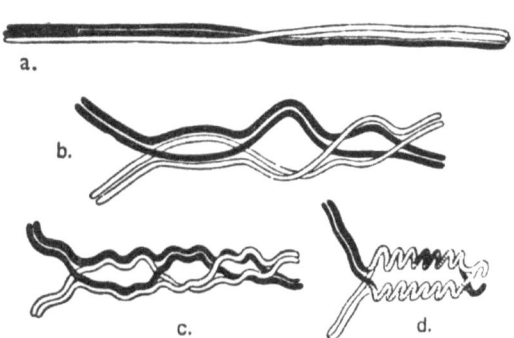

ne stage, both homologous chromosomes are next to one another, in b the chromatids are locally repelled, in c coiling of the strepsitene stage starts to become visible, whereas the contact between the chromatids from different chromosomes becomes closer and clearly shows

a.

b.

c.   d.

Fig. 125. Scheme of cytological crossing-over based on chiasma breakage (after SAX, 1930).

chiasmata, and in d both these chromatids are repelled again after partial exchange. Hence, in this contemplation the chiasma is primary; the break follows after the occurrence of chiasmata, and as a final result the repair of 'combination-chromatids' takes place.

This mechanism was accepted as possible by MORGAN and held by SAX as a karyologically probable theory. However, nowadays this proposal of such a mechanism meets a number of difficulties, which are brought forward especially by the school of DARLINGTON.

One of the many processes which seem to be in opposition to the hypothesis of SAX is that of the so-called *'interlocking'* of chromosomes (figure 126).

This phenomenon occurs, if two pairs of chromosomes during the amphitene stage make contact with one another in such a way that one or both members of the one pair is clasped between both the homologous of the other pair of chromosomes. In the first case there originates a so-called 'single interlocking'; in the second case, a 'double interlocking'. Both these interlockings remain visible upto metaphase. The double interlocking does indeed seem to be inconsistent with the hypothesis of SAX because if this occurs (figure 126 A) and homologous chromatids undergo chiasmata with reciprocal fusions, the picture will originate which is represented in figure 126 B. The interpretation of this picture on the basis of the hypothesis of SAX seems to be im-

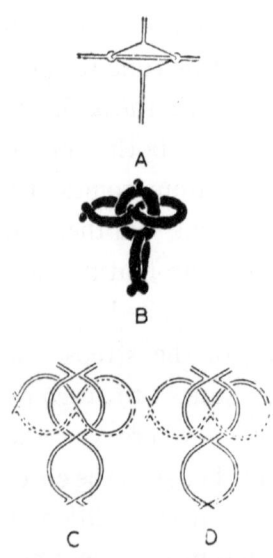

Fig. 126. Interlocking of chromosomes (after MATHER, 1933).

possible (126 C) and therefore an explanation on the basis of the figure 126 D remains. Such examples of interlocking are also known in other objects; extremely nice figures are given by HUSKINS and SMITH (1935) for *Trillium erectum* and by STRAUB (see OEHLKERS, 1937) for *Gasteria trigona*, both of which are botanical material. Figure 126 D is based on a totally different aspect of the nature of the chiasmata which was a provisional one by BELLING (1933) and which was worked out by DARLINGTON (1935a, 1936) especially.

DARLINGTON associates crossing-over with the tensions which occur as a consequence of the coilings. As a result of the winding of the chromosomes (which still remains from the preceding mitotic contraction) both homologues lay themseles in the same direction in such a manner that both their internal spirals balance with one another. This is possible on the basis of the supposition that specific forces act during this chromosome conjugation which prevent the conjugating chromosomes from winding themselves around one another. Apart from this process, each chromosome splits into two thinner and easily breakable threads: the chromatids. The span of their coils is the cause of one of the chromatids breaking. This results at once, in a rather complicated reorganization of forces acting at this point. The chromosome situated opposite is hit and one of the chromatids also breaks at the analogous point. Crossing-over occurs because the broken ends of the one spiral chromatid stretch themselves in the opposite direction and here they meet the broken ends of the other chromatid with which they fuse reciprocally. The span of the coils is then decreased by the crossing-over. The chromosomes show a longitudinal cohesion by means of which the reduction in tension has spread itself on both sides of the chiasmata.

The consequence of this is that crossing-over does not occur so very easily in these parts of the chromatids, and a secondary effect is the origin of chiasma as a consequence of the breakage.

Both hypotheses, the simple one of SAX and the more refined one of DARLINGTON to-day stand next to one another; the first one is the more obvious, the second makes higher demands on the physico-chemical conditions which are necessary; strictly speaking the latter is more probable than its primitive predecessor, however, even this hypothesis seems not to be as well grounded as it should be. The critisisms which OEHLKERS (1937, 1940) exercised, do indeed have merit.

Both hypotheses nevertheless agree in the following point: that parts of the chromatids are indeed mutually exchanged.

There is still another possibility to which WINKLER (1930) has pointed, i.e., his *hypothesis of conversion.*

The hypothesis of DARLINGTON seems attractive but it represents not more than one of the possibilities even when the evidence necessary for this hypothesis will be produced, and undoubtedly, between possibility and validity there is a great distance especially in this field of science. Moreover the occurrence of truly karyological crossing-over does not completely exclude the possibility of other processes playing a rôle.

WINKLER recognized this very clearly: 'By no manner of means can we see in this a denial of the crossing-over hypothesis, but only a denial of the assertion that the phenomena mentioned could be explained only by crossing-over' (1930, p. 163). We note that WINKLER is extremely careful in this and gives a dampener to the great over-enthusiasm of a number of investigators who considered the chiasmatypy hypothesis as the only valid one.

WINKLER gives the following hypothesis: 'In the heterozygotes (but not in the homozygotes) dominant factors are transformed into recessive-ones during the formation of the gametes in a definite percentage. The reciprocal also holds'. He calls this phenomenon conversion.

Here are several possibilities:

1) the conversion takes place in both of the homologous chromosomes (*digenic conversion*) viz.:

a) the frequency of the conversion is equal in both the homologous chromosomes; the conversion occurs in the same percentage, i.e., *equal digenic conversion*;

b) the frequency is greater in one chromosome of a pair of homologues, than in the other one; i.e., *unequal digenic conversion*;

2) the conversion in one of the homologous chromosomes excludes conversion in the other: *monogenic conversion*;

a) the frequency of conversion for each of the homologous chromosomes is equally distributed over the gamete mother-cells (*equal monogenic conversion*);

b) conversion in one homologous chromosome occurs more frequently in the gamete mother-cell, than does conversion in its chromosome partner (*unequal monogenic conversion*).

Digenic conversion upholds the heterozygous nature of the gamete mother-cell; monogenic conversion changes heterozygosity into homozygosity .

WINKLER elaborated this conversion hypothesis very carefully in a number of cases for which we will refer to his book (1930). This theory has also in every respect almost as much reason for existence as the crossing-over theory. This also was put explicitly by STERN. He says (1930 a, p. 611): 'In this way WINKLER has clearly shown that the hypothesis of crossing-over is not the only one possible, by which the *normal* (printed in bold type by STERN) results of linkage can be explained, but that the hypothesis of conversion does the same', and further (1932, p. 375): 'WINKLER succeeded in explaining the results obtained with heteromorphous pairs of chromosomes formally and completely with his hypothesis of conversion, when he assumes the existence of genes, which govern the form and the coherence or separation of chromosome parts on the spot where they are located'.

But for the crossing-over hypothesis, as well as for the validity of the conversion theory, evidence has to be produced. However, final proof is still lacking. There are some indications in the literature which support the supposition of a 'conversion'. Perhaps in the future a final proof can be made for heterozygotes, by isolating the four gametes or spores originating from a single cell (*tetrad analysis*) and by using these for marked fertilization.

Such material is available to us in species where an obvious haploid-diploid life-cycle occurs, as is the case in hepaticae, musci, fungi and algae. The segregation expected for a monohybrid Aa will be, that two out of the haploid gametes will carry the dominant alleles and two gametes carry the recessive alleles.

Indeed it seems that data exists which indicates that this is not always the case. In a series of investigations concerning the sex factor of *Coprinus fimetarius* (Aa and Bb) BRUNSWIK (1924, 1926) found four exceptions to the rule: No. 29 of his cultures yielded the combinations AB, Ab, AB and ab, i.e., 3 A :1a, No. 41: AB, aB, ab,

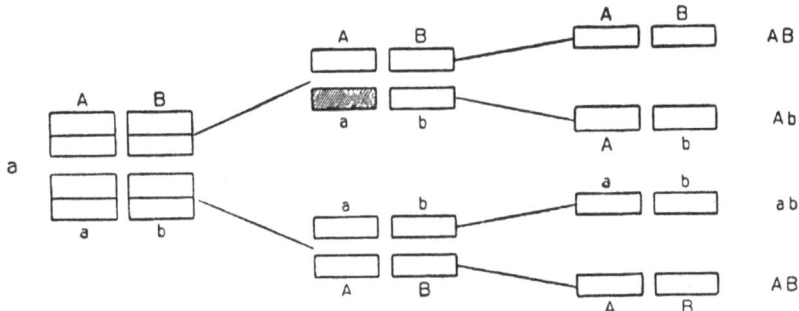

Fig. 127. Conversion of a factor 'a' into 'A' (after WINKLER, 1930).

and ab, i.e., 1 A: 3 a; No. 59: AB, Ab, aB, and aB, i.e., 3 B: 1 b, and No. 81: aB, ab, aB, and Ab, i.e., 1 A: 3 a.

BRUNSWIK describes these phenomena as due to mutation, or to quote him 'In – sich – selbst – Mutation' (mutation within itself) because a new characteristic does not originate. Nevertheless, it is obvious that this process is the same as the 'conversion' of WINKLER. Figure 127 gives a representation of the possible course of BRUNSWIK's No. 29 in which the ratio of spores expected is changed from 2A: 2a into 3A: 1a by conversion of an a-factor into an A.

Such results can be found in other investigations with fungi and hepaticae. Reliable data in this direction is rather scarce so that a complete proof of the possibility of conversion may not yet be considered as having been produced. Despite this there is still good reason for the existence of conversion next to the karyological hypothesis of crossing-over and it certainly may not be denied that both processes may play their rôle. Whatever the decision will be, the name of WINKLER will remain with merit, at least it was he who fluttered the dove cotes and compelled the geneticists, who were swayed by the attractive scheme of MORGAN to study more closely the real background of the crossing-over process.

Apart from these two hypotheses, which attempt to explain the exchange mechanism, there have been attempts to hold the physico-chemical processes as responsible (GOLDSCHMIDT, 1917; RENNER, 1920; AEBLY, 1924, KARCZAG, 1928 and others). For the time being these points of view may be considered as both uncertain and unprovable, although they comprise many interesting facts.

Thus, in the question of a mechanism for gene exchange we have at our disposal three hypotheses:

1) The chiasmatype in its original definition and in its later altered forms, for which the final proof of karyological crossing-over, as being a normal process of exchange, may be considered as having been obtained.

2) The conversion hypothesis stating that dominant factors can be converted into their recessive state if they occur heterozygously and conversely for conversion of the recessive factor into the dominant one. There are indeed indications in some cases that this hypothesis is valid but conclusive evidence has not yet been produced.

3) The supposition of physico-chemical forces as the causes of an exchange of chromosome parts or genes. This hypothesis is partly based on karyological observations, but these are not yet sufficiently developed. This view has the advantage, that it considers the chromosomes as colloid-chemical systems. Nevertheless proof is still lacking.

It is probable that the three processes act in a complementary sense.

CHAPTER XIV

THE NATURE OF GENES

The very suggestive character of MORGAN's theory concerning the basis of heritable characteristics contains the danger that the morphological aspect of it may be accentuated too much and that too much attention may be given to the localization, and too little to the nature of these factors. The physiological side of the question is not taken into consideration at all.

Nevertheless, there are some indications that the 'factors' are in reality very special chemical substances and that the whole problem of the heritable factors is of a physico-chemical nature. When considered in this way a great distance still exists between the field of activity of MORGAN's school and the background of these phenomena which, in recent years has begun to be gradually bridged by more precise investigations. We now set foot into a sphere of great uncertainty, and intuitive genius has led to interesting deviating results alongside all criticisms. Let us start with the point of view to which EAST gave expression at the International Botanical Congress in Ithaca, 1926: 'The genes are units useful in concise descriptions of the phenomena of heredity. Their place of residence is the chromosomes. Their behavior brings about the observed facts of genetics. For the rest, what we know about them is merely an interpretation of crossover frequency. In terms of geometry, chemistry, physics, or mechanics, we can give them no description whatever' (EAST, 1929b, p. 895). In the lecture, which followed this, MULLER showed the audience that the total number of genes in *Drosophila* must lie between 1400 an 1800 and that, at the utmost, they have a size of 1/50 micron.

Genetics can take comfort in the development of the atom-conception, which had played an important rôle in chemistry for over a century, before a clear cut insight was obtained into the nature and structure of the atom. LOEB was the first person who wanted to see a definite chemical substance considered as the basis of the phenomenon of inheritance. But even LOEB could not reach a conclusion from the following: 'Either morphological circumstances or its chemical nature are responsible for each character of an organism' (1906, p. 262), but in 1909 (p. 232) he arrived at the hypothesis that the nuclei of the cells of different organisms differ in chemical composition, and that these differences cause the multiformity of the organisms. Each cell nucleus should contain an enzyme for its own synthesis so that it is able to multiply. In raising hybrids, two different enzymes come together, which split to different poles during the formation of gametes and which are then distributed over the reproductive cells so that each gamete contains either one or the other enzyme. In the same year BATESON (1909, p. 266) gave his views as follows: 'What the physical nature of the units may be we cannot yet tell, but the consequences of their presence is in so many instances comparable with the effects produced by ferments, that with some confidence we suspect that the operations of some units are in an essential way carried out by the formation of definite substances acting as ferments'.

Later HAGEDOORN (1911), BEYERINCK (1917) and TROLAND (1917) expressed their views in the same way, though in rather more detail. However, it is true that the differences are still based principally on the fact that genes with particular autocatalytic substances can be identified with enzymes.

If we accept the view that factors are definite chemical substances multiplying themselves (autocatalysts or enzymes for instance), then the immediate consequence will be obvious, that laws of physical chemistry are applicable here.

On the basis of a chemical process it can be shown, how heritable factors can be identified with chemically known substances in certain cases. The question here is of two genes C and R, which in co-operation (e.g. in *Lathyrus odoratus*) (see p. 147) give rise to the formation of anthocyan; i.e., the gene C (see WHELDALE and BASSETT, 1914) is here a well-known chemical substance belonging to the group of flavones, whereas R is probably an oxygen transferring enzyme (i.e., an oxidase). In this case the chemical process could be considered as due to the co-operation of both heritable factors.

For a precise study into the problem of gene-enzyme relationship we are indebted to GOLDSCHMIDT (1916a, 1927, 1928a), who brought the quantitative nature of the genes to the foreground. A great number of investigations into sex determination and into the processes through which the 'factor' passes during ontogenetical development of the individual, have led to the view that not only can the gene be considered as an enzyme, but also as a well defined quantity of enzyme. We will return later (chapters XVIII and XXI) to these investigations, but we now discuss the conclusions which were made by GOLDSCHMIDT from his investigations: 'The gene is a particle of material which not only (as is obvious) has a specific quality but also a typical quantity at the starting-point of development, typical but different for different genes or groups of genes. The mechanism of the chromosomes, which is responsible for the presence of the right combination of genes at the starting-point of the development, shall also possess the necessary equipment to control the typical quantity of each gene. Every gene now is a substance, which firstly plays a rôle in the activation at the moment of fertilization and secondly takes part in a reaction or a chain of reactions, the specific quality of which is controlled by the quality of the gene and its working sphere (i.e. chiefly the eggplasm). The reaction-velocity however is proportional to the quantity of the gene. As a concrete idea the substance of the gene may be thought to belong to the group of autocatalytical substances. If so, the gene catalyzes a reaction with a velocity, proportional to its quantity ...... A system of gene quantities in the right dosage and the course of catalyzed reactions with proportionally measured velocities brings about that the results of these reactions, the points of determination in the development of the organism will follow in the right succession and at the right intervals' (GOLDSCHMIDT 1927, p. 40–41).

The idea that each gene contains a definite quantity of a particular enzyme, and that these enzymes therefore should be unbreakable units, has many attractions but on the other hand there are some other hypotheses which in their turn do not consider the gene to be a multiple of unchangeable enzyme molecules but a complex of mutually equal or unequal quantities which are connected in a way that can be analyzed. Building on the preliminary supposition made by CORRENS (1919) to explain the results of his investigations with

Capsella bursapastoris albovariabilis EYSTER (1925, 1928, 1929) worked out his hypothesis of *'genomery'*. His investigation had special reference to the colour pattern of variegated maize kernels. In a group of maize plants with red pericarps, a plant was found which had a kernel very much lighter in colour than the starting material. This plant with light red (dilute red, orange) kernels turned out to be inconstant with regard to the colour of the kernel and yielded an offspring which varied from pure white to deep cerise-red, and moreover, the progeny comprised some plants in which the pericarp showed a coloured white-pattern.

The colour intensity turned out to be absolutely heritable; the different colour tints of orange therefore cannot be considered to be modifications. A similar heredity was also observed in the colour pattern; light-variegated kernels yielded almost solely light varie-gated plants; deeper variegated kernels produced heavily variegated progeny.

It was also striking that in heterozygous orange-white individuals with orange as the basic colour, dark orange-red bands occurred next to white ones in almost equal widths (figure 128).

By comparison of the colour of the kernels on the same cob it became evident that important differences exist, which were always grouped in widths along the axis and whose intensity is always characteristic of the cob.

On the grounds of a large number of data, EYSTER could conclude that a somatic segregation of factors had taken place, so that the factor O, which causes the orange colour of the pericarp could not be a unit, but a complex of different elements. He said: 'The simplest conceptions of the gene which would permit quantitative segregation in the course of somatic development was to regard it as a compound structure composed of a constant number of gene elements, or geno-meres. It is assumed that a particular gene is composed of a constant number, k, of genomeres which may or may not be of the same chemical, physical, or physico-chemical nature. Granted this simple assumption, all of the known facts concerning variegations can be satisfactorily explained' (1928, p. 675).

This hypthesis was worked out by EYSTER as follows: the normal red maize type contains a gene which is composed of k genomeres C. If a change of one of the genomeres from the colour-forming type

C into the colourless type c takes place, the construction of the gene will be (k-1)C + c. If this change takes place in n genomeres a gene of the composition (k-n)C + nc will originate. As a result of nuclear division (somatic as well as generative) a gene of the structure (k-1)C + c, can yield either two genes identical to (k-1)C + c, or two different genes: (k-2)C + 2c and kC. A whole series of genic

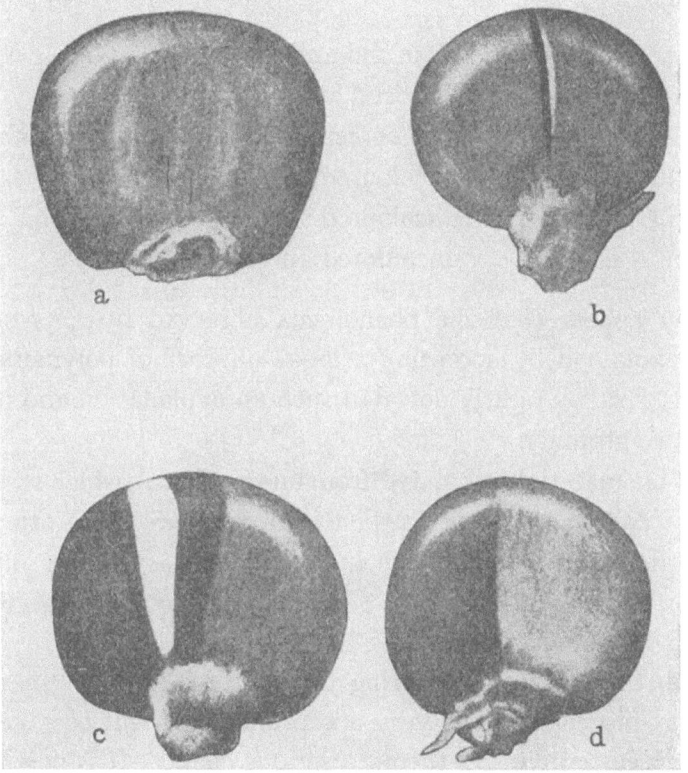

Fig. 128. Somatic colour aberrations in maize kernels (after EYSTER, 1925).

types can be derived between the absolutely stable form kC(red) to (k-1)C + c (lighter red and unstable) to (k-n)C + nc, the lightest observable colour-forming gene and likewise unstable. From these nonstable colour-forming genes, genes with greater and smaller intensities can originate. If the number of C-genomeres decreases even more, a second series follows (k-n-1)C + (n + 1)c which is really colourless but still bears possibilities of formation of genes in the somatic or generative cells which belong to the type (k-n)C + nc and

upwards, next to genes belonging to (k-n-2)C + (n + 2)c and down-
wards. Therefore, the members of these series are able to develop
colour patterns in the pericarp whose intensity depends on the number
(m) of the C-genomeres which were replaced by c-genomeres. Hence,
the extreme of this series is formed by the gene kc (when n + m = k),
that is to say, a complete lack of colour forming ability. This gene is
stable. The compositions of the series will be clear from the following
summary:

kC	plain red and stable
(k-1)C + 1c	light red and non-stable
(k-n)C + nc	just perceptible colour and non-stable
(k-n-1)C + (n + 1)c	uncoloured, but able to cause variegation
(k-n-m)C + (n + m)c	uncoloured with light variegation
kc(as n + m = k)	uncoulored and stable

It may appear that the phenomena observed by EYSTER could
also be explained by accepting a great number of polymeric genes.
However, EYSTER rightly defeated such an explanation and he based
this on two grounds:

1) All his material originated from one single cob which showed the
divergent colour of the pericarp, therefore, there is no other possibility than to accept a change in one single factor only, and 2) a
theory on grounds of polymeric genes is not able to explain the variability in the somatic cells.

Indeed, the idea of considering the gene in EYSTER's material as
a complex of genomeres is very acceptable, and will be a very constructive idea to give this theory a serious chance. It is possible that
the complex nature of the gene allows an exchange of units (genomeres) between two alleles and this can perhaps in some cases replace
an explanation based on polymery.

A number of investigations (see STURTEVANT, 1925) concerning the
'Bar'-factor, which greatly influences the number of facets of the
eye in Drosophila, led to a more or less analogous representation of
the nature of the gene. Whereas the normal eye has an average of
about 800 facets a type was discovered in 1913 ('Bar') in which the
eye was reduced to a high degree. It turned out that the factor 'B$_2$'
in question was not completely dominant over the normal factor but
it still decreased the number of facets in the heterozygotes.

The homozygously normal types $\dfrac{B}{B}$ had as their average the number 779·4; the heterozygous $\dfrac{B_2}{B}$ females 358·4 and the homozygous $\dfrac{B_2}{B_2}$ females only 68·1.

Besides being sex-linked (present only in the X-chromosome and absent in the Y-chromosome, and hence, heterozygous XX-females could originate but no heterozygous XY-males) the B_2-factor does not give any peculiarities.

However, another type was described later by ZELENY (1920) as *ultra-bar* (this was called *double-bar* by STURTEVANT) which occurred in a homozygous culture of Bar flies. The reduction of the number of eye facets occurred here in a high degree. Heterozygous with bar, gave 45·4; in a homozygous state it gave 25·0. The following table is borrowed from STURTEVANT (1925) and fig. 129

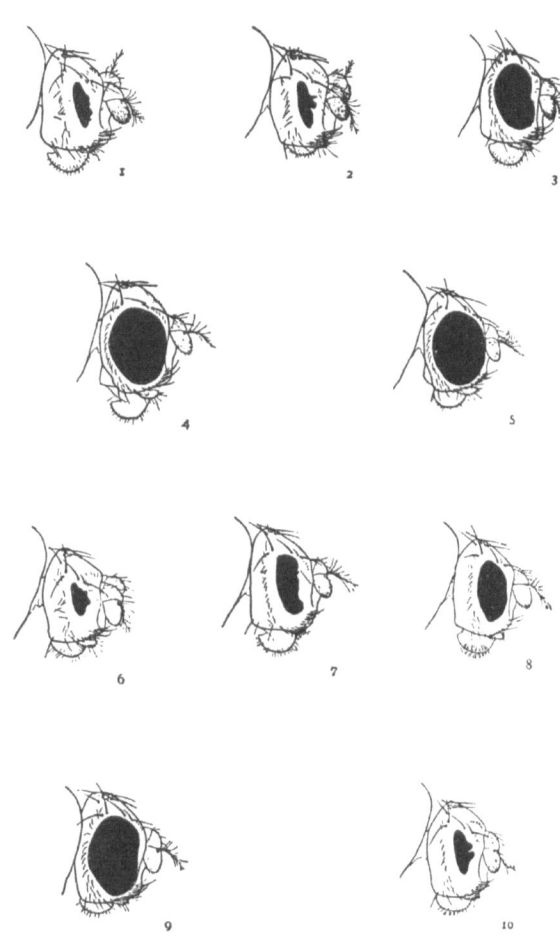

Fig. 129. Eyeforms in *Drosophila*. $1 = \dfrac{B_2}{B_2}♀.\ 2 = B_2♂.$

$3 = \dfrac{B_2}{B}♀.\ 4 = \dfrac{B}{B}♀.\ 5 = B♂.\ 6 = B_2B_2 ♂.$

$7 = \dfrac{B_1}{B_1}♀.\ 8 = B_1 ♂.\ 9 = \dfrac{B_1}{B}♀.\ 10 = B_1B_1 ♂.$

(after STURTEVANT, 1925).

gives a clear survey of the different eye-shapes of these types and of the heterozygous combinations. In this table the normal type is indicated by B, the Bar-factor by B_2, the ultrabar (double bar) by B_2B_2. The difference between the normal eye shape and the Bar type seems to exist only in

	$\male$	$\female$					
		B	B_1	B_2	B_1B_1	B_1B_2	B_2B_2
B	738.8	779.4					
B_1	478.1	716.4	348.4				
B_2	91.0	358.4	73.5	68.1			
B_1B_1	46.0	200.2	138.2	38.3	38.2		
B_1B_2	29.7	50.5	37.8	37.0	27.9	26.7	
B_2B_2	29.0	45.4	41.8	36.4	26.7	24.1	25.0

the number of facets and is therefore of a purely quantitative nature. On the other hand another factor B_1, *infrabar*, became known which not only decreases the number of facets (though less strongly than does the bar-factor itself), but moreover causes a much less well defined arrangement of the facets in the eye, so that the eye takes on a rougher appearance. An intensified form of this B_1 factor was discovered later, and was called *double-infrabar* or B_1B_1.

It is evident that this material was extremely important from several points of view: not only because the quantitative action of these factors could be determined precisely, but especially because the 'double' types (double-bar or ultra-bar and double infra-bar) originated in homozygous bar-, respectively infra-bar cultures. Apart from these aberrant forms with an intensified nature of the bar gene a higher percentage of reversions to the normal eye shape occurred in these cultures. These processes seem to be more or less opposite to one another and bar $\rightarrow$ double-bar and bar $\rightarrow$ normal could be perhaps ascribed to the same process. STURTEVANT believes that this is indeed the case and supposes an unequal cross-over between both the X-chromosomes as the cause of origin of these types. From the normal animal $\frac{B}{B}$ a heterozygous bar-normal individual $\frac{B}{B_2}$ originated in one or another unknown manner, and from this again, the homozygous $\frac{B_2}{B_2}$.

However, crossing-over now occurs between two factors located on both sides of the bar-factor (forked and fused) and the breakage place of both the chromatids ties in the one case between "forked" and "bar" and in the other case between "fused" and "bar". Hence the result will be that from two chromatids of the following constitution

$$\frac{\text{Forked / B}_2 \text{ Fused}}{\text{forked B}_2 \text{ / fused}}, \text{ after crossing-over we will obtain:}$$

$$\frac{\text{Forked} \qquad \text{fused,}}{\text{forked B}_2\text{B}_2 \text{ Fused}} \text{ as chromatids, that is to say, one chromatid with}$$

$2B_2$-genes and one without a single B_2-gene.

Such an explanation is indeed possible; it gained much in probability by the investigation of BRIDGES (1936) who could show that in salivary gland chromosomes the same part in which the Bar-factor is supposed to be located (according to experiments sector 16A of the

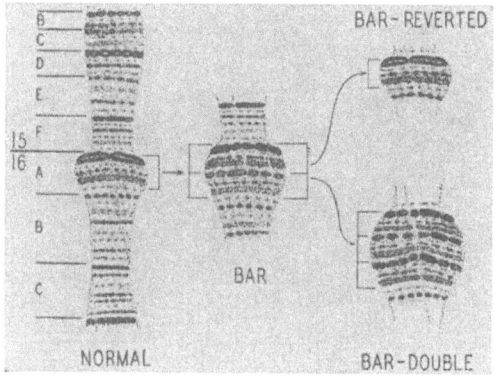

Fig. 130. Chromosome sectors with the Bar-factor in normal, Bar and double Bar individuals of *Drosophila melanogaster* (after BRIDGES, 1936).

X-chromosome), occurs once in normal and reverted (Bar-reverted) flies, twice in Bar-flies, and three times in double-Bar flies (figure 130).

STURTEVANT's hypothesis of *'unequal crossing-over'* as the cause of Bar-flies was given a sound experimental basis by this work of BRIDGES.

A third series of investigations which led to the theory of considering the gene as a complex was made by SEREBROVSKY and his school (AGOL, 1931). This theory was based on the presence and distribution of the bristles of *Drosophila* (especially on the head and the thorax). It is known from entomological taxonomy that these appendices of the insect body are very important as characteristics for determination. Hence, the complete set of hairs of a *Drosophila* fly are described according to their position (see figure 131) with the following names: orbitalia (1–3), ocellare, verticalia (1–2), post-verticale, humeralia (1–2), noto-pleuralia (1–2), prae-suturale, supra-alaria (1–2), dorso-centralia (1–2), post-alaria (1–2) and scutellaria (1–2); more to the ventral side there are still two sterno-pleuralia and some others in the region of the supra-alaria.

Since PAYNE first discovered the type of *Drosophila* in which the number and size of the hairs was decreased, more such aberrations have been found. To the factor scute [1] of PAYNE, a series of others,

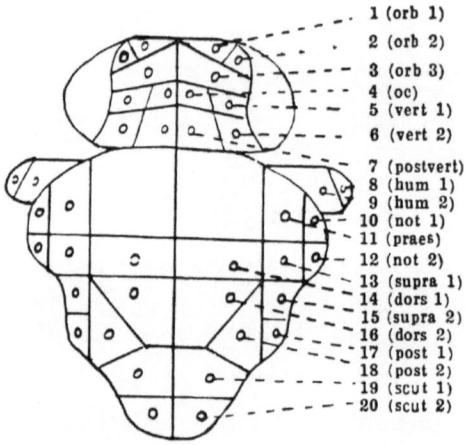

1 (orb 1)
2 (orb 2)
3 (orb 3)
4 (oc)
5 (vert 1)
6 (vert 2)
7 (postvert)
8 (hum 1)
9 (hum 2)
10 (not 1)
11 (praes)
12 (not 2)
13 (supra 1)
14 (dors 1)
15 (supra 2)
16 (dors 2)
17 (post 1)
18 (post 2)
19 (scut 1)
20 (scut 2)

Fig. 131. Location of the bristlehairs on the thorax of *Drosophila* (after GOLDSCHMIDT, 1931c).

viz., scute[2]–scute [13] have been added mostly by SEREBROVSKY and his school.

These scute-genes form a series of multiple alleles. But each individually influences a number of hairs or bristles in different combinations. Moreover, in pure breeding lines with a definite scute-gene there are some animals which are not affected by this gene which influences the hairs. This is due to the fact that the scute-genes are sex-linked and therefore their effect on male animals is usually different to that on female animals. The following table lists these effects. The data are borrowed from AGOL (1931) and DUBININ (1932) (females with scute [3] are lacking because they are not viable):

	sc¹		sc²		sc³	sc⁴		sc⁵	sc⁶	sc⁷	sc⁸	sc⁹	sc¹⁰	sc¹¹
	♀	♂	♀	♂	♀ ♂	♀	♂	♀	♀	♀	♀	♀	♀	♀
ocellare . .	5.0	2.0	100.0	100.0	—56.0	0.0	0.0	100.0	0.0	0.0	100.0	0.0	100.0	100.0
postverticale	39.0	9.0	94.0	87.5	— 3.0	19.0	11.8	100.0	27.0	15.0	100.0+ 5	100.0	100.0	100.0
orbitalia 1–2	18.0	1.0	100.0	98.0	—16.5	0.0	0.0	100.0	1.5	0.0	100.0	0.0	100.0	100.0
orbitale 3 .	100.0	100.0	100.0	99.0	—52.0	0.0	0.0	100.0	93.0	84.0	100.0	95.0	100.0	100.0
verticale 1 .	100.0	100.0	100.0	99.0	—18.0	100.0	100.0	100.0	100.0	100.0	100.0	100.0	100.0	100.0
verticale 2 .	100.0	100.0	100.0	100.0	—59.0	0.0	0.0	100.0	100.0	100.0	100.0	100.0	100.0	100.0
dorsocentralia 1–2 . .	100.0	100.0	100.0	100.0	—24.0	100.0	100.0	100.0	100.0	100.0	95.0+14	100.0	28.0	100.0
dorsocentralia 3–5 . .	100.0	100.0	100.0	100.0	—24.0	100.0	100.0	100.0	100.0	100.0	96.0+ 9	100.0	2.0	3.0
scutellare .	40.0	19.0	39.0	19.0	—95.0	23.5	0.0	80.0	100.0	0.0	100.0	0.5	100.0	100.0
humerale .	100.0	100.0	75.5	74.5	—80.0	17.0	0.0	100.0	100.0	100.0	99.5	98.0	100.0	100.0
praesuturale	100.0	100.0	100.0	98.0	—30.0	0.0	0.0	100.0	100.0	100.0	99.0+ 4	100.0	100.0	100.0
notopleurale 1	7.0	2.0	100.0	100.0	— 6.0	57.0	9.0	100.0	62.0	100.0	100.0	12.0	100.0	100.0
notopleurale 2	100.0	100.0	100.0	100.0	—83.0	100.0	96.0	100.0	100.0	100.0	100.0+ 4	100.0	100.0	100.0
supra-alare .	100.0	100.0	100.0	100.0	—47.0	100.0	100.0	100.0	100.0	100.0	66.0	97.0	69.0	100.0
postalare . .	100.0	100.0	100.0	100.0	—75.0	70.0	61.6	100.0	99.0	71.0	100.0+49	95.0	100.0	100.0
coxale . . .	0.0		100.0		—	0.0		100.0	0.0	0.0	100.0	0.0	100.0	100.0
sternopleurale . . .	17.5		17.1		—	15.8		18.2	18.2	12.1	27.0	12.5	15.5	14.9
sternitale .	32.0		16.5		—·	30.4		25.2	56.1	18.5	65.3	17.4	50.4	52.4
ventrale . .	100.0		100.0		0.0	0.0		100.0	100.0	100.0	100.0+28	100.0	100.0	98.0

The double figures scute [8] means that not only does this gene reduce the bristles, but it also increases their number in some areas of the body surface. A sample of the phenotype of these animals is given in fig. 132, where, as well as the normal, also scute [1]–scute [7] are represented.

	Dorsocentralia.	Supra-alaria.	Verticale 1.	Notopleurale 2.	Verticale 2.	Praesuturale.	Orbitale 3.	Notopleurale 1.	Ocellare.	Orbitalia 1 en 2.	Coxale.	Postverticale.	Abdominalia.	Scutellaria.	Humeralia.	Postalaria.
Scute – 1.	+	+	+	+	+	+	+	—	—	—	—	—	—	—	+	+
Scute – 2.	+	+	+	+	+	+	+	+	+	+	+	—	—	—	—	+
Scute – 3.	—	—	—	—	—	—	—	—	—	—	—	—	—	—	—	—
Scute – 4.	+	+	+	+	—	—	—	—	—	—	—	—	—	—	—	—
Scute – 5.	+	+	+	+	+	+	+	+	+	+	+	+	—	—	+	+
Scute – 6.	+	+	+	+	+	+	—	—	—	—	—	—	+	+	+	+
Scute – 7.	+	+	+	+	+	+	+	—	—	—	—	—	—	—	+	—
				a		b	c	d		e		f		g	h	i

A mutual comparison of the data given in the table shows that the effect of the different alleles is very different, but certain regularities can still be found. This regularity, which according to SEREBROVSKY and DUBININ (1930) is evident in the grouping of the

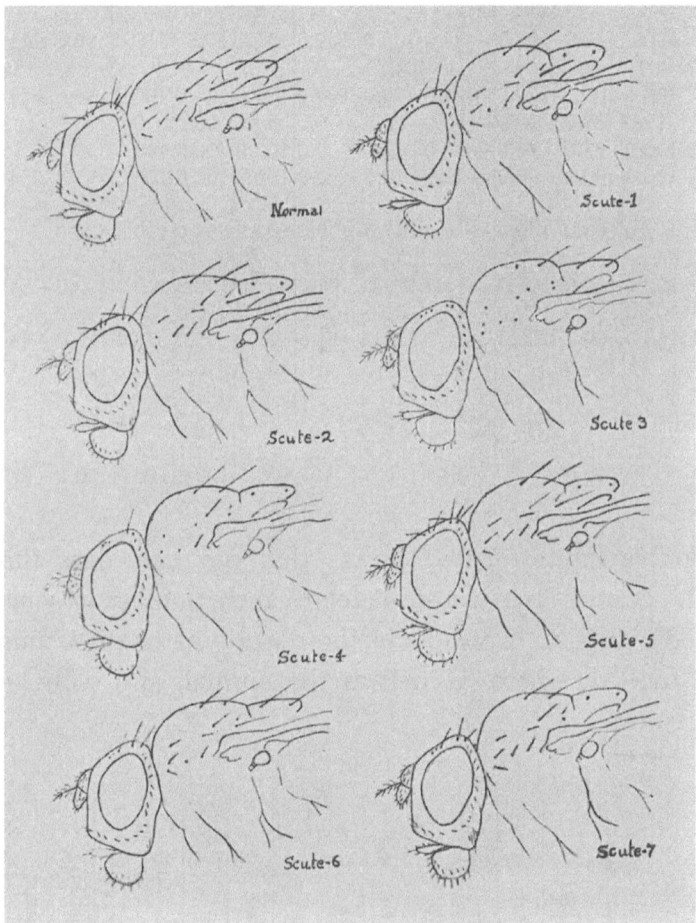

Fig. 132. Types of reduction of the bristle hairs of *Drosophila* (after SEREBROVSKY and DUBININ, 1930).

different hairs, is shown in the second table on p. 317 (completely developed hairs or bristles indicated by +, incompletely developed hairs by —).

It is obvious from this table, that scute [3] has the greatest effect and practically completely reduces the length of all hairs; next to this is scute [4] and then the others, each causing a definite combination of normal and reduced hairs. These phenomena are described by the school of SEREBROVSKY (AGOL, DUBININ, LEVIT, and others) and are called *step-allelomorphism (Treppenallelomorphe)* by them.

But after this MULLER and his co-workers (MULLER and PROKO-FYEVA, 1935; MULLER, PROKOFYEVA and RAFFEL, 1935) were able to show that breakages at different places in or near the 'scute-locus' in the chromosome could be held responsible for the different types of action of the scute-gene so that for these phenomena a purely mechanical cause could take the place of the hypothesis of step-allelomorphism.

By all appearances these investigations have made the oneness of the gene a more or less academic matter of dispute. Hypothetical contemplations concerning its complexity seem to make way for indications in favour of the more probable oneness although this oneness turnes out to be breakable.

However, such a postulation is not valid. From many aspects there is still evidence for a complexity of the gene, but we have to admit that the final proof for such a complexity is not yet at hand. We already mentioned the approach of DUBININ (1929) and SEREBROVSKY (1930) into this subject; their suggestion of step-allelism could not be approved of, but in recent research, cases of varying locations of identical or almost identical genes appear again. In order to get an intelligable terminology it is however, necessary to make a sharp distinction between several related phenomena.

We have mentioned the occurrence of *multiple allelism*. It is of great importance that we stick to the original definition that is to say, a group of factors from which each member is able to form a pair of alleles together with another member of the group. It is obvious that such a group of factors forms an interchangeable series, that is to say the localization of each factor of the series is the same.

Another kind of allelism is suggested as *step-allelism*. According to the original literature we will define this as: 'the phenomenon that the activities of the various members of a series of multiple alleles partly do not coincide or completely do not coincide. This partial absence of coincidence is shown by the dominance of a part of the characteristics produced by one allele. This dominance leads to a partial reversion to the wild type, which is as constant as the complete

reversion, which occurs when two genes of different loci are combined'
(DUBININ, 1929, p. 337). In this definition it becomes clear that we
deal with almost multiple alleles, but instead of their being similarly
localized in the chromosome, the alleles overlap one another. This is
not only true of their location but also of their physiological effect.

A third valuable concept is that of *polymery* or *polygenism*, that is
to say, more than one (i.e. two, three, etc.) genes influence one
characteristic (cf. p. 129 ff.). These factors cooperate physiologically
and may or may not segregate independently from each other, i.e.
may or may not be located in the same chromosome.

It seems a pity that genetics was not able to limit itself to these
three concepts, which may look sufficient to cover each phenomenon
concerning the problem of "allelism" in all its detail.

The main difficulty arises from the fact that the concept of the
gene can be approached from two sides: the purely morphological or
karyological side, and on the other hand the biochemical aspect. As
PONTECORVO formulated (1952, p. 218): "The gap between the
morphological and the biochemical approach to biological problems
is due mainly to our inability to grasp the significance of the organization in space of biochemical processes and to attack it experimentally". And this is certainly true for the gene.

A clear exposition of these problems has been given by RAFFEL
and MULLER in 1940, and this seems to be still valid to day.
We may quote the following from their paper (1940, p. 570): 'In
genetic theory, genes have been considered as (1) crossover units
— hypothetical segments within which crossing-over does not occur;
(2) breakage units — again hypothetical segments within which
chromosome breakage and reattachment do not occur (at any rate,
not without destruction of one or both fragments); (3) mutational
and functional units — those minute regions of the chromosomes,
changes within one part of which may be so connected with changes in
the functioning of the rest of that region as to give rise to the phenomenon of (multiple) allelism; or (4) reproductive units — the
smallest blocks into which, theoretically, the gene-string could be
divided without loss of the power of self-reproduction of any part.
A category of auto-attractive units might also be added.'

'Although it seems often to have been assumed, there is as yet
no empiric evidence, and only doubtful theoretical ground, for as-

suming that the lines of demarcation between genes, as defined on any one of these systems, would coincide with those on any of the others, or even for assuming, in the case of some, given one of these systems (especially the mutational one) that such lines of demarcation are necessarily invariable, non-overlapping, well-defined and absolute. The minimal gene numbers and maximal gene sizes hitherto arrived at have to be considered in relation to the particular method used in arriving at them, one value having been a limit obtained through crossover studies, another through studies of mutation (allelism), and the most recent one through breakage studies. Hence they do not necessarily represent the same thing, even though it might be simplest to assume that they did'.

'In view of the above consideration, it is not far fetched to imagine that the 'gene for scute', as recognized by the test of allelism of its mutations, may nevertheless consist of an undetermined number of parts, some of which may become separated from others by breakage without such separation resulting in the loss of the reproductive power of any of these parts (although, perhaps, interfering with their functioning). These parts, then, might themselves be denoted 'genes', and the whole a 'gene-complex', or the parts might be called 'sub-genes' or something equivalent, and the whole 'gene', depending upon the taste of the writer and upon the criterion which he prefers (that of 'allelism' or that of breakage) for defining the limits of a 'gene'.'

Because of this difficulty we find references in the literature of many other terms, from which we will take for discussion: 'pseudo-allelism' and 'position pseudo-allelism'. LEWIS's investigations (1939, 1951) into the problem of the 'Star' and 'asteroid' genes in Droso-phila melanogaster are of significance in this matter. 'Star' (S) refers to rough eyes with a rumpled hairy surface. This dominant factor was, as a gene, located at locus 1·3 on the second chromosome. LEWIS (1939) obtained a recessive mutation ('star-recessive') with small rough eyes when homozygous. The gene involved was also located at locus 1·3 on the second chromosome. Flies heterozygous for 'Star' and 'star-recessive' had very small, rough eyes with many of their facets fused. It appears therefore, that the dominant character 'Star' does not show up in the hybrid. The phenotypical expression of both these factors, and their similar mapping distance indicate that these factors behave as alleles, that is to say, mutants for the same gene. LEWIS was able to prove this hypothesis by showing that flies heterozygous for 'Star' and 'star-recessive' give rise to two types of germ-cells. This can be expected in the case when there is a simple segregation of alleles.

Hence, up to this point it appears that this case points to the existence of a series of multiple alleles; however in a very careful and extensive study LEWIS (1951) showed that these two mutant factors were separable by means of crossing-over. With regard to the crossing-over frequency he calculated that the distance between these two genes was 0·02 units (20 micro Morgan). The difference in location led to the alteration of the name 'star-recessive' to 'asteroid' (ast.). By means of combining both types into double heterozygotes $\dfrac{S^+ \text{ ast.}}{S \text{ ast}^+.}$ and $\dfrac{S \text{ ast.}}{S^+ \text{ ast}^+.}$ he found remarkable differences in the phe-

notypes: $\dfrac{S^+ \text{ ast.}}{S \text{ ast}^+.}$ (trans-arrangement) had very small rough eyes and

other abnormalities. The cis-arrangement $\dfrac{S \text{ ast.}}{S^+ \text{ ast}^+.}$ gave flies which

are similar to those heterozygous for 'star' only. The effect of 'Star' is dependent on whether ast. or ast$^+$. is located next to it. From a theoretical point of view this kind of *position effect* has significance for the origin of new genes.

New genes may arise in a process composed of two steps. A gene may duplicate into two similar genes. By means of mutation of one of these two a deviation of function can be established. Such a process however, could lead to a related function (and hence to a probable formation of a chain of reactions) of the mutated gene towards the old one. LEWIS assumes such a reaction-chain to occur on the chromo-somal surface for normal eye development in *Drosophila*:

$$\begin{array}{cccc} A \text{ ast}^+ & B \text{ S}^+ & C & \\ \text{------}\!\!\to & \text{------}\!\!\to & \text{------}\!\!\to & \text{--}\text{--}\!\!\to \end{array}$$

In the trans-configuration this chain is blocked in each chromosome and hence an extreme effect can be expected. In the cis-configuration the chain is blocked in one chromosome only and it can proceed unchecked on the other.

LEWIS refers to the case of 'star' and 'asteroid' as a case of *position pseudo-allelism*. However, McCLINTOCK (1944) put forward that the term '*pseudo-allelism*' presupposes a knowledge of some special alteration which accompanies the expression of allelism. When this knowledge is not present no 'pseudo' modifies the term allelic. Various causal factors produce mutations and are responsible for allelic expression. Mutants giving allelic expression need not be located at comparable positions in homologous chromosomes and they need not be inseparable by the mechanism of crossing-over.

Hence there is no need for the term '*position pseudo-allelism*' and we may classify the case of 'Star' and 'asteroid' in *Drosophila* under the heading pseudo-allelism which implies a special alteration in this case, a position effect.

Pseudo-allelism can be defined as follows: When an apparent multiple allelic series is shown to be fractionable by crossing-over, and when a special chromosomal alteration accompanies the expression of allelism (i.e., a position effect due to the occurrence of a trans-location or to a deletion) they are called pseudo-alleles. From the case of LEWIS it became evident that the adjacent loci interact in position effect, which may suggest a close relationship in develop-mental effect of the genes involved. However, this is not so much a characteristic of pseudo-allelism, rather it is a possibility arising from the occurrence of position effects. Hence the definition of SRB and OWEN (1952, p. 416) postulating the necessity of an interaction of adjacent loci in (position) pseudo-allelism refers to a rather special case of pseudo-allelism which includes interaction of position effects. SINNOT, DUNN and DOBZHANSKY (1950) refer to pseudo-allelism as being due to several genes with related effects as those common to members of an allelic series, which have turned out not to be alleles but to be due to alterations in separate genes which are so close together that in the great majority of gametes they behave as alleles. In spite of its incorrectness, because this definition is also valid for a special case of a close linkage group, it describes a very significant fact. In this case we are dealing with a group of closely linked genes, which are all similar, or almost similar, in their phenotypical ex-pression. From the chemo-genetics of *Neurospora*, to which we return in chapter XXI, we will borrow two examples to explain the signifi-cance in more detail. These are not the only existing examples, cases are known in other fields as well (for example corn, *Drosophila*, *Aspergillus*).

BONNER (1951) describes a case dealing with the "Q" locus in *Neurospora crassa*. Strain 3416 requires niacin for growth (fig. 133) and is blocked in its ability to convert 3-hydroxyanthranilic acid into niacin (nicotinic acid). This strain therefore accumulates quinolinic acid. In other mutation experiments allelic strains were obtained, that is to say biochemically allelic, requiring niacin and non-utilizing 3-hydroxyanthranilic acid. However, these strains are not absolutely identical. One strain turned out to be temperature sensitive that is to say, requires niacin only above 30°. The biochemical identification could be proved. By means of crossing these strains niacin-independent cultures were obtained; the frequency of occurrence depends on the cross involved. The five strains used can be subdivided into three classes with respect to the frequency with which niacin-independent cultures are found; those which yielded no niacin-independent cultures (A), those which yielded about 10% niacin-independent cultures (B) and those

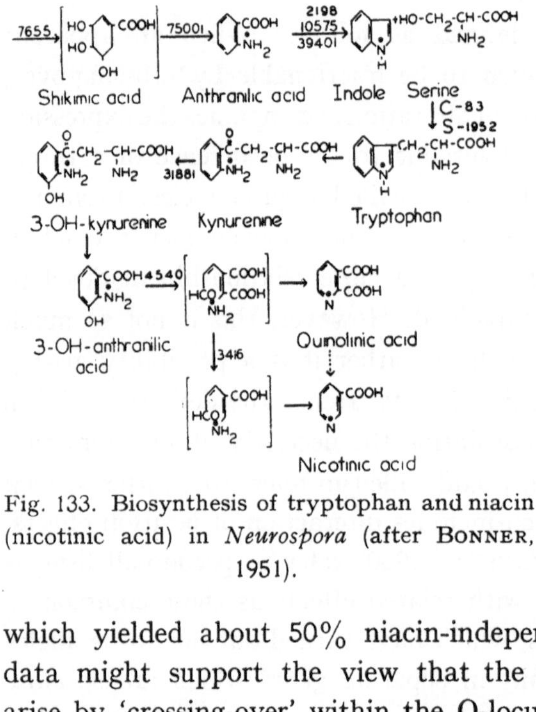

Fig. 133. Biosynthesis of tryptophan and niacin (nicotinic acid) in *Neurospora* (after BONNER, 1951).

which yielded about 50% niacin-independent cultures (C). These data might support the view that the niacin-independent strains arise by 'crossing-over' within the Q-locus.

Another case of presumed crossing-over within the locus can be borrowed from the studies of YANOFSKY (1951) and WEIJER (1954) with regard to the tryptophan desmolase locus in *Neurospora*.

Several strains are known which are biochemically identical, that is to say, are blocked at the same step in the synthesis of tryptophan. However mutual crossings give rise to tryptophan-independent cultures. The significance of this phenomenon seems to be of some importance for the gene concept. Providing that the strains involved, are karyologically free from rearrangements and providing that the rate of back mutation is very low, there is a possibility that a 'crossing-over' within the locus exists.

It is not easy to rule out chromosomal rearrangements. From the work of WEIJER (1954) it became evident that the genetical behaviour of these strains is frequently very abnormal. Crossing to a morphological mutant 'fluffy' does not give a regular pattern of a Mendelian segregation as could be expected, but gives several classes which are incomplete in number. There are several causes which may be responsible for such an abnormal segregation. However, since chromosomal rearrangement is often the cause of a low germination we may as well account this as being due to a translocation or another kind of rearrangement. The karyological approach is, in the case of *Neurospora*, very difficult, however, without a direct karyological investigation we are able to obtain more knowledge of the chromosomal structures involved. WEIJER was able to mark the rest of the chromosome by a morphological marker and in this way could establish the successive crossing-over frequencies between this marker and the centromere. From this data it is possible to derive some conclusions with regard to a probable existence of translocations, deletions etc.

It is also necessary to form an idea of the spontaneous reversion rate of the stocks involved. In none of the cases is the spontaneous reversion rate higher than the frequency of crossing-over observed in the crosses.

In both examples, i.e., in the Q-locus as well as in the tryptophan desmolase locus, we face the problem that there is some evidence that these strains, while having similar biochemical blocks, represent alterations of genetically separable genic material.

BONNER (1951, p. 151) states that: 'these 'alleles' may represent three closely linked units, all of which must be present together in the same nucleus for expression of the normal Q-locus in *Neurospora*. These units appear genetically as three closely linked genes. Biochemically, however, they appear as three component parts of a larger unit which is necessary for the performance of a single specific reaction characteristic of niacin formation. Thus the nature of a gene controlling a biochemical reaction appears very different when viewed from a biochemical standpoint'.

BONNER refers to this phenomenon as a case of '*pseudo-allelism*'. However, pseudo-allelism is used here erroneously in the sense of closely linked genes. At first sight this may appear very unimportant but on closer examination of the data it becomes clear that BONNER (1951) on the grounds of crossing-over data only, comes to his conclusion of the complexity of the gene, that is to say, a gene in some cases can be considered as a group closely linked units which all need to be present for a normal genic expression. HOROWITZ and LEOPOLD (1951) on the other hand stick to the one gene — one enzyme relationship and denies a greater complexity in the gene. Between these extremes WEIJER supports the view that more genes are involved in the synthesis of one enzyme. These genes may be different (qualitative) or they may be similar (quantitative) in effect or their expression may even depend on the presence of other genes. However, there is no reason, and no evidence, for a complexity of the gene itself. At this moment this phenomenon is fully explained by the postulation of its being due to a strongly linked group of factors. In this sense we almost approach the borders of the phenomenon of *polymery*. It is even possible to denote this process as being polymery of a special kind i.e., the factors involved are strongly linked. According to the biochemical tests we are inclined to say that these factors all have a similar effect and therefore we could speak of isomeric factors although whether we deal with cumulative or non-cumulative isomeric factors is hard to say.

At this moment it is hard to foresee what the truth will be. It is possible that this whole phenomenon is based on the occurrence of minute chromosomal rearrangements however, it may also be correct that the gene does not form a unity inseparable on crossing-over and that hence the gene is composed of separable units.

The question arises: are these units similar in their activity (forming a quantitative series) or are these units each responsible for a step in the synthesis of a biochemical product or even requires one gene for normal action full integration of all its parts, parts which are separable by crossing-over? We donot know, but intuitively, we feel that there is more beyond the gene than we previously thought. PONTECORVO (1952) in his studies with *Aspergillus nidulans* concludes that in all these cases we are dealing with heterogeneous groups of phenomena which are not as yet fully analysed and when analysed it has been with different techniques for different cases.

The full information necessary should therefore include the following requirements:

1) We should know in which kinds of heterozygous arrangements the normal effect breaks down. The crucial comparison in the case of two genes, is that between the cis-arrangement and the trans-arrangement (interaction of position effect).

2) We should have valid proof that the apparent recombinants actually do originate from normal crossing-over (ruling out chromosomal rearrangements).

3) We should try to discover, by chemical and biological tests, the possible non-identity of the elementary action of 'pseudo-alleles' (closely linked isomeric genes) (PONTECORVO, 1952).

Apart from this fundamental problem which is of great importance but is, on the other hand, difficult to approach in its present form, there is one other, no less essential problem concerning the *nature of the gene*, a problem ripe for development.

Two paths led to an insight into this very interesting part of genetics; a direct method which touches the gene in its being and tries to open up the character itself, and an indirect method, which tries to probe the gene as it acts and wields itself in the phenotype. We will limit ourselves preliminarily to the first point. Later we will discuss the second one, when we will deal with the development of the individual with respect to its hereditary dispositions (chapter XXI).

The data at our disposal in the former direction (i.e., that of direct attack on the gene) is borrowed from the results obtained by TIMO-FÉEFF-RESSOVSKY (1937, 1940b), and with his co-workers DELBRÜCK and ZIMMER (1935), concerning the influence of X-rays upon the process of so-called gene-mutation, that is to say, the origin of the changes in the gene itself as wielded by the X-rays.

The work of TIMOFÉEFF in this way forms a part of a greater whole, which comprises the biological effect of the different rays, in other words comprises *radiation biology* (see DESSAUER, 1954, DUGGAR, 1936, LEA, 1943 and ZIMMER, 1937). In a later chapter (XIX) we will discuss this phenomenon in more detail. The following points are of importance with regard to the problem of the nature of the gene:

1) The frequency of spontaneous mutation is small, the average for *Drosophila* can be estimated to be about 0·0005%, from a fixed curve of variability on which each gene has its own place.

2) This mutation frequency can be increased considerably by treatment with rays of short wave-length (X and gamma rays).

3) In general, the same new types occur through this kind of mutation caused by radiation, only a few irradiation mutants are unknown in their spontaneous form.

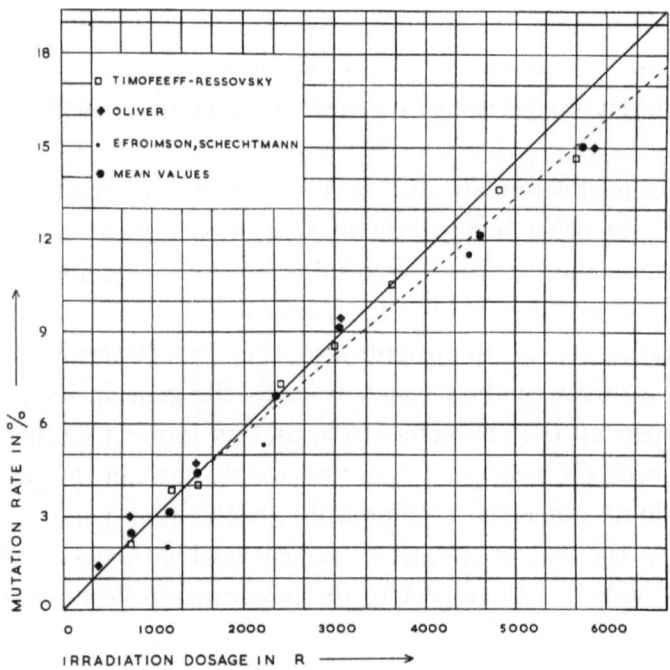

Fig. 134. Relation between the total irradiation dosage and mutation frequency (after ZIMMER, 1934).

4) The irradiation directly affects the genes, bearing in mind that the properties of the tissue, the physiological state of the cells and other circumstances of minor importance will have some influence upon the mutation frequency.

5) The mutation frequency provoked by irradiation is directly proportional to the total summation of r-units received (fig. 134), this means that the distribution of these doses over shorter or longer time has no significance, provided that the total dosage remains unchanged (fig. 135). Also the type of irradiation (soft or hard X-rays

or gamma-rays) is of no significance providing the total dosage of units remains unchanged (fig. 136).

TIMOFÉEFF-RESSOVSKY related these results to the work of DESSAUER (1922, 1933) concerning his '*Punktwärmetheorie*', which developed out of the '*Collision theory*' in Physics, which states, that the biological effect of the rays is caused by the irradiation (or its secondary products) within a clearcut sensitive area. Hence the effect of irradiation is due to the action of the electrons, provided that the energy is concentrated at very localized points and with great strength. In these cases the energy is not converted into heat in the irradiated area. From the fact that rays act by the direct effect of their quantums (or of their secondary products) and not by means of indirect physiological plasmic alterations upon the genes, it is clear that they cause definite changes in the very limited area of activity. ZIMMER (1934) then produced evidence, that the relation between the irradiation dosage and the frequency of the genemutation is given by the

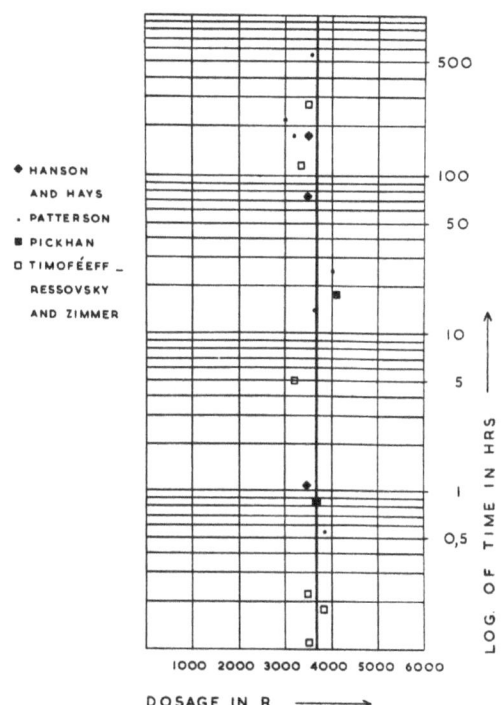

Fig. 135. To show that the mutation frequency is independent of time (after TIMOFÉEFF-RESSOVSKY, 1937)

formula $x = a(1 - e^{-kD})$ where x = the number of mutated genes, a = the number of irradiated genes, D = irradiation dosage, k = a constant of velocity and e the base of the natural log. (the exponential function).

In 1922, BLAU and ALTENBURGER put forward a general formula for damage, based on the theory of DESSAUER, in which n appears. This represents the number of impacts necessary to produce damage. It turned out that the formula of ZIMMER could be derived from the previous formula provided that n = 1. From this it would seem to be evident that only one impact is sufficient to provoke one gene-mutation. On this basis TIMOFÉEFF-RESSOVSKY, ZIMMER and DELBRÜCK (1935) concluded that a gene-mutation is a shifting of atoms. That is to say, an atomic reaction within the molecule which should act like a gene. Thus a gene should have the form of a definite and separate stable atomic complex which should have a certain independence in its activity and relations to other genes (see DEHLINGER, 1937; SOMMERMEYER and DEHLINGER, 1939; DEHLINGER and WERTZ, 1942). We now have a definite model of a gene which can be discussed. By means of this model a link was made with the conception of the chemical structure of the chromosome as a gene complex, which will be discussed later on in the book. It was not only with respect to the

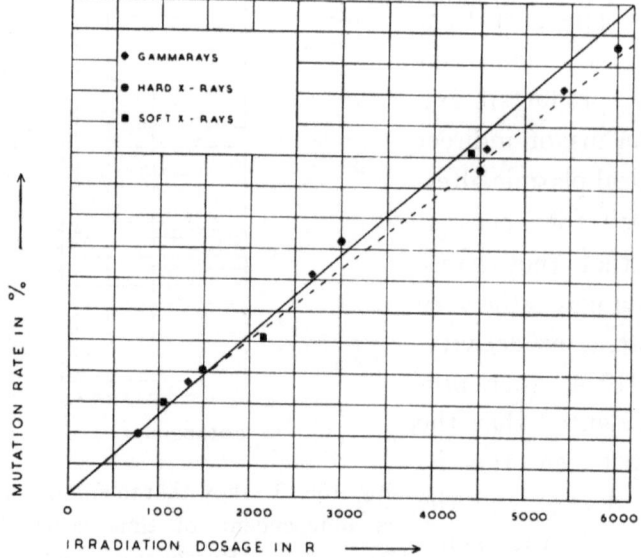

Fig. 136. To show that the mutation frequency is independent of the type of irradiation (after TIMOFÉEFF-RESSOVSKY, 1937).

structure of the gene but in other directions also that the work of TIMOFÉEFF-RESSOVSKY opened up new avenues of investigation: namely the size of the genes. In the beginning of this chapter we saw that MULLER demonstrated that the size of a gene should be at the most 1/50 micron (1 micron = 10^{-3} mm). In 1943 he came to the conclusion that genes are rod-shaped with a length equal to 6–30 times their diameter, and a length of 0·12 micron and a diameter from 0·02–0·004 micron. This conclusion was based on the localization of the salivary gland chromosome, and the proportions of length between these chromosomes and the normal ones.

These calculations are stimulating but not irrefutable. Nevertheless they are supported by calculations of LEA (1940 a and b) based on his X-irradiation-experiments, who found as an average for the area of impacts for *Drosophila* a volume of $2·9 \times 10^{-20}$ cm^3. This led, after some conversions to dimensions analogous to those obtained by MULLER. The order of volume is very important and may be

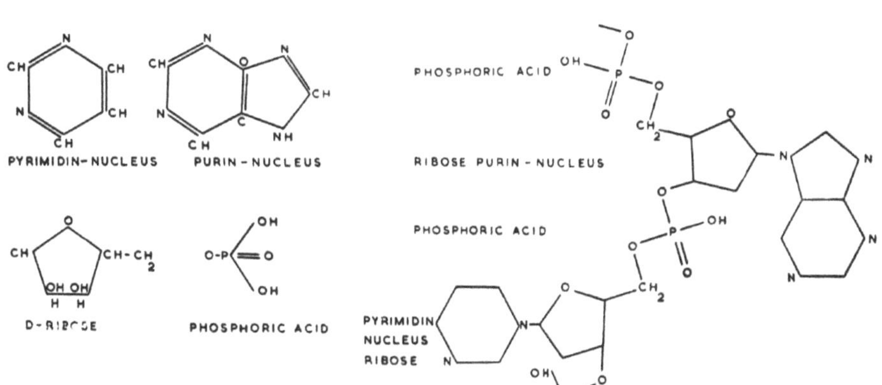

supposed to exist in genes equal to those in which the biggest protein molecules are located. This kind of investigation, comes nowadays more to the fore as *irradiation genetics*, and forms a field of science on its own (see Chapter XIX). The contributions of conferences concerning this sphere (Symposium 1941 b and Symposium 1948) and the recent reviews of CATCHESIDE (1948) give a clear picture of the progress.

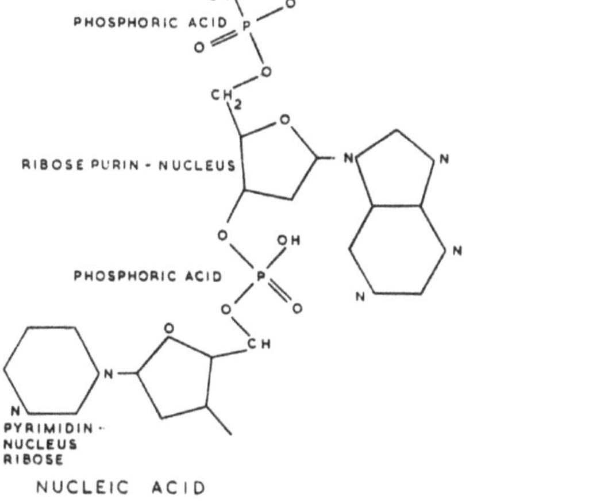

Fig. 137. Structure of a nucleic acid molecule.

It appears possible that the analysis of the chromosomes into the genes which compose it will undergo further progress with the aid of the electron microscope. Some statements of PEASE and BAKER (1949) have a bearing on this.

Hence if the gene is to be considered as a separate and definite atomic complex, then the question arises as to how it will be possible to build a bridge between this and the morphological conception of its structure. Properly, the morphological picture is no more than the

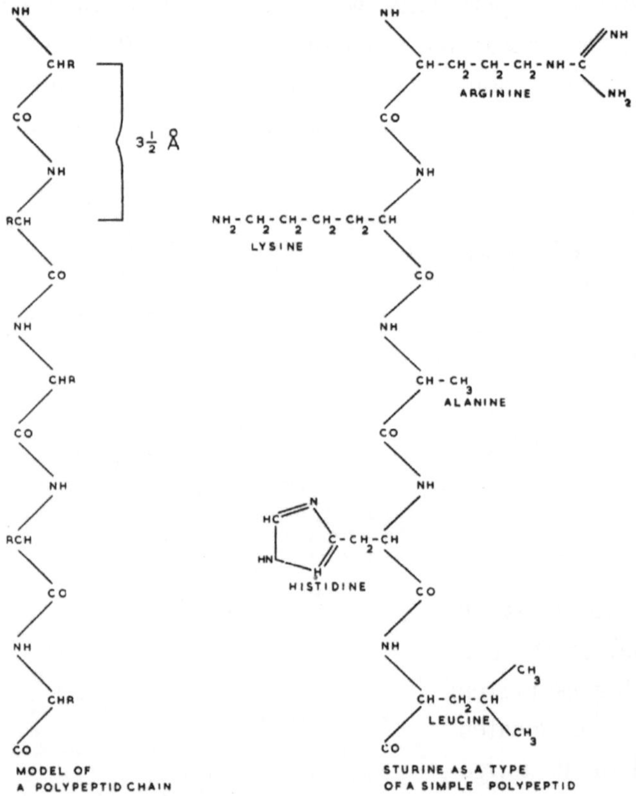

Fig. 138. Structure of a protein molecule as a chain of polypeptide.

drawing of an architect without knowledge of the materials used. It is obvious that this knowledge is indispensable. On the one hand it is hard to obtain because of the small sizes of the chromosomes, which are often no more than a few microns, and even at that size contain several thousand genes. The chemical investigations were entirely dependent on the cell nucleus as a whole. In their classical work with spermatozoa of fish, MIESCHER and KOSSEL were able to ascertain that 96% of the components are nucleo-proteins, that is simple systems of *nucleic acid* and *proteins*. This could be distinguished by means of Feulgens reaction. The fish spermatozoa were such a succesful object because they are nearly wholly composed of cell nuclei without protoplasm.

Attention was drawn to both these organic groups. An insight into their properties is necessary for our further knowledge of the structure of nucleus and chromosome. *Nucleic acids* (figure 137, cf. DAVIDSON, 1953, CHARGAFF and DAVIDSON, 1955) are chainstructures which are usually composed of a number (often 4) of short rod-shaped molecules, which are called rod-nucleotides. The nucleic acid molecules are compelled to fall apart into a molecule of phosphoric acid, a molecule of pentose and a heterocyclic ring of the pyrimidine or purine groups. Generally speaking, two groups of nucleic acid are known: *zymonucleic* acids (*ribonucleic acids*), which occur especially in the protoplasm and contain d-ribose as sugar and which bear adenine and guanine from the purine group as side chains in the nucleotides, and cytosine and uracil from the pyramidine group, and *thymonucleic* acids (*desoxyribonucleic acids*, DNA) with an analogous structure in which the d-ribose is replaced by d-2-ribodesose (at the bottom left H instead of OH) and in one of the four nucleotides the uracil is replaced by thymine. The structure of the thymonucleic acid is thus as follows:

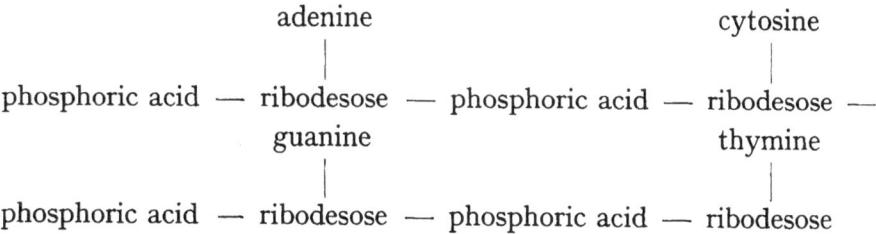

by means of which a ring can still occur, because group 8 again joins group 1. *Proteins* (figure 138), at any rate the relatively simple protamines and histones, which seem to play the principal rôle in the structure of the chromosome, are chains of polypeptids composed of spirally built, folded successions of amino acids, which are coupled together as *amino acid groups*. The side chain in the figure indicated by **R**, can represent all kinds of compounds (glycine, lycine, serine, valine, leucine, arginine, lysine, histidine etc.). These amino acid groups are measurable; each has a length of 3·5 Å (1 Ångström = = 0·1 millimicron = 0·0000001 millimetres); a molecule of a simple protamine, like that of sturine from the sperm of the sturgeon (which contains twenty amino acid groups) has a length of 70 Å. A rather more complicated one such as the clupeine from the herring has 32 amino acid groups and is nearly 100 Å (or 10 millimicron). Long chains of polypeptides have a regular periodically repeating structure. It seems evident that the fibroine of silk (a higher amine) is composed of 2592 amino acid rests, with a length of about 900 millimicrons, in which glycine lies at second position, alanine at fourth place, and tyrosine at the sixteenth place. This series is repeated over and over. Similarly, polypeptide chains can be mutually bridged so that they form flat shapes and these in their turn can be glued together mutually by side chains into three-dimensional bodies.

The chemical investigation of these substances was much facilitated by the discovery of the *salivary gland chromosome*, because we now had at our disposal structures visible in the microscope which can be treated by microchemical methods. CASPERSSON (1936) carried out a number of very original investigations by treating chromosomes with lanthanium compounds, in which the thymonucleic acid was converted into lanthaniumthymonucleate. If after this procedure the chromosomes were treated with trypsine, the proteins were fermented, whereas the lanthanium salt of nucleic acid was unassailable. Hence he could show, that the fibrils between the chromomeres were wholly composed of proteins and form a continuous thread through the chromomeres, whereas the nucleic acids are limited to the chromomeres. Also, other differences became known: protein-chains are kationic, nucleic acids are anionic, protein-chains are optically positive, nucleic acids are optically negative, and later investigations of CASPERSSON (1940 a/b) (ultra-violet absorption curves) made it possible to distinguish nucleotides, protamines, histones and higher proteins of the globulin type one from another. Through this, it became possible to ascertain that the *euchromatic parts* of the salivary gland chromosomes contain nucleic acids and a few histones, the intermediate slices contain proteins of the globulin type. The *heterochromatic* parts contain nucleic acids and a high amount of proteins of the histone-type. The conclusion seems to be justified, that the chromonema forms a continuous thread of protein, and that on this thread, nucleic acid is accumulated locally in the chromomeres.

On this basis there developed a *physico-chemical-picture of the chromosome structure* (BRACHET 1945, FREY WYSSLING 1948, GULICK

1944, MIRSKY 1943, DE ROBERTIS 1948, Symposium 1947a, Symposium 1949a, Symposium 1950a). The question now arises as to what rôle the polynucleotids play. The earlier conception based on their ability for staining, that they should act as the substance of heredity, is no longer held as true. Polynucleotides periodically appear in mass, expecially during the division period of the nucleus, they disappear again in the resting stage, but they remain present in the heterochromatic parts. There are two opposite conceptions. FREY WYSSLING (1938, 1948, p. 152) believes that they play a rôle as protectors of the proteins, as some hydroxyl groups in sugars are protected by phosphatides during the dis-assimilation. He states that nucleic acids are formed in the time that the nucleus prepares itself for division and that they there-after are accumulated in the chromosomes. In the chromosomes they protect specific side chains of proteins during the splitting.

Another conception is held by ASTBURY (1939) and CASPERSSON (1940b). The latter had already found in 1936, that nucleic acids play an important rôle in every biological synthesis of proteins. In all cases of rapid increase of proteins inside the cell investigated, considerable increase in the concentration of zymonucleic acid was also observed to be present in the cytoplasm. It also appeared that there exists a correlation between the number of chromosomes and the amount of thymonucleic acid. ASTBURY put forward the supposition that a possible rôle of the thymonucleic acid should be a binding together and a stretching of a polypeptide chain in preparation for the proper reproductive process of the genes. CASPERSSON used this conception and worked it out further. The opinion of L. W. JANSSEN (1939) was almost analogous. It is interesting to note, in this connection, the fact that a number of viruses which cause plant diseases, turned out to be nucleoproteins (see BAWDEN, 1950); for instance, the tobacco mosaic virus is a nucleoprotein, i.e. like a rod-shaped molecule with a diameter of 15 and a length of 200 millimicrons. Their measurements are of the same order as those of proteins and genes. From this chemical investigation concerning the structure of the chromosome a conclusion can be broadly outlined as follows:

The principal element, the *chromonema*, could be a series of polypeptides with clear-cut side-chains as in the genes. A vague picture was made in 1928 by KOLTZOFF and later developed by HAASE BESSELL (1936). Following the track pointed out by chemistry, she supposed that the chromonema should be the separate bearer or *'pheron'* with colloidal characteristics to which the active groups were attached in the form of a symplex-system called *'agon'*. In 1938, these conceptions were worked out more precisely by FREY WYSSLING and more especially by GOLDSCHMIDT, in the following sense, that the whole chromosome is one long protein molecule to which the genes are connected as side-chains almost like the beads on a necklace. Such a scheme can still only be of elementary importance as a working hypothesis, but it may be true. Even in the case when this hypothesis turns out to be unacceptable, it still provides us with a picture of the gene as a part of the chromosome, which will perhaps make it possible that we will understand some of the processes observed morphologically in the chromosome 'life'. Before we can draw this picture much time must yet pass.

It is indeed an ascertained fact that genetics and karyology meet one another in a sphere, which approximates in its deepness very much with the question posed by SCHRÖDINGER "What is life?"

CHAPTER XV

THE ROLE OF THE PLASM

The very promising character of MORGAN's localization hypothesis resulted in a wide circle of genetical investigators giving attention to the karyological phenomena and especially to the life-cycle of the chromosomes, on the other hand the possible independent role of the plasm was not considered of interest. We must state however, that such a point of view is incorrect; the plasm is undoubtedly of importance, not only as a substrate for the phenotypical expression of the genes (which are located in the chromosomes) but also in some cases as an active ruler of the genotype. But the nature of the cell as an indivisible one-ness of the nucleus and the plasm does not make it easy to distinguish separately the functions of both these component parts. Without plasm the genes, which are present in the nucleus, could not express themselves and without the nucleus, the plasm will not be viable or, at least not capable of development. With regard to the rôle which the plasm plays, some definite statement can be given although a number of questions remain undecided. For a long time the view of the geneticists over cytoplasmic inheritance was one of indifference, and only a few investigators wrote papers on this subject (CORRENS 1937, SIRKS 1938b, WETTSTEIN 1937). However, in the past fifteen years more interest has been shown in the rôle of the plasm and hence SONNEBORN could write in 1950: 'Cytoplasmic inheritance is one of the capital facts of biology' (1950, p. 31). Recent reviews and discussions have been given by CASPARI (1948), EPHRUSSI (1953), MICHAELIS (1954), OEHLKERS (1949, 1953), SONNEBORN (1953).

There are many different ways in which evidence can be produced concerning the question of the rôle of the plasm. The first method of investigation is that of the so-called *'defect-experiments'*. If we are to remove the nucleus from an egg-cell artificially and we fertilize this egg-cell with a spermatozoon of another species, we have eliminated the maternal nucleus substance completely, and therefore plasmatic influences are the only explanation of a possible occurrence of maternal characteristics in the fertilization product. This merogony-method was used by BOVERI (p. 224) and later by a great number of zoologists (BALTZER, HADORN, G. and P. HERTWIG, HÖRSTADIUS, JOLLOS, PETERFI, SCHLEIP, see RAVEN 1954).

There were several different attempts to make egg-cells without nuclei, as follows; — fragmentation of the egg-cell, by shakings eggs of sea-urchins, by making a ligature of fine hair round *Triton*-eggs, removal of female nuclei with the aid of a micro-manipulator and the destruction of the nucleus of the egg by chemical means, ultra-violet light, radium and X-rays. From the illustrative point of view of the methods, these investigations were very important and their significance in the developmental physiology was great, but up to now, these zoological investigations could not give the final result for the genetical problem of the plasm. The haploid individuals which originated, suffered a decrease in their viability, and hence never reached the fullgrown stage (see BALTZER 1940, FANKHAUSER 1937).

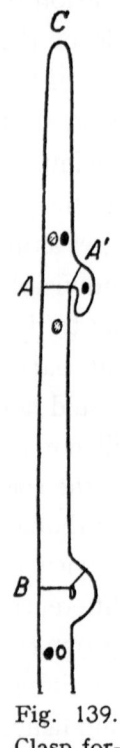

Fig. 139. Clasp formation in *Schizophyllum* (after HARDER, 1927b).

However, the *micro-surgical method* has led to some results in definite species of Fungi: HARDER (1927 a and b, 1929) has obtained results in a series of interesting experiments, which are very important for a critical review of the rôle of the plasm. The Basidiomycetes (*Pholiota, Schizophyllum*) yielded suitable material. The copulating cells of both parents have a considerable quantity of plasm. After copulation, a union of both nuclei fails to appear, they remain separate inside the zygote and divide. The zygote grows into a cell thread by means of cell-division (each cell contains a paternal and a maternal nucleus). This procedure seems very complicated, because only one of both nuclei goes at once into the cell which is situated directly under the top-cell, whereas the other nucleus still remains provisionally inside the top-cell. This nucleus is later transferred to the second cell from the top. However, this transference of the second nucleus does not occur directly, but takes place through a cell, which is separated laterally. This is represented schematically in figure 139: the one parental nucleus is black, the other striped. At A', clamp-forming takes place as a side-cell of the top-cell. The clamp unites with the second cell from the top and the nucleus is moved into this cell; in diagram B, this process is completed. With the help of a micromanipulator, HARDER succeeded

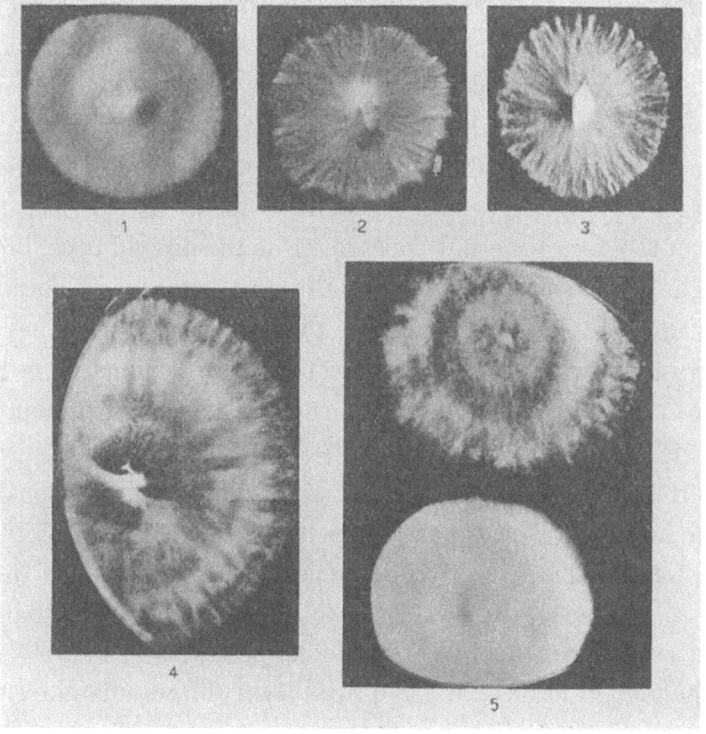

Fig. 140. Mycelia of *Pholiota*. 1) Haploid PhA$_1$. 2) and 3) Haploid PhB$_2$. 4) Diploid hybrid. 5) Op$_{42}$ (after HARDER, 1927a).

in pricking the top-cell, the clamp and the third cell from above so that they burst. Only the cell situated between A and B, containing both parental plasmata, but only one nucleus remains alive and is able to develop into a mycelium.

Different lines originating from one-spore-cultures of *Pholiota* showed very typical differences in phenotype. One, PhA$_1$ forms a homogeneous felt-like, structureless, haploid mycelium, of which the advancing edge creeps over the substrate and forms a dense and extensive air-mycelium (see figure 140). The haploid mycelia of the other line PhB$_2$ are less dense, more transparent with hyphae arranged like spokes, and in bundles. These bundles jut out at the margin and are the cause of an uneven outline. The diploid mycelium which originates through the crossing of both forms, is about intermediate between both these types and has a rich air-mycelium, but is loosely arranged with bundles at the margin which are arranged very weakly in radial spokes. From this diploid hybrid material six operation-lines (Op) were obtained which are therefore haploid-bi-protoplasmic. By crossing these Op-lines, evidence could be produced that they had the nucleus of PhB$_2$. Because these nuclei are differentiated sexually, it turned out that all six Op-lines could copulate with the original PhA$_1$ but not with the other parent PhB$_2$. The conclusion therefore is justified that they posses B$_2$-nuclei. These Op-lines have, in the first period after the operation, the same external appearance as the diploid type, but if the cultures are enabled to grow continually, this external appearance disappears and is replaced by a definite phenotype which was never that of the B$_2$-type (which could be expected, since the nucleus is B$_2$) but in 4 cases that of A$_1$. In one number there is a weak resemblance with B$_2$, but nevertheless it is clearly different from this and, in another number the characteristics of A$_1$ and B$_2$ occurred simultaneously with the same intensity. The apparent resemblance with the diploid mycelium-type can therefore be considered as an after-effect of the influence of the diploid nucleus, but the later occurrence of a clear A-habitus in four of the six Op-strains with B$_2$ nuclei produces evidence for HARDER's theorem that here the A-protoplasm differs characteristically from the B-plasm, and that in this example, heritable characters are localized in the plasm. It is obvious that the plasm of the Op-lines is a mixture of the plasm of A + B, but it may be concluded that an almost complete dominance of the A-habitus over the B-habitus exists, since the A types are present in a preponderance among the haploid mycelia (which originate as segregation products of the hybridization of PhA$_1$ × PhB$_2$.)

Apart from these defect experiments, *reciprocal crosses* between varieties, or between species, can also yield important data for the question of plasmatic inheritance. Very different results were obtained in these crosses. Their analysis often gives an insight into the processes as given in Chapter X, but several other reciprocal differences are only open for explanation, if we assume a plasmic cause.

If we consider the plasm as a substrate in which the phenotypical effect of the genotype (which is formed by the genes located in the chromosomes), expresses itself, then it will be obvious that this plasm in its turn has an influence on the realization of the phenotypical appearance. In this process it often plays a passive rôle, but still it may be able to exhibit its own characteristics.

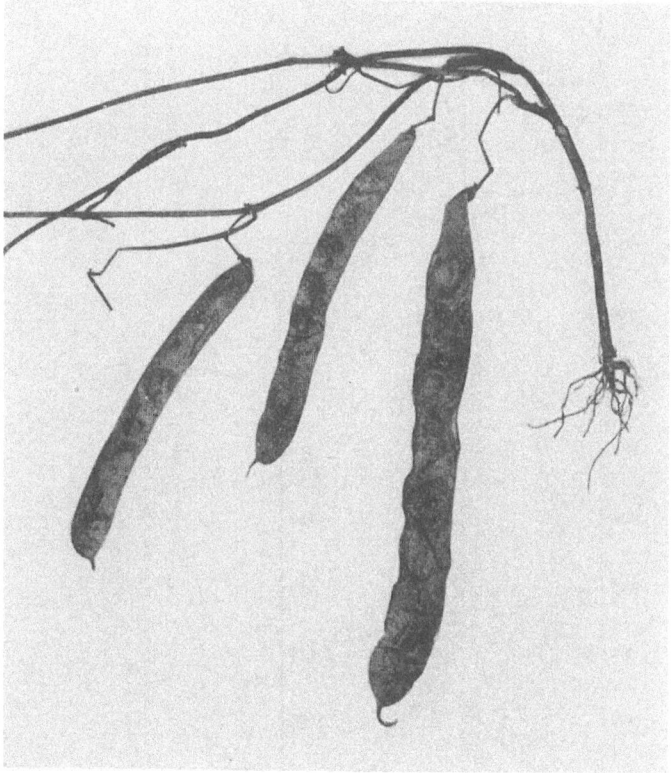

Fig. 141. Bud variations in French beans; beside a normal pod (III), two smaller ones (I and II) (after SIRKS, 1938a).

The most obvious possibility is that external circumstances give rise to *modifications* in the maternal organism which are handed over and embodied in the egg-cells originating from this maternal individual. These modifications usually have after-effects in one or more generations.

Such a case was already referred to earlier (p. 220) in which an after-effect was caused by the application of a dye. However typical examples of this phenomenon are known which seem to arise from absolutely natural causes or an unknown physiological change. In his investigations with the ichneumon fly, *Habrobracon juglandis* in which the colour intensity is strongly correlated with the environmental temperature (see p. 7) KÜHN (1927) found that these influences are already embodied in the unfertilized egg-cells; that is to say, an animal reared at a high temperature will still for some days after it has been

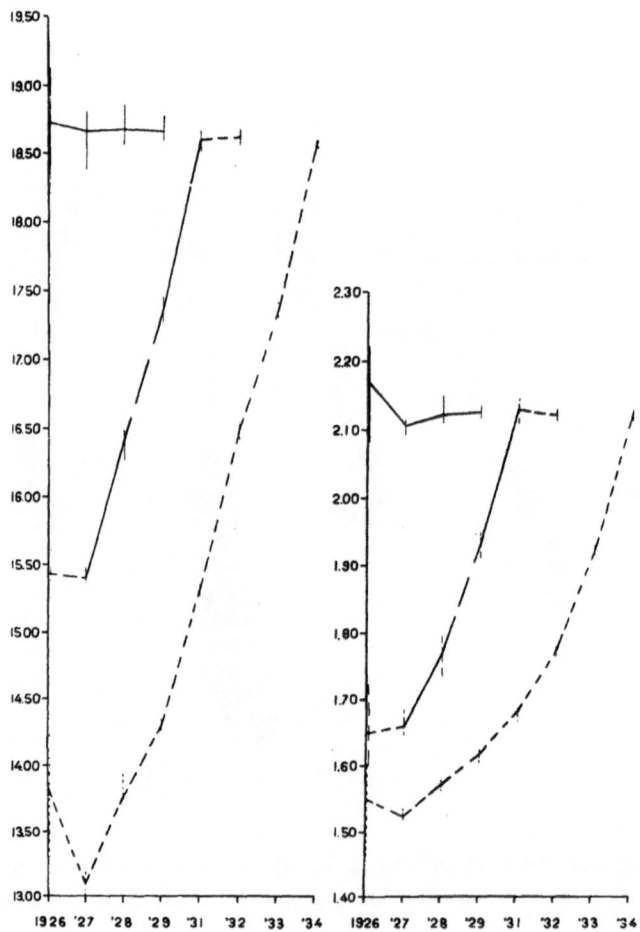

Fig. 142. Graphs showing lengths and widths of pods of the descendance of seeds inside the pods of fig. 141.D₁–D₈ (years 1926–1934) successive generations obtained by selfing (after SIRKS, 1938a).

transferred to a lower temperature, produce eggs from which animals will originate which are much lighter in colour than they ought to be according to the low temperature. The influence of the higher temperature is still perceptible in offspring reared at the lower temperatures.

KÜHN called this phenomenon '*predetermination*'. An analogous process, but one which has a longer after-effect and whose original cause is unknown, was observed in haricot beans (SIRKS 1938a). On a plant two pods (I and II) appeared as bud variations which were strikingly smaller than the normal type (figure 141); the offspring from the seeds of both these aberrant fruits had mainly shorter and narrower fruits than those of the original race, but the difference turned out not to be permanent in successive generations. In the second generation (obtained by self-pollination) a recovery of the former measurements started, which was continued in the later generations. Finally in the eighth generation (D_8) the original size of the pods was reached again (figure 142). From crosses between plants originating from the normal and from the aberrant pods, it was evident that the aberrant form only inherits in the maternal line. Descendants of the cross large ♀ × small ♂ give rise to big fruits, those of the reciprocal cross to small ones. The after-effect was therefore a purely plasmatic one. One or another physiological disturbance in the plasm of the original plants must be supposed as a possible cause.

A second plasmatic cause of the difference between reciprocal crosses can be found in the possibility that the genotype of the mother-individual itself (in which an egg-cell with a high amount of protoplasm is formed) wields an after-effect, which remains visible over a shorter or a longer period of time. In HARDER's *Pholiota*-investigations we have already learned about such an after-effect. We will also mention an almost classic case taken from the botanical investigations of CORRENS, concerning the differences in *Matthiola*. CORRENS (1900b) observed differences in colour between the reciprocal crosses of *Matthiola incana* (with blue embryonic epidermis) with *M. glabra* (in which the epidermis in the embryo is yellow). Seeds obtained from a cross between *M. incana* as mother and *M. glabra* as father, were nearly as blue as those of the pure *incana* race. Seeds of the reciprocal cross, *glabra* ♀ × *incana* ♂ deviated from the pure *glabra* form in the sense that they were also more or less blue coloured, which can be ascribed to a dominance of the factor for a blue embryonic epidermis. But, both reciprocal crosses still differ considerably in the intensity of the blue. From the cross with the blue *incana* mother, all seeds were a rather dark blue; the seeds from the yellow *glabra* as mother formed a series of transitional types which varied from almost pure yellow to blue.

In genotype, both seed-types were completely similar. It is rather obvious that we may suppose a plasmic difference here, that is to say, the *incana* plasm with a heterozygous genotype immediately produces a blue colour, whereas the *glabra* plasm reacts less strongly on this genotype as a result of an after-effect of the factors for yellow in the mother.

A nice analogous case taken from zoological investigations can be found in the experiments concerning the dextral and sinistral shape of the snail-shells of *Limnea peregra* (BOYCOTT and others, 1923, 1930; DIVER and others 1925, 1938). The principle results of the cross between a coloured dextral race and an uncoloured sinistral can be summarized as follows (see BOYCOTT, 1930, p. 61):

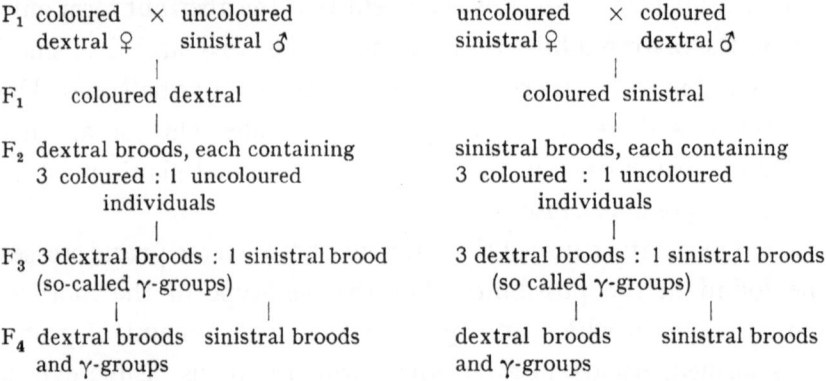

The interpretation of the segregation coloured : uncoloured will not give any difficulties as opposed to the dextral and sinistral shape. It was STURTEVANT (1923), who considered the dextral shape to be dominant over the sinistral. This view was based upon preliminary data from BOYCOTT's school, and this opinion was also held by BOYCOTT himself. Nevertheless the phenomenon that the F_1 generation from dextral mothers is wholly dextral and those of sinistral mothers wholly sinistral, whereas the F_2 generation from both reciprocal crosses is wholly dextral needs a separate explanation. Such an explanation could be derived from the behaviour of the F_2–F_4 generations. The F_3 generation forms so-called γ groups which are composed of 3 dextral broods : 1 sinistral brood. Hence, the segregation is here observed in mutually absolutely uniform broods and not in the individuals. The explanation of this apparently complicated behaviour is rather easy fundamentally. The convolution direction of the snail-shell depends on the genotype of the mother animal and is embodied in the plasm of the egg-cell. The influence of the fusion product of the egg-cell nucleus with the nucleus of the spermatozoon occurs too late to express itself in the egg-cell and hence in the animal which originates from this egg-cell. Indeed, a perceptible after-effect exists from the maternal genotype in this case, but BOYCOTT still rightly believed in grouping this after-effect under *'delayed inheritance'* and not under *'maternal inheritance'*.

The plasm shows more independence towards the genotype in other cases in which the nature of the plasm controls the genotype.

DE VRIES (1913) already communicated that the reciprocal crosses between the homozygous *Oenothera Hookeri* and the complex hetero-zygous *O. Lamarckiana* yielded different offspring. If *O. Hookeri* is used as mother, then the descendants *Hookeri-laeta* and *H. velutina* are normal green. The reciprocal cross with *O. Hookeri* as father gives green *laeta* as well as yellowish and weak *velutina* individuals. This fundamental idea caused RENNER (1924a) and KRUMBHOLZ (1925) to give a more detailed investigation. The result was that this phenomenon occurs rather frequently among the *Oenothera. Velutina* individuals, which originate from a cross *Lamarckiana* × *Hookeri* are for the most part yellowish and hence die young; almost 15% of the seedlings show green spots on a yellow background in the beginning of their development. These individuals are viable, because the green tissues increase considerably, they grow up into yellow variegated or sometimes to wholly green fully-grown plants. *Velutina*-individuals from the reciprocal cross were indeed green from seedlings up, as was observed by DE VRIES, however two exceptions were observed, in a batch of some hundreds of individuals, which had yellow spots on a green background.

An explanation of this reciprocal difference was put forward by RENNER as being due to the difference between the plastids (future chlorophyll) of *O. Lamarckiana* and those of *O. Hookeri*. It appears to be an ascertained fact that the pollen tube in *Oenothera* can carry some plastids, but this number is much less than those present in the egg-cell. The diploid nature of the nucleus of the *velutina*-types is the combination of the *Hookeri*-genome with the *velans*-complex and it is therefore presumed that the *Lamarckiana* plastids do not react upon genes present in this hybrid nucleus which would otherwise result in chlorophyll formation (as is the case with *Hookeri*-plastids). Therefore a *Lamarckiana* egg-cell, fertilized by *Hookeri*-pollen contains many plastids which remain yellow, and only a few which have the ability to become green. On the other hand, the reciprocal cross contains many green coloured *Hookeri*-plastids and only a few *Lamarckiana*-plastids which remain yellow.

On the grounds of these results and also of numerous others concerning the heredity of chlorophyll characteristics, RENNER (1934) concludes that the chlorophyll or plastids spread around in the plasm have their own heritable disposition, which is independent of the genotype. As an analogy with the heritable gene-property which was called 'genome' by WINKLER (see p. 250), references to the heritable disposition of the plastids are grouped by RENNER under the heading of '*plastome*'.

Plastids are indeed situated in the plasm, but they are still not an inherent part of its composition. Therefore it is important, that with regard to other properties differences were also ascertained in the ability of the plasm to react towards the genotypical constitution.

In investigations mentioned earlier in this book (p. 138) concerning *Vicia Faba* (SIRKS 1931), it was established that the same series of multiple allelomorphs G_4–G_1 can act as basic factors for the size of different organs. The species *Vicia Faba* contains two subspecies, which can be distinguished systematically: *V. F. major* and *V. F. minor*. The crosses between races, which differ only in this basic factor which are however identical with respect to possible co-factors, yielded the result that the stem length was more strongly developed in minor plasm than in plants with major plasm, with the same genotypical constitution. Hence greater intensity of reaction-ability of the minor plasm upon these G-factors may be presumed. No differences between either of the reciprocal crosses could be observed in the leaves, but in the fruit the reverse phenomenon took place; here the measurements of fruits carrying the minor plasm were obviously smaller than those containing the major plasm with the same genotypical constitution. In the F_2-generations segregations occurred in different genotypical groups as could be expected. Nevertheless, characteristic differences between plants with major plasm and plants with minor plasm were still preserved in this generation (as was obvious from the graphical representation in fig. 143). These reciprocal differences could be attributed to a different reaction-ability of minor plasm and of major plasm upon the same genotype. But, how are we to explain physiologically the decrease of the reaction ability in the minor plasm during the ontogenetical development (stems, leaves, fruit) and on the other hand, the increase of the reaction-intensity of the major plasm. These questions remain unanswered.

Analogous results were obtained by SCHLÖSSER (1935) in tomatoes.

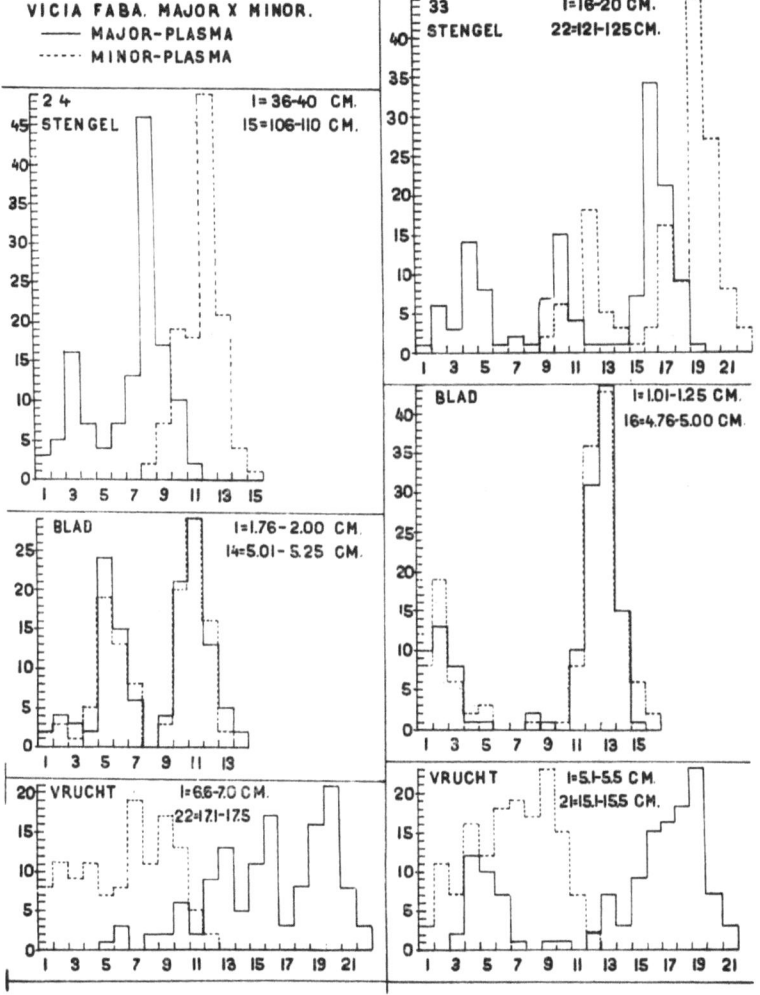

Fig. 143. Plasmic influences upon the length of the stem (stengel), length of leaf (blad) and length of pods (vrucht) in *Vicia Faba* (after SIRKS, 1931).

Differences in lengths of the plants and leaves in two races turned out to be due to genes, however, the effect of the same genotype is stronger in the one race than in the other.

The investigations concerning the hybridization products of *Epilo-*

bium-species (cf. MICHAELIS, 1954) are very important for the latter case. It was found that an important difference occurred between the species *Epilobium hirsutum* and *E. parviflorum* on the one hand and species like *E. luteum* and *E. roseum* on the other hand. Reciprocal crosses between one of the first mentioned species and one of the latter always gave very differing results. The stem of *E. hirsutum* ♀ × *E. roseum* ♂ is striking in its habit. This stem is erect as is that of the mother plant, whereas in the reciprocal cross, the F_1 resembles the variety used as mother (*E. roseum*) which is bent at the top. For the rest, the reciprocal differences lie only in quantitative properties. They express themselves mostly in the phenomena of inhibition in growth (height of the plants, measurements of the leaves and those of the

corolla, of the ovary, pollen fertility, etc.). Growth inhibition is evident from the flowers of *E. hirsutum* ♀ × *E. luteum* ♂, whereas flowers of the reciprocal F_1 *E. luteum* ♀ × *E. hirsutum* ♂ are developed normally (fig. 144). These inhibition phenomena always act in concert, but to different degrees. Different races of *Epilobium parviflorum* (or of *E. hirsutum*) often give different results when crossed with the same father-plants. The inhibition phenomena differ considerably in the F_1-offspring. We can therefore speak of very strong (inhibiting) races of

Fig. 144. Flowers of F_1-individuals of the cross *Epilobium luteum* × *E. hirsutum*. Upper: *luteum* plasm, lower: *hirsutum*-plasm (after MICHAELIS, 1933).

E. parviflorum and of weak races. Mutual crosses between *E. hirsutum* and *E. parviflorum* show such an inhibition in many cases, and the same phenomenon occurs (though to a lesser degree) if weak and strong races within the same species are crossed with one another. If was evident from very detailed investigations of MICHAELIS (1940–1948) and his co-workers that the species *E. hirsutum* comprised a great number of geographical races which, by mutual hybridization give different results. The race 'Jena' behaves in an especially striking

Fig. 145. Reciprocal differences between hybrids of geographic races of *Epilobium hirsutum*:

a. Right: Jena ♀ × Insel candidum (Müncheberg) ♂; middle: Insel candidum self; left: Insel candidum ♀ × Jena ♂.

b. Right: Jena ♀ × Beuren ♂. Left: Beuren ♀ × Jena ♂.

c. Right: Jena ♀ × München ♂. Left: München ♀ × Jena ♂.

d. Right: Jena ♀ × Lissabon ♂. Left: Lissabon ♀ × Jena ♂.

e. Right: Jena ♀ × Jassy ♂. Left: Jassy ♀ × Jena ♂.

way. Fig. 145 represents five reciprocal crosses. F_1-individuals with the Jena-race as father were normal, those with Jena plasm, that is to say Jena as mother, showed a series of transitions between a stronger growth of the hybrid (fig. 145a) and all sorts of degrees of perceptible inhibition phenomena (fig. 145 b–e).

It is evident that these inhibition phenomena must have an interesting physiological background. ROSS (1941–48) studied this side of the question and found that the race Jena excels in the tendency to activate oxydase ferments, which are located in the plasm and are possibly due to definite characteristics of the plasm colloids. This activation causes inhibition of the function of growth substances and should be excited by a great number of genes present in different geographical races.

Among the investigators who have studied these different *Epilobium* crosses, there existed a sharp opposition in interpretation at the very beginning. On the one hand RENNER and KUPPER (1921) supposed a pure plasmic influence of the maternal egg-cell plasm, on the other hand LEHMANN and his students held heterogamy as responsible for the phenomenon observed (as was found in *Oenothera*). Both these extreme points of view gradually grew nearer to one another: RENNER (1929, p. 26) concluded 'that the characters of the hybrid flowers are principally brought about by the interaction of both the genomes present, as is the case in all other organisms. The nature of such an interaction is specifically influenced by the maternal plasm', whereas LEHMANN (1936, p. 662) concludes: 'According to these observations the reciprocal differences caused by the plasm remain, though these differences cannot be explained with the knowledge available'. The difference brings us to the question as to which part of the cooperation between plasm and genes is due to the plasm and which rôle is due to the genotype.

Investigations on mosses are of particular significance for the study of plasmatic inh:ritance: these were published by F. WETTSTEIN (1924–28, 1926) and later amplified by his students (DÖRRIES-RÜGER and others, 1929–32). In crosses between different races of the same species no reciprocal differences were observed. However, if the parents were less closely related and belonged to different species or to different genera, some difference could be observed. A nice example from his studies deals with the cross between *Funaria mediterranea* (indicated as Me) and *F. hygrometrica* (Hy). The species *F. mediterranea* has small setae with a high, pointed operculum. The haploid moss-plant has leaves of which the top has lengthened to a long thread, the leaf-rib suddenly ends below the leaftop and the paraphyses are composed of a coiled series of oval cells. The other parent *F. hygrometrica* has great setae with a broad operculum, plants with flat very contracted leaves, protruding leaf-tops. The leaf-ribs end at the top of the leaf, and the paraphyses are composed of straight series of more or less bulb-shaped cells (see fig. 146). The diploid stage, which originates after hybridization of these species is different in both reciprocal combinations. The setae of Me ♀ × Hy ♂ (MeHy) are smaller with a high and pointed operculum, similar to the mother (Me), those of HyMe are much bigger with a flat operculum, like the *hygrometrica* mother. Greater differences still came to light in fully grown diploid hybrid plants obtained by regeneration of the seta-tissue. It turned out that the connection of the leaf-rib with the leaf-top is completely ruled by the mother-type. In MeHy, the leaf-rib ended far below the top; in the reciprocal HyMe the leaf-rib continues up to the leaf-top. Also, the thread-shaped ends of the leaves appear only in MeHy-individuals. The HyMe-plants have pointed leaves. The paraphyses in MeHy were smaller or more spiral, but in HyMe they are straight. Fig. 146 gives clear representation of these phenomena.

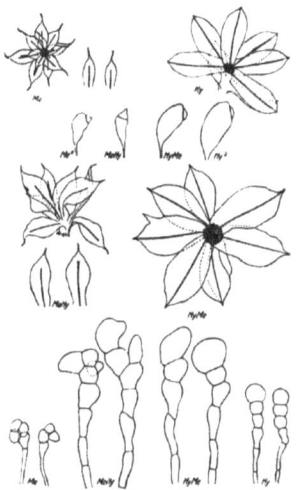

Fig. 146. Leaf-rosettes, leaves, setae and paraphyses of *Funaria mediterranea* (Me), *F. hygrometrica* (Hy) and their reciprocal crosses Me–Hy and Hy–Me (after WETTSTEIN, 1926).

An investigation of the haploid descendants, obtained from spores showed that the maternal influence had indeed asserted itself in all three characteristics, but to a very different degree. As to the length of the leaf-rib, very little variation was observed in the offspring, providing we recall that in all MeHy-plants the leaf-ribs did not end at the top of the leaf, while the opposite was the case in the HyMe-descendants without exception. This characteristic of the maternal type was preserved through succeeding generations. The haploid offspring formed by segregation varied more in the leaf-shape. The

proportion between leaf and hair-shaped appendix was determined
and amounted in Me to a mean of 2·47, in Hy to 27·72. If now the F_1-
haploids were classified, then the following distribution was found for
both reciprocal crosses.

MeHy

1·5	2·0	2·5	3·0	3·5	4·0	4·5	5·0	5·5	6·0	6·5	7·0	7·5
5	24	62	82	33	36	15	6	2	2	–	2	

HyMe

7	8	9	10	11	12	13	14	15	16	17	18	19	20	21	22	23	24	25	26	27	28
1	–	1	–	1	2	4	4	6	5	–	–	1	8	1	3	1	12	2	2	461	

It is evident from this, that the F_2 segregation products of the MeHy
vary between 1·5 and 7·5, so that the paternal type is not attained. The
haploid individuals of HyMe vary between 7 and 28 with an overwhelming majority in the class between 27 and 28.

The segregation of the shape of the paraphyses was much stronger.
Here, in both crosses both parental types were found to be present,
though a preponderance of the maternal shape could be observed.

The only acceptable explanation for these phenomena is found in
the supposition that the plasm plays a very important rôle here in the
expression of the characters. Evidence for this supposition was produced by WETTSTEIN in the following way: a plus variant with respect
to leaf-type from the cross MeHy (which had thus Me-plasm and which
genotypically could be supposed to be Hy, written as (Me)Hy, if the
plasm-type indicated between ( ) was back-crossed with *F. hygrometrica*. This species as father gives spermatozoa Hy and the hybridization product must thus be (Me)HyHy. That is to say, the true-breeding
(Me)Hy-type. The egg-cells of *F. hygrometrica* of course are (Hy)Hy
and the descendants of this with (Me)Hy as father must then be pure
(Hy)HyHy-plants. Such a test gave affirmative results and evidence
for the influence of the plasm had been obtained. Thus the plasm has a
heritable nature by which it was enabled to react upon the genotypical
constitution of the cell-nucleus according to its own disposition and
even to develop distinct characteristics by means of its own power. To
describe this genetical element of the plasm, WETTSTEIN introduced the
term *'plasmon'* in contradistinction to the term *'genome'* of WINKLER,
which delineates the totality of genes present in the chromosomes.

The three characters investigated form examples of three different
intensities of the plasmon. The leaf-rib seems to be almost entirely

governed by the plasmon. The leaf-shape is influenced to a large degree by the plasmon, but is also dependent on the nucleus-genes. The genes play an overwhelming rôle by establishing the shape of the paraphyses, whereas the influence of the plasm is only slight. WETTSTEIN believed that in all the three cases, plasm and gene play a rôle. That is to say that in the first case an extremely strong predominant plasm (*antecedent*) repels the influence of the (*recedent*) genes. In the latter case, the genes should therefore be extremely antecedent and the plasm recedent.

The realization of a definite phenotype in reciprocal crosses is thus dependent on the genotypical constitution of the genome so far as it is made possible by the heritable composition of the plasm. Such a plasmatic heredity is not only limited to morphological characteristics; physiological ones can also be influenced by the protoplasm; e.g. resistance to disease in *Epilobium* (MICHAELIS, 1935), intensity of the plasm-permeability for definite substances and viscosities (V. DEL-LINGHAUSEN, 1935, 1936), osmotic values (SCHLÖSSER, 1935). It turned out that those properties were dependent on heritable characteristics of the plasm.

Alongside these botanical data concerning plasmatic inheritance are the investigations of GOLDSCHMIDT (1920b, 1934) with the butterfly *Lymantria*, and those of KÜHN (1927) with the ichneumon-fly *Habrobracon*. The common butterfly *Lymantria dispar* is, as a caterpillar, plain grey in colour. The Japanese *L. japonica* has bright yellow spots in each segment after the first peeling. Their reciprocal hybrids show a strong matrocliny, but this decreases after the first peeling, so that both hybrids become more or less similar with a prevalence of the dispar-type. This could point to a predetermination process, but the behaviour of the F_2 generations prove that this is not the only factor which plays a rôle. A monohybrid segregation (1 dispar : 2 heterozygotes : 1 japonica) occurs in the F_2 generation, but the colour of the F_2 animals with the japonica appearance is dependent on the original mother used in obtaining the F_1 generation. Those descended from $(J \times D)^2$ remain light-coloured, but those from F_1 individuals $(D \times J)^2$ become much darker. The genotypes of all F_1 animals are equal. Hence in this case predetermination is out of the question, so that a difference in plasmatic reaction upon the same genotype has to be accepted.

23

Analogous to this are the investigations of KÜHN (1927) concerning *Habrobracon*. These deal with races which differ in colour intensity at a mean temperature (25°C); the one light, the other dark. Moreover, in both races, the males are darker than the females. The F_1 females are intermediate, though both the reciprocal crosses differ according to the mother races used. This difference remains after back-crossing with light-coloured males. It also occurs in males which arise in this species from unfertilized egg-cells. Here again a striking influence of the mother is observed in three successive generations (P, F_1 and F_1R). Such an influence can only be ascribed to a plasmatic origin, though the genotype is also of importance for the colour-intensity. As we have already mentioned, pre-determination also plays a rôle in *Habrobracon*.

A very important question is; how far are these heritable constitutions of a genome and of plasmon mutually and lastingly independent, or, how far can a change of genome permanently affect the plasm?

A quantity of data is available to help us answer this question, but up till now it has not enabled us to reach a final conclusion. WETTSTEIN found as a result of his investigations into mosses, that the plasmon is to a high degree independent of the constitution of the genome, and that this latter is not able to influence permanently the constitution of the plasmon. In order to investigate this question, WETTSTEIN again made use of his method of crossing species and genera. He tried, for example, in the genera cross *Physcomitrium piriforme* (Pi) × *Funaria hygrometrica* (Hy) to obtain a solution in three different ways: firstly by continuous selection of the individuals which resemble the father Hy in which the plasm remained Pi; secondly by the formation of polyploid individuals in which an accumulation of paternal nucleus-material had taken place; that is to say, a method of considerably intensifying the influence of the paternal nucleus upon the plasm, and thirdly by continuous back-crossing in successive generations of (Pi) Pi-segregation products with Hy-spermatozoa (see fig. 147). The first method did not lead to any result and this was due to the considerable sterility in the individuals with Hy-characteristics. By means of the second method, plants with more than one Hy-genome could be obtained, that is to say, plants which had the constitution $PiHy^2$ and $PiHy^3$. However, intensification of the number of descendants resembling the father was not observed. It was the

third manner of investigation especially which produced evidence
that the Pi-plasm in all successive generations was independent from
any definite influence of the Hy-genome.

A true-breeding Pi-plant, which
originated from a PiHy-spore, was
crossed again with a true-breeding
Hy-father and this was repeated in a
number of successive generations.
The segregation of the offspring was
rather complicated, as could be expected in a cross of little related
individuals. Nevertheless, it was
possible to make a classification by
paying attention to some eleven
characteristics, in which the parents
differed (leaf-shape, leaf-margin,
paraphyses, seta-length, seta-colour,
seta-grooves, operculum, seta-
margin, peristome and rows of
interstices) and hence, the proportion
between the number of Pi-characteristics and the number of Hy-
characteristics could be determined.
Out of the 113 individuals (F_1-
segregation products), 57 (50%) were
pure haploid Pi-individuals with a
ratio of 11 Pi-characteristics : 0
Hy-characteristics. There were also
4 plants with a ratio 10 : 1; 3 with
9 : 2; 2 with 8 : 3; 2 with 7 : 4 and
2 with a ratio of 6 : 5. The remainder were either diploids, or could
not be classified for one or another
reason (fig. 148). Back-crossing of a

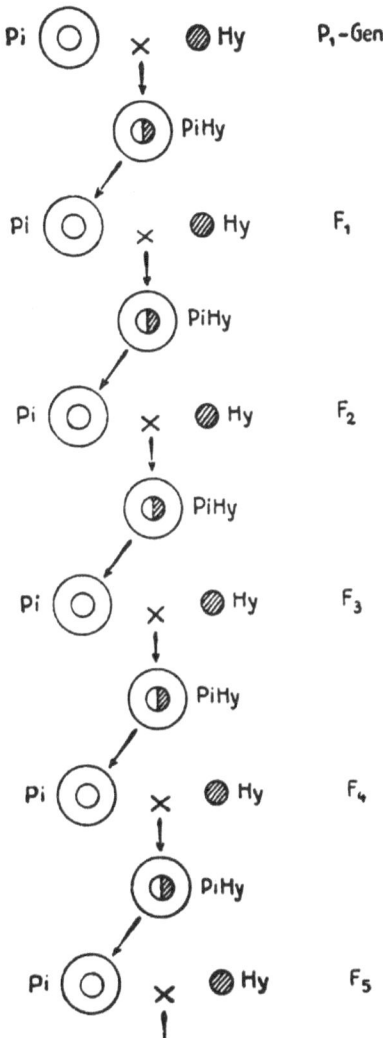

Fig. 147. Scheme of the crosses and
continuous backcrosses of *Physcomitrium piriforme* × *Funaria hygrometrica* (after WETTSTEIN, 1930).

pure haploid Pi-individual with Hy yielded an F_1R of 129 individuals
of which 88 were pure Pi-plants (68%). In this way, WETTSTEIN
obtained successively higher and higher percentages of pure Pi-plants.

The very high number of pure Pi-individuals, which occurred in these generations, combined with the high percentage of sterile spores, and the fact that never less than 6 Pi-factors were observed, pointed clearly to the fact that the Hy-genome is not viable in the Pi-plasm, even when the Hy-genome is mixed with Pi-factors.

The successive generations of back-crosses provide evidence that the Pi-plasm is not changed at all in the direction of the Hy-plasm by repeated inbreeding of a fresh Hy-genome. The viability of the Hy-genome in the Pi-plasm did not increase. On the contrary the per-

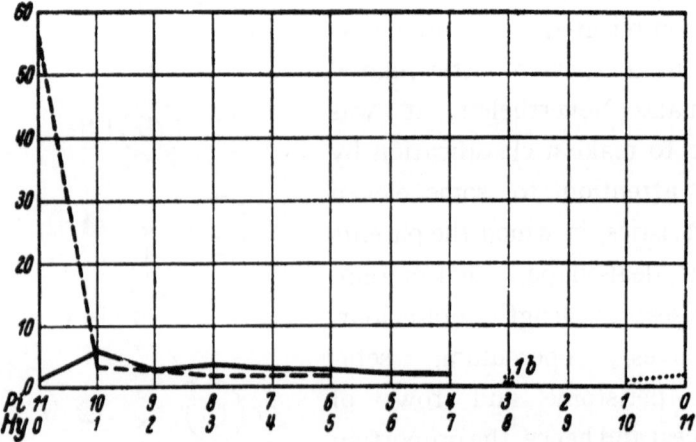

Fig. 148. Graph showing relationship between Pi and Hy-characteristics in the F_1-generation of *Physcomitrium piriforme* × *Funaria hygrometrica* (broken line, Pi × Hy haploid; continuous line Pi × Hy diploid; dotted line Hy × Pi (after WETTSTEIN, 1928).

centage of pure Pi-plants was highly increased over the course of the years.

If we could hold that a change of Pi-plasm was responsible, then we have to presume that such a change would take place in the reverse direction to that which one would expect. That is to say, the 'Pi-nature' of the Pi-plasm is intensified rather than weakened over the course of successive generations.

This same question concerning the possibility of a permanent influence of the genotype upon the plasm, was also discussed in the investigations concerning *Epilobium* mentioned earlier. It is evident that this material is not so easy to approach in order to obtain an exact interpretation; the technique does not permit a breeding of haploid descendants. Hence, in this aspect, the opinions differ. RENNER accepts the point of view of WETTSTEIN and believes, that in *Epilo-bium* the plasm cannot be changed by the genome. LEHMANN on the other hand accepts the opposite point of view.

For investigations parallel to those of WETTSTEIN concerning con-tinual back-crossing with the father we are indebted to MICHAELIS (1929, 1932, 1938). His crosses were made principally between *Epilo-bium hirsutum* and *E. luteum*. The F_1-generation which had hirsutum as mother (h × l) had completely sterile pollen. The reciprocal l × h had a pollen percentage of 26·4. On the other hand the germination of the F_1-seeds in the h × l was 77·3%, in l × h it was 24·7%, and the viability respectively 85·5 and 8·33. Sterility and decreased viability play an important rôle in these crosses, therefore it is not surprising that the interpretation of the results is less certain than in the mosses of WETTSTEIN. The reciprocal products h × l and l × h show important differences in quantitative characteristics as was described above. The leaf-width is 20·82 in h × l and 23·67 in l × h, the proportion lengths : widths of the corolla were 4·59 : 3·37 and 10·66 : 9·21. MI-CHAELIS had tried another way to give a conclusion to the problem of a permanent change of the plasm under the influence of the non-specific genome. The h × l hybrids as mother back-crossed with *luteum* as father gave such a great decrease in germination of the third generation that investigations had to be broken off. The l × h hybrids could be back-crossed 14 times with the father h, so hence the plants of the constitution $(l \times h^1) \times h$ or lh^2 up to lh^{14} could be ob-tained. In order to trace whether the luteum-plasm was changed permanently by the influence of back-crossing with *hirsutum*-pollen, MICHAELIS (1928) investigated some combinations. On the grounds of this investigation he concluded 'that the characteristics of the plasm remained principally the same after five generations' (1929, p. 309).

A few years later the intensity of the pollen sterility was investigated again by MICHAELIS who at that time concluded that a slow shifting of the pollen percentage seems to be the probable result of a continual back-crossing of the l × h-hybrid with *hirsutum*-pollen. Fig. 149 gives a representation of this phenomenon, MICHAELIS concludes; 'these investigations show that the *luteum*-plasm is changed in a charac-teristic way under the influence of *hirsutum*-nucleus which is unrelated to this species' (1932, p. 98).

In his recent review (1954, p. 327–328), in which he compiles a rich documentation, MICHAELIS again is very careful in his conclusions: 'Section IV described detailed experiments which demonstrated the constancy of plasmon, even when the cytoplasm constituents are exposed to the influence of different genomes and environments. In discussing alterations of cytoplasmic inheritance, therefore, it must be emphasized that in general, in all experiments, the constancy of plasmon has been well established. Alterations of reciprocal differences

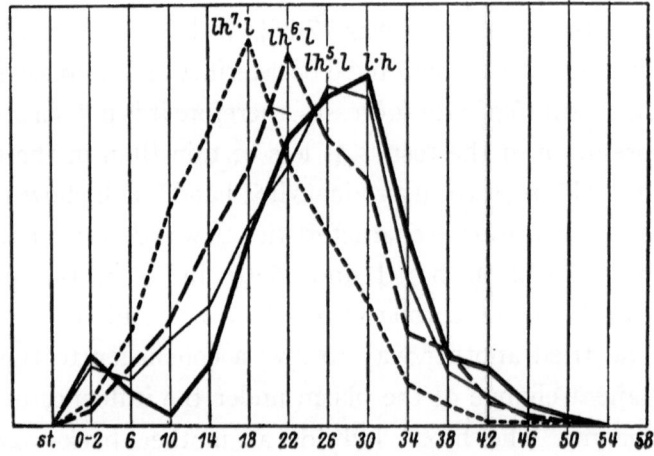

Fig. 149. Backcrosses of *Epilobium luteum* × *E. hirsutum*, showing shift of the pollen percentage (after MICHAELIS, 1932).

have been observed only in single, exceptional plants or in quite definite genome-plasmon combinations'.

It is not at all surprising, that the same investigator comes to two conflicting conclusions on the basis of such complicated material. An investigation dealing with the possibility of a permanent change of the plasm under the influence of nuclear genes in Phanerogames gives rise to many other possible explanations (contrary to the conclusions derived from the investigations into mosses by WETTSTEIN), that it will be extremely difficult to exclude all contingent sources of error. Even MICHAELIS cannot give a final conclusion at this moment.

Of the grounds of WETTSTEIN's investigations with mosses it was concluded that the plasm has an independent nature, which cannot be affected by nuclear genes. On the other hand it seems that the *Epilobium*-crosses pointed to the fact that the plasm may change under the influence of a non-related genome.

In connection with this the statement of AVERY, MACLEOD and McCARTY (1944) seems to be very important. They pointed out that it may be possible to transfer an isolated desoxyribonucleic acid (that is to say a nucleic substance) from strain III into strain II of *Diplococcus pneumoniae*. Strain II is now compelled to synthesize the plasmic substance, a polysaccharid characteristic for strain III, whereas on the other hand the desoxyribonucleic acid III introduced can maintain and even multiply in strain II.

The reverse question is also interesting: does plasm have such an active rôle in the process of heredity, that the genotype of the individual is permanently influenced by the plasmon? Such an influence must therefore have the consequence that the course of a segregation is ruled by the nature of the plasmon.

Crosses between the sub-species *Vicia Faba major* and *V. F. minor* have given different results in a reciprocal direction (SIRKS, 1931) as we already saw. These differences not only come to expression in a difference in reaction towards the same genotype, but as much in an elimination of certain zygotes under influence of the plasm. If we compare the percentages of *zygote-sterility* (that is to say egg-cells which do not develop into seeds) in the parents and the cross in question, then we will obtain the following results:

Cross	10	24	25	33	54	62	71	87
Major-parent . .	5·04	5·04	5·04	2·70	2·47	8·67	5·04	5·00
Minor-parent . .	4·08	3·65	3·55	4·08	4·41	4·41	4·41	4·41
F_1 Major-mother .	3·69	5·08	2·19	8·59	9·17	10·55	2·42	9·26
F_1 Minor-mother .	26·68	27·58	27·65	26·46	30·71	32·29	17·62	35·30

From these numbers it is evident that the plasm of the *minor*-parents causes considerable elimination of zygotes, which can be estimated to be on an average, 25%. This zygote sterility was attended by an apparent true to type breeding with regard to a number of heterozygous characters in the F_1, as is evident from the following table:

Factor	A		E		M		O		B		T	
	A	a	E	e	M	m	O	o	B	b	T	t
F_2 Minor-mother	931	297	212	81	92	38	867	295	235	88	337	113
F_2 Minor-mother	944	—	161	—	121	—	944	—	187	105	294	144

In this cross we deal with the dominant factors A, E, M and O and the recessive factors b and t originating from the *minor* parent and the recessive ones a, e, m, and o and the dominant B and T of the *major* parent. The numbers obtained pointed to the fact that the homozygous combinations of the factors, which are contributed by the *major*-parent were not viable in the *minor*-plasm. Therefore, the double recessives of the characters A, E, M and O were eliminated which resulted in an apparent breeding true to type for the dominant factor, whereas, for the factors B and T, double dominant individual were not viable. This caused a segregation of 2 : 1 instead of the expected 3 : 1 segregation. A striking peculiarity was that the factors in question mutually form one linkage group. Hence we may presume that the combination of two paternal *major*-chromosomes in the *minor*- plasm gives rise to *zygotic sterility*. Nevertheless the real cause of the elimination of zygotes could not be concluded from the data available.

This sterility of zygotes with a certain genotype needs not be so radical that the development of the individual is prevented so early. It is possible that the sterility does not occur at once after fertilization but in a much later stage of development of the zygote. In crosses between two races of flax which differ highly in their habit, the phenomenon of sterility of the anthers was observed by BATESON and GAIRDNER (1921) in one of the reciprocal crosses. These observations were the cause of a closer investigation (CHITTENDEN, 1927a, CHITTENDEN and PELLEW, 1927, GAIRDNER, 1929). An individual of the species *Linum usitatissimum* of unknown origin differed considerably in habit from the normal type. Instead of being unramified, this plant formed at its basis a great number of side-branches which apparently lie flat on the earth, but more or less straighten themselves in the flowering stage. Other differences were observed between this 'procumbens'-type and the normal one with regard to the colour of the style, shorter slips of the sepal and later flowering time. After self-pollination this aberrant form behaved as a true-bred for many years. Crossing of this race with a normal high erect unramified ('tall')flax gave reciprocal differences in both directions. Procumbens ♀ × tall ♂ yielded an erect F_1 intermediate in stem length, and an F_2 generation which was extremely complex in stem lengths but which showed a simple segregation with regard to the fertility of the anthers: 75% of the individuals were normal hermaphrodite (☿), whereas in 25% all the anthers shrivelled

and contained no pollen (ms = male-sterile). The reciprocal cross:
tall ♀ × procumbens ♂ gave normal F_2-individuals. By means of a
great series of systematic back-crossings (see fig. 150) it could be proved
that male-sterility depends on a recessive m-factor, which only wields
its action in procumbens-cytoplasm. This m-factor was present in the
'tall'-races used. Hence, MM, Mm and mm-individuals with 'tall'-
plasm are always normal and completely hermaphrodite. In pro-

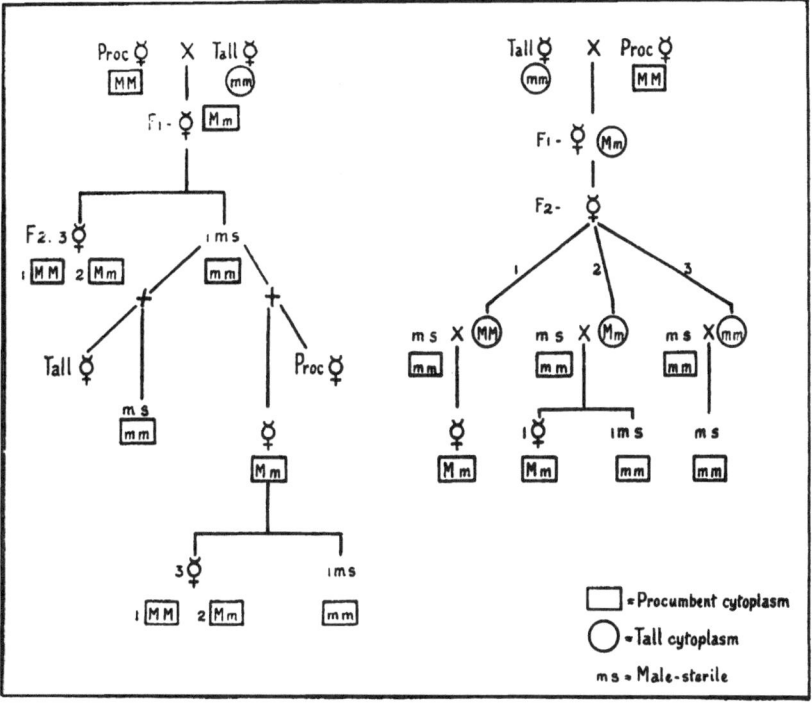

Fig. 150. Male sterility in the flax cross tall × procumbens
(after GARDNER, 1929).

cumbens, only MM- and Mm-plants are able to produce normal pollen,
whereas mm-individuals are male-sterile. Is is possible to interpret
these results in this way, that mm-individuals are stunted in their
normal development by the procumbens-plasm. This inhibition finds
its expression in the sterility of a definite organ (the anthers). The
active sterilizing influence of this plasm-type is therefore, in this case,
a delayed one.

Analogous results on plasmatic-conditioned flower abnormalities
and male sterility have been obtained by OEHLKERS (1938–1941) in the

cross *Streptocarpus Rexii* × *S. Wendlandii* and by KOOPMANS (1951, 1952–1955) in the cross *Solanum chacoense* × *S. rybinii*.

Finally a consequence of plasm activity upon the genotype can be found in the *elimination of the gametes*. Crosses between different species of *Aquilegia* brought SKALINSKA (1930) to the conclusion that in some cases the segregation seems to be absolutely normal (*A. californica* × *A. flabellata*) and in both reciprocal crosses comparable F_2-generations were obtained, whereas in other cases (*A. chrysantha* × *A. flabellata*) a number of types which resemble the paternal flower with respect to spur, shape and colour of flower are lacking in the F_2-generation. In the latter cross, the F_1-plants were highly sterile. This sterility turned out to be due to the fact that gametes succumb after the reduction division. It could be proved by back-crossing the F_1-individuals with the parents, and reciprocally, that this elimination of gametes under the influence of the plasm of the mother is correlated with the non-occurrence of individuals of the paternal type. It was presumed that in this cross, the foreign maternal plasm is the cause of the destruction of gametes with a genome which is partly or completely similar to that of the original father.

Similar results were also obtained with the variegated leaf-types of *Vicia Faba* (SIRKS, 1931), and other investigations. Hence, it is evident that the plasm is able to eliminate certain types of gametes. Nevertheless, we will put forward that such a gamete-sterility and also a sterility of zygotes does not always have to be the result of a direct plasmic influence. For instance if pure somatic causes in the maternal body (e.g. a too small seed-coat, see p. 219) give rise to disturbance in the development of the embryo, then this will naturally result in a sterility of the zygotes as well (see LAIBACH, 1931).

It is evident that the desirability of trying to penetrate into the still vague concept of '*plasmon*' is forced upon us. An analogy of the genome, as being localized in the chromosome, could be perhaps found also in this plasmon. The investigations up to now in this direction have, for a great part been carried out by WINGE and his co-workers (e.g. 1939, 1940, 1949, 1952, 1954), by LINDEGREN (1945, 1946, 1949, 1952, 1953), by SPIEGELMANN (1945) and by EPHRUSSI (1953) with yeast (*Saccharomyces*), alongside those with *Paramaecium* by SONNE-BORN (1945, 1947, 1950, 1954; cf. BEALE, 1954). The experiments of

L'Héritier (1948) dealing with the sensitivity of *Drosophila* for CO_2 must not be overlooked.

The data with regard to *Saccharomyces* is concerned with the so-called adaptive enzymes, which are indeed formed under the influence of certain genes, but only when the suitable sugar is present in the medium. Both yeast species *Saccharomyces cerevisiae* and *S. carlsbergensis* differ in the following way: the first species is not able to ferment melibiose, whereas the second can with aid of its enzyme melibiozymase. By means of hybridization Winge was able to show that this difference was based on two genes. It is of especial importance to note the fact that this ability to ferment is determined in the plasm. The cells, originating from such a cross do not have the gene in question but are still able to achieve this fermentation process.

From this the conclusion follows, that one or two enzymes can be formed by the dominant genes, but only when the sugar melibiose is present in the medium. It also follows that these enzymes are fixed in the plasm and can multiply there under favourable conditions, independent of the presence of the dominant genes in question.

Partly analogous to these yeast investigations are the results of the studies of Sonneborn and his associates over the so-called 'kappa'-factor in *Paramaecium*. A certain line of this animal has the characteristic of synthesizing and of secreting a substance, which is a deadly poison for nearly all other lines of the same species. This race is known as 'killer' and is immune to this poison whereas all other races are susceptible. A dominant gene 'K' gives rise to this poison but only if a cytoplasmic factor is also available. By itself the K-gene is non-active; if the individual does not contain kappa, the dominant K-gene remains harmless. Hence there are four combinations: no kappa + K (harmless); no kappa + k (also harmless); kappa + K makes the animal dangerous; in individuals with kappa next to the recessive k-gene (kappa + k) the dangerous characteristic is present, but gradually disappears. The plasm-compound 'kappa' does not seem to be able to multiply. It is also evident that the dominant gene K does not have the ability to give rise to the killing kappa-substance, but only to increase an already present quantity of kappa. This is one of the very few cases, in which the plasmatic factor, i.e. the kappa-substance is clearly visible as separate particles in individuals of the killing strain (fig. 151).

The sensitivity for CO_2 of certain races of *Drosophila* expresses itself in an analogous way; these races do not tolerate CO_2 and die after some seconds in a CO_2 atmosphere in contrast to normal ones. A cross of a sensitive female with a resistant male yields only sensitive descendants. This is true also of successive back-crosses with resistant males. The reciprocal cross; resistant female × sensitive male gives rise to a mixture of sensitive and resistant individuals. The sensitive males originating from this cross produce in their turn only resistant descendants. The explanation of this is not easy. It is possible that there is indeed a rather particular virus, which is transferred by the cytoplasm. But the phenomenon can also be interpreted as true cytoplasmic inheritance.

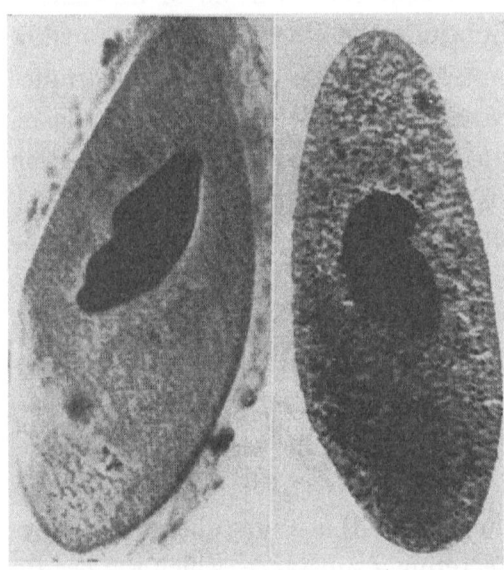

Fig. 151. Left: sensitive *Paramaecium*, right: Killer-*Paramaecium* with plasmatic Kappa-particles (after SONNEBORN, 1950).

From this and other data we can derive how far the cytoplasm has its own materialistic heritable basis. WRIGHT (1941, 1945) introduced, as the first conception '*plasmagene*', and supposed that a gene, during the release by the chromosome in the interphase (between two successive divisions of the nucleus) changes its nature in the cytoplasm (mutates) or attaches itself to the 'hapten' which are already present in the plasm or are produced there by an enzyme. In this way the gene should become a '*plasmagene*'. LINDEGREN has suggested an analogous supposition (1945, 1946) in considering the locus in the chromosome (called '*chromogene*' by him) as the attaching place for the '*cytogene*', which is transferred from the chromosome into the cytoplasm. Here in the cytoplasm the cytogene should multiply completely independently. On the other hand, DARLINGTON (1944, 1949) uses the term '*plasmagene*' only for small bodies which multiply by themselves and are present in the cytoplasm. It is tempting to look for a material basis for the active rôle of the plasm such as the chromosomes for the genes of the nucleus. So far as plant characteristics, connected with plastids are concerned, the problem is easy; the plastids certainly form permanent compounds of the cytoplasm, and the conception *'plastome'* seems to be realized in a material form. For other activity of the plasm the question of a material basis is much less distinct. We know of the existence of plasm-particles in the killer-strain of *Paramaecium*, mentioned above. We know a number of plasmic compounds which appear to belong principally to two groups: *chondriosomes* (that is to say *mitochondria* and *chondriokontes*) and *chromidia* (see DE ROBERTIS and others, 1953 and MONNÉ, 1948). These compounds could be distinguished one from another and also from the substances of the nucleus by several methods, as is evident from the following table:

	absorption ultra-violet light of 2600 Ångström	Feulgen-reaction	Pyronine-staining
chromatin	absorption	positive	negative
chromidia	absorption	negative	positive
mitochondria	no absorption	negative	negative

Chromatin is organized in the chromosomes; the chondriosomes are easily recognizable as cytoplasmic compounds; the chromidia are scattered as very small units (microsomes) in the cytoplasm, but by treatment with Sodium-azide, it became known that they are ranged along cytoplasmic fibrils. These fibrils are parallel threads, upon which chromidia containing ribonucleic acid and interchromidia which do not contain ribonucleic acid alternate (figure 152). We now enter a sphere where a lot of uncertainty still occurs, but it still can not be denied, that such a picture is extremely suggestive and the thought of a chromosome structure is called to mind by this. But how far the possibilities for the explanation of the plasmic activity lie here is, for the present, still only a subject of speculation (see BEADLE, 1949, SONNEBORN, 1949, SYMPOSIUM 1945a, 1946, 1949d).

A summary of the rôle of the plasma can be resolved into the following essential points:

1. the independent nature of the chlorophyll (plastids), which is characterized as *'plastome'*;

2. the after-effect of the modifications located in the egg-cell, or any after-effects of the genotype of the maternal organism in the egg-cell (i.e. *pre-determination, delayed inheritance*). Such an after-effect results in a fixation of characteristics in the embryo, before the new combination of genes (which is present after fertilization) expresses itself in the phenotype;

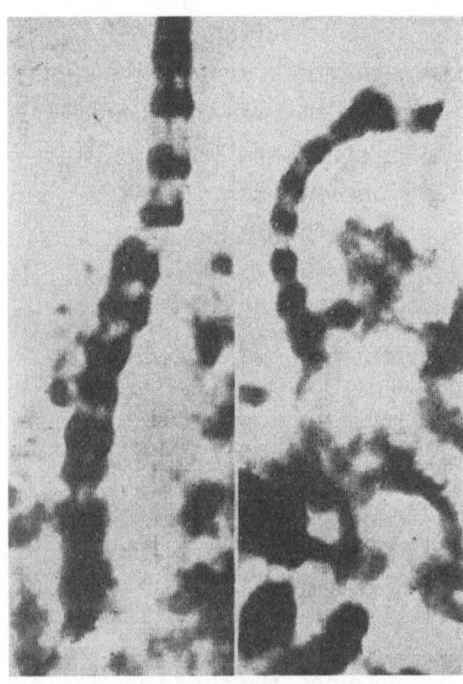

3. the independent accomplishment of properties by the cytoplasm without any action of the genotype;

4. differences in reaction of different plasmata upon the same genome;

5. in most cases it seems that the plasm type (*plasmon*) is always constant, and independent of the genotypical constitution (*genome*); in a very few cases a change of this plasmon due to the influence of the genome appears to have been observed;

Fig. 152. Cytoplasmic fibrils under the influence of Sodiumazide (after MONNÉ, 1948).

6. the plasm can cause elimination of zygotes or of gametes with certain genotypical constitutions;

7. not much can be said with certainty with regard to the material basis of the plasmon or over the existence of so-called plasma-genes.

CHAPTER XVI

THE MOMENT OF FACTOR SEGREGATION

The problem of localization of heritable factors or characters is closely connected with the answer to the following: at what moment in the life of the organism does the segregation of heritable factors take place? From the original Mendelian point of view, only one answer can be given: the moment of segregation of the factors in the hybrid lies in the instant of the formation of the gametes. MENDEL believed this on the strength of his experimental data. When the principal difference between the divisions which take place in the nuclei of the body cells, and the important division (i.e. the reduction-division) during the formation of the gametes, became known from karyological investigations it seemed that the battle was already over. The principal difference (as discussed in Ch. XII) is that, during the divisions of the nucleus in the body cells, all chromosomes are split longitudinally, so that in MORGAN's concept, each new body cell originating will receive a complete set of factors, whereas in the reduction division, in which gametes are formed, the chromosomes are not split, but from each pair of chromosomes one is carried to the one daughter cell, the other, to the other daughter cell. On the grounds of the localization theory of the heritable characters, it is acceptable that all body cells contain the same set of heritable factors in their nuclei, while the gametes contain incomplete sets.

So far as is known, the proof of this theorem is based upon the analysis of the offspring of a certain cross composed of diploids. The conclusion that a monohybrid gives rise to two types of male and two types of female gametes is properly only obtained by indirect means. Therefore it became important that a more direct study into the genotypical constitution of gametes (or spores) originating from the reduction division should be carried out. After HARTMANN (1912) pointed to the great importance of such a study, NEWELL (1915) published an unfortunately short summary of his studies with bees. It appeared to be an ascertained fact (see chapter XVIII) that in bees the male individuals (drones) are born from unfertilized egg-cells, whereas the females (workers) owe their origin to fertilized egg-cells. Male animals can be considered as somatic haploids, and the consequence of this is that the composition of the males in an F_2-generation gives a picture of the gametotypes, which are produced by the F_1-female. NEWELL crossed an Italian queen (*Apis ligustica* ♀) with a 'krainer' drone (*A. carnica* ♂) and found a complete dominance of the Italian type in the F_2-females. The F_2-generation on the other hand, after mutual mating of F_1-individuals, as well as after back-crossing with *A. ligustica* or with *A. carnica*, was composed of 50% Italian and 50% krainer types so far as the drones were concerned. With this evidence he put forward that the F_1-females do indeed give rise to two sorts of egg-cells and that the segregation takes place during the reduction division.

Further investigations were made into the possibility of a morphological recognition of the genotypically different haploid cells. Above (p. 180) we have seen that RENNER (1919a and b) succeeded in distinguishing both types of pollen grains, which are formed in a complex, heterozygous *Oenothera*. This investigation was soon followed by proof that the nature of pollen grains, with different factors for reserve-food could be shown by chemical reaction. PARNELL (1921) analysed the pollen grains of a hybrid between two races of *Oryza sativa*, in which one contained starch as its reserve and the other one amylodextrine. Among the pollen grains of the F_1-plants, two types could be distinguished, which were differently stained by iodine. In 18 anthers originating from three plants, the starchy pollen varied from 43·2 up to 51 8% with an average of 48·1%, so that both pollen types are formed almost in equal numbers.

Hence these results prove, that both the types were formed in equal numbers as products of the reduction division. However, evidence was not obtained that two of the daughter cells, originating from one mother cell during the formation of the tetrad, belong to the one type and two to the other type. An analysis of the tetrad, which originates from one mother cell is necessary for irrefutable proof of the fact, that the reduction division is the cause of the formation of genotypically different generative cells.

Such a tetrad analysis on the grounds of the data provided by diploid descendants will be hard to obtain and it will give no answer to the inseparable question: which of the two divisions that give rise to the formation of gametes (or spores) is the proper division? For this problem genetics only can derive a complete answer from data dealing with tetrad analysis of organisms, in which the product is not only distinguishable as a generative cell, but also as an individual. The theoretical basis of the problem is as follows: if the separation of the allelomorphs in a di-heterozygote takes place in the first of the two reduction divisions, then the result will be, that the four cells (which together form a tetrad) are similar in pairs. However, if separation takes place in the second division, then a tetrad can be formed consisting of four genotypically different cells. The exactitude of this conclusion can be derived from the following scheme:

	Di-heterozygote	First division	Second division
1.a.	AaBb....	............AB..	..............AB
			..............AB
		............ab...	.... ab
			.... ab
1.b.	AaBb....	.... Ab..	..............Ab
			..............Ab
		.... aB..	..............aB
			..............aB
2.a.	AaBb....	...........AaBb..	..............AB
			.... ab
		...........AaBb..	..............Ab
			..............aB
2.b.	AaBb....	...........AaBb..	..............AB
			.... ab
		...........AaBb..	..............AB
			.... ab
2.c.	AaBb....	...........AaBb..	..............Ab
			..............aB
		...........AaBb..	..............Ab
			..............aB

Therefore if segregation of allelomorphs takes place in the first division, then only tetrads of the combinations (2AB + 2ab) or (2Ab + 2aB) (case 1.a. and 1.b.) will originate. Segregation of the allelomorphs in the second division, gives rise to a tetrad 1A B + 1 Ab + 1aB + 1ab (case 2.a.) as well as to the possibilities (2 AB + 2 ab) and (2 Ab + 2 aB) (cases 2 b and 2 c).

In karyology there is an analogy with this question in the problem of pre-reduction or post-reduction (see EKMAN 1927, BRIEGER 1933). Pre-reduction may certainly be considered to be the most common case and therefore it was discussed here (p. 231) as the normal case. Pre-reduction supposes that homologous chromosomes separate in

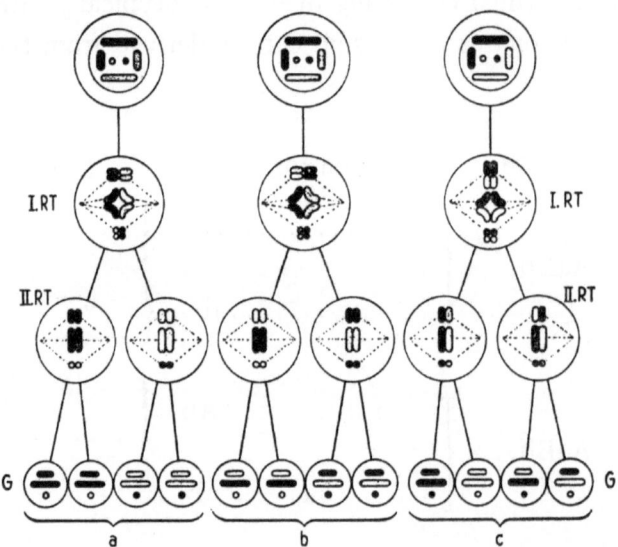

Fig. 153. Scheme of possibilities of pre-reduction (a and b) and post-reduction (c) (after BĚLAŘ, 1928).

the first division (I.RT in figure 153) of the reduction divisions. Finally in the second division both split-halves of each homologous chromosome separate. Post-reduction presupposes that in the first division of the reduction division, both split halves separate and in the second division (II.RT) the homologous chromatids separate. Figure 153 gives a clear representation of these processes: a and b represent pre-reduction and show that the future gametes (G) are similar in pairs; c gives the case of post-reduction, in which the four future gametes (G) are different. However, the conception that pre-reduction should be the only process that occurs, meets with difficulties (see Bĕlař 1928, Wilson 1928). The investigations of Carothers especially (e.g. 1931) may be mentioned here. These appeared to show that post-reduction plays as much of a rôle as does pre-reduction. In some of the species investigated by her, only pre-

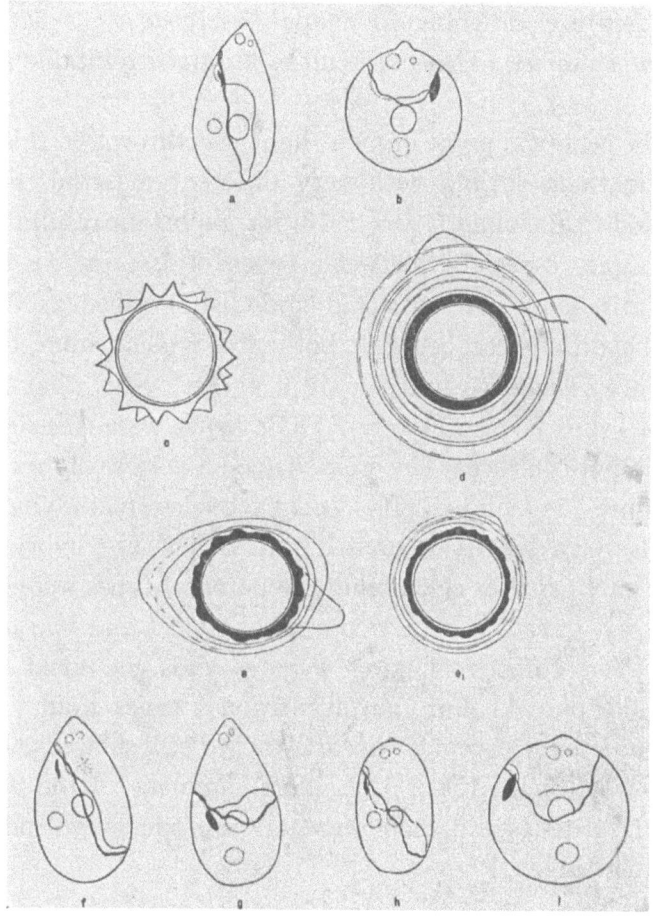

Fig. 154. Two races of *Chlamydomonas* (a with c, b with d), their hybrids (e and e₁) and four types of individuals originating from the hybrid (after Pascher, 1916).

reduction occurred, whereas in other species, only post-reduction. There were also species in which one pair of homologues can behave differently from the other pair. In *Trimerotropis citrina* (a grasshopper species), it was observed that for one of the chromosomes, pre-reduction occurred in 95% of the animals and post-reduction in 5%. For another chromosome of the same animal these numbers were 10% pre-reduction and 90% post-reduction. An analogous botanical case is mentioned by RESENDE (1936) in his karyological studies concerning the genus *Aloe*. However, the karyological investigation of both these possibilities is very limited because the investigation is only completely successful if both the homologous chromosomes are different in shape (*heteromorphous chromosomes* and *sex-chromosomes*). These difficulties are not irrefutable as we will soon see.

From the genetical point of view, light was thrown on this problem by investigations dealing with very different material. Firstly we are indebted to PASCHER (1916, 1918) for important results in crosses between algae, especially between types of *Chlamydomonas*. Two types of this genus were selected and distinguished as Chl. I and Chl.II. The differences between both are evident from figure 154 (a and b are vegetative individuals, c and d cystes, that is to say, zygotes of I and II respectively). These types were sufficiently true breeding in pure cultures. The cross yielded a very easily recognizable zygote (figure 154 e and e_1). PASCHER succeeded in studying the four individuals which directly originated from each of the 8 hybrid zygotes: in 5 cases only gametes resembling the parental types were observed, that is to say, in each zygote two cells of type I and two of type II occurred. The three remaining zygotes did not yield cells resembling the parents, but four transitional types from which two had more or less the parental appearance, but still showed characteristic differences (figure 154 f–i). A short summary of the differences between the parents and their segregation products will be given in the following table:

	body shape	membrane	papilla	chromato-phore	eyespot
Chl. I	pear-shaped	tender	—	lateral	linear
Chl. II	spherical shaped	tough	+	basal	spot-shaped
f.	pear-shaped	tender	—	lateral	linear
g.	pear-shaped	tender	—	basal	spot-shaped
h.	spherical shaped	tough	+	lateral	linear
i.	spherical shaped	tough	+	basal	spot-shaped

If these characters are considered on their own, it is certain that each characteristic is present twice among these 4 cells, which therefore points to a Mendelian segregation. But if we take the cells as a whole, four types can be distinguished. These results therefore, may account for the fact, that in the five zygotes first mentioned, the reduction of homologous chromosomes takes place in the first division, whereas in the last three mentioned zygotes reduction may have taken place in the second division. However we will soon see that such an interpretation does not need to be true.

Many investigations concerning the problem of segregation factors were carried out with the help of fungi. Ascomycetes as well as Basidiomycetes are used in this kind of research and yield excellent

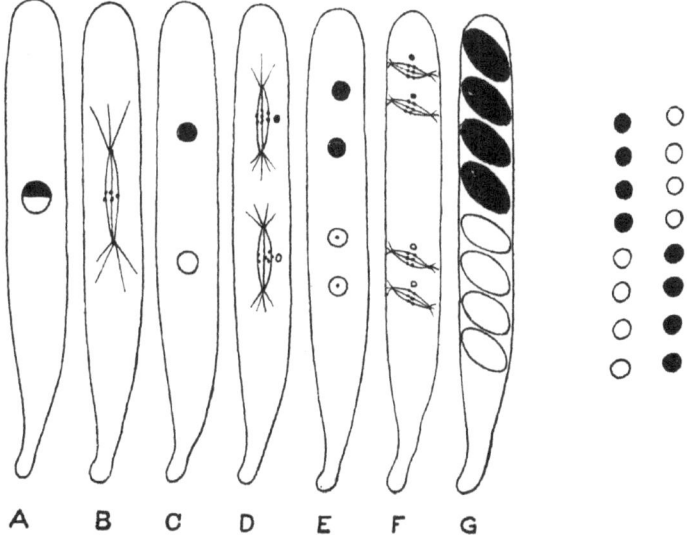

A   B   C   D   E   F   G

Fig. 155. Scheme of the spore formation for the case in *Neurospora sitophila*, that separation of sex-factors takes place in the first nuclear division (after KNIEP, 1929).

results (see KNIEP 1928, 1929). The Ascomycetes, as we know, have an ascogeneous cell with one nucleus. This nucleus successively undergoes a division three times and hence eight nuclei originate, each of which together with an amount of plasm forms an ascospore. These eight ascospores lie in a row inside the protracted ascus. If the individual is heterozygous for one or more factors (for instance sex) then by means of single spore cultures, the distribution of genotype in the ascospores which lie next to one another in the ascus, can be studied. SHEAR and DODGE (1927) pointed to the importance of this kind of research. Figures 155 and 156 give a summary of the position in the ascus of both sexes of the ascospores as may be expected if a segregation of the heterozygotes takes place in the first division or in the second division. In the first case all four spores of the same sex lie next to one another, in the latter they are scattered in groups of two identical ones over the ascus (C after the first division, E after the second division, G after the third division). On the basis of these suppositions e.g. DODGE (1931) carried out investigations with single spore cultures. These investigations dealt with two characters, that is to say both the 'sex' types (S) A and B and the formation

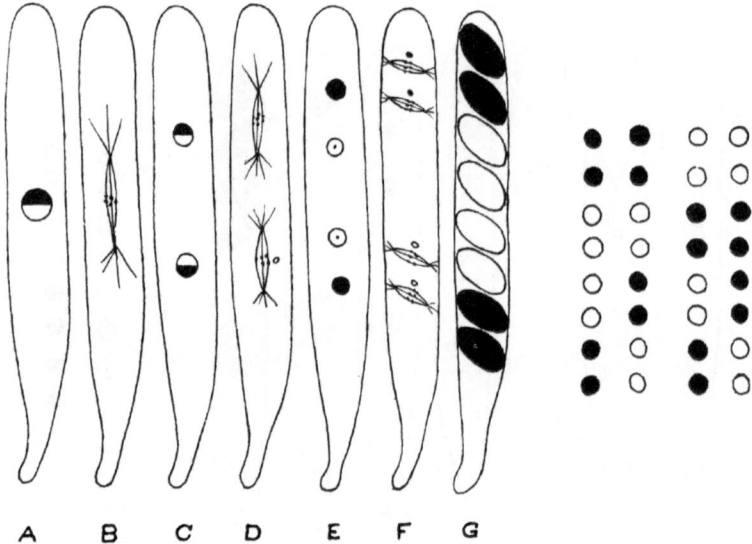

Fig. 156. Scheme of the spore formation for the case in *Neurospora sitophila*, that separation of sex-factors takes place in the second nuclear division (after KNIEP, 1929).

or absence of yellow conidia (+C and —C). The following table represents the results of DODGE:

Ascus Nr / Spore Nr	A S	A C	B S	B C	B' S	B' C	C S	C C	D S	D C	D' S	D' C	L S	L C	M S	M C	N S	N C	O S	O C	P S	P C	Q S	Q C
1	B	—C	B	+C			B	—C	A	+C			B	±C	B	—C	A	—C	A	—C			A	—C
2	B	—C					B	—C	A	+C			B	±C	B	—C	A	—C	A	—C	A	+C		
3	A	+C			B	—C	B	—C	A	+C	B	±C	A	±C	A	+C	A	+C	A	±C	A	+C		
4	A	+C					A	+C	B	±C	A	±C	A	+C	B	+C	A	±C	A	+C				
5	B	+C	A	—C					B	—C					A	+C	A	+C	B	+C	B	—C	B	+C
6	B	+C	A	—C	A	+C			B	—C	A	+C			A	+C			B	+C	B	—C	B	+C
7	A	—C	A	—C	A	+C					A	+C			B	—C			B	+C			B	+C
8			A	—C	A	+C					A	+C			B	—C			B	+C	B	—C	B	+C

It is evident from this table, that the spores 1 and 2, 3 and 4, 5 and 6, 7 and 8 always mutually resemble one another in character as far as they germinate (the non-germinating ones are indicated by —). This indicates that the third division gives rise only to similar products. The sex factor sometimes segregates in the first division (asci B, B', C, D, D', O, P and Q) and in other cases, in the second division (asci A, L, M). Ascus N behaves irregularly with respect to the sex-distribution (3A: 1B: 1A: 3B) which was explained by DODGE as due to an experimental error. Likewise, for the probable possession of conidia (first division asci B, B', C, D, P, Q; second division asci A, M, N) both segregating divisions are held to be responsible. However this character is somewhat dubiously expressed in the asci D', L and O. In those cases some abnormal conidia ±C develop on a mycelium, which is normally without conidia.

DODGE concludes, that the first two divisions both can serve as the segregating division for sex and for the origin of conidia, and that for both characters, this takes place independently from one another. Further we will discuss HÜTTIG's (1931) experiments carried out with species of a large group of the Basidiomycetes, the *Ustilagineae* (smut fungi). In this case the four spores from the zygote germinate each into a promycelium, which consists of four cells (see figure 157) of which each cell forms a sporidium, which can be isolated and give rise to a mycelia. HÜTTIG comes to an analogous conclusion

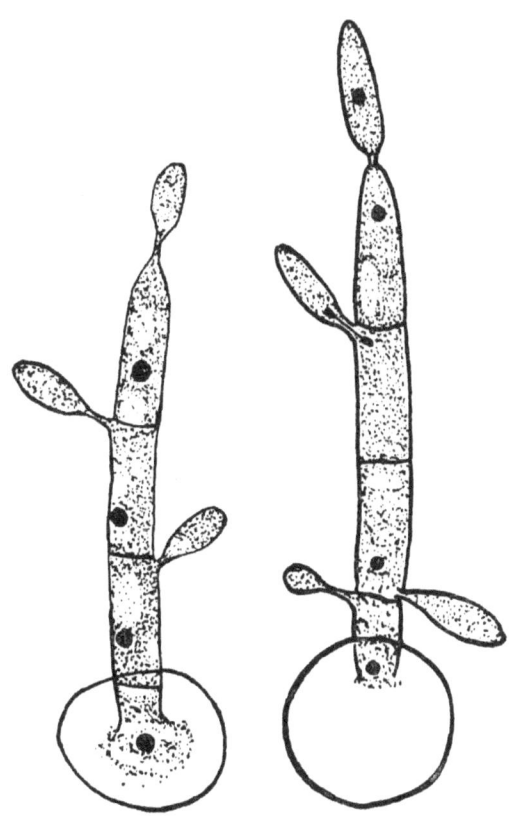

Fig. 157. Promycelium with four sporidia of a smut-fungus (after HÜTTIG, 1931).

to that of DODGE, viz., that the factors involved in his crosses segregate independently, sometimes in the first and sometimes in the second division. It appeared that these crosses are temperature sensitive; 33% of the sexfactors segregate at 10°C in the first division and 67% in the second division of the promycelia. For temperatures of 20°C and 25°C these numbers were 31% and 69%; and 9% and 91% respectively.

WETTSTEIN also gave important contributions to the problem under discussion in his investigation concerning mosses. He made a tetrad-analysis (WETTSTEIN 1924b, p. 95; see also 1928, p. 17 note) with relation to four pairs of factors (G = macrospora, g = microspora; B = latifolia, b = angustifolia; P = globosa, p = cylindrica and C = aurea, c = ochracea) combined in a cross GBPC × gbpc. By means of narcosis he caused four spores of one single tetrad to remain temporarily together, hence he could make cultures of these four isolated spores. Two groups of 24 and 11 respectively of tetrads analysed yielded the following results:

16 and 6 tetrads segregated into 2 spores GBPC and 2 spores gbpc
 5 and 1 tetrads segregated into 2 spores GBPc and 2 spores gbpC
 0 and 1 tetrads segregated into 2 spores GBpC and 2 spores gbPc
 2 and 0 tetrads segregated into 2 spores GBpc and 2 spores gbPC
 1 and 1 tetrads segregated into 2 spores GbPc and 2 spores gBpC
 0 and 2 tetrads segregated into 2 spores GbpC and 2 spores gBPc
―――   ――
24   11

From these results it is obvious that spores originating from one tetrad are similar in pairs. Therefore we may conclude that the segregation of the factors has to take place in the first division. Moreover WETTSTEIN was able to support this hypothesis because he succeeded in suppressing the second division of the cells in a sporogonium with chloralhydrate. Instead of four cells, only two cells were formed. These spores differed in heritable tendency, which proves a segregation of factors in the first division. On the grounds of these investigations the hypothesis can be built up that, in mosses, the first of the two divisions is the proper reduction division.

However, the results of ALLEN (1924, see 1935) later amplified by KNAPP (1936) from studies with the liverwort *Sphaerocarpus* do not agree with the hypothesis derived by WETTSTEIN. In 7 crosses (by means of tetrad analysis) both pairs of factors, male and female and polycladus (polylobate, p) and non-polycladus (np) segregate as follows: out of 70 tetrads analyzed there were 39 with 2 spores ♀ and np + 2 spores ♂ and p; 19 tetrads with 2 spores ♀ and p, 2 spores ♂ and np; 12 tetrads with 1 spore ♀ and np, 1 spore ♀ and p, 1 spore ♂ and np, 1 spore ♂ and p. In this case there is no doubt that tetrads with 4 types of haploid spores originate. Therefore these analyses give evidence for the view that segregation can take place in the second division.

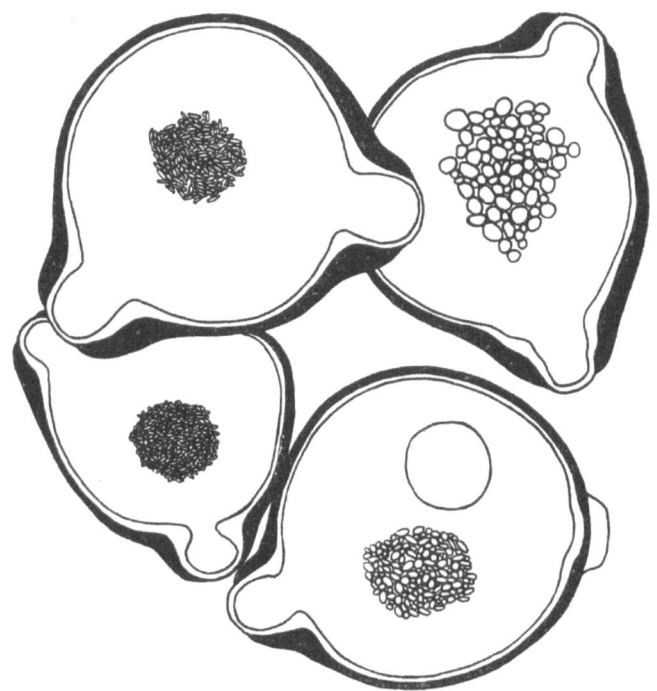

Fig. 158. Four pollen-grains from one tetrad of the hybrid *Epilobium hirsutum* × *E. luteum* (after MICHAELIS, 1931).

Finally we will mention the example of a tetrad analysis in Phanerogamae viz. the investigation of MICHAELIS (1931). He crossed two species of *Epilobium* (*E. hirsutum* and *E. luteum*) which differ considerably in the shape of starch in the pollen grains: the first has rod-shaped (bacillaris) and big (augens) starch grains, the latter egg-shaped (oviformis) and small (diminuens) ones. He was able to observe the four possible combinations in the same tetrad, that is to say, big egg-shaped starch grains (fig. 158 top right), small egg-shaped (bottom right), big rod-shaped (top left) and small rod-shaped (bottom left). If our theoretical supposition is true then evidence has been produced that the second division is the division in which segregation takes place.

There is a tendency in genetical literature to take the occurrence of four genetically different haploid cells in one tetrad as proof of the significance of the second division. However, such a significance has not been proved, because alongside the theoretical supposition, other possibilities exist. Classic contributions to this very important problem have been given by GRÉGOIRE (1905, 1910). Unfortunately his studies in this sphere are almost forgotten today. A first indication of the direction in which we will find a possible solution was given by the case of two factors (sex and polylobate) investigated in *Sphaerocarpus* in which ALLEN was able to show that partial linkage existed between these factors. It is evident that four types of gametes can originate in the same tetrad after crossing-over between the linked factors located in two chromatids of the homologous chromosomes. This is also true in the case that reduction takes place in the first division only.

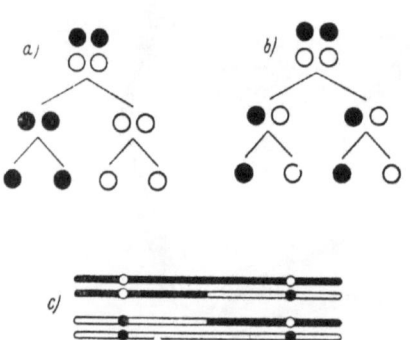

Fig. 159. Scheme of pre- and post-reduction (resp. a and b) in connection with crossing-over (after KNAPP, 1937a).

The previously mentioned amplifying study of KNAPP (1936, 1937a) has turned this possibility into a great probability; tetrad analysis of a great number of linked genes in *Sphaerocarpus Donnelli* showed that during the reduction division, crossing-over takes place between two of the four chromatids as could be expected. However, a certain part of the chromosome, near the centromere, always shows pre-reduction, whereas other genes are submitted to post-reduction if crossing-over took place between the centromere and the locus of the gene in question. The conclusion is obvious: non-exchanged genes give pre-reduction and exchanged genes give post-reduction. Figure 159 will explain this.

It looks as if the karyological data, which seems to support the post-reduction (p. 371) can also be explained by crossing-over between the centromere and the heteromorphous chromosome part. This resulted in a loss of significance for the whole debate concerning pre-reduction-post-reduction. Pre-reduction remains the only process for the number of chromosomes, and for the genes pre-reduction is the principal process. Post-reduction can thus be considered as a secondary process and a consequence of crossing-over.

Apart from this important problem, as to which of the two successive divisions (which gave rise to the formation of gametes or spores), is the proper reduction division; a second problem comes to the foreground. Is the segregation of allelomorphs a monopoly of the reduction division or is it possible that this can also take place in somatic cells? This question can also be formulated in the following way: are all descendants which originate by vegetative multiplication (these are called *clones* by WEBBER, 1903), mutually identical in heritable characters?

In the vegetable kingdom the phenomena of 'budvariations' (also known as budmutations or budsports) are important. Horticultural literature is especially rich in its records of observations, which all contain the fact that a part of the plant (usually a branch with leaves, or an inflorescence) has a striking appearance, different from the rest of the plant. These observations are scattered throughout all the horticultural and botanical periodicals, and it was therefore a hard task for CRAMER (1907, 1954) to collect together all this data: a study of the data makes it obvious that this phenomena is very common among plants and especially among cultivated ones.

Three organisms are of classical value in the battle which was waged concerning the significance of the phenomena, they are the bizarria fruits, which have partly the appearance of oranges and partly that of lemons, while they also show internal differences, the *Cytisus adami* and the *Crataegomespilus*-types.

The best known form, *Cytisus adami* is characteristic because it generally unites the different characteristics from two other *Cytisus* species, *C. laburnum* and *C. purpureus* and especially because nearly every well-developed specimen with typical flesh-coloured flowers of *C. adami* shows branches with pure yellow flowering inflorescences and also (though less common), branches with short, compact purple racemes. With regard to the flowers and leaves, the first branches are not distinguishable from those of *C. laburnum*. The other branches are similar to *C. purpureus* in flower and leafshape. The problem is: what gives rise to the origin of these budvariations of *Cytisus adami*. This remained unanswered for a long time. The most probable explanation was that budvariations originated by means of grafting. A conclusion was reached when anatomical investigations of BUDER (1910, 1911) showed that *Cytisus adami* could always be sharply dissected into two parts by the structure of the different

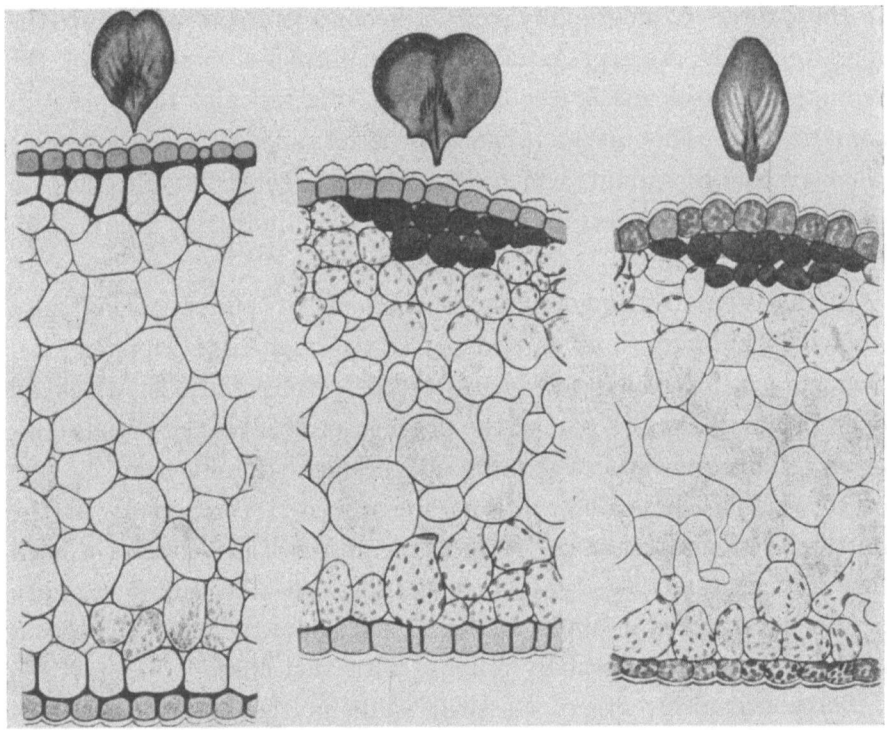

Fig. 160. Corolla leaves and their sections in *Cytisus purpureus* (left), *C. adami* (middle) and *C.laburnum* (right) (after BAUR, 1911).

tissues viz. an internal part which completely resembles the corresponding part of a *C. laburnum* plant and around this, a cylindrical cell layer, which completely resembles the epidermis of *C. purpureus*. This followed immediately from sections of the corolla leaves (figure 160). Therefore *Cytisus adami* can be compared with a hand in a glove; the hand is composed of *C. laburnum* tissue, the glove of cells originating from *C. purpureus*. BAUR (1909a) suggested the name *'periclinal chimeras'* (enveloping chimeras) for these buds. Thanks to his investigation we are able to explain the origin of budvariations. *Laburnum* tissues can form the outside layer if the enveloping *purpureus* epidermis is injured, that is to say, the *laburnum* tissue which is the next deeper layer comes freely to the surface and provides for the regeneration. An absolutely *purpureus* branch is only possible if the epidermis develops into a multicellular tissue by means of repeated divisions and if in this multicellular tissue a growing point is formed. The supposition that these periclinal chimeras owe their origin to grafting remained hypothetical for a long time, but was supported by the investigations of WINKLER (1907, 1916) who proved experimentally the possibility of such an origin. He succeeded in making plants artificially, which were chimeras, by very prudently arranged experiments. These plants were composed of tissue partly taken from one species and partly from another species. His results with graftings of two species of *Solanum* (the tomato, *S. lycopersicum* and the black nightshade, *S. nigrum*) are especially important. Both these species resemble one another sufficiently to give a good chance of success, and on the other hand,

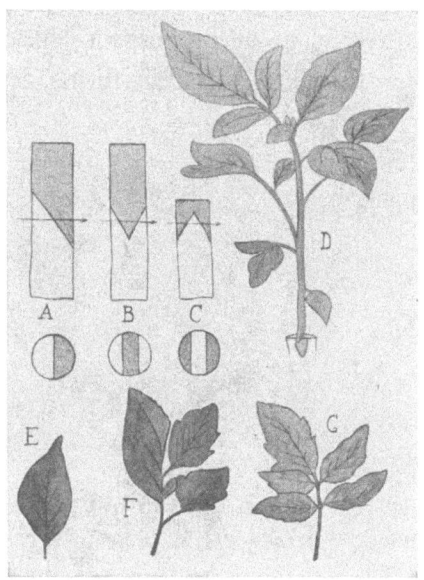

Fig. 161. Representation of the graft experiments of WINKLER (1907).

differ sufficiently to be distinguished at first sight. The chromosome number also differs (*S. lycopersicum* 2n=24; *S. nigrum* 2n=72) and by means of this (in the growing points) the tomato cells can be distinguished from nightshade cells. Each of the two species were grafted upon one another and after the graft thrived, the stem was cut at the grafting place. Therefore, at this section both tissues were situated next to one another, i.e. tissue originating from the tomato and from the nightshade (see figure 161). All which is speckled in this drawing is nightshade tissue, the unspeckled part is tomato. At the ends of the stems which are cut off, adventitious buds grew out of the callus tissue. The buds usually give nightshade or tomato plants. However, some other buds grew branches which had tomato tissue and nightshade tissue lying next to each other in 'sectors' (see figure 161D). Hence such branches yielded, apart from pure tomato leaves (G) and pure nightshade leaves (E), leaves which were partly tomato and partly nightshade. This branch when grafted gives a plant which was characterized as a 'chimera' by WINKLER. In this case we are dealing with a '*sectorial chimera*', because the different tissues are situated next to one another like sectors. However, these results cannot yet be compared with the case of *Cytisus adami*. Later on, buds comparable to those in *C.a.* did originate in WINKLER's experiments which represent in every respect 'enveloping chimeras' or '*periclinal chimeras*' in four different forms. In the growing points of these plants

Fig. 162. Flower, leaf and fruit (from left to right) of *Solanum nigrum* (nightshade), *S. tubingense, S. proteus, S. gaertnerianum, S. koelreuterianum* and *S. lycopersicum* (tomato) (after Bonner Lehrbuch).

both kinds of tissue could be distinguished by chromosome counts and therefore it was evident, that the four new forms were differently built (figure 162): *Solanum tubingense* is composed of a central mass of nightshade tissue encircled by a single layer of tomato cells. *S. proteus* contains nightshade tissue and around this a double layer of tomato tissue. *S. koelreuterianum* has an internal structure of tomato tissue and a single layer of nightshade cells as an envelope and *S. gaertnerianum* is a mass of tomato cells enveloped by a double layer of nightshade cells. From WINKLER's investigations it was obvious that it is possible to obtain periclinal chimeras by means of grafting two different forms one upon the other. All forms mentioned give occasional budvariations, composed of either tomato or nightshade tissue.

The problem of the budvariation in *Cytisus adami* and these graft hybrids is now almost completely solved, however it is not so much a genetical question. From a genetical point of view, the importance of chimeras is not so great in the case when the origin of these chimeras is the result of grafting. On the other hand however, it is of great importance if such sectorial and periclinal chimeras can be brought about without grafting. This question is the main point which we

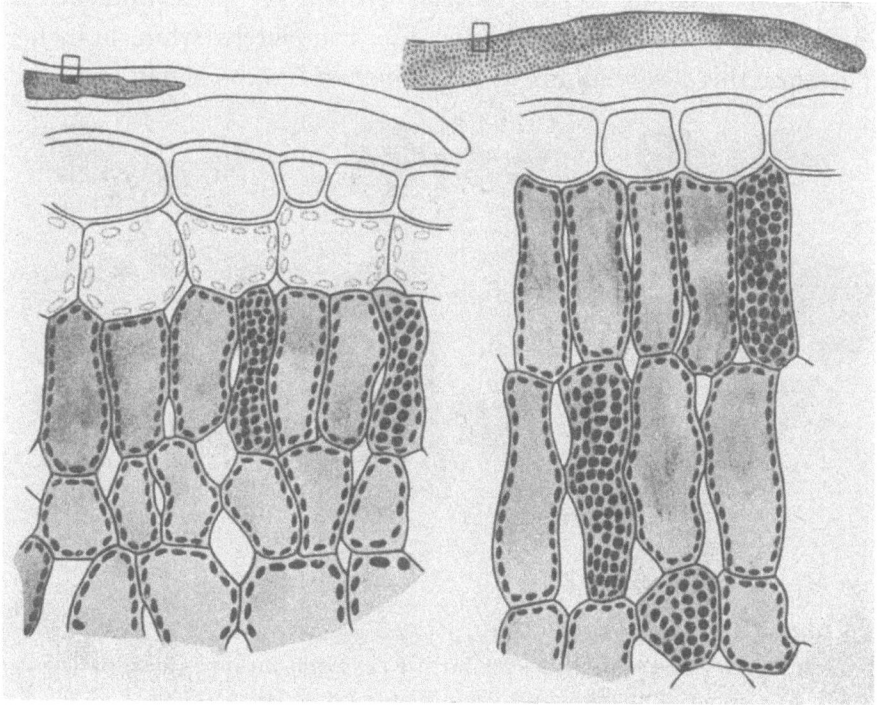

Fig. 163. Sections of white-edged (left) and green *Pelargonium* (right), the squares indicated in the upper portion are enlarged in the lower part of the diagram (after BAUR, 1909b).

have to answer before we recognize the reduction division as the only moment in which segregation takes place.

The first and most appealing phenomena is the occurrence of pure green or pure white branches on variegated plants. In connection with this we must mention a study of BAUR (1909b) with variegated leaves of pelargoniums (*Pelargonium zonale*). The material with which he worked was composed of a number of plants with leaves that were only green in the central part, encircled by a white margin. Anatomical investigation (Figure 163) showed BAUR, that this white margin was due to the fact that in these variegated plants not only the epidermis is lacking chlorophyll (as in all, including the green plants), but also the sub-epidermal layer (which lies directly under the epidermis). The leaf thickness decreases towards the margins and as a consequence of this colourless cell layers of the under and upper side do not enclose more chlorophyll containing parenchyme, and hence the leaf margins appear to be completely white. These plants with white margins regularly bear absolutely green branches as budvariations and also, occasionally, completely white branches. Therefore this case reminds us very much of *Cytisus adami* and other

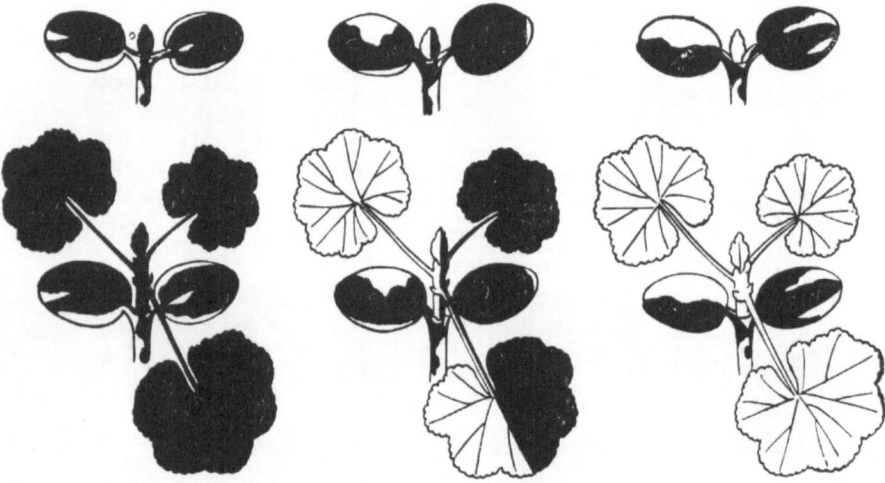

Fig. 164. Young variegated plants of *Pelargonium zonale*, green × white branches, with three possibilities for further development into green, variegated sectorial chimera and white plants (after BAUR, 1909b).

periclinal chimeras. However, no grafting was carried out here and hence plants with white margins must thus have some other origin.

This became evident from sowing experiments carried out by BAUR. All descendants from variegated branches of *Pelargonium* with white margins turned out to be white seedlings (like those of the white budvariation), but they died after about 10 days. This is easily understandable, if we bear in mind the fact, that the sub-epidermal layer especially, is the layer from which pollen grains and egg-cell originate. Therefore the possession or absence of chlorophyll in these cells decides the nature of the descendants. In accordance with this, there is also the fact that all the seedlings of plain green budvariations are green without exception. However if a cross is made between a flower from a variegated, or from a white branch, with a flower of a branch with green leaves, then the seedlings obtained will be green-white marbled in different nuances ranging from almost pure green, up to and including, almost pure white. The further development of these variegated seedlings depended on the position of the growing point between the two seed-lobes. For instance if this growing point accidentally lies wholly in a green part of the young seedling, then the growing point (and as a consequence of this the fully grown plant developing) is also plain green (figure 164 left). But if this growing point lies in a white part, then the plant becomes white, whereas a growing point situated in the white and green part gives rise to a variegated plant in which the white and green tissues lie next to one another, in sectors.

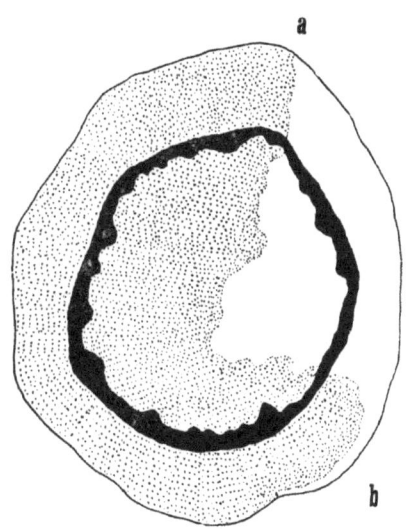

Fig. 165. Section through a sectorial green-white branch, with a peculiar boundary of the tissue at b (after BAUR, 1909b).

These investigations have further consequences. In green-white variegated plants (which were divided sectorially into both tissues) the boundaries between these two parts were not always running radially straight from the centre to the margin, but often followed all sorts of paths. Figure 165 represents a section through the main stem of a sectorial green-white plant, in which at a, a border between the (speckled) green and the (unspeckled) white part is situated perpendicularly to the margin, but at b, it runs parallel with the margin for a long distance. This peculiar course is the cause of the origin of plants with white margins. In all cases in which BAUR observed the development of leaves at such margins of the sector, it turned out that these leaves had white margins. Buds originating in the axil of such leaves grow out into branches with wholly white margined leaves. A growing point originating in the green part

25

(situated directly under the sub-epidermal layer) during development takes the sub-epidermal layer and the epidermis from the white upper layer with it and hence develops into a periclinal chimera, that is to say, a green content enveloped by a cell layer without chlorophyll. This is the reason that green branches, sectorially green-white and periclinally green-white branches (figure 166) can occur on the same plant.

Differences of opinion with regard to the interpretation of these phenomena still exist. BAUR holds that the segregation process is due to a mixture of green plastids, originating from the one parent, and white plastids originating from the other parent, and a redistribution of these different plastids. This point of view, that all cells are already differentiated at a very early ontogenetical stage into green (containing only green plastids), is upheld by the investigations of KRUMBHOLZ (1925) and ROTH (1927) but is not upheld by others (e.g. CORRENS 1928b). This con-

Fig. 166. *Pelargonium zonale* plant which shows green, sectorially variegated, and periclinal variegated branches (after BAUR, 1909b).

ception however, remains the most probable one, but it is still evident that not all forms with variegated leaves can be explained in this way (see CORRENS 1937, DE HAAN 1933, RENNER 1934). However, in spite of these differences with regard to the differentiation of cells the genetical behaviour of the plants investigated by BAUR produced evidence for the justification of the following conclusion: by means of processes of segregation in the growing parts of the plant, without any co-operation of a reduction division, sectorial, as well as periclinal chimeras can originate. The two parts of such a chimera differ in genetical predisposition. However, BAUR's investigations have reference to a special group of heritable characters; i.e. to those of

Plate V

Budvariations

chlorophyll, which form a more or less independent *plastome* (see p. 346). Therefore this problem reaches still further for a positive answer. But the question is now, has a somatic segregation of hybrids also been proved with regard to phenomena other than inheritance of plastids, that is to say heritable characters which may be supposed to be located in the cell-nucleus?

Unlike the former question, this one cannot be answered conclusively at the moment. We are still looking for a solution. The phenomenon is too frequent not to be of importance for genetical studies. One of the most typical cases of budvariation in which no cytoplasmic inheritance is involved is shown in Plate V: here is shown a hyacinth (King of the Blues), in which the raceme is sectorially divided into two parts in a well defined manner, the original purple-blue has one sector in which all bells are pink and the stem is partly purple and partly green. The bells situated just at the border between these two parts give rise to perianths which are partly purple and partly pink to varying degrees. Another case occurs in the *Dahlia* (Helvetia) in which plain red flowers as well as normal red-white marginally striped flowers occur. In the potato (Blue Eigenheimer) white budvariations occur on the same plant with normals.

A number of cases have also been found in animals. These cases became known as '*mosaic phenomena*' and are comparable to the botanical '*budvariations*'. However it is difficult to detect the cause of the underlying process which gives rise to this phenomenon since only a few observations are recorded. Several possibilities can be entertained: 1) modification of a part of the organism by means of nutrition or other external influence, 2) changes in the chromosome set by an irregular behaviour of the chromosomes during a somatic division, 3) mutation in somatic cells from a homozygous nature into another heterozygous or homozygous constitution, and 4) vegetative segregation of characters with respect to which the organism is heterozygous. It is not surprising that investigation of this problem will present many great difficulties. The study concerning budvariations and mosaic animals was, up to now only a descriptive one; experimental data in this sphere is less easily available. Budvariations in plants have the advantage that they are able to give a genetical solution to the problem concerning the behaviour of budvariation after vegetative multiplication and after generative reproduction, combined with an analysis of the offspring of the original type and with a study of the karyological processes. The mosaic phenomenon in animals is for the most part situated in other parts of the body than the reproductive organs, so that a genetical investigation is excluded, however, karyological data gives an acceptable supplement in these instances.

As a typical example of the influence of nutrition with a long *after-effect* on a part of the plant we may mention the *arborea* variety of the ivy, *Hedera helix*. This variety differs very much in external appearance and structure from the non-flowering branches and its origin is due to the fact that cuttings were taken from flowering branches of the ivy, which differ considerably from non-flowering branches in structure and habit. It can be seen at once that in this case no genotypical differences exists; seeds from *H. helix arborea* produce normal individuals. The case mentioned earlier (p. 341) of a budvariation in beans also belongs under this heading of modification, however it turned out that the modification has an after-effect which continues over a number of generations.

Changes in chromosome constitution by means of *changes in the number of chromosomes* during the somatic cell division can be expected because cells are found in different plants which have another number of chromosomes than that typical for the individuals (see p. 253). The fact that budvariations can indeed originate is proved by the case of *Primula kewensis* cited earlier (p. 256), and also by the investigations of WINKLER (1916) with regard to the *gigas* forms of *Solanum*-species. The formation of *polyploid somatic cells* seems to play a greater rôle in animals and plants than was previously supposed (see GEITLER, 1941).

No certainty exists concerning the possibility of the occurrence of true somatic mutation in homozygous material. The cases given in the literature cannot be controlled because the so-called mutation arose in vegetative tissue, from which no separate offspring can be cultivated. If the individual which produces this '*somatic mutation*' has a heterozygous nature (which is often the case) we have to be extremely careful in our interpretation.

A very important source for the occurrence of budvariations and mosaic forms is found in the *heterozygous nature of an individual*. The following theorem: 'that in the preponderant majority of cases a relationship exists between the heterozygous nature of the individual and the occurrence of budvariations' finds strong support in the fact that budvariations originate so frequently on cultivated plants, which are multiplied vegetatively. These plants usually are of a heterozygous nature. But an irrefutable proof can only be obtained if the budvariation is self-fertile or can be backcrossed

Fig. 167. Speltlike ears of wheat (after ÅKERMAN, 1920).

with a recessive individual in which homozygosity is an ascertained fact, so that the progeny can be compared with those of the same crosses with the rest of the plant on which the budvariation occurred.

A fundamental investigation in this field is that of ÅKERMAN (1920) which deals with speltoid chimeras, or spelt-like characters in wheat plants. In a cross between Sun-wheat 11 and Rumanian Balan-wheat, an ear was found in which nearly all the spikelets were spelt-like, at any rate in the upper part of the ear. On the lower part of the ear, the lower eight glumes on the one side were spelt-like but on the other side of the ear they were typical wheat glumes (see fig. 167). From six seeds, situated near the spelt-like glumes, 5 spelt-like plants and 1 normal plant were obtained, the seeds of four of these spelt-like plants gave a segregation in the ratio 49 normal: 533 spelt-like plants. The normal plant yielded only normal descendants. But on the other hand, seeds situated near the normal glumes yielded normal wheat plants and normal offspring. The rest of the ear (which thus bore only spelt-like glumes) gave 5 descendants, and their offspring again segregated into 50% typical and 50% speltlike plants. It seems as though a cross between wheat and spelt took place here (this occurring spontaneously), and that these F_1-plants have formed some glumes of the pure wheat type during the ear-formation at some given moment, together with the corresponding wheat gametes.

Material for investigations in this sphere is also given in numerous forms of *Coleus hybridus*. These plants not only form a great number of budvariations, but their high fertility also makes a study of the genotype of the separate branches possible. After STOUT (1915) and KüSTER (1917) published extensive studies concerning the occurrence of vegetative phenomena, a genetical investigation into the subject was added by BEYER (1923). *Coleus hybridus* has leaves in which green and yellow alternate and a red colour occurs in spots in the epidermis and the sub-epidermal layers. The distribution of yellow and green can be distinguished as 5 forms: yellow-margins, yellow spotted, green margins, plain green and plain yellow. The epidermal red occurs in four forms: coarsely spotted, finely red spotted, plain red and uncoloured. The sub-epidermal red can be found in 3 distinct forms: red veins, red in the main vein only, or lacking in red. BEYER could now prove that cuttings originating from the same plant can differ in heritable pre-disposition. He found the following relation with the epidermal red in the composition of the offspring:

Parental type	Descendants		
	plain red	red spotted	uncoloured
coarsely red-spotted . .	15	47	50
finely red-spotted . . .	2	45	29
plain red	62	16	22
uncoloured	0	0	388

Though it was not possible to draw up the factor constitution on the grounds of this data, it becomes evident that there is an analogy between the origin of budvariation and the heterozygous nature of the plant, so that the same fundamental cause may be presumed for these two phenomena.

A second example of this is found in a publication by CLAUSEN and GOODSPEED (1923), however they do not give a decisive explanation. The cross of a carmine red *Nicotiana tabacum purpurea* with a white *N. tabacum Cuba* yielded carmine red F_1-plants, which segregated in the F_2 into the ratio 9 red : 3 pink : 4 white. Two pairs of factors may be assumed in this case: one pair A–a, and one other pair B–b, which is cryptomeric in absence of A. Red plants possess both A and B, pink ones A only (AAbb or Aabb) and white plants may have B only or both factors in a recessive state (aaBB or aaBb or aabb). One of the red F_1-individuals had a branch with pink flowers. Red flowers of this plant backcrossed with white produced a segregation of 64 reds, 29 pink and 64 white (expected 3 red : 1 pink : 4 white), while the pink flowers backcrossed with white yielded a segregation of 40 : 37 : 72 (expected 1 : 1 : 2). These segregations seem to be fairly in accordance with our expectations. So the supposition seems justified that the pink budvariation is indebted for its origin to a somatic hybrid-segregation.

This assumption is in line with the results of vegetative propagation. After clones were obtained from these budvariations by means of cuttings, it turned out that these were all different depending on the origin of the cuttings. Stem sprouts gave only pink individuals and root cuttings gave only red plants. This is understandable if we consider that the dermatogen participates in the origin of stem buds, and that the anthocyane is contained in the epidermis cells, which belong to the dermatogen. Root buds on the other hand, are formed from the internal part of the plant.

There is indeed data on hand which indicates that budvariations can be attended by a somatic segregation of a hybrid. In this case the conclusion is justified that the heterozygous nature of the individuals is also the cause of budvariations.

But how can we explain the cause of budvariations, which give rise to an offspring which is completely identical to the original parents? The explanation may be that somatic segregation is only concerned with somatic cells and not with the reproductive cells.

There are cases available which seem to prove that the segregation took place at the moment of the forming of the epidermis cells, and in such a case the genotype of the epidermis is that of the budvariation, the genotype of the subepidermal cells, which are the origin of the reproductive cells however is that of the original plant.

Finally such phenomena can still find their origin in irregularities during the fertilization. We already encountered the mosaic-like sea-urchins (p. 225) which HERBST obtained by fertilization of sea-urchin eggs in which the egg-nucleus had divided before its fusion with a sperm nucleus. From this, individuals were born in which the body was composed of cells which had partly a maternal nucleus and partly a fertilized nucleus. We are indebted to the investigations of TANAKA (1917) for similar material (fig. 168). This work was later continued by GOLDSCHMIDT and KATSUKI (1927–31). The mosaic formation can be related either to body properties (such as the transparency or non-transparency of the skin), or to the reproductive organs (gynandromorphs) only or to both. It seems that the ability to form mosaic-individuals by itself is determined by a recessive factor, which is submitted to a delayed inheritance as in the manner of the direction of convolution in snail shells (see p. 344). The F_1 of the cross: mosaic line × normal line was always completely similar to the mother; the F_2-generations contained normal animals only, and in the F_3's there occurred a segregation of 3 normal broods:

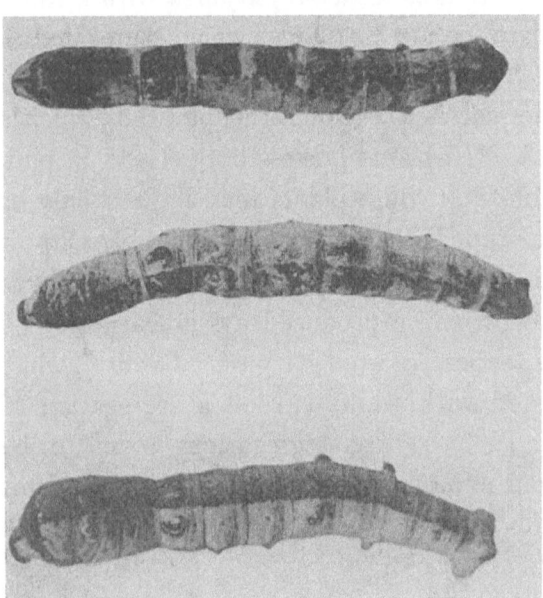

Fig. 168. Three different forms of mosaic silk-worms in which both halves of the body showed different racial properties (after TANAKA, 1917).

1 mosaic. GOLDSCHMIDT and KATSUKI now believe that we may explain the origin of these mosaic animals in the following way, that, by the action of a certain gene, two nuclei remain behind in the egg-cell after the reduction division. Both these nuclei should be suitable for fertilization, so that the individual which grows up should be composed of two genotypically different halves.

Experimental investigations thus show that there are four possible causes: 1) Permanent modifications, 2) Chromosome aberrations in somatic cells, 3) Somatic segregations of heterozygotes (so that a somatic segregation plays a rôle alongside the proper reduction). This kind of vegetative segregation takes place either in somatic cells only (that is to say it does not concern the genotype), or the somatic segregation also involves the reproductive cells. In the latter cases somatic segregation causes a difference in offspring between the two genotypically different parts of the individual. 4) Formation of an egg-cell with two nuclei and fertilization of both these.

With this experimental data the question arises as to how far these possibilities are supported by karyological data on the grounds of MORGAN's localization theory. Such proof is not necessary for cases of nutritive modifications, since no genotypical change takes place. With respect to the second possibility, i.e. the tendency to *polyploidy (euploidy*, see p. 250) of the somatic cells, it may be noted that this is a common phenomenon in animals and also occurs to some extent in plants as is evident from the summary of GEITLER (1941). It appears that this process is really limited to somatic cells and does not penetrate to the germ-line (the tissue from which the reproductive cells will later originate). In plants (where the reproductive cells are not grouped as a separate germ-line) cases of chromosome duplication in a branch and in the reproductive organs formed on it, have been observed. The example of *Primula Kewensis* represents this kind of phenomenon (p. 255). However, we have to bear in mind that this tendency to polyploidy in somatic cells does not always result in a constant number of chromosomes. In many cases a reversion to lower chromosome numbers was observed (VAARAMA, 1949; KOOPMANS, 1951). The reverse of the tendency to euploidy is the decrease of the chromosome number in the body-cells, whereas the number of chromosomes in the reproductive cells remains at the same level. Alongside the process described in *Sciara* (p. 286) there are still more of such cases, e.g. in the gall-gnat *Oligarces paradoxus* in which REITBERGER (1940) finds the chromosome number of 66 in the germ-line, whereas this number is reduced to 10 in the body cells; the reduction takes place in two phases of which the first (from 66 down to 11) takes place in the third cleavage division of an egg-cell, the second (from 11 down to 10) occurs several cleavage divisions later. A combination of chromosome-increase and chromosome-decrease is found to be due to a rather frequent occurrence of non-disjunction in somatic processes of nuclear divisions. SVÄRDSON (1945) described this in the Salmonoidea *Coregonus lavaretus* in which there can originate alongside somatic cells with the normal somatic chromosome number, somatic cells with rather different numbers. The fourth possibility deals with the fertilization of an egg-cell with two nuclei mentioned above for which a karyological basis was found, however, this phenomenon may prove to be an exception.

It is very difficult to find the karyological basis for the process of somatic hybrid-segregation; the course of mitosis (the nuclear division of body-cells) seems to be an ascertained fact, and the splitting of the chromosome into similar halves during this division is so clear that something special must be presumed as the explanation for the origin of two genetically different body-cells from one and the same cell. In hybrids which originate from a cross between less related individuals this is not always the case: in the cross *Godetia nutans* × *G. Whitneyi* for instance, HÅKANSSON found a striking variability in chromosome numbers in the somatic cells. SVÄRDSON (1945) described this phenomenon in hybrids between species of Salmonoidea. Such a somatic hybrid-segregation also occurs in other material. EMERSON (1915, 1921) left the choice open of explanation between somatic hybrid-segregation and somatic mutation, but he preferred the first and upheld the theory that the cause may lie in a non-disjunction of the chromosome in which the factors in question are located, hence by means of somatic division of the nucleus. In this case the one nucleus should obtain $2x+1$ chromosomes and the other $2x-1$.

In another investigation, JONES (1936, 1937) showed that this explanation cannot be exact, because the linked genes, which are situated in the same chromosome, behave independently during the formation of such mosaics, that is to say they do not always move together into the one daughter nucleus. After careful consideration of all the possibilities JONES reaches the conclusion given by STERN earlier (1936), concerning the mosaic phenomena in *Drosophila*, viz., that the basis for this must be presumed to lie in a process of somatic crossing-over between two chromosome halves out of a group of four, from which a homozygous combination of originally heterozygous genes can be obtained. This explanation of STERN necessarily assumes that the homologous chromosomes should pair up in somatic nucleic divisions also. This was observed in Diptera (the group of insects to which *Drosophila* belongs); HUETTNER (1924) was the first to record the fact. KAUFMANN (1934) ascertained chiasma-like bridges between homologous chromosomes and PETO (1935) also recorded the occurrence of chiasmata in roots of barley after irradiation. Oddly enough a fundamental difference appears to exist between the results obtained in Diptera and in maize although both are organisms in which the process of mosaic formation by means of somatic hybrid-segregation has been thorougly investigated. Maize data is available which indicates that non-homologous chromosomes are able to pair in somatic divisions of the nucleus. However in Diptera (e.g. *Drosophila*) it seems that it is the homologous chromosomes only which can show crossing-over in the nuclear divisions of the body-cells.

Hence the possibility of a somatic hybrid-segregation lagged behind the hypothesis of a somatic mutation for a long time because this latter hypothesis was easier to imagine, but not nearly as easy to prove. At the moment however, the former gains more and more in probability.

CHAPTER XVII

TRUE-BREEDING HYBRIDS

Those geneticists who investigated hybridization of individuals belonging to different species before 1900 did not have an easy time. Their observations led them to the conclusion that descendants of hybrids were changeable and of different types, however, on the side of the florists, the opinion was held that fertile hybrids were uniform and could maintain themselves as a true species. It was difficult for GAERTNER to take sides in these alternatives. On the one hand he postulates: 'Changeability in descendants of hybrids with regard to shape as well as to fertility is one of the most common properties' (1849, p. 551). However, on the other hand he supports 'the hypothesis of some plant-physiologists that fertile hybrids which breed rue to type can be transferred into stable species if these hybrids have confirmed and intensified their type and their fertility during a series of successive generations, by the law of habit forming (to which all living things are submitted) and that by this means, the vegetable kingdom will be replenished and enriched'. Support is given by the hybrid between *Dianthus armeria* and *D. deltoides:* 'this hybrid has maintained itself down to the tenth generation without a change in type and sowed itself in the first eight generations, but its fertility decreased year by year, until by the tenth year, the plant had completely disappeared' (p. 553). GAERTNER however, upholds the other side as well: 'But with the exception of these few examples, the fertile hybrid in the second generation and even more in the following generations do not yield similar types in their offspring, but give variants of the types or aberrant forms. These latter partly revert to the maternal and partly to the paternal type'. GAERTNER has thus not been completely convinced about the true-breeding of species-hybrids. NAUDIN also was not convinced, as is evident from statements given earlier in this book. Both are followers of the thevry of constant hybrids.

In 1865 a contribution was published by WICHURA concerning artificial hybrids of willows. These hybrids had a value for everyone who liked to hold the concept of hybrid constancy. The florists such as FOCKE, very much wished the concept to be true and approved WICHURA with respect to two points: 'WICHURA, in contradistinction to GODRON, corroborated the data of KOELREUTER, HERBERT, GAERTNER, NAUDIN and so many others, that hybrids are many times fertile with their own pollen, and also found that the offspring of the willow hybrids are true-breeding, a view which differed from that of NAUDIN. He disproved the principal errors of both his direct predecessors'. FOCKE, a good florist and a connoisseur of hybrids was at once willing to accept WICHURA's results as a proof for the constancy of species-hybrids. The problem was: Do true-breeding hybrids exist, or put into other words, do plant or animal hybrids exist which appear to be real hybrids in the F_1-generation and which in their subsequent generations give rise to offspring (F_2, F_3 etc.) which are uniform and completely similar to the F_1? This question is one of great importance for genetics. For this reason we will discuss the experimental data available which seem to support this thesis.

Firstly, the main pillar of the theory of the constancy of hybrids: *Salix*-hybrids of WICHURA. It was singularly unfortunate that this publication which played such a great rôle was interpreted so wrongly and without criticism. We, in our time will not find support for the hybrid-constancy in WICHURA's study. This is obvious from a quotation taken from this publication: 'From the observations of GAERTNER and myself, it seems to be an ascertained fact that descendants of hybrids originating from the pollen of hybrids are less uniform than those from the pollen of true species' (WICHURA, 1865 p. 54, see SIRKS, 1915). A clearer statement could not be expected before 1900. There were several underlying reasons why WICHURA could not formulate his opinions more sharply, but we will not discuss them at this time. It is enough to state that one may not consider him as a follower of the constancy concept.

The investigations of WICHURA were repeated by HERIBERT NILSSON (1918–1937) and IKENO (1918–1922); their results showed experimentally that willow hybrids do not give any support to this hypothesis of direct constancy. Both came to the conclusion that there is no constancy of the hybrids and that an F_2-generation of a hybrid between two willow species is multiform. IKENO made several hybrids between the species *Salix purpurea, S. purpurea multinervis, S. purpurea sericea, S. gracilistyla, S. opaca and S. viminalis* and investigated especially the F_2-generations of the cross *S. purpurea multinervis* × *S. gracilistyla*, which was found to segregate in the F_2 with regard to very differing characteristics such as stem shape, hairy leaves, the existence of stipules, colour of the stigma, shape of the catkins. HERIBERT

NILSSON worked with different species of *Salix*. He studied especially the hybrid *Salix viminalis* × *S. caprea*. He studied an F_1 of 26 individuals, which were mutually similar and of more or less intermediate leaf-shape between the two parents (see fig. 169) and an F_2-generation, which originated from two female F_1-individuals and

Fig. 169. Leaves of *Salix caprea* (left), *S. viminalis* (right) and their F_1 (after HERIBERT NILSSON, 1918).

comprised 106 + 51, i.e. 157 individuals. We need hardly to point out that the investigation and the classification of these F_2-individuals was extremely complicated. An obvious segregation occurred, firstly with regard to the habit of growth which was almost transitional between the two parents, but which sometimes resembled one of the parents, and sometimes even showed completely new forms as well. Secondly, a segregation with regard to the leaf-size and shape was observed. Not only did transitional forms between both parents occur, but there were also descendants with very large wide leaves and some with short, small leaves (fig. 170). In fig. 171 leaves of 3 F_2-plants are shown; the left one has the width of *viminalis*-leaves but only half the *viminalis*-length, in the middle is an individual with almost complete *caprea*-leaves and on the right is one with very big leaves which have the length of the *viminalis*-leaves and the width of *caprea*. Some F_2 individuals more closely resemble *Salix aurita*, others more closely resemble *S. cinerea*. Therefore these plants when more closely examined may more likely be considered as hybrids of one of these species, than as F_2 individuals of the cross *S. caprea* × *viminalis*. Alongside an F_2, obtained by mutual crossing of F_1 individuals, HERIBERT NILSSON also made all sorts of back-crosses between F_1 individuals and one of the parents, and also crosses between the F_1 individuals and a third species, and crosses of mutually different F_1 plants (e.g. *S. (cinerea* × *purpurea)* × *(purpurea* × *viminalis))*. He followed this unusual path,

Fig. 170. Two plants of *Salix caprea* × *S. viminalis* (after HERIBERT NILSSON, 1918).

because it was his intention to repeat the investigations carried out by WICHURA, who worked along this direction. As it could be expected, there also occurred in these offspring a strong segregation into several types. The interesting cytological phenomena of these hybrids were investigated by HÅKANSSON (1938). Thus the most classical example of direct constant hybrids was shown to be invalid by the work of HERIBERT NILSSON.

The cross *Dianthus armeria* × *deltoides* as quoted by GAERTNER, a cross which should breed true to type down to the tenth generation, was no better. To test GAERTNER's results, BAUR started by breeding an F_1 gereration between both these species, however, his experiments were taken over by WICHLER (1913). In contradistinction to the uniform F_1 generation, 300 plants of the F_2 generation were mutually quite different, so that a mere superficial observation distinguished numerous different types. Both parents were different in 15 easily recognizable characteristics, with respect to the vegetative parts of the plants, as well as to the flowers. It hardly need be pointed out that the F_2 generation yielded good material for analysis of the factor in question. From some points of view, the F_2 did not consist of enough individuals to show all the possible combinations, so that in the F_3 and F_4, some types were still found, which did not show up in the F_2. The leaves of the corolla, both in their shapes and stripes (see figure 172) are an example of such a

Fig. 171. Leaves of three F_2-plants of *Salix caprea* × *S. viminalis* (after HERIBER NILSSON, 1918).

very striking segregation. A hasty glance shows quite obviously, that there is no question of constancy, because this hybrid also segregates to a high degree after selfpollination.

In both cases the original statement with regard to the constancy of the hybrid and the uniformity of their offspring could not be supported. Since that time, many other publications concerning species-hybrids have been published, which also prove the possibility of segregation in the F_2 and subsequent generations. For a review in the botanical sphere we will refer to summaries by RENNER (1929) and by OEHLER (1941). On the zoological side, the result of the cross between the fish-species *Xiphophorus strigatus* and *Platypoecilus maculatus* (GERSCHLER, 1914) is valid as a classical case, which was later followed by others (see P. HERTWIG 1936).

Nevertheless, we are not permitted to draw from this the conclusion that all hybrids between species should behave in every respect in complete accordance with the laws of MENDEL. In the first place numerous species-hybrids are more or less sterile due to lack of similarity between the chromosome sets of both parents. The formation of gemini often fails to occur, which means that gametes originate with very different chromosome numbers, which are only partly fertile (see TISCHLER 1925, BLEIER 1934, SAX 1935). This sterility nearly always causes irregularities in the segregation of the offspring, which are obtained from the few fertile gametes. On the other hand we have also to acknowledge that F_1 hybrids were obtained between plant species, which yielded a completely uniform offspring.

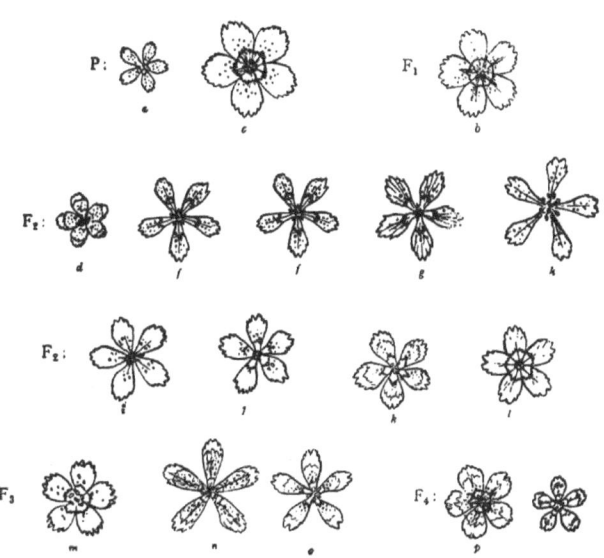

Fig. 172. Flowers of *Dianthus armeria* (a), *D. deltoides* (c), their F_1 (b) and some forms of the F_2, F_3 and F_4 (after WICHLER, 1913).

Constant hybrids exist but the constancy is caused by special circumstances. MENDEL himself found examples of this. Apart from his experiments with peas and beans he also made numerous crosses between species of *Hieracium*. The following data were obtained: 1) 'With regard to the shape of the hybrids we must mention a striking phenomenon that up to now the forms originating are not mutually the same. . . . . If we compare the hybrids with their parents, according to the sum of the characteristics, then both forms *Hieracium praealtum* × *H. aurantiacum* are intermediate though they differ mutually. On the other hand we can see that the hybrids from the cross as *H. auricula* × *H. aurantiacum* and *H. auricula* × *H. pratense* differ very much, so much that one of them resembles the one parent, the other the second parent, and in the latter cross there is a third type which is intermediate between the two. We may suppose that we are dealing with some members

26

only of a yet unknown series, formed by direct action of the pollen of one species upon the egg-cells of another species. 2) With a single exception, the hybrids discussed form good viable seeds. Some were completely fertile, others fairly so and again others are partly fertile; only one was sterile. 3) Up to now the descendants of the hybrids originating from self-pollination were constant; they agree mutually in their characteristics with the hybrid plant from which they originate' (MENDEL, 1869, p. 29–30).

If a sharp observer such as MENDEL, comes to such conclusions which are in contradiction with his results with peas, then there must indeed be some substance behind them. The discussion in Chapter IX will explain MENDEL's first conclusion. The cause of this was the heterozygous nature of the parents. Nevertheless, both the others and especially the third conclusion are very important as exceptions to the rule of segregation which appeared to be an ascertained fact.

Therefore it was no wonder that attention was drawn by later investigators to the genus *Hieracium* which has become a very important subject of study. The important difference in opinion was elucidated by the work of OSTENFELD, RAUNKIAER, and ROSENBERG (see ROSENBERG, 1930). In the genus *Hieracium* apomixis occurs, that is to say egg-cells are formed in the ovary but although they resemble egg-cells, they are not true egg-cells, because they donot pass through the so-called reduction division (see Chapter XII). They have the double number of chromosomes and can develop without fertilization into a plant embryo. The seed in such plants originates only from the mother plant and hence pollen does not take a part in the origin of this seed. Therefore *apomixis* is a form of asexual reproduction emanating from a single, pseudo egg-cell. There are some species of this genus which are completely normal, so that their flowers must always be fertilized in order to form seeds. Others are absolutely apomictic and always give rise to seed without the help of pollen. Again other species have true egg-cells in the same inflorescence next to pseudo egg-cells. The former must be fertilized and the latter develop into seeds independently, that is, without the aid of pollen. This third group of species is the most interesting one: *Hieracium aurantiacum, H. excellens* and *H. pilosella* belong to this group. Thanks to the presence of true egg-cells, these species can be used for hybridization experiments; their hybridization always yields an F_1-generation in which two types of plants are next to one another: individuals which originated by means of apomixis and completely resemble the mother and recognizable hybrids formed by the interaction of true egg-cells and pollen-grains. In the case that both the parents of the cross are partly apomictic, then the recognizable hybrids in the F_1-generation were either apparently fertile (that is to say, completely apomictic), or completely sterile. In the first case, an F_2-generation was obtained which completely resembled the hybrids of the F_1-generation, so that a constant, non-segregating hybrid was obtained; in the second case an F_2 cannot be obtained.

Not only were the exceptions to the rules found by MENDEL himself corroborated by this finding, but the explanation of apparent constancy in *Hieracium* was also discovered: the seeds of this hybrid do not origina te from the usual means i.e. fertilization of the true egg-cells with the aid of pollen, but by an abnormal process of the development of pseudo egg-cells, without the cooperation of pollen, that is to say, completely by their own power. Hence the uniform F_2 is an asexual continuation of the F_1-generation. This uniformity persists in the successive generations despite the fact that the plants are all heterozygotes.

The fact is now established that artificially produced hybrids in the vegetable kingdom can propagate themselves apomictically, that is to say, can form seeds which contain vegetatively formed embryos. This throws light on a number of other cases of constant hybrids, which are in fact pseudo-hybrids.

In the first place this is valid for the experiments of LIDFORSS (1905, 1907, 1914), who carried out a series of systematically arranged hybridization experiments in the extremely polymorphic genus *Rubus* and met very peculiar and apparently unexplainable results. In spite of careful extirpation of the anthers before they opened (proved sufficiently by non-pollinated control flowers), each artificial cross yielded alongside a number of true hybrids, a certain percentage of plants of the purely maternal type. The recognizable hybrids always segregated and gave a multiform F_2-generation. The other '*false hybrids*' (which resembled the mother plant) yielded a similar offspring after self-pollination for several generations. If the parents show much resemblance we shall usually obtain true and false hybrids in equal numbers (*Rubus acuminatus* ♀ × *R. caesius* ♂), or as is also possible, a single true hybrid among a great number of false hybrids (*R. thyrsanthus* ♀ × *R. caesius* ♂) or alternatively, only false hybrids (*R. polyanthus* ♀ × *R. caesius* ♂).

The F_1-generations of true hybrids are sometimes completely homogeneous, and from this fact we may conclude that the father species (*Rubus caesius*) was homozygous; at other times they are highly heterogeneous (*Rubus plicatus* ♀ × *R. caesius* ♂). *R. plicatus* used here as mother is therefore (according to the multiformity of the F_1) a hybrid, but a hybrid which is able to give rise to a number of pseudo egg-cells under the influence of stimulating pollen, as well as to give rise to different types of true egg-cells. The behaviour of the blackberry differs in two points from the *Hieracium*-hybrids discussed. The hybrids segregate and form pseudo-egg-cells only in the apomictic manner provided that a definite stimulation by the pollen is exercised. The relationship between these pseudo egg-cells and the pollen does not extend further. Hence this occurrence of a constant offspring after hybridization (that is to say an offspring similar to the mother) can be called '*induced apomixis*' or '*pseudogamy*' (FOCKE, 1881, p. 225). GUSTAFSSON (1930) proved that true apomixis does not occur in the genus *Rubus*, only pseudogamy. Over the course of years (see ERNST 1918, WINKLER 1920, KUHN, 1930) it became known that pseudogamy also plays a rôle in other genera.

With the recognition of apomixis, whether *autonomous* (i.e. of its own accord) or *induced* by means of foreign pollen, to form an exception to the generality of the rule of segregation of hybrids, some light was dawning on a problem which before 1900 was rather mysterious. This problem was that of the 'fausse hybridation', which for example was obtained in strawberries by MILLARDET (1894). In the years 1884–1893, this French investigator carried out hybridization experiments with different species of strawberries, viz. the European *Fragaria vesca* and *F. elatior* and the American *F. chiloensis* and *F. virginiana*. Mutual crosses of the European species gave nothing particular, neither did mutual crosses of the American species. The hybridization of a European and an American species yielded in a very few cases less clearly recognizable hybrids, in most cases however a number of plants identical with the species used as mother. Occasionally some plants occurred which completely resembled the father, MILLARDET called the first type *metromorphic*, and the last type, *patromorphic* and these two groups together he termed '*faux hybrides*' or '*hybrides sans croisement*', as opposed to normal, true hybrids.

For a long time these results formed the theme of varying contemplations. After SOLMS-LAUBACH (1907) had partly corroborated these, the problem of the metromorphic and patromorphic *Fragaria* hybrids was experimentally and karyologically investigated by others (see SCHIEMANN, 1931, 1937, 1943). At present there is enough karyological data which points to the fact that pseudogamy can occur in *Fragaria* crosses, that is to say, plants originate with maternal characteristics, and moreover that *patrogenesis* (a better term for this is *androgenesis* as advocated by WILSON, see p. 270) can play a rôle. However, it appears that in the latter case the F_1-individuals were not identical to the father from a morphological point of view, a fact that was already observed by MILLARDET (1894, p. 360). MILLARDET probably rightly supposes an influence on the part of the maternal plasm.

The unexpected occurrence of plants with paternal chromosomes only in *Fragaria* is no longer a lone fact; we have already mentioned the investigations of CLAUSEN and LAMMERTS (1929) and those of KOSTOFF (1929) with *Nicotiana*. In these studies, descendants originated by means of *androgenesis* which had a haploid chromosome number. The *Fragaria* hybrids go a step further, in that the paternal chromosomes have doubled. Hence the F_1-hybrid is completely similar to the male in its chromosome configuration. With this fact, the main point of the 'faux hybrides' seems to have been solved. The constant (true-breeding) hybrids which resemble the mother originate by means of pseudogamy. Those which resemble the father originate by androgenesis.

With both these processes the possibilities of the origin of constant hybrids are not exhausted. We discussed earlier (p. 258) investigations of FEDERLEY (1913) in relation to the cross between the butterflies *Pygaera anachoreta* and *P. curtula*. From these experiments it became evident that this hybrid gave rise to unreduced sperm-cells. From the chromosome numbers x = 30 and x = 29 of both parents, animals originated which had 2x = 59 as their somatic number. However the chromosome number of the spermatozoa (x) was 59 as well, hence constancy in the offspring is caused in this case by lack of a reduction division in the male hybrids. Nevertheless, the diploid sperm-cells are able to fertilize egg-cells of the parental species.

We will refer for a botanical analogy for this phenomena to the case of the *Rosa Wilsoni* a hybrid from the cross between *R. pimpinellifolia* ♀ (x = 14) and *R. tomentosa* ♂ (x = 7). This hybrid has in its somatic cells 42 chromosomes, and in its gametes 21 chromosomes (see BLACKBURN and HARRISON 1924). Another case is the cross *Nicotiana glutinosa* (x = 12) × *N. tabacum* (x = 24). In general, the F_1-plant is sterile, however one plant was selected with a branch which yielded seed, and an F_2 was obtained which was mutually similar and which had an x of 36 (CLAUSEN and GOODSPEED, 1925, see also WINGE 1932a).

In the classic *Aegilops* × *Triticum*-cross (so extensively discussed in older literature) a constant hybrid was produced with a double number of chromosomes. This only occurred in certain combinations, viz. *Aegilops ovata* × *Triticum dicoccum, T. dicoccoides* or *T. durum.* Crosses with *T. vulgare* (common wheat) did not give any results up till now. We are indebted to TSCHERMAK for the first of such *Aegilotricum* types. Both crosses of *Aegilops ovata* × *Triticum dicoccoides var. spontaneo-villosum* and *A. ovata* × *T. durum Arraseita var. Hildebrandtii* have given rise to almost true-breeding *Aegilotricum* species. Karyological work was carried out by BLEIER (TSCHERMAK and BLEIER, 1926; BLEIER, 1928b). Later TAYLOR and LEIGHTY (1931), obtained the same true-breeding type in the cross *A. ovata* × *T. dicoccum.* KIHARA (1930–37) confirmed this behaviour using a cross between *T. dicoccoides* var. *Kotschyanum* × *A. ovata.* Karyological investigations have led to the conclusion that TSCHERMAK's *Aegilotricum fertilis* has 56 chromosomes in its somatic cells (both parents have x = 14), and that a completely normal reduction division takes place during the formation of the gametes so as to give x = 28. The morphological investigations of TAYLOR and LEIGHTY confirmed that this 'species' *Aegilotricum* is, in the main, constant (see fig. 173). Nevertheless, some aberrant descendants are known. The karyological results of KIHARA have corroborated BLEIER's work. These aberrant forms which are as yet exceptions, owe their origin to irregularities in the reduction division (gametes are formed, which have lost one or more of their 28 chromosomes).

This origin of a new constant 'species' by means of hybridization of two other species is of great importance as is especially evident from studies of HERIBERT NILSSON (1928, 1931) who succeeded in obtaining both species *Salix laurina* and *S. cinerea* by hybridization of *Salix caprea* × *S. viminalis*. These species had been known for a long time in systematics. It therefore appears that species-hybrids play a great rôle in nature (ALLAN 1937).

Fig. 173. *Aegilotricum* (A) and its parents *Aegilops ovata* (B) and *Triticum dicoccum* (after TAYLOR and LEIGHTY, 1931).

Whereas in the above mentioned examples, *chromosome accumulation* or doubling of their number leads to the origin (at least in the main) of mutually similar gametes in the hybrids, there are also hybrids which form gametes with a very different number of chromosomes, and only those gametes of a certain type are suitable for further development, for use as a pollinator or to be pollinated (*elimination of gametes*). For instance this seems to be the case in the cross of two species of tobacco: *Nicotiana sylvestris* and *N. tabacum*. GOODSPEED and CLAUSEN (1917) came to the following conclusion on the basis of their investigations: '1) *N. sylvestris* when crossed with various varieties of *N. tabacum* gives F_1 hybrids which are replicas on a large scale of the particular *tabacum* variety concerned in the cross. 2) The F_1 hybrids of *sylvestris* and *tabacum* produce a small number of functional ovules which represent the *sylvestris* and *tabacum* extremes of a recombination series, the great majority of the members of which fail to function because of mutual incompatibility of the elements of the two systems. 3) Backcrosses with *sylvestris* give *sylvestris* and aberrant forms, and of the two the *sylvestris* alone are fertile and breed true to type. On the other hand, back crosses with *tabacum* apparently produce only *tabacum* forms of which some are completely fertile and continue to produce only *tabacum* forms' (1917, p. 99).

In this example gametes of definite types succumb and therefore a constancy of hybrids can be imitated. On the other hand embryos with definite genotypical constitutions can also degenerate simply and solely because of their constitution (*elimination of zygotes*). We discussed previously (Chapter VIII) the two types of generative cells, (i.e. egg-cells as well as pollen-grains) formed by the genetically well known *Oenothera Lamarckiana*. The consequence of the hybrid nature of *O. Lamarckiana* naturally should be, that alongside heterozygous descendants (velans-gaudens-combinations) homozygotes should be formed (in this case the homozygotes velans-velans and gaudens-gaudens). This however does not happen; all descendants of an *O. Lamarckiana* plant are heterozygous and mutually resemble the mother plant. The anticipated homozygotes do not occur, but on the other hand, RENNER was able to show, that from the young embryos formed in the ovaries of a self-pollinated *O. Lamarckiana* half of the embryos succumb before developing into ripe seed, which is capable of germinating. Conclusion: this 50% of degenerative embryos are composed of the 25% gaudens-gaudens expected and the 25% anticipated velans-velans combination. This view of RENNER is shown in fig. 174 and it appears that this might be the true solution. Hence *Oenothera Lamarckiana* is a hybrid which always forms two types of ovules and two types of pollen-grains, which does not segregate in a striking scheme with regard to the two types (excluding some products of exchange between the velans and the gaudens complexes) and therefore remains constant as a result of the elimination of the homozygotes formed.

If we summarize the discussion concerning the possible constancy of hybrids, we are able to say:

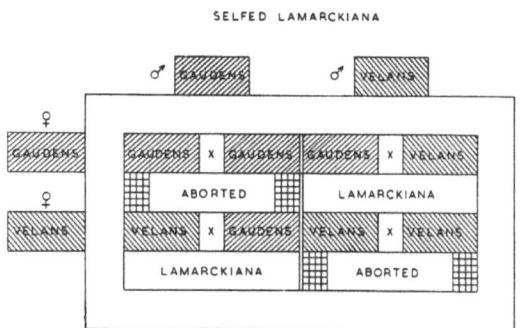

Fig. 174. Scheme of selfed *Oenothera Lamarckiana* and the origin of aborted homozygotes.

Species-hybrids segregate in many cases in spite of earlier statements, though in a more complicated manner than hybrids between varieties.

Really true-breeding hybrids can occur and are then the result of:

a) apomixis, development of pseudo egg-cells, which do not participate in a reduction division, either autonomous, or induced under the influence of pollen-grains (pseudogamy); however, fertilization is completely lacking;

b) androgenesis, fertilization by normal gametes and after which the nucleus of the female gametes succumbs;

c) absence of a reduction division, followed by fertilization of the non-reduced egg-cells;

d) duplication of the number of chromosomes in a hybrid followed by a normal reduction division;

e) formation of different types of gametes, from which only one or only a few are viable;

f) formation of two types of ♀ and ♂ gametes with normal fertilization, however the homozygotes formed degenerate.

CHAPTER XVIII

SEX AS A GENETIC CHARACTERISTIC AND A GENIC EXPRESSION

Throughout the ages was no lack of explanations for the cause of sex determination, and the influences which make one organism male, the other female, and yet another bisexual. The search was principally applied to man himself and has been therefore for a long time completely statistical. GEDDES and THOMSON (1889) estimated the number of theories put forward before the beginning of the nineteenth century as 500, and in the nineteenth century itself many investigators added new ones to this number. All these different hypotheses (none were more), could very well be classified into three groups (HAECKER, 1902), i.e. *progamous, syngamous* and *epigamous sex-determination*. The advocates of the first group believe that some circumstance or other is present in the egg-cell by which it is predestined to become either a male, female or bi-sexual individual. The *syngamous sex-determination* should take place at the moment of fertilization, at which moment, by means of the concurrence of tendencies (present in the egg-cell and spermatozoon), it should be decided which of these tendencies is winning. The supporters of the *epigamous sex-determination* on the other hand, considered the fertilized egg-cell as a neutral area in which the decision with regard to its sex should be confined to later influences of its environment.

The influence of genetics however completely changed the character of the problem of sex-determination, that is to say it became a problem of general biology. Hypotheses and statements may not only consider man or species of animals in which we are able to distinguish male and female individuals, but they have to embrace the varying phenomena of sex-differentiation expressed by nature in toto and find a relationship between these expressions (see SIRKS, 1942). The realisation of this immediately enlarges the question, and the classification of the hypotheses into three given groups becomes obsolete. Hence the choice between the possibilities now lies between: heritable, that is to say genotypical causes, and non-heritable, or a phenotypical expression of sex. Over and above this is the fact that in organisms which show an alteration of a diploid and a haploid life-phase (such as mosses and ferns), the difference between male and female can also occur in the haploid phase. Therefore both groups must subdivide again into two parts: *haplogenotypical* and *diplogenotypical* alongside *haplophenotypical* and *diplophenotypical sex-determination*. This newer classification covers the older one: the progamous and syngamous hypotheses belong to the genotypical group, and the epigamous belongs to the phenotypical one.

At this moment the state of affairs is such that with respect to man and the large majority of other organisms in which the sexes are normally obvious, no influence is credited to environment. We are able to put forward arguments which are so strong and well-founded that a diplogenotypical explanation must be accepted as true. So-called *identical (monozygotic) twins* which occur in man (and cattle) and are striking in their mutual resemblance (see NEWMAN, 1917, 1942, SIEMENS 1924, DAHLBERG 1926, KOMAI 1928), may prove that the sex of the individual is determined at the moment of fertilization. Both organisms originate from the same fertilized egg-cell, which divides at a given moment into two halves. Each half grows up into a completely normal individual. These twins, originating from one egg are either both girls or both boys. This is a very strong argument against epigamous sex-determination. Though exceptions among mammals, such identical or monozygotic twins are a general phenomenon among other groups of animals. We are indebted to the investigations of FERNANDEZ and of NEWMAN and PATTERSON (see PATTERSON, 1927) and others for discovering that the four or eight young which occur together in the uterus of the armadillo (*Dasypus* or *Tatusia*) are of the same sex and are moreover all formed from the same fertilized egg-cell. Through very detailed embryological studies it was proved that during each ovulation only one egg-cell leaves the ovaries, and moreover that this fertilized egg-cell is able to give rise to 4 or 8 embryos. It is obvious that these embryos have a joint amnion (figure 175).

This phenomenon of polyembryony occurs to a higher degree among certain species of insects. Prior to this MARCHAL (see 1904) had already found that the egg of the ichneumon-fly *Encyrtus fuscicollis*, which develops inside the body of the moth *Hypomoneuta* falls apart into a series of young germs (up to 120). All these germs can become individuals and all these individuals born from one egg are of the same sex.

Hence the idea that feeding has an influence which can induce sex phenotypically has been disproved. We therefore have to concede a genotypical determination of sex.

From one point of view the numerous statistical data have given us an important fact, that is, that as a rule the numbers of male and female individuals of a definite species are almost equal (see PARKES 1926, CREW 1937, MAYR 1939). Sex ratios are usually expressed as follows: the number of female is fixed at 100, and the number of males calculated per hundred females. The

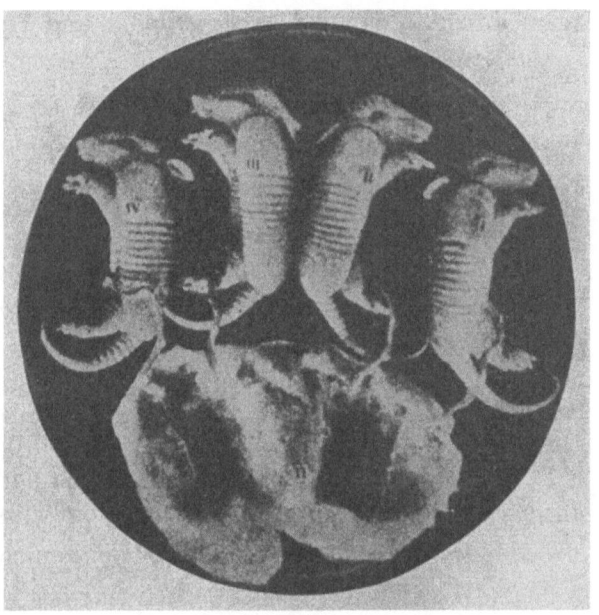

Fig. 175. Four embryos of *Dasypus novemcinctus* originating from one egg-cell (after NEWMAN and PATTERSON, 1909).

ratio between them generally fluctuates around 100 : 100. Deviations from this are, as a rule very small. In man this ratio is very near 100 : 100, however it can differ slightly due to the races and countries in question. Nevertheless there is a small surplus of boys at the time of birth, about 105 : 100. The same is true in plants where, by means of counts, a ratio of 1 : 1 was obtained (see CORRENS, 1928a, p. 97). Important differences are sometimes observed between local races, and some species show striking deviating ratios. These exceptions are not detrimental to the rule that male and female individuals occur in equal numbers.

This fact once established brings us at once to the problem of the causes of heritable sex-determination. In the simple case of a mono-hybrid backcrossed with a true-breeding recessive one, an offspring is obtained which is always composed of two groups: heterozygotes and recessives. The pink *Mirabilis jalapa* of p. 88 originating from the cross of a white with a red can be backcrossed with a white plant and will give an offspring composed of 50% pink and 50% white individuals. The pink monohybrid gives rise to 50% 'red' egg-cells A and 50% 'white' a. The first 50% again yield Aa (i.e. pink) plants when crossed with white pollen-grains. The last mentioned a-egg-cells when crossed with white a-pollen-grains yield aa, i.e. white plants.

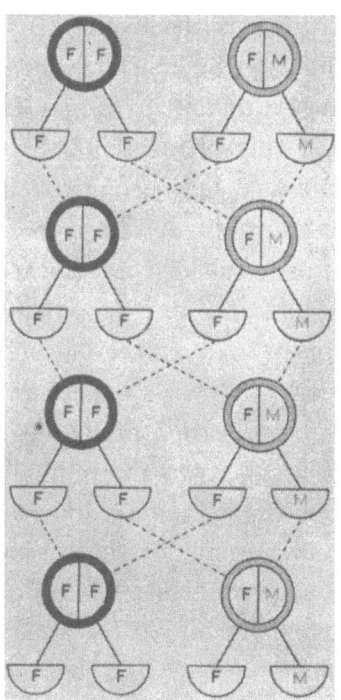

In the same way we are able to derive the origin of equal numbers of male and female individuals if we suppose that one of the two, let us say the female, is always homozygous for a certain pair of allelomorphs and the male, heterozygous. Let us take for example (fig. 176), the case when a heritable factor M is responsible for the expression of the male sex, and this factor is dominant over a recessive m, (which can also be written as F), then the formula for all male individuals will be MF, and that for all females will be FF. Hence 50% of all sperm-cells will contain the factor M, the other 50% the factor F, all egg-cells F. Therefore in each 'backcross' of a male with a female individual (50% M × F and 50% F × F) we will find that half are male and half are female individuals. It is possible to explain the sex-determination by a simple Mendelian scheme by supposing that both sexes differ in one factor only, and further that one sex is heterozygously dominant, and the other homozygously recessive.

Fig. 176. Mendelian scheme of sex-determination.

The fact that a supposition which affords such a simple explanation of an apparently complex problem is not without proof needs no discussion. The data obtained from experimental investigations with regard to factors, which are probably linked to a sex-factor, is so abundant that it appears that no other explanation need be put forward. Therefore we may consider it as proved that a heritable factor is necessary for the expression of sex.

It was also shown by karyological means that organisms of the one sex are homozygous and the organisms of the other sex are heterozygous. All gametes formed by the one sex are the same in their chromosome configuration, whereas those of the other sex belong to two types. The first who observed inequality in chromosome numbers between two gametes was HENKING in 1891. He found that in the bug *Pyrrhocoris apterus* the reduction division (through which sperm-cells are formed), was abnormal. Hence gametes with 11 as well as with 12 chromosomes were formed. However, neither he nor a number of later investigators succeeded in recognizing the significance of this phenomenon. All sorts of names were given to these supernumerous chromosomes such as X-*chromosome, accessory chromosome, heterochromosome, idiochromosome*, etc. In 1902 McCLUNG (1902 a and b) came to the idea of integrating this chromosome with the determination of the sex: 'Upon the assumption that there is a qualitative difference between the various chromosomes of the nucleus, it would necessarily follow that there are formed two kinds of spermatozoa which, by fertilization of the egg, would produce individuals qualitatively different. Since the number of each of these varieties of spermatozoa is the same, it would happen that there would be an approximately equal number of these two kinds of offspring. We know that the only quality which separates the members of a species into these two groups is that of sex' (1902 a, p. 225).

Nowadays on the grounds of this hypothesis the chromosome in question is called a *sex-chromosome*. Karyological work carried out since 1900 by many investigators, has shown irrefutably, in a number of species of animals, that the male organism gives rise to two types of sperm-cells which differ in the number of chromosomes. This is represented by figure 177 in which the four spermatids (future spermatozoa) of *Ancyracanthus cystidicola*, a worm, originate from one single cell. Both right hand ones in the figure contain six, both the left ones five chromosomes each. One of these six chromosomes is the '*accessory chromosome*' in which the gene responsible for sex is thought to be located.

However, the difference between the two types of sperm-cells rarely lies in the number of chromosomes. In most cases these numbers are equal, but one of the chromosome pairs is composed of two parts which differ in such a manner in shape, size or behaviour that because of this we are allowed to consider these chromosomes individually. They are nowadays called *gonosomes*, sometimes *allosomes*, in contradiction to the remaining chromosomes or *autosomes*.

Fig. 177. Four spermatids of *Ancyracanthus cystidicola*,
originating from one cell (after MULSOW, 1912).

This can also be observed in man; the male gives rise to two kinds of sperm cells, the one with a large chromosome (X), the other with a small (Y) homologous chromosome (PAINTER, 1923; MACLEAN, EVANS and SWEZY 1929; SHIWAGO and ANDRES 1932; see fig. 178).

The representation of the course of events which the advocates of the hypothesis of heterochromosomes as sex determining chromosomes present is clearly given in the figure borrowed from LOEB (1916) (fig. 179) in which we find two egg-cells (above) each with 6 autosomes and one heterochromosome (x), together with two sperm cells, one (b) with 6 autosomes + 1 heterochromosome and one (c) with 6 autosomes only. The two possible combinations are represented in this figure: a fertilized egg-cell (d) with 12 autosomes and 2 heterochromosomes (xx), hence a female individual will grow up from this zygote; and a fertilized egg-cell (e) with 12 autosomes and 1 heterochromosome (x) will develop into a male individual. In this way, the hypothesis of the heterochromosomes can find a schematic representation.

Hence, in general we can indicate the homozygously female individual (which has in its body-cells 2 X-chromosomes) by XX; the male will then be XO. In this latter case there is only one X-chromosome and this X-chromosome does not have a partner for pairing. However, when a homologous heterochromosome is present we can indicate the constitution by XY. Some investigators use the indication XO and XY arbitrarily. This was done by MOR-GAN with respect to *Drosophila melanogaster* as discussed earlier. The female fly has two similar (XX) idiochromosomes, the male, on the other hand has two different idiochromosomes (see figure 82). MORGAN proposes this on the grounds that the bent Y-chromosome in the male organism may not contain factors and hence this chromosome may be "empty". It is however evident from the study of STERN (1926c) that this view is incorrect. He showed that some genes were still present in the Y-chromosome.

Fig. 178. Chromosome constitution of man with clear sex-chromosomes (X and Y) (after SHIWAGO and ANDRES, 1932).

The data concerning the '*sex-linked inheritance*' obtained experimentally can be satisfactorily explained by the above mentioned hypothesis (see p. 203 and further, also FÖYN 1932). The cross of *Drosophila* ♂ with white eyes and a ♀ with red eyes yielded, as discussed, male and female flies both of which had red eyes. The F_1-animals give red-eyed female descendants and red-eyed as well as white eyed males by mutual pairing. The explanation of this segregation should be based upon the fact that, according to the chromosome hypothesis of heredity and the hypothesis of the 'sex chromosomes' the factor for red as well as the sex factor is located in the X-chromosome. Hence, the homozygously red-eyed female has the factor for red eyes in both X-chromosomes. The white-eyed male lacks the factor for red in the only

X-chromosome he possesses. The explanation of both reciprocal crosses as presented in figure 68 (p. 209) now needs no further elucidation. In this figure XO stands for males which give rise to two types of sex determining sperm-cells: X and O. XX are females which produce only egg-cells with X.

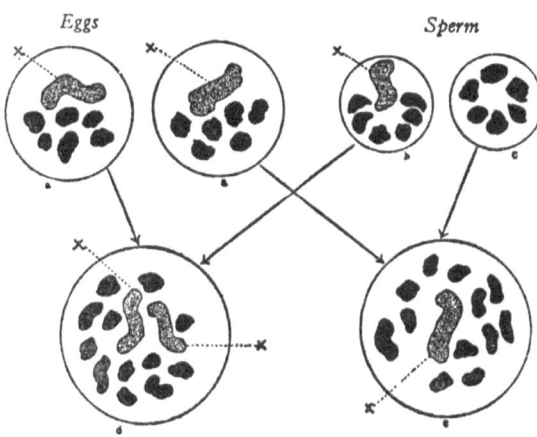

Eggs   *Sperm*

Fig. 179. Origin of both sexes by means of heterochromosomes (after LOEB, 1916).

It is also clear that there should be an agreement between the data of the experimental investigation concerning sex-linked inheritance and the hypothesis of the heterochromosomes. It is obvious that MORGAN accepted this agreement gratefully as a support for his localization theory of inheritance.

It might seem that the followers of this karyological hypothesis of sex-determination overrated their results and hence their view has to be regarded as unsettled. The general aspect of the question was considerably strengthened since a parallel with sex-linked inheritance could be shown. More support for this supposition was obtained from experimental investigation, when BRIDGES (1913, 1914) discovered the phenomenon of the socalled non-disjunction in *Drosophila* (see p. 271). The occasion for such a discovery was presented by the occurrence of some very unexpected aberrations in his crosses. The regularity which could be expected was not always obtained: about 5% of the individuals of the F_1 (white ♀ × red ♂) in BRIDGES' experiments behaved differently; these F_1-females had white eyes and the F_1-males red eyes. This result must have had a cause and BRIDGES put forward the explanation that in a certain number of the egg-cells, originating from the originally white-eyed mother fly, the separating process of both X-chromosomes during the reduction division fails to appear (*non-disjunction*). This supposition was also confirmed later on (BRIDGES 1916) by karyological investigation. In almost $2\frac{1}{2}$% of the egg-cells, which originate from the white-eyed female, two X-chromosomes were present: in the other $2\frac{1}{2}\%$ X-chromosomes were lacking.

The consequence of this abnormal behaviour of X-chromosomes was that in the F_1-generation alongside heterozygously red-eyed females (see figure 68, p. 209) and white-eyed males, other types also occurred. Egg-cells with 2X-chromosomes (both with factor 'white-eyed') could be fertilized by sperm-cells with the X-chromosome (with factor for red eyes). From this cross a female originated with 3 X-chromosomes. One of these chromosomes has the factor for red and therefore the ♀ fly had red eyes. However, egg-cells with two white X-chromosomes which are fertilized by Y-sperm-cells must develop into females with white eyes, because there are two X-chromosomes present it is a female. None of these X-chromosomes contains the factor for red and therefore these females have white eyes. The other abnormal types of egg-cells (those without X-chromosomes) can also be fertilized: fertilization with an X-sperm-cell (factor for red present) gives rise to an individual with red eyes and a single X-chromosome of male sex. The fertilization of an egg-cell without an X-chromosome with a Y-spermatozoon seems to lead to none viable individuals. Thus, apart from XX- and XY-individuals other combinations can be formed which can be indicated by the formulae XXX, XXY, and X.

Supposing that these flies again form reproductive cells, then an XXY female will produce four types of egg-cells: XY, X, XX, and Y. By backcrossing these combinations with a normal male still more genotypical formulae can be derived. A white-eyed XXY-female crossed to a red-eyed X^1Y-male (X^1 contains the factor for red) gives the 8 possible combinations XX^1Y (red ♀), XX^1 (red ♀), XXX^1 (red ♀), XXY (white ♀), X^1Y (red ♂), XYY (white ♂), XY (white ♂) and YY (non-viable). However all these combinations do not occur in equal numbers, because the XXY-female gives rise to more egg-cells with XY or with X, than with XX or with Y but this is a question of minor importance. For a long time the karyological investigation did not go hand in hand with the experimental results. Apparently BRIDGES could find only flies with a typical XXY-constitution (fig. 180). Later he succeeded in obtaining karyological evidence of the occurrence of XXX- and XYY-flies. Moreover he established the fact that a female fly existed which had four sex chromosomes XXYY. It is possible that this animal is indebted for its origin to the mating of an XXY-female with an XYY male fly. This phenomenon is a very important support for the hypothesis of sex determination by specific heritable factors, localized in the X-chromosome. These factors express male sex if present singly, and female sex if present in duplicate in those species of animals which are homozygous as female and heterozygous as male. This is also the case in man: the inheritance of colour-blindness (see p. 211) and of haemophily is in complete accordance with this.

As discussed in Chapter X, the opposite is the case in a number of other animals such as *Abraxas grossulariata*, (cf. p. 203). On the basis of the results of breeding experiments with *A. grossulariata*

Fig. 180. Chromosome configuration of an XXY-individual (after Morgan a.o., 1915).

and its pale variety *lacticolor*, it was concluded that it is the heterozygous disposition of the female and the homozygous disposition of the male with regard to the heritable factor which is responsible for the sex. In accordance with this Doncaster (1914b, 1915) also found that in the female currant geometer 55 chromosomes are present in the body-cells, whereas the egg-cells formed by this animal have partly 28 and partly 27 chromosomes. The male contained 56 chromosomes per somatic nucleus and only gives rise to spermatozoa with 28 chromosomes (see Cockayne, 1938).

By a series of other investigations (see Witschi 1929, Crew 1933, Guyénot 1935, Bridges 1939) it has now been ascertained that heterozygosity with regard to sex occurs in the male individuals of almost all groups of animals, including mammals, and in females of butterflies and birds. With regard to fish some contradictions still exist (see Goldschmidt 1937a, Kosswig 1939, 1941). But we may still ask, has the question concerning sex determination been solved by this hypothesis of heterochromosomes? If so heterochromosomes must also occur in diploid plants. Several cases of sex-linked inheritance are known (p. 215) in diploid dioecious plants. The research into the chromosome constitution of dioecious plants was energetically carried forward. After *Melandrium dioicum* (Blackburn 1923; Winge 1923b; Belar 1925, see figure 181) and *Rumex acetosa*

(KIHARA and ONO 1926) a number of other plant species followed, in which the occurrence of heterochromosomes was ascertained (see ALLEN 1940).

However rich our series of data concerning morphological differ-

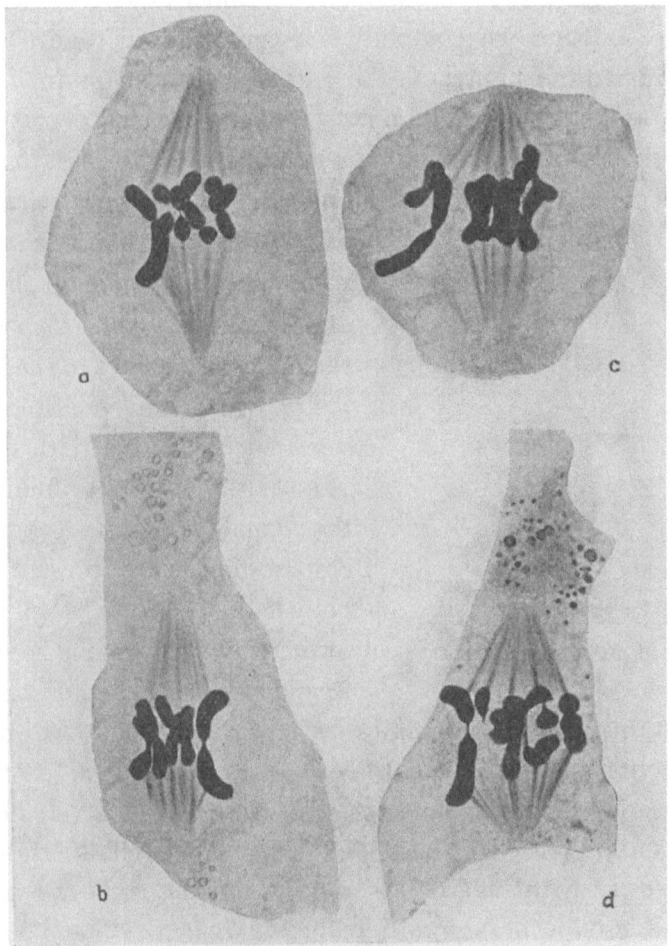

Fig. 181. *Melandrium dioicum*. a, ♂ with XY-pair left;
b, ♀ with XX-pair right; c and d, hermaphrodites with
XY-pair left (after BELAR, 1925).

ences between both sex chromosomes may be, there still remain a number of gonochoristic animals and completely dioecious plants in which no pair of chromosomes is striking in its shape or behaviour and in which the sex chromosome as such, cannot be recognized. There is no doubt however, that in these organisms also, sex determination takes place according to the scheme previously developed. It has been experimentally proved in the so-called *sub-dioecious plant species* that the female individual is homozygous and the male is heterozygous for the sex factor. Sub-dioecious species are composed of female plants which have a few male flowers alongside their female flowers (*subgynoecious individuals*), while the male plants in exceptional cases carry some female flowers (*subandroecious individuals*). This situation makes it possible to get seed from these almost female plants by self-fertilization and similarly from the sub-androecious individuals as well. It was pointed out by KUHN (1939) that the offspring of subgynoecious plants of *Thalictrum* and of *Mercurialis* is composed only of females and sub-gynoecious individuals; but that of subandroecious plants segregated into one homozygous male, two heterozygous males and one female plant. From this it is evident that the subgynoecious individual is homozygous and the subandroecious one is heterozygous for the sex-factor. Thus the conclusion also follows that the combination of genes for male sex does not necessarily make the plant completely male, neither does a combination of genes determine absolutely the female sex. We will soon discuss this precariousness in detail.

With the above type of sex determination, that is to say, determination by means of genes in the diploid stage (*diplogenotypical*), is associated the parallel phenomena in haploid organisms. First of all we may mention here the *Hymenoptera* (bees, wasps etc. see WHITING, 1935). Through a series of investigations, and after much contesting (see ARMBUSTER, NACHTSHEIM and ROEMER, 1917) it was ascertained that the queen bee lays two sorts of eggs, fertilized and unfertilized. The fertilized eggs have the potentiality to become female individuals (i.e. workers or queens), the unfertilized will develop into male drones (see p. 368). Even this theory has a karyological basis; it was pointed out by some investigators, such as NACHTSHEIM 1913, that female bees (either workers or queen) have 32 chromosomes in their body-cells, and 16 chromosomes in their reproductive cells, whereas the male individuals have only 16 in their somatic cells, as well as in their gametes. It is an inevitable condition for the origin of the female sex, that the egg-cells have been fertilized; when fertilization does not take place, then the egg-cell gives rise to a male individual. The sex determination is therefore *diplogenotypical* for the female, *haplogenotypical* for the male.

A more eloquent example of a *haplogenotypical sex determination* is found in groups of mosses. From the studies of ALLEN (1917, 1919) sex chromosomes were discovered in some species of the dioecious liverwort *Sphaerocarpus* (see figure 182). This was later repeated in a number of other mosses and liverworts (ALLEN 1935, TATUNO 1934–38, LORBEER 1927, 1934, 1941). As was already known, the reduction division takes place in these organisms during the formation of spores and the spores which originate have the ability to develop into plants. It is therefore in the cells of a moss-plant that we find the reduced chromosome number. This number is 8 for *Sphaerocarpus Donelli*, in the female

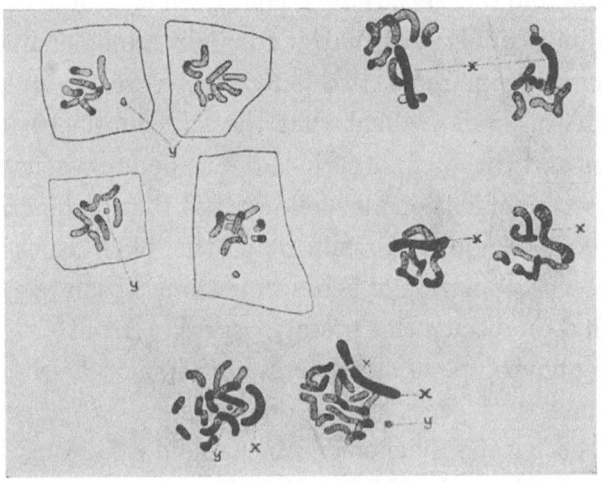

Fig. 182. *Sphaerocarpus*. Left above: 4 nuclei of ♂ haploids; right above: 4 nuclei of ♀ haploids; below: 2 diploid nuclei (after LORBEER, 1927).

plants one of the eight chromosomes is much longer and bigger (x), than the remaining seven, which mutually resemble one another, to a high degree. In the male plants, we find one striking small chromosome (y) alongside the seven similar chromosomes. The female set is found again in the egg-cells, the male set in the spermatozoa. After fertilization the small y-chromosome becomes the partner of the long one in the female cells. Therefore there are 14 autosomes and both the idiosomes in this fertilized egg-cell; the reduction division, that is to say, the formation of spores takes place when the partners separate, so that each sporogeneous cell gives rise to two spores which develop into ♂ plants, and two spores which develop into ♀ plants.

Apart from this form of haplogenotypical sex determination in which the genes responsible are located in easily observable chromosomes, we meet yet other possibilities among the lower organisms, which demand a haplogenotypical explanation of sex determination, however, these possibilities are not very explicit. In the first place we find an apparently physiological sex determination when no morphological differences can be observed between the individuals which act as male and female. The important fact of the occurrence of these so-called plus and minus races (names which represent the current terms

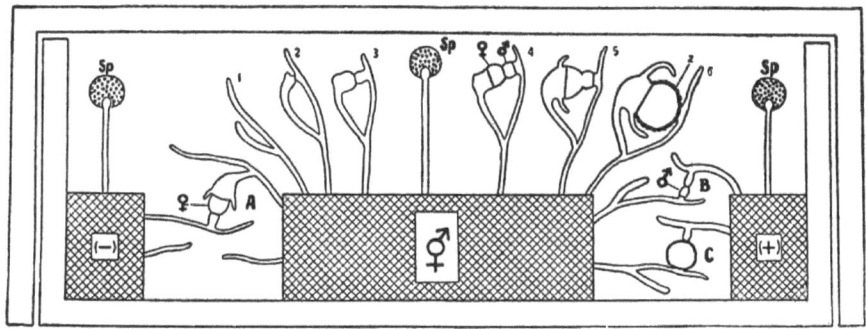

Fig. 183. Cross between (+) and (-) strains of *Mucor* with the anisogamous-dioecious *Absidia spinosa* (after BLAKESLEE, 1915).

for female and male in this special case) was discovered by BLAKESLEE (1904) in fungi from the group of *Mucorineae*. After this the idea was worked out in greater detail by a number of other investigators (see HARTMANN 1929). Plus-races must fuse with minus-races in order to give rise to zygotes. Combinations of two equivalent races are sterile. It became evident from work of BLAKESLEE (1915) that the difference of plus- and minus-races is analogous with sex-differentiation in higher organisms. He made crosses between a fungus species *Absidia spinosa* and *Mucor* races. *Absidia* is an anisogamous-bisexual form, that is to say, two sorts of copulating cells are formed on the same mycelium. These can be recognized morphologically (see fig. 183: 1–6). The content of the male cell (antheridium) is transferred to that of the more voluminous female cell (oogonium), after which the zygote is formed as a fertilization product. If mycelia of *Absidia* are crossed with the + and the — races of a *Mucor*-species, it was found that the female *Absidia*-cells fuse with the minus mycelium, and the male fuses only with the plus mycelium (fig. 183 A, B and C respectively). From this behaviour, we can derive the conclusion that the minus-races of *Mucor* have the nature of male organisms, and that the plus-races may be considered as female.

Since BLAKESLEE's discovery, the occurrence of sex differentiation has been found many times with fungi and algae, where formerly no differentiation was suspected on morphological grounds; some such cases have been discussed already, and we will refer to the summaries of KNIEP (1928, 1929), and of HARTMANN (1929).

The second group of rather complicated phenomena is encountered in the *Basidiomycetes*: in some species (especially *Coprinus*-species), a sexual differentiation can be shown in the same sense as in the above-mentioned *Mucorineae*, so that only two types of sexes are present. These are based upon the activity of one single pair of allelomorphs A-a. The four spores which are formed in one single basidium, are then 2A : 2a. A number of other Basidiomycetes show a more complex behaviour; it seems that the reproduction does not depend on a single pair of allelomorphs but on two pairs: A-a and B-b. Copulation between haploid mycelia is in this case only possible if they differ in both factors, that is to say, between AB and ab or between Ab and aB. This so-called *tetrapolar sexuality* is found in *Schizophyllum commune*,

		AB				ab				Ab			aB		
		1	8	9	11	2	3	5	6	7	12	13	4	10	14
AB	1	−	−	−	−	+	+	+	+	−	−	−	−	−	−
	8	−	−	−	−	+	+	+	+	−	−	−	−	−	−
	9	−	−	−	−	+	+	+	+	−	−	−	−	−	−
	11	−	−	−	−	+	+	+	+	−	−	−	−	−	−
ab	2	+	+	+	+	−	−	−	−	−	−	−	−	−	−
	3	+	+	+	+	−	−	−	−	−	−	−	−	−	−
	5	+	+	+	+	−	−	−	−	−	−	−	−	−	−
	6	+	+	+	+	−	−	−	−	−	−	−	−	−	−
Ab	7	−	−	−	−	−	−	−	−	−	−	−	+	+	+
	12	−	−	−	−	−	−	−	−	−	−	−	+	+	+
	13	−	−	−	−	−	−	−	−	−	−	−	+	+	+
aB	4	−	−	−	−	−	−	−	−	+	+	+	−	−	−
	10	−	−	−	−	−	−	−	−	+	+	+	−	−	−
	14	−	−	−	−	−	−	−	−	+	+	+	−	−	−

as is evident from the preceding table, which represents the combination possibilities between 14 haploid mycelia which originate from one single mycelium (+ fertilization, — no fertilization).

The four possible types of mycelia are formed in equal numbers. They can be distinguished morphologically by means of small differences in their manner of growth, germination rate of the spores and other characteristics. KNIEP succeeded in proving that in one single basidium of *Aleurodiscus polygonius*, the four spores represent the four different types, 1 AB : 1 Ab : 1 aB : 1 ab. Evidence was produced that this kind of tetrapolar sexuality had a genotypical basis. Moreover GREIS (1942) showed, that the A-factor is a real sex-factor, whereas the function of sterility must be ascribed to the B-factor.

Finally it appears that yet another complication occurs in the sex differentiation of some species of algae. This phenomenon is described as *relative sexuality* (see HARTMANN, 1930, 1932). By this we understand the presence of sexual differences within one single sex, that is to say, differences in intensity of the sex in question, which can, in extreme cases lead to the fact that a normal female gamete (A) behaves as a female towards a male gamete B, but as a male gamete towards a more strongly female gamete C.

We will find this relative sexuality particularly in the lower organisms especially algae. An excellent review is given by MOEWUS (1941–1943). We will choose as an example the algae species *Chlamydomonas eugametos*, an isogamous species, that is to say the copulating gametes cannot be distinguished by shape or size, but a differentiation into male or female exists. These gametes are originally scattered equally through the cultures, but if two cultures are grown together then copulation takes place and gametes are formed in groups. We are able to distinguish four degrees of formation of groups by gametes: 0) no copulation, 1) joining of a few gametes into groups, 2) joining of 10–20 gametes, 3) group formation of many gametes (100). This phenomenon finds its explanation in the fact that alongside a physiological sex determination a relative sexuality exists which can be indicated successively as f^4, f^3, f^2, f^1, — m^1, m^2, m^3, m^4. Hence a series from strong (4) to weak (1) female, and from weak (1) to strong (4) male. The combinations of these 8 types give copulation intensities which are summarized in the following table:

	f⁴	f³	f²	f¹	m¹	m²	m³	m⁴
f⁴	0	0	1	2	3	3	3	3
f³	0	0	0	1	3	3	3	3
f²	1	0	0	0	3	3	3	3
f¹	2	1	0	0	2	3	3	3
m¹	3	3	3	2	0	0	1	2
m²	3	3	3	3	0	0	0	1
m³	3	3	3	3	1	0	0	0
m⁴	3	3	3	3	2	1	0	0

This relative sexuality is based upon the formation of so-called fertilization substances which belong to the group of *carotinoids* (MOEWUS 1938), viz., cis- and trans-crocetine-demethylester. The gametes of *Chlamydomonas* produce both these compounds but in different ratios: f^4 makes 95 cis/5 trans, f^3 produces 85 cis/15 trans etc., as follows:

	f⁴	f³	f²	f¹		m¹	m²	m³	m⁴
cis	95	85	75	65		35	25	15	5
trans	5	15	25	35		65	75	85	95

A copulation is then only possible provided that the cis- and trans-percentages of both gameto-types differ by at least 20. As a consequence of this, the type f_I behaves as female towards all male gametes and as a male towards types f^4 and f^3.

This review showed how many ways lead either haplogenotypically or diplogenotypically to a heritably based sex determination. Yet this does not alter the fact that the problem of sex determination which is especially complicated is not solved. In the first place it turned out that the explanation of sex determination by one single factor (with respect to which one of the sexes should be homozygous, and the other heterozygous) was much too simple even for those species with very obvious sex differences. Moreover there are still many phenomena in this field which can barely be explained. It appears that the chromosome-hypothesis is inadequate. The multiformity of sex differentiation animals was demonstrated by MEISENHEIMER (1921); for the higher plants we will refer to the review of CORRENS (1928a), and for lower plants to the work of KNIEP (1928).

In the animal kingdom, a difference is drawn between *gonochorism* (i.e. distinct differences between male and female individuals) and *hermaphroditism*. Of the latter there are two forms: *simultaneous hermaphroditism* (both sexes occur simultaneously in the same individual) and *consecutive hermaphroditism* (both sexes form successive phases in the life-cycle of the animals). We can also distinguish abnormal types of hermaphroditism such as *gynandromorphs*, that is to say, animals composed of a mosaic of cells with a different sex-chromosomal constitution, and *intersexes*, by which we mean consecutive hermaphrodites which occur as abnormal exceptions among normal gonochoristic species.

In plants, we are unable to make this difference between gynandromorphs and intersexes; abnormal types which occur in pure dioecious species are termed *hermaphrodites*.

The occurrence of *gynandromorphic* individuals (bilateral or a scattered mosaic) is most successfully and strikingly investigated in insects, although it is also known in other groups of animals. Their origin can be based on different processes of which there are four distinguishable cases: *Bombyx*, *Drosophila*, *Apis* and *Habrobracon*.

1) The case of *Bombyx*, the silkworm. Individuals regularly occur in certain strains which differ bilaterally in external appearance. This was mentioned earlier in the discussion over mosaic animals (p. 392, fig. 168). In this case the external asymmetry is often attended by a difference in sex; one half of the animals is male and the other half is female. GOLDSCHMIDT and KATSUKI (1927–1931) ascertained that the realisation of such a gynandromorphism is based upon the fact that a polar-body remains behind in the egg-cell. This is the cause of the cell obtaining two nuclei. In the case (the female is heterozygous for the sex factor) that one nucleus contains the X-chromosome, and the other contains the Y-chromosome, provided that both nuclei are fertilized by X-spermatozoa, the result will be that the fertilized egg-cell contains two fertilized nuclei, the one, XY ($\female$) and the other XX ($\male$). A cell-division between these two nuclei and a further regular course of development of the animals will result in a completely bilateral gynandromorph.

2) The case of *Drosophila* (MORGAN and BRIDGES, 1919, fig. 184). The chromosome configuration of the female is here XX and that of the male is XY. After fertilization of an egg-cell by an X-spermatozoon one single X-chromosome presumably remains behind in the equatorial plane during the first cleavage. Hence one daughter nucleus receives 2X, the other one X. In consequence of this, the one half of the

body develops into a male, whereas the other part is wholly female.

3) The case *Apis* (BOVERI 1915, MEHLING 1915, fig. 185). The most probable explanation in the bee (female diploid, male haploid), is that the egg-nucleus divides once again after both reduction divisions under the influence of refrigeration (7–13°). However, only one daughter-nucleus is fertilized (XX) and hence gives rise to female tissue. The other non-fertilized one remains, and is the origin of the male tissue.

4) The case of *Habrobracon* (WHITING 1927). This is presumably based on the fact that during the second reduction division, the polar body remains in the egg-cell. However this polar body is not fertilized, and cells which originate from this polar body preserve the male characteristic due to haploidy.

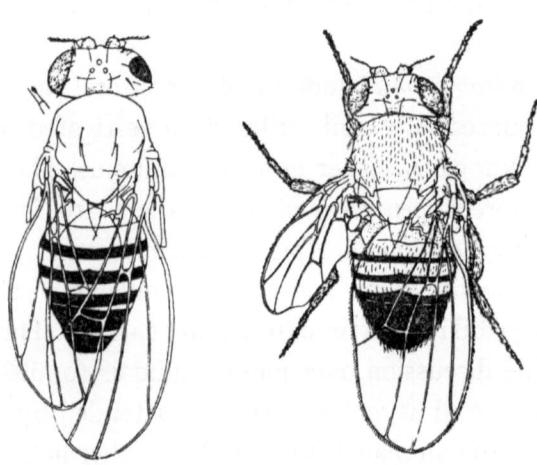

Fig. 184. Bilateral gynandromorphy in *Drosophila* (after MORGAN and BRIDGES, 1919).

More important than these gynandromorphs for an investigation into the sex factors are the intersexes. *Intersexuality*, that is to say a change in sex at a certain stage in the development of an organism, can be found throughout the whole animal kingdom. For a general review we will refer to the publications of CREW (1926–1927) and GOLDSCHMIDT (1931b, 1937b). A summary concerning mammals is given by KREDIET (1942), over birds and other vertebrates by DOMM (1939) and VAN OORDT (1942). From the richness of the phenomena we will choose four, which are rather different in character. Firstly, intersexuality occurs as a characteristic of certain races within the obviously gonochoristic species *Rana esculenta* and *R. temporaria* (WITSCHI, summaries of 1929, 1939, 1934); GOLDSCHMIDT called this type *transitory intersexuality*. Secondly, it was found that intersexuality originated in the offspring of crosses between European and Japanese races of the butterfly species *Lymantria dispar* (*L. japonica*) (GOLDSCHMIDT, summaries of 1931a and 1934) and is interpreted as a *factorial intersexuality*. Thirdly, an intersexuality is known in *Drosophila* (BRIDGES 1922, 1925b) which is based upon a chromosome irregularity which may therefore be called *chromosomal intersexuality*. Fourthly, the *hormonal intersexuality* of the vertebrates.

Let us consider firstly the *transitory intersexuality*. In both species of frogs which he mentions, WITSCHI draws attention to differences between differentiated types, which under normal circumstances can be separated into females and males on the fifteenth day after metamorphosis, and undifferentiated ones which develop provisionally into

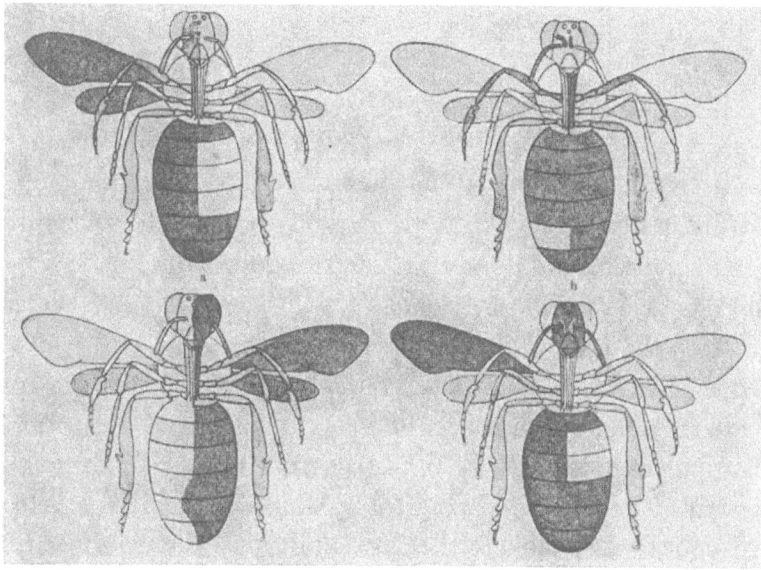

Fig. 185. Mosaic gynandromorphy in bees (after MEHLING, 1915).

females. In the course of the first two years of their life 50% of these latter females are changed into males. By means of this process the 1 : 1 ratio between the sexes is corrected for the undifferentiated animals. Moreover some animals develop into true hermaphrodites, but these are exceptions. The moment of differentiation of the apparently female into male animals is fixed for each local race, and because of this the ratio of numbers of the sexes is found to be different when the various races are compared at any given time after fertilization. This ability to differentiate is based upon typical 'potentialities' in the gametes which in their turn can again be ascribed to quantitative differences in intensities between the genes in question.

So WITSCHI obtained the following results after fertilization of the eggs of a single female of the differentiated race 'Bonn' with sperm of different races of R. *temporaria*. They are judged one month after metamorphosis.

Bonn ♀ ×	Real females	Provisional females	Herma- phrodites	Males
Freiburg ♂	190	176	14	—
Elsasz ♂	167	135	29	2
Berlin ♂	167	68	16	82
Bonn ♂	124	—	36	129
Davos ♂	135	—	2	139

From this it is evident, that there is a considerable variability in early differentiation from the undifferentiated race 'Freiburg' up to the highly differentiated race 'Davos'. It is obvious that it is not sufficient to explain such results by saying they are due to one single sex-factor localized in a heterochromosome. We will soon arrive at the general assumption these investigations are leading to.

Factorial intersexuality. The complex character of the process of sex determination finds even more expression in a long series of investigations carried out by GOLDSCHMIDT with *Lymantria dispar* and *L. japonica*, a species of butterfly in which the female is held to be heterozygous with regard to the determining sex-factor. The essentials of the phenomena observed by GOLDSCHMIDT (summary of 1934) led to the conclusion that pure breeding within the race always gave rise to normal females and normal males in the normal ratio, but that by mutual crossing of these races intersexes occurred with a definite regularity. The conception of intersexuality was defined by him as follows (1920b, p. 179): 'An intersex is an individual which develops upto a given development critical instant as a female (or male respectively) and completes its development as a male (or female respectively) subsequent to this critical moment. The higher the degree of intersexuality, the sooner this moment of change occurs'.

Those intersexual forms which originated in GOLDSCHMIDT's experiments formed a complete series of transitional forms (see fig. 186). A distinction between female intersexes and male intersexes is possible. Female intersexes are XY-individuals, which as a matter of course should be female. Male intersexes have two X-chromosomes. A whole series of female intersexes can be found starting with normal females through successive types of intersexuality (with an increase of male characteristics) upto a complete change of sex, viz., normal function-

Fig. 186. Below: normal ♂ and ♀ of *Lymantria japonica;* above a series of intersexual male animals (after GOLDSCHMIDT, 1912).

ing males with XY chromosomes. A corresponding series of male intersexes begins with normal males upto XX-individuals, which are functional females. The phenomenon expresses itself in all types of bodily properties: shape of the wing, colour of the wing, abdomen, antennae, reproductive and copulative organs etcetera, and sometimes in the pupa as well. If the above-mentioned critical moment occurs in the very early development of the individual, then the individual obtains the opposite sex to a high degree (that is to say, opposite to the sex that was determined by its genotype). However, when this moment of change takes place much later, then the original heritable disposition shows up, and in this case we obtain weak intersexual animals. The moment of change when the sex ascerts itself, depends on the races used for the cross. There are Japanese races which give nothing but normal males and females when crossed with European races, but there are also those races which yield only intersexual forms when used for crossing with European individuals.

It was found that the races used differed considerably in the 'potential' of their sex-determining factors, so that races can be distinguished which possess successive degrees of potential. We meet here a similar difference in potential (intensity, tendency, valency) between the sex determining genes, as we have already seen in the section dealing with relative sexuality (p. 425). The reciprocal crosses between a strong and a weak race, with their F_2-generations and all possible backcrosses yielded the following results (I = intersex, N = normal):

weak ♀ × strong ♂	females	males	strong ♀ × weak ♂	females	males
F_1a	100% I	100% N	F_1b	100% N	100% N
F_2a	50% N + 50% I	100% N	F_2b	100% N	50% N + 50% I
weak ♀ × F_1a ♂	50% N + 50% I	100% N	strong ♀ × F_1b ♂	100% N	100% N
weak ♀ × F_1b ♂	50% N + 50% I	100% N	strong ♀ × F_1a ♂	100% N	100% N
F_1a ♀ × weak ♂	100% N	100% N	F_1b ♀ × strong ♂	100% N	100% N
F_1a ♀ × strong ♂	100% I	100% N	F_1b ♀ × weak ♂	100% N	100% I

How can we bring these remarkable results into line with the theory of genic sex determination? GOLDSCHMIDT tried as follows: Each fertilized egg-cell normally has two sorts of heritable factors. Activity of both these factors is necessary for the expression of the one or the other sex. These heritable sex factors have a certain potential. He called the factors for female sex F, and those for male sex, M. The F-factor should only be inherited from the maternal line, and the M-factor presumed to lie in the X-chromosome. Therefore the homozygously male individuals have the formula FMM and the heterozygously female, FMm. The male factors should be counteracted by the factor for female F. In this way it becomes possible to explain the phenomenon of intersexuality in such a way that these 'male' and 'female' factors differ in strength in different races. Hence we can express this potential by numbers and assume that $F = 80$ for the weak race, and that $F = 120$ for the strong one. M_w in the weak race is 60 and hence $M_w M_w = 120$. For M_s in the strong race, 80, and hence $M_s M_s = 160$. The recessive m should then be equal to 0. It is obvious that one must have a surplus of 'F'-units over those bearing the Mm characteristics to produce a normal female individual (FMm). The same surplus has to exist between MM and F in order to give an individual the chance to develop into a normal male. If these surpluses are below the minimum and the opposite sex continues to wield an influence, then intersexual forms will originate. By means of the occurrence of such a mechanism it is possible to form an idea of the underlying causes which lead to the following results:

weak ♀ × strong ♂
F_w M_w m (N) × F_s M_s M_s (N)
80 60 0(+20)   120 80 80(−40)

	female		male	
F_1a	F_w M_s m (I)		F_w M_s M_w (N)	
	80 80 0(±0)		80 80 60(−60)	
F_2a	F_w M_w m (N)	F_w M_s m (I)	F_w M_s M_w (N)	F_w M_s M_s (N)
	80 60 0(+20)	80 80 0(±0)	80 80 60(−60)	80 80 80(−80)
weak ♀ × F_1a ♂	F_w M_w m (N)	F_w M_s m (I)	F_w M_s M_w (N)	F_w M_w M_w (N)
	80 60 0(+20)	80 80 0(±0)	80 80 60(−60)	80 60 60(−40)
weak ♀ × F_1b ♂	F_w M_w m (N)	F_w M_s m (I)	F_w M_s M_w (N)	F_w M_w M_w (N)
	80 60 0(+20)	80 80 0(±0)	80 80 60(−60)	80 60 60(−40)
F_1a ♀ weak ♂	F_w M_w m (N)		F_w M_s M_w (N)	
	80 60 0(±20)		80 80 60(−60)	
F_1a ♀ × strong ♂	F_w M_s m (I)		F_w M_s M_s (N)	
	80 80 0(±0)		80 80 80(−80)	

strong ♀ × weak ♂
F_s M_s m (N) × F_w M_w M_w (N)
120 80 0(+40)   80 60 60(−40)

	female		male	
F_1b	F_s M_w m (N)		F_s M_s M_w (N)	
	120 60 0(+60)		120 80 60 (−20)	
F_2b	F_s M_s m (N)	F_s M_w m (N)	F_s M_s M_w (N)	F_s M_w M_w (I)
	120 80 0(+40)	120 60 0(+60)	120 80 60(−20)	120 60 60(±0)
strong ♀ × F_1b ♂	F_s M_s m (N)	F_s M_w m (N)	F_s M_s M_s (N)	F_s M_s M_w (N)
	120 80 0(+40)	120 60 0(+60)	120 80 80(−40)	120 80 60(−20) ,
strong ♀ × F_1a ♂	F_s M_s m (N)	F_s M_w m (N)	F_s M_s M_s (N)	F_s M_s M_w (N)
	120 80 0(+40)	120 60 0(+60)	120 80 80(−40)	120 80 60(−20) ,
F_1b × strong ♂	F_s M_s m (N)		F_s M_s M_w (N)	
	120 80 0(+40)		120 80 60(−20)	
F_1b × weak ♂	F_s M_w m (N)		F_s M_w M_w (I)	
	120 60 0(+60)		120 60 60(±0)	

Figure 187 represents GOLDSCHMIDT's view. The time taken for development is along the horizontal axis, the vertical axis gives the effect produced by the gene corresponding with its strength. The curve represents the development of the sex; the line S-S indicates that the development has been completed. The curves M_8M_8 and F_8 show that the male develops regularly, those of F_8 and M_8m show the same for the female. By replacing the factor M by weaker ones for instance M_1 (in order of potential M_1, M_2 and M_3) the total effect of the genotype M_1M_1 is different. At a certain point the genotype M_1M_1

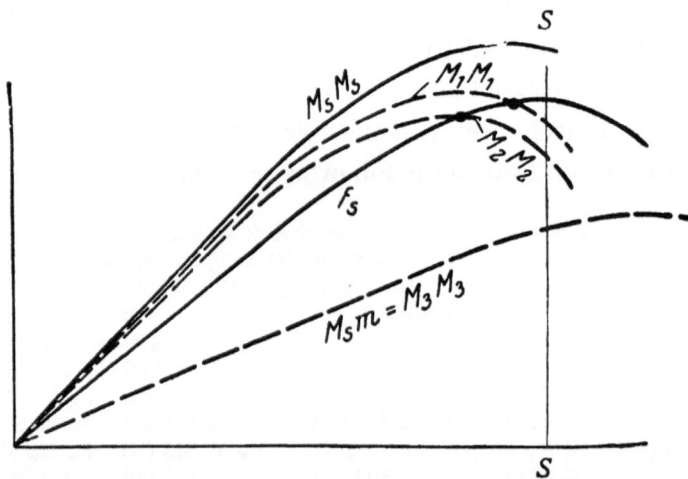

Fig. 187. Schematic representation of intersexuality in the butterfly *Lymantria* (after GOLDSCHMIDT, 1931 b).

is outweighed and typical female characteristics develop. This moment takes place earlier in M_2M_2 and in M_3M_3 even so early that the individual becomes female from the beginning (though it is nevertheless genotypically male).

In the main this explanation of GOLDSCHMIDT, concerning the behaviour of the intersexual *Lymantria* is plausible, but still open for discussion (see BALTZER 1937 GOLDSCHMIDT 1938b, 1946 and SEILER 1946, 1949). The greatest difficulty lies in the localization of the F-factor. GOLDSCHMIDT believed that this factor is of a cytoplasmic nature; WITSCHI, on the other hand accepts that the F-factor (s) are localized in a pair of autosomes and therefore formulates the constitution as FFMm (♀) and FFMM (♂). WITSCHI presumes analogously that in animal species in which the male is heterozygous, two M-factors are located in the autosomes and the F-factor in the X-chromosomes, hence the constitutions MMFF (♀) and MMFf (♂) can be drawn up. We have to admit that GOLDSCHMIDT's solution of the phenomenon of intersexuality is a very original one and on the whole appears acceptable. It is possible to present this supposition as the explanation of similar phenomena in other groups of animals, such as birds, *Daphnia's* and others (see CREW, 1927, KREDIET 1942, VAN OORDT 1942). On the other hand it cannot be denied that a certain amount of criticism is not without justification.

For analogous investigations dealing with plant hermaphrodites we are indebted to SHULL (1910, 1911b), whose work was mostly concerned with the very dioecious *Melandrium dioicum*. He discovered hermaphroditic individuals in this species. His crossing experiments point to a homozygous nature of the female plants and a heterozygous nature of the male plants. This is in accordance with the experiments dealing with sex-linked inheritance and karyological investigations already mentioned. SHULL's principal results were as follows:

1) From normal crossings of male and female plants he found that male and female descendants occurred in equal numbers.

2) Two types of hermaphrodites can be distinguished: somatic and genetic hermaphrodites. The difference between them will appear after crossing.

3) Genetical hermaphrodites give rise to equal numbers of hermaphrodites and female descendants after self-pollination.

4) Normal female plants pollinated with pollen of genetical hermaphrodites yield hermaphrodites and female plants in equal numbers.

5) Normal female plants with pollen of somatic hermaphrodites give rise to male and female plants, 50% of each.

6) Genetical hermaphrodites with pollen of normal ♂ individuals give 50% ♂ and 50% ♀ plants.

7) Crosses of normal female plants with pollen of hermaphrodites in some cases give a few plants which are apparently male. The percentage of these is very low.

8) Crosses of normal female with male plants sometimes give rise to some hermaphrodite individuals.

SHULL concluded from these results that the hermaphroditic individuals in *Melandrium* take the place of male individuals and that hermaphrodites can change into male plants and the converse. The interpretation that SHULL gave to his observations did not completely agree with GOLDSCHMIDT's view. The later investigations of G. and P. HERTWIG (1922) have corroborated the results and extended them viz., female plants have the formula FFMM, male plants have the formula FfMM, and the hermaphrodites have the same constitution as the male individuals, with the condition that the F-factor is changed into the more intensive F'-factor and the f into the f'. Hence there are three possible types of females, the normal FFMM, F'FMM, and F'F'MM; there are two types of hermaphrodites, F'f'MM and F'fMM, and four types of males FfMM, Ff'MM (sterile), f'f'MM and f'fMM (the last two are not viable). The karyological investigations also showed that there is no chromosomal difference between male and hermaphrodite plants (see fig. 181). In many cases the results obtained from investigations into botanical hermaphrodites in species which are otherwise dioecious, can be explained following the theory of GOLDSCHMIDT.

The third type of intersexuality, based upon aberrations in the chromosome configuration (chromosomal intersexuality) was observed for the first time by BRIDGES (1922, 1925) in *Drosophila melanogaster*. The sex character of all sorts of chromosome combinations was determined by him and later investigators. These are summarized in the following table in which a set of autosomes is represented by A. From this data it is evident that not so much the individual presence of either X or Y chromosomes determines the sex, but that the ratio between the quantity X and the quantity of the autosomes is a defin-

Chromosome configuration	sex type	ratio X : A
2 A + 3 X	hyper-female	3 : 2 = 1.5
4 A + 4 X	female	4 : 4 = 1.0
3 A + 3 X	female	3 : 3 = 1.0
3 A + 3 X + 1 Y	female	3 : 3 = 1.0
2 A + 2 X	female	2 : 2 = 1.0
2 A + 2 X + 1 Y	female	2 : 2 = 1.0
2 A + 2 X + 2 Y	female	2 : 2 = 1.0
3 A + 2 X + 1 Y	intersex	2 : 3 = 0.67
3 A + 2 X	intersex	2 : 3 = 0.67
2 A + 1 X + 1 Y	male	1 : 2 = 0.5
2 A + 1 X + 2 Y	male	1 : 2 = 0.5
2 A + 1 X	male	1 : 2 = 0.5
3 A + 1 X	hyper-male	1 : 3 = 0.33

ing influence. If this ratio is 1 0 or higher, then the animal will be female or sometimes hyperfemale. If the ratio is 0'5 or lower then the male sex is determined whereas when the ratio lies between 1'0 and 0'5, an intersex occurs.

The fact which comes to the fore from this table is that the autosomes are not as neutral as formerly believed. This is evident from the ratios of numbers given. The conclusion seems to be justified that in all individuals the male character is located in the autosomes as the factors MM, that the sex factors FF in homozygous state surpass the effect of their antipodes MM, but in a heterozygous state Ff is not sufficient to convert the male character into the female one. This hypothesis is supported by all sorts of other investigations with different materials. If the M-character is enhanced by triploidy (MMM) then FF is no longer completely able to surpass the effect of MMM. Hence the normal scheme comes down to the genotypical formula MMFF for the female, and MMFf for the male. Sex can be considered as being a result of the counteracting effect of M in the autosomes and F in the X-chromosome. Factors in the Y-chromosomes do not wield any influence.

An amplification of the investigations carried out with *Drosophila* is the possibility of making polyploids of dioecious plants (see KUHN 1942). Almost simultaneously, polyploids of the dioecious *Melandrium* were obtained by WARMKE and BLAKESLEE (1939) and ONO (1940) by means of colchicine treatment, and by WESTERGAARD (1940) through the influence of certain temperatures; their results corroborated one another completely. Alongside the normal diploids $(2 A + 2 X = ♀; 2 A + X = ♂)$, tetraploids as well as triploids were obtained. Very extensive studies, especially those of WESTERGAARD, yielded the following possibilities (A = autosome):

female	male	hermaphrodites
44 A + 4 X	44 A + 2 X + 2 Y	44 A + 3 X + fragm. Y
33 A + 3 X	44 A + 4 X + Y	43 A + 3 X + Y
	44 A + 3 X + Y	31 A + 2 X + Y
	44 A + 2 X + Y	
	33 A + 2 X Y	

The genetical conclusion, which follows from this, is that in the dioecious ragged robin the sex of the male plants is caused by the overbalance of a very strong male element (located in the Y-chromosome) over other elements, which determine the female sex and are supposed to be located in the X-chromosome. With regard to this point there is a great difference between *Melandrium* and *Drosophila*, in the latter, the Y-chromosome does not have any function as a sex-determiner. From the shortage of autosomes in two of the three types of hermaphrodites we may presume that accidental genes for the male sex are present in the male autosomes.

In connection with this chromosomal intersexuality or hermaphroditism on the one hand, we would like to point to the relationship which exists in mosses between the polyploids and a possible separation of sex (heteroecy as an opposite to monoecy in the haploid stage) on the other hand. We will return later to these relations in the discussion of the normal-monoecious character of most of the plant species.

As the last group we will take the *hormonal intersexes* (see DANFORTH 1939, WILLIER 1939, BURNS 1949). We find a classical example in the so-called *free-martins*, in which a non-identical twin (a genetical cow-calf) is found side by side with a bull-calf. The cow-calf is a sexual deformation which is going to develop into a male direction. A basis of explanation for the occurrence of free-martins was laid down in the work of KELLER and TANDLER (1911, 1916, 1920) and of LILLIE (1917); all data points to the fact, that the bull-calf produces substances (hormones), which influence the cow calf through the placental anastomosis. The female calf obtains masculine features, but not in sufficient quantities to revert the animal into a bull-calf, because it remains sterile. Such a hormonal intersexuality does not properly belong within the sphere of genetics; indeed it is better to classify it among the physiological-chemical problems. Nevertheless, this process of intersexuality is extremely suited to experimental physiological genetical work: DANTSCHAKOFF (1938, 1941) showed that, in a large series of experiments with chickens (female sex heterozygous, male homozygous) and guinea-pigs (♂ heterozygous, ♀ homozygous) that the hormone of heterozygous sex administered in an early embryonic stage acts as an intensifier in this sex and when administered to the homozygous sex, it even changes this type of sex, whereas the hormone of the homozygous sex acts lethally in both types of embryos. This points to the fact that a relationship exists between the moment at which the hormone in question will be produced in normal life, and the homo-heterozygous nature of sex determining genes. In the heterozygous sex, the hormone in question is secreted at a very early embryonic stage, a long time before the proper gonads are formed. In the homozygous sex, the hormone is produced at a much later stage of development, the tissues are then adapted to this hormone. An artificial supply of this hormone in an early stage of development leads to lethality (see also PONSE, 1949), since the tissue is sensitive to such a compound.

All this data which can be brought to bear on the haplo- or diplogenotypical sex-determination problem can also be related to the great complexity of problems which the bi-sexual nature of the higher plants presents.

In this way we will discuss the crosses between species of which one is ex-

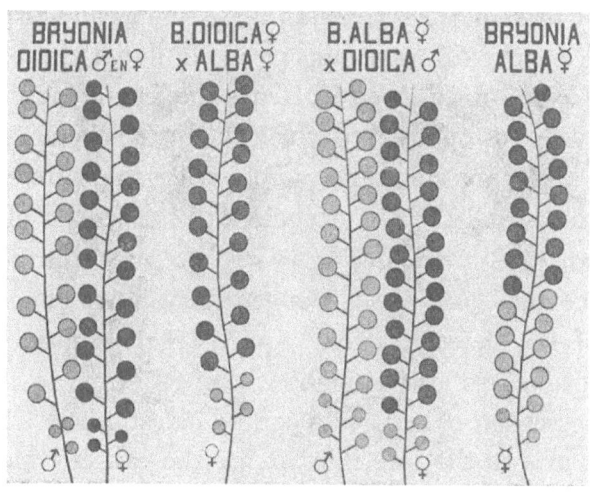

Fig. 188. Selfings and hybrids of *Bryonia dioica* and *B. alba* (after CORRENS, 1913).

clusively *dioecious*, the other exclusively *bisexual*. Investigations concerning this meet many difficulties, principally as a result of the sterility of the F_1-individuals and hence the impossibility of obtaining offspring. The most important study in this sphere is concerned with the cross of *Bryonia dioica* with *B. alba* (CORRENS 1907a, CORRENS and GOLDSCHMIDT 1913). *B. dioica* forms its male inflorescences on different individuals to that of its female inflorescences (this is indicated in fig. 188 by light and dark circles). The second species *B. alba* has mixed inflorescences on all plants, the lower part in all cases is composed of male flowers and the upper part of female flowers. Hence if the female flowers of *B. dioica* are pollinated with pollen of the same species equal numbers of male and female plants will originate (fig. 188, left). But after pollination by pollen of *B. alba* the seed gives only plants with female flowers (excluding a few incompletely developed flowers on the lower part of the inflorescences, see fig. 188, second from the left); the reciprocal cross (*B. alba* ♀ × *B. dioica* ♂) gives equal numbers of male and female plants. Selfed plants of *B. alba* yield hermaphrodites exclusively.

It is not easy to give a correct interpretation by means of formulae of these data of CORRENS, but it seems to follow that: firstly, dominance of dioecy over monoecy exists, and secondly, that one may presume that with regard to the sex factor all the egg-cells produced by the female *Bryonia dioica* are similar and also one may presume the occurrence of two genotypically different groups of pollen grains in the ♂ plant of this species. Hence the female individual be termed homozygous as opposed to which the male is heterozygous.

This investigation corroborates the heterozygosity of the male plants and the homozygosity of the female plants, however, it does not give a decisive answer concerning the heritable nature of the monoecy of *B. alba*. The general validity of each theory in the realm of sex determination requires that *monoecious* and *hermaphrodite* species are also brought within its scope so that their questionable possession of sex genes can be investigated. The problem which now comes to the fore is what are the causes which are able to convert dioecy into monoecy or hermaphroditism and the converse (i.e., what are the causes which are able to convert monoecy into dioecy). From the investigations into this, three groups of influences were discovered: a) change of the number of chromosomes, b) the effect of certain genes, c) the activity of environment.

The first possibility comes to expression in a comparison of normal haplo-diploid mosses, with others which are diplo-tetraploid in their life-cycle. There are two aspects of approach to the problem: 1) observations in free nature and 2) experiments. On the grounds of data available concerning Hepaticae and Bryophyta, HEITZ (1939) formulated his rule of a '*polyploidy-monoecy relationship*' that is to say alongside haplo-diploid dioecious mosses which occur in nature, related species are found which are polyploid (diplo-tetraploid) and monoecious. From this rule one would expect as a matter of course that a change in monoecy should be attended by a tendency to polyploidy in dioecious mosses. This rule is corroborated by an experiment performed by WETTSTEIN in 1937: regeneration of diploid sporangiae of the dioecious species *Bryum caespiticium* (in which the moss plant is either male or female) yielded diploid moss plants which bore antheridia as well as archegonia. Nevertheless these diploid monoecious moss plants were at first highly sterile but finally one of these diploid types attained complete fertility thereby maintaining its monoecy. Hence it is proved that the polyploid combination of the chromosomes gives rise to monoecy and holds both male and female factors together which were originally separated in the normal *Bryum caespiticum*, and hence this chromosome accumulation in dioecious individuals makes them monoecious.

The reverse case can also take place, that is to say, that certain genes make diploid organisms change from monoecious into dioecious. A clear example of this was given by JONES (1934) in maize. Maize is monoecious, that is to say male and female inflorescences are formed by the same individuals, however they are separated from one another. Genes have now been found by JONES, which prevent the development of anthers in the male inflorescences and give rise instead to female inflorescences in these plumes; and other genes which allow male flowers to grow normally, but which sterilize the female cobs. Though both types of inflorescences remain morphologically unchanged, these genes still appear to deform the monoecious plant into a purely male or a purely female plant.

As a third group of causes for such changes, we will examine the action of environment. We will find an example in the investigations of CORRENS (1907b) dealing with *Satureia hortensis*. This species is gynodioecious, that is to say, alongside the purely female individuals, plants are also found which form a number of hermaphrodite flowers side by side with female flowers (so-called gynomonoecious individuals). Specimens with hermaphrodite flowers only were not observed. Fig. 189 gives a representation of different types of flowers: A and B are two forms of hermaphrodite flowers, C represents flowers with shrunken anthers and D-J represents purely female flowers with corolla leaves of different sizes.

The extent to which the origin of these flower types depends on external or internal non-heritable factors is evident from fig. 190.

All flowers were counted which had appeared on a total of 390 plants (the total of flowers was over 20.000) over a period of ten weeks; the four curves in the figure represent the course of the countings; a) purely hermaphrodites, b) hermaphrodites with partly shrunken

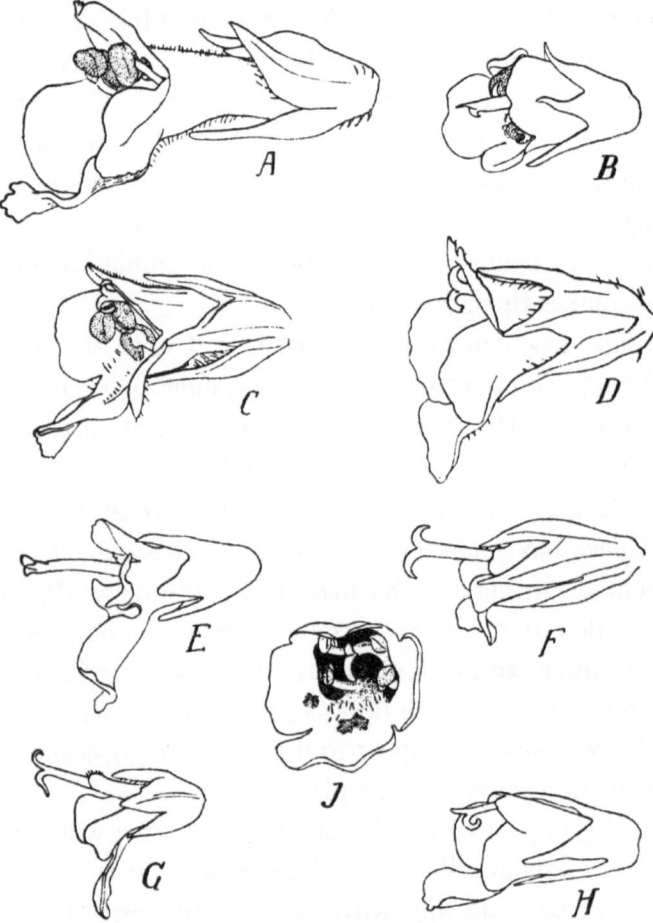

Fig. 189. Different flower types of *Satureia hortensis*
(after CORRENS, 1928 a).

anthers, c) hermaphrodites with completely shrunken anthers, and d) pure female flowers without anthers. It is obvious from this that the hermaphrodite flowers have the upper hand at the beginning of the flowering season ($\pm$ 80%) whereas at the end of the season the large majority of the flowers are purely female. This change in sexual nature of the flowers is called a *phenotypical determination of sex* by CORRENS. The heredity of sexual forms, which belong to the species

Satureia hortensis, seems to be completely maternal: descendants of selfed gynomonoecious plants are again all gynomonoecious. Descendants of female specimens pollinated with pollen of gynomonoecious plants were all pure female. At present it is still not possible to decide what the cause of this phenomenon is. Different explanations have been put forward of which the most obvious is that the plasm of the female and that of the gynomonoecious individuals are different and

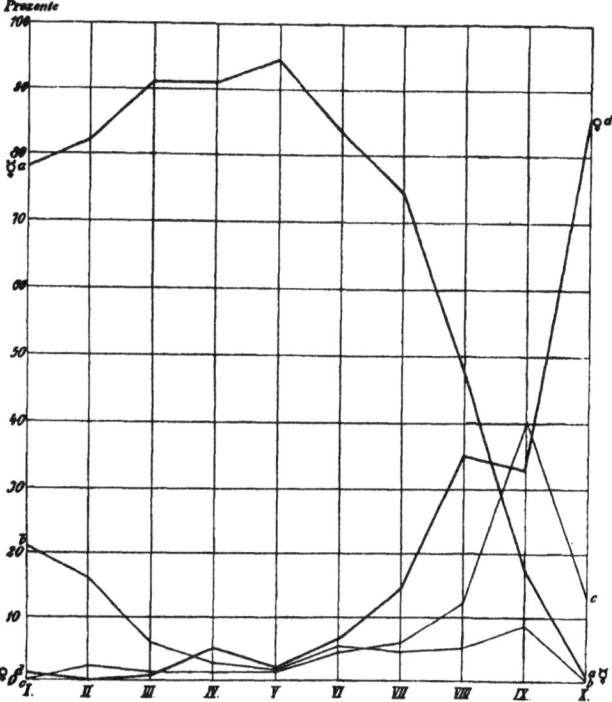

Fig. 190. Dependence of the frequencies of flower types on external conditions in *Satureia hortensis* (after CORRENS, 1907 b).

that its nature determines the nature of the descendants. CORRENS postulates that we may draw an analogy between these results and those obtained in crosses between erect and recumbent plants of *Linum usitatissimum* concerning the occurrence of male sterile plants (see p. 361).

We thus meet for the first time an example of *diplo-phenotypical sex determination*; in the diploid phase (the flowering plant), one of the two possibilities occurring in the individuals (i.e. male or female) is so favoured by environment, that the phenotype becomes purely female. Such a phenotypical sex determination seems to occur more frequently in lower organisms in the haploid phase, though generally the proofs are not above criticism. Among the higher diploid plants

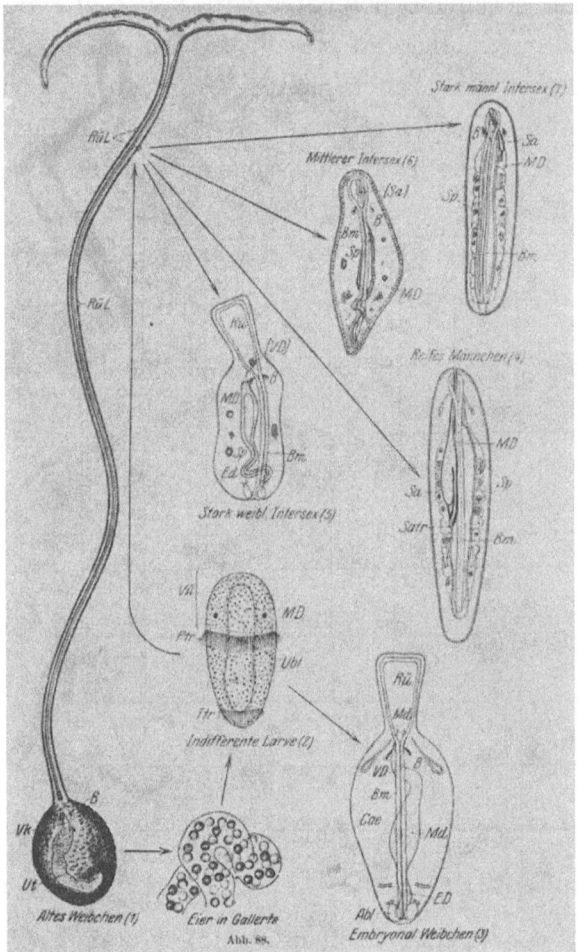

Fig. 191. The development of *Bonellia* (Altes Weibchen = old female; Eier in Gallerte = eggs in jelly; Indifferente Larve = indifferent larva; Embryonal Weibchen = embryonic female; Reifes Männchen = mature male; Stark weibl. Intersex = strong female intersex: Mittlerer Intersex = intersex of medium type: Stark männl. Intersex = strong male intersex. All figures are 60× as compared with female (after BALTZER, 1925).

the phenomenon apparently occurs frequently (see CHODAT, 1943) but among animals a diplo-phenotypical sex determination is scarce. A few striking cases are known however, for which no other explanation seems acceptable, and in which environment favours one of the two sex potential present in a bisexual individual. Famous in this sence is the *Bonellia* case (see fig. 191) which for years appeared to be an insoluble puzzle, until the work of BALTZER (1914, 1925, 1928 and later the work of BALTZER and ZURBUCHEN, 1931, 1937) uncovered the explanation.

Both sexes of the worm are strikingly different in shape, the female has a round body, about 4 cms long with a long proboscis (fig. 191,1), the male is at the utmost 1 mm. in length and of an oblong-cylindrical shape (fig. 191,4). The larva (fig. 191,2) born from the fertilized egg-cell is undifferentiated with regard to sex (i.e. neutral). If such a larva attaches itself to the proboscis of a fully-grown female, then it will not increase in size, but its shape will become more narrow and changes inside the body will take place so as to give rise to a male; however, if the larva remains free in the sea-water then it will grow and sink and attach itself to the rocks. After this it develops into a female (fig. 191,3). By interfering with the life of a larva (which had attached itself to the proboscis) by dislodging, it will turn out that the duration of attachment to the proboscis determines the sex. If this happens soon after attachment, then the animal will become an intersex with a strong female tendency (fig. 191,5); when the larva remains longer on the proboscis then the male characteristics increase, the female ones decrease, and the sex will become intermediate (fig. 191,6). If there is a long time between attachment and encroachment, then the intersex will become preponderantly male (fig. 191,7). On the grounds of his experiments, BALTZER concluded that certain substances are synthesized in the proboscis of the female which will make the indifferent larva into a male, and that the extent to which the larva will develop into an intersex or into a pure male, depends on the supply of these substances. Later investigations of HERBST (1928–40) state more precisely the effect of definite chemical elements on this kind of sex determination.

The general theorem, which follows from this is that the diploid organism is bisexual in nature and hence has the potential for both sexes, of which one of the two can be brought into development by environment i.e. the sex is determined phenotypically.

This theorem can even be extended to the genotypical sex determination. The concept originated from CORRENS, HARTMANN (see 1943) and WETTSTEIN (1936) that each diploid organism, also those with a genotypical sex determination, is fundamentally bisexual and that the effect of the sex gene remains limited to the *realization of one of the two given potentials*. Hence, sex determining factors are not, as was originally believed, *determinators*, which have an absolute power; it is better to consider them as factors which *realize* the sex by making a choice between the two possibilities available. If we approach the matter in this way, the effect of the sex-factors (as being located in the gonosomes) becomes more easily understandable. The rather vague and mysterious character associated with the gene, when viewed as a sex determiner wholly disappears when we consider the gene as a 'sex realizer'.

CHAPTER XIX

CHANGE IN THE HERITABLE PREDISPOSITION OF THE INDIVIDUAL

There is a process which takes place not only artificially by means of rearing and breeding but also in free nature, which at its face value seems to contradict all heredity; viz., the origin of new forms which are heritably different from their parents. If heredity is to maintain its reputation as phenomenon of life worthy of investigation and a vital point in many scientific problems, then genetics must be aware of this contradiction and give attention to this apparently insoluble problem. What is the cause of the origin of these heritably different forms? Is the genotype variable for each given individual?

The manner in which an attempt was made to answer this question varied during the history of genetics, depending on appreciation of the methods which led to the answer. As in nearly all the problems of general biology, there were three methods: 1) purely contemplative, 2) by observation and 3) by experiment. Each of these led to a number of solutions. The first method, although flourishing in large circles of authors is practically without significance from a scientific point of view, as is also the method of pure observation due to the subjective element present in every observation, and to the interpretation of these observations.

The studies of LAMARCK started with his 'Système des animaux sans vertèbres' (1801) and followed by his 'Organisation des corps vivants' (1802), his 'Discours d'ouverture' of 1806 and his ,,Philosophie zoologique'' of 1809. These studies are of paramount historical importance. He opened these publications with a discussion on the question of the *heredity of acquired characters.*

Piling observation upon observation, LAMARCK derived two rules:

1) 'In every animal which has not reached the bounds of its development, one or another organ is gradually developed by frequent and

continued use and an ability is given to this organ proportional to the
duration of its use. On the other hand a continual neglect of an organ
results in an almost imperceptible weakening and gradual decline
until final disappearance occurs'.

2) 'Everything that nature gains or loses in individuals under
the influence of environment (to which the race was exposed over a
long period of time and thus through the continued use of an organ,
or by a continued non-use of a part of the body) will be preserved in
individuals which originate by means of reproduction provided that
these changes occur jointly in both sexes, or at least in those individuals which have given rise to the new ones' (1809,1. p. 235).

The main point of LAMARCK's contemplations embodies the hypothesis that the change of environment gives rise to new properties
or organs which should be useful to the organism and make these
properties develop further by frequent use; at the other hand properties can gradually disappear by falling into disuse. Thus new forms
of plants and animals come into the world which differ in heritable
aspect from their parent, simply and solely as a result of a reaction of
these parents to the influence of environment. LAMARCK is absolutely convinced of his views and the reliability of the 'known facts'. He
therefore states very positively that 'this is an ascertained fact and
later on I intend to give convincing proof of this'.

All LAMARCK's arguments are based on observations in nature
among both plants and animals, and his contemplations deal with the
changes of races under the influence of changes in cultivation. In the
time at which LAMARCK published his views they had great influence,
but they are less recognized today because the value of pure observation as a reliable basis for deduction is questioned. After LAMARCK
there were numerous other investigators in the course of the nineteenth
century who uttered more or less speculative and rather Lamarckian
views concerning the changes of heritable predisposition. HERBERT
SPENCER whose writings we discussed earlier in this book gave an
interesting discussion in the Contemporary Review (1893, 1894) of
his point of view in a paper war with WEISMANN.

From the pen of comparative anatomists alongside palaeontologists,
plant-geographers and ecologists, more or less Lamarckian theories
dealing with the change of the genotype under influence of environment were published on a large scale and these found their ex-

treme consequence in the Mneme-hypothesis, stubbornly held by SEMON (1904, 1909).

Both declining appreciation for accurate observation so typical of the nineteenth century and also exacting conditions dictated by modern genetics as a basis for hypotheses have wielded their influence on this topic. Therefore a more stable foundation for the problem of change in heritable predisposition was sought experimentally. It became obvious, as WEISMANN remarked, that a distinction must be drawn between the different investigations on account of the contrast which exists between uni-cellular and multi-cellular organisms. Whereas in the former there is no distinction between the reproductive cells and body-cells, in the multi-cellular organisms this difference between reproductive cells and body-cells is of prime importance because the special reproductive cells are for the most part situated in definite organs, surrounded by a number of tissues of differentiated body cells. We have to take into account the fact that in uni-cellular organisms some always propagate exclusively in an asexual manner whereas in most other species a frequent asexual division and a periodical sexual reproduction occur side by side.

There is still a great deal of uncertainty concerning sexual reproduction of bacteria: DUBOS (1946, p. 181) calls the data 'unconvincing', KNAYSI (1946, 1949, p. 137) speaks of 'unsatisfactory', but BISSET (1950, p. 76) says that 'sexual fusion is a normal part of the nuclear cycle'. TATUM and LEDERBERG (1947) also produce evidence by means of indirect demonstrations, when they speak of 'gene recombinations' and LEDERBERG (1947) even speaks of 'linked segregations' in *Escherichia coli*. These cases of linked segregation are different in character from results of experiments dealing with the unicellular *Paramecium* with its well-known though complicated, sexual reproduction. Studies with higher animals and plants must be considered from yet another point of view.

In order to interpret data obtained in bacteriological investigations the technique must satisfy strict conditions. In the first place the cultures used, the so-called 'single cell cultures' must originate from one single isolated bacterium. Cultures which do not satisfy this condition will prove misleading and unreliable. A second point to be satisfied lies in the fact that a source of errors for bacteriological research (or rather its interpretation) exists in even the most accurate of the existing techniques. This error is based on absence or obscurity of sexual reproduction. It is not possible to prove homozygosity in bacteria. No way is known of proving that the bacterial cell is homozygous, in the same sense that we use this term when speaking of higher organisms. Therefore, we do not have 'pure lines' among microbes which will fit the description given by JOHANNSEN (1915, see p. 40). Because of this, evidence has never been produced to show that environment plays a rôle in the realization of the difference between two strains obtained from one single bacterial cell whose strains differ in a lasting manner. This is due to the fact that firstly proof is lacking that the original cell was a homozygote and hence both cells could perhaps already be different from the very beginning of the strains. Secondly the necessary criterion viz., sexual reproduction and its consequences are therefore uncertain in this case. Thirdly, we must be very careful in accepting as permanent any change in microbes as a consequence of differences in culture conditions. JOLLOS (1914) pointed out that the after-effect of a deviating environment after a sexual propagation has taken place, can last an exceptionally long time in bacteria. We have learned earlier of the phenomenon of an after-effect of a former genotype (p. 343); similarly, an after-effect of aberrations in the phenotype brought about by environment frequently occurs. JOLLOS rightly held the view that such durable after-effect changes, are not changes in heritable predisposition. There is no important or vital difference when compared to those non-permanent changes in higher organisms (modifications) which are due to the influence of environment and remain limited to the individual subjected to it, and to its clones (asexual offspring, vegetative offspring). JOLLOS therefore believed that all cases of such changes in bacteria must be considered as modifications. He gives the name 'Dauermodifikation' to this long after-effect.

It is of great importance to separate the conception of 'Dauermodification' from that of a 'mutation', by which is meant a change in the genotype of the individual. As we have already seen it is impossible to study the genotype in detail because of the unsuitable disposition of the material which means that investigations with bacteria are improper for use in any problem concerning the changeability of the heritable predisposition of the individual; hence, it is also not permissible to speak of bacterial mutations (see HÄMMERLING 1929, JOLLOS 1939).

On the other hand experiments with uni-cellular organisms have significance if we understand the nature of their sexual reproduction. *Paramaecium caudatum* is an especially suitable object, because this organism gives parthenogenesis alongside vegetative propagation and generative reproduction (by *parthenogenesis* we understand a further development of cells destined for sexual processes without fertilization).

JOLLOS (1913, 1921), using *P. caudatum* obtained results which are

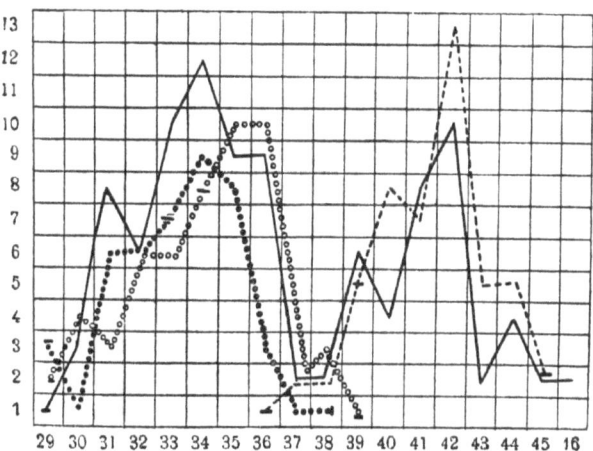

Fig. 192 Frequency curves of *Paramaecium*. Horizontal axis: measurements of length. Vertical axis: numbers. Continuous line: original race (between 37 and 46) and new race (between 29 and 37); dotted line = new race; open dotted line = new race after conjugation; interrupted line = original race by itself (after JOLLOS, 1913).

of great value in answering this problem. His experiments diverge in two directions: action of abnormal temperatures and influence of ions of certain metals (by diluted salt solutions) upon the characteristics of *Paramaecium*.

From the first series he appeared to obtain a positive result: in a culture starting with one single individual cultivated over a long period of time at 31°C, alongside Paramaecia of normal size (between 37 and 46 measure-units) a high number of smaller individuals (between 29 and 37 measure-units) were obtained after 9 weeks. The whole culture gave, from exact measurements, a typical two-topped frequency curve (fig. 192, continuous lines) as we met earlier. This pointed to the fact that two races were represented, and this proved to be the case; some of the smallest individuals were cultivated separately and gave rise to a culture in which the frequency distribution had only one top (fig. 192, dotted line) whereas the individuals, originating from separate cultivation of some of the largest animals, also yielded a one-topped distribution (interrupted line). This latter distribution coincides completely with that of the ancestral race (a), the smaller however (a_1) did not appear before in JOLLOS' cultures. Apart from size, this new race also differs from the original one by the characteristic that it could be transferred safely from a temperature of 31°C into a medium at 39°C, whereas the ancestor race could not tolerate such a sudden increase in temperature.

Both these properties (small size and resistance to temperature) remained unchanged in the new race a_1 , even after being cultured for several months at 31° as well as at lower temperatures. They did not disappear even in cultures originating from individuals which had passed through the fertilization process (conjugation). Originally, this was accepted as a criterium for changed heritable predisposition. After conjugation, the frequency-distribution had moved to the right (fig. 192 open dot line), that is to say, had moved into the direction of size increase. JOLLOS did not attach great significance to this because such movements can always be observed after fertilization in ancestor cultures. Prolonged experiments however, have shown that the continuance of such artificial changes after conjugation may not be considered as a criterium for a lasting change in the genotype. Cultures of *Paramaecium* in which the ratio of divisions was delayed by the influence of Ca-salts, preserved this change in character for seven months (December '17 – June '18) after their return to normal environment. In the eight month (July '18) a natural rate of division reappeared. Seven successive parthenogenetical reproductions had established this reversion to normal division rate in March '18 or alternatively, two conjugations. Therefore it is not so much a qualitative difference which exists between the three forms of reproduction but more a quantitative difference: conjugation has thus lost its fundamental importance. A parthenogenesis is in general as active as 30–34 vegetative divisions, and a conjugation in its turn wields the same influence as 3–6 parthenogenetical processes.

JOLLOS' work has taught us that apparently positive results

with regard to changes of the genotype in uni-cellular organisms must be viewed with the greatest scepticism. The uncertainty of results obtained from research dealing with more complicated plants and animals (as opposed to research dealing with uni-cellular organisms) is still greater. In uni-cellular organisms we may consider the nucleus of the cell, the principal depository of heritable factors, more easily open to influences from the outside and, moreover this type of cell is both body-cell and reproductive cell at the same time. In higher organisms this is completely different; a definite distinction between body and reproductive cells exists, the latter are situated in definite organs. This is also the case in plants, however, here it is not so sharply localized as in animals. Hence, if an external influence is to act upon the heritable pre-disposition of the individual, then such an influence can take place primarily by acting upon the body, that is to say, the changed environment causes changes in the body and these penetrate into the reproductive cells (so-called *somatic induction*, action by way of the soma, fig. 193, top left). In the second place, the changed environment can act directly upon the reproductive cells, whereas the body remains untouched and therefore unchanged (*direct induction*, top right). In the third place, the changed environment can wield an influence upon the body-

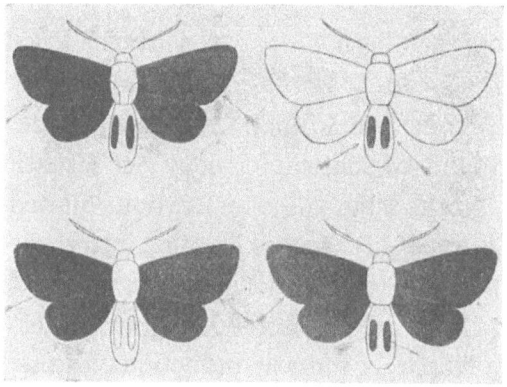

Fig. 193. Schematic representation of the four possibilities, mentioned in text: somatic induction (top left), direct induction (top right), no change in the reproduction cells (bottom left) and parallel induction (bottom right). The dark coloured parts are influenced by means of the altered environment (after ZIEGLER, 1910).

cells, but this influence does not extend further, and *parallel* with this influence a direct *induction* acts upon the reproductive cells. In the latter case, both inductions occur parallel to one another (fig. 193 bottom right). And as a fourth possibility it is possible that the environment only acts upon the body, and the reproductive cells remain unchanged (fig. 193 bottom left).

We have to bear in mind the fact that these four points are possibilities and no more. It is nevertheless important to realize their existence, because it immediately brings to the fore a most difficult question, that of producing conclusive proof in a positive or negative direction.

Alongside a number of apparently positive arguments are a much higher number of investigations in which the results are negative and sources of error could be detected in most of the positive arguments by those who did not believe in inheritance of acquired characteristics. The assumption of direct induction gives its advocates the opportunity of denying the significance of negative proofs on the grounds of the possibility that the reproductive cells were not in the necessary sensitive period at the time of the experiment. So the fight goes on. With regard to this problem in higher organisms (see ZIMMERMANN 1938) we see periodically first advocates, and then opponents in the majority. Since 1948 the point of view of 'inheritance of acquired characters' has been propagated by the Russian government in the form of the so-called 'Michurinism'. Modern genetical research in the sense of MENDEL and MORGAN is thereby made completely impossible on the grounds of political ideology (see Situation 1949 and HUXLEY 1951).

In the main, all investigations quoted by advocates as positive (see SEMON 1911) can be classified into three groups: those due to injuries, those due to conditions of climate, and those due to chemical or physical influences (poisons, X-ray and others). The first may be considered to be purely somatic inductions, the second in some cases as parallel inductions, the third perhaps as direct induction.

From these groups of external factors (by which a change in genotype can be supposed) we may omit injuries because experiments dealing with this exhibit such a large number of weak points that their significance in a critical review is undermined.

It is not so easy to detect errors in the series of climate and nutritive changes and their consequences. The majority of investigations in higher plants and animals belong to this group which is due to the fact that these results can be used in a practical way: acclimatization of animals and plants which have been transplanted in new environments. A problem very important in forestry is that dealing with the origin of seeds of fir used for reforestation. This is very closely related to the problem of the influence of climate upon heritable constitution.

The experiments of SCHUBELER (1862, 1873) have been considered for years as classical proofs in this sphere. These experiments were chiefly concerned with the shortening of the vegetation time (i.e. from sowing to harvest) in the imported German summer wheat after it had been grown for 3 years in Norway. In Norway, the vegetation time amounted to 103 days (1857) at first and successively decreased to 93 days (1858) and finally to 75 days (1859). Half this seed was brought back to Breslau and it turned out that the vegetation time amounted to 80 days in Germany (this was 75 in Norway). Thus for a large part, the artificially produced shortening appeared to have been preserved.

As a once famous zoological counterpart to this, we mention the temperature experiments of E. FISCHER (1901, 1907) with the butterfly *Arctia caja*. FISCHER made the butterflies much darker by refrigeration to —8° C. and this darker colour was retained in the descendants reared at normal temperatures.

Against the unassailability of these and other data as proof that a genotype is changed by influence of the environment, modern genetics has brought a number of arguments to the fore. These arguments completely nullify the proof of the latter experiments.

We must first take the heritable purity of the material into account. The early investigations, formerly considered to be pillars for acclimatization, start with a number of more or less phenotypical similar individuals as opposed to one single individual. These groups of individuals (i.e. populations), may not consist of genotypically similar organisms despite the phenotypical resemblance. In this lies the first source of error. One may not exclude the possibility that the changed environment caused a selection of individuals with a certain genotype, and that all other genotypes were eliminated. Thus we can see that this causes a lasting change in the composition of the population and hence these results do not give a ghost of a proof in favour of a change in genotype of the individual. If genotypical differences occur among the individuals of a population, then in nearly all cases heterozygous individuals will also be encountered, especially if the characters investigated are based upon the action of polymeric factors. Hence, the chance will be very great that selection by a changed environment will also result in a shifting of the population composition. FEDERLEY (1920) showed the significance of polymeric factors in the wings of butterflies by means of a very detailed experiment. This meant that FISCHER's experiments with *Arctia caja* are not evidence for the change of genotype of the individual. Wild specimens often show an extremely great variability in the degree of their light and dark markings. Such a variability is also sometimes encountered in one and the same culture under the same normal circumstances (see fig. 194). The injustice of positive evidence interpreted on the grounds of SCHUBELER's and FISCHER's work as shown by this, can no longer be denied. Genotypes which seldom occur under normal circumstances and are thus very rare, can be highly favoured by other environments so that they finally form the majority of the population.

In this kind of experiment it is prescribed to start with individuals, in which homozygosity is a proven fact. It is also clear that we can start with the offspring of such a homozygous individual if we can show that this offspring forms a pure line (see chapter XX).

Alongside this first source of uncertainty, (that is to say the lack of genetical purity of individuals under investigation) there is yet another source: i.e. the possibility of dealing with 'Dauermodifications'. It is true that the occurrence of this phenomenon of 'Dauermodifications'. may be more frequent in uni-cellular organisms than in higher plants and animals (in which the modification is usually limited to the 'individual' itself or to its clones, that is to say, to an individual considered in a broader sense as being produced by means of asexual propagation), however some clear-cut cases are known to exist in *Drosophila* (JOLLOS 1931a, 1932). Several aberrations were caused in the full-grown animal by the influence of high temperatures (36° C) upon the larvae in the 5–6 days stage, viz., altera-

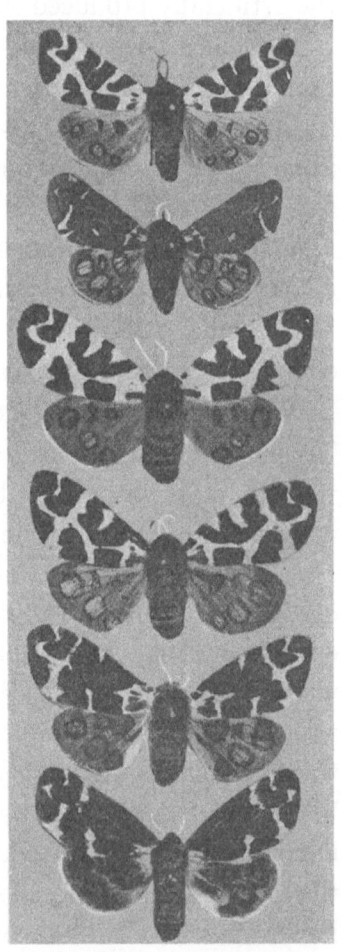

Fig. 194. Above: two ♂ butterflies captured in free nature, one very light and one very dark. Below: ♀ four butterflies from the same brood (after FEDERLEY, 1920).

tions in the wing-muscles, which causes the wings to be folded together in an abnormal manner. This type of animal phenotypically resembles another type, which is known as 'Aeroplane', however, the alterations in this type of fly are based upon genotypical differences. Hence GOLD-SCHMIDT (1935) calls such phenotypically aberrant forms *'phenocopies'*. In most of JOLLOS' cultures, this modification was limited to one generation only, however, in some cases the descendants of these abnormal flies, reared at normal temperature, also showed the same aberration.

From the data given by JOLLOS it can be seen that only abnormal mothers give rise to abnormal offspring. Abnormal fathers do not appear to have any influence. In the F_1 and F_2 generations almost all the flies are abnormal, however, in the F_3, the majority are abnormal but a large number of normal flies also occur. The F_4-generation is composed of only a few flies which show the abnormal characteristic (in only a weakened form). The F_5-generation is completely normal. Because abnormal males produce normal descendants only, we deduce that the genotype of *Drosophila* was not changed, and that the abnormalities are an after-effect caused by the egg-plasm through a limited number of generations. Hence nothing is changed genotypically and the aberration is limited to a 'Dauermodification'.

For an accurate interpretation of the data obtained in studies concerning a change of genotype due to an altered environment it is absolutely necessary that we take account of the following possibilities: Firstly, the presence of different genotypes in the material with which we start the experiments (including heterozygotes) and selection of certain genotypes under the influence of environment. Secondly, an after-effect of the phenotypical change caused by the environment. In the light of these prescriptions it appears doubtful if any seeming evidence which apparently concerns a genotypical change as the result of 'climatic' circumstances can stand up to this test.

Alongside the influence of 'climatic' circumstance (under which is included temperature), we mentioned a third group of investigations. These concern a method by which we can change the genotype of the organism or the germ-plasm by means of chemical substances, radium and X-rays. GAGER (1908, 1911) and MACDOUGAL (1911) were the first botanical investigators in this field. Their studies seem to indicate that a change in heritable constitution, or at least a change in the reproductive cells can be produced by irradiating the ovary or by injecting salt-solution into the ovary of plants. We are however able to point out many weaknesses in these groping experiments and hence we may not consider the work as satisfactory for providing a solution to the problem involved. Insufficient consideration is given in these studies to the genotypical constitution of the original material. We will refer to the summaries of KOERNICKE (1922), IVEN (1925), P. HERTWIG (1927a) and G. HERTWIG (1928) for earlier investigations in this field. In these older studies, the heritable moment of the aberrations still plays an inferior rôle. Of the toxic compounds, which can cause a change of genotype, we have mentioned colchicine (see p. 253). The doubling of the chromosomes set up by the application of this compound can indeed be considered as a *genome-mutation*.

In later investigations, the influence of radiation proved very successful. In connection with this, such material as *Drosophila* and *Antirrhinum* yielded many interesting results. STEIN (1926–1936) studied the influence of *radium* as an agens upon the genotype of *Antirrhinum*. She obtained striking aberrations in the irradiated individuals, which she described as *phytocarcinomes*. Seeds of an *Antirrhinum*-line which had been selfed since 1912 (in which we may consider the heritable purity as proven) were irradiated. The viability of the embryos present in the seeds was considerably reduced, and hence only 10% of the radiated seeds yielded full-grown plants (normally, without radiation 76%). The radiated plants were, for the most part different from the normal type, either dwarf plants or possessing smaller leaf-width, or differed in connection with their shape and colour.

Such 'radiomorphoses', obtained by direct irradiation, are very complex in type. Some parts of the plant show one aberration and other parts show completely different type of aberration. STEIN preliminarily thought that these abnormalities were not heritable, because the descendants obtained by selfing were not identical with the irradiated parental plants. However, in more recent publications of STEIN the opposite conclusion comes to the fore. A microscopic investigation proved that the disturbance in the tissues was not always located in the sub-epidermal layer (from which the reproductive cells are formed) but some plants were obtained in which such a disturbance really did occur in this layer of cells. One of these plants, which showed only slight aberrations, produced an offspring which had the abnormal features, sometimes even to a higher degree than in the irradiated parental plant. Crosses between these obviously aberrant individuals and normal untreated plants yielded almost normal offspring in both reciprocal ways, whereas the next generation appeared to be subjected to a Mendelian segregation.

Despite the fact that the disturbance phenomena are extremely complex, it still seems that in this case a type was produced and isolated in which aberrations are due to radium irradiation and in which the reproductive cells of both sexes, and hence their descendants are also altered in their heritable constitution.

The field of *radiation genetics* was greatly extended by the application of *X-rays*. After the first publications of MULLER (1928) his work and that of numerous followers (see CATCHESIDE 1948, DOBZHANSKY 1936a, GOODSPEED and UBER 1939, HERTWIG 1937, MULLER 1950b, OLIVER 1934, PACKARD 1931, SCHULTZ 1936, STADLER 1936, STUBBE 1937, 1938, TIMOFÉEFF-RESSOVSKY 1934a, 1937, 1940a and Symposium 1941b, 1948, 1950c) has proved convincingly that irradiation can be of importance for the genotype of the individual, or at any rate, for the genotypical structure of the reproductive cells produced by this individual. Alongside all kinds of processes which have a greater influence on the course of a reduction division (for instance 'genome *mutation*' by means of non-disjunction), or by which smaller or greater parts of the chromosomes are involved (*chromosome mutations* i.e. deficiency, translocation, inversion) separate genes also can be influenced by irradiation ('*gene mutations*', '*point mutations*', '*trans-genations*', BRIDGES 1923b). The boundary between chromosome mutations and gene-mutations is not at all sharply defined. In spite of the beautiful material as given in the salivary gland chromosomes, it is still possible that a chromosome defect comprises such a small part that even in these giant chromosomes it is impossible to detect whether the gene itself or its enveloping chromosomal cover is hit by the radiation applied. Earlier in this book (p. 327) we have referred to the main points of mutation investigations in connection with discoveries over the nature of the gene.

The method known as the "ClB-method", first applied to *Drosophila* by MULLER (1928a, b) was so ingenious that all other explanations than only real gene mutation can be ruled out. This technique is as follows (fig. 195): the ClB-females have the dominant Bar-factor in their X-chromosomes (see p. 313) and by means of this it is easily possible to distinguish flies bearing this gene. This bar-factor occurs alongside a long inversion which prevents crossing over between both X-chromosomes (C) and the X-chromosomal lethal gene (1), which when present homozygously prevents the development of the female larvae and when present singly will disturb the development of the male larvae. In the case when males of another strain (not possessing the B-factor) are irradiated (the irradiated chromosomes are indicated by a thick line in the figure) and after this are mated with females of the ClB-strain, then four kinds of flies ($F_1{}^1$, $F_1{}^2$, $F_1{}^3$, $F_1{}^4$) can be expected from this cross theoretically. Since one kind is not viable ($F_1{}^3$ ♂†) we will only find one kind of male fly ($F_1{}^4$ ♂), that is to say all the males which bear the combination ClB are lacking because this combination causes lethality in the male larva.

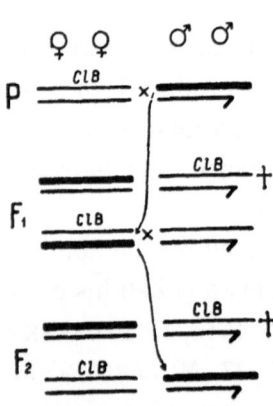

Fig. 195. CIB-method for detection of gene-mutations resulting from X-rays irradition (after TIMOFÉEFF - RESSOVSKY, 1937).

The females originating from this cross will either be ClB over irradiated X-chromosome ($F_1{}^2$ in fig. 195) or normal over irradiated X-chromosome ($F_1{}^1$ in fig. 195). By mating the F_1 ClB-females with normal non-irradiated males, it will be clear that all the males of the F_2 will bear the irradiated X-chromosome ($F_2{}^4$) since the ClB males are not viable ($F_2{}^3$). The females originating from this cross will be (concerning their X-chromosome) irradiated over normal ($F_2{}^1$) or ClB over normal ($F_2{}^2$).

Hence possible gene mutations, which were established in the original parental fly (P ♂) are now expressed in the grandchildren.

The first results MULLER published showed that, on the one hand the offspring of irradiated flies did indeed possess a highly increased mutation percentage when compared with the non-irradiated control cultures (viz., 16·3%, instead of 0·11%). However, the majority of these mutants (almost 90%) were lethal, and hence the cause of death of these flies appears dubious.

However, 10% of the flies are still viable, and therefore a corrected mutation percentage of 1·6% (to 0·11% in the control culture) can be calculated. Repetition of these experiments showed that there is a striking qualitative agreement between spontaneous mutability and provoked mutability, and one can establish parallels between both these types of gene mutation.

However, on the other hand, differences of a quantitative nature between the several gene-types seem to occur. In some genes, the mutation frequency can be greatly increased by irradiation, but for other genes this may not be the case, or only a slight increase may occur. In this connection, the investigation concerning the behaviour

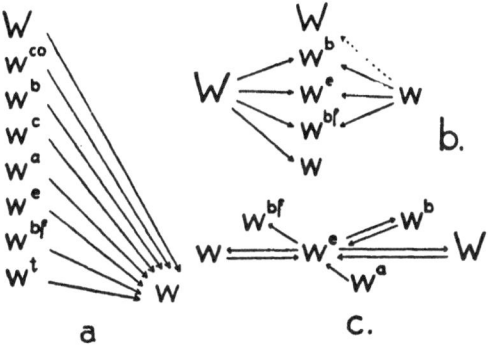

Fig. 196. Mutations in a series of multiple
alleles concerning eye colour of *Drosophila*
by means of X-ray irradiation
(after TIMOFÉEFF-RESSOVSKY, 1937).

of a series of multiple alleles under irradiation are of particular interest. In *Drosophila*, a multiple series of alleles concerning the eye colour is known of decreasing intensity, which is composed of at least nine members, viz., wild colour (dark red, W), decreasing to coral (w^{co}), blood (w^b), cherry (w^c), apricot (w^a), eosine (w^e), buff (w^{bf}), tinged (w^t) and finally white (w). Not only could it be shown that these genes can mutate into lower members of the series of multiple alleles and even to the end type, white (the wild colour (W) especially, mutated into lower members), but also that *back-mutation (reversion)* is possible. The w-gene can revert into w^{bf}, into w^e and into w^b, but not directly into W. Nevertheless, this latter case was possible in two steps, that is to say, firstly by reversion from w into w^e, and secondly into W from w^e. Figure 196 gives a summary of these possibilities. In spite of the fact that the mutation frequency of the dominant type into the recessive type is much higher (almost ten times higher) than that for the reverse process, it seems however to be an ascertained facr that back-mutation of the recessive gene obtained by irradiation into the original dominant type is possible. From this, one may draw the conclusion that X-irradiation does not destroy anything, or rather need not destroy anything and that a reverse process can be produced. These reversable processes can be explained as being due to a shift of atoms and this proposition can again be related to our modern conceptions concerning the nature of the gene. Analogous investigations with regard to the study into the effect of X-rays, radium, ultraviolet light and temperature shocks are found in plants; maize, oats, barley and wheat (STADLER 1928, 1936), *Nicotiana*-species (GOODSPEED 1929, 1937), *Antirrhinum* (STUBBE 1930–1938), the liverwort *Sphaerocarpus* (KNAPP 1935, 1937b). After irradiation of young plants, normal descendants were usually obtained, however their offspring yielded a rather considerable number of 'mutants' in the F_2-generation. Irradiation of the inflorescences, that is to say, of generative tissue gave rise to a great number of abnormalities even in the F_1-generation. These aberrations either concerned alteration of shape (which reminds us of those obtained by STEIN, known as radiomorphoses) or chromosome mutations or definite genes. We will classify the latter as gene mutation. The mutants obtained were for the greater part breeding true to type and the mutants were always recessive in their mutated characters.

MULLER's conclusion that the frequency of the mutants produced is exactly proportional to the energy of the doses absorbed will immediately raise the question as to whether or not natural radio-activity may be a cause of the origin of gene mutations which occur spontaneously in *Drosophila* and other organisms. Some experiments (BABCOCK and COLLINS 1929; HANSON and HEYS 1930) carried out at places where the natural radio-activity is considerably higher, than in the laboratory, seemed to show an increase in mutation frequency in the cultures involved, however, this increase was not more than 2˙5 and 2˙09 × the probable error so that statistical certainty could not be obtained. MULLER and MOTT-SMITH (1930) in a later publication postulated, on the grounds of calculations, that perhaps natural radio-activity does have some influence, but is certainly not the sole source of explanation for the frequency of spontaneous mutations.

This will certainly not exclude any possibility that spontaneous mutations are caused by natural radiation. It is clear that natural short wave irradiation can hardly be held responsible, however the proposition can be put forward that cosmic rays may play a rôle in the mutation. Such a hypothesis is hardly to prove under experimental conditions because we are not able to construct a place on the earth which is free from these kinds of rays. In 1936 FRIESEN, in Russia, and at the same time JOLLOS, in the United States, tried to increase the mutation frequency by sending a *Drosophila* culture attached to a hydrogen balloon into the upper atmosphere. FRIESEN obtained an absolutely negative result, however, JOLLOS' results pointed in a positive direction. Nevertheless, the question still remains as to whether the magnitude of intensity of the cosmic rays (110 times that at the earth's surface) was the main source; also the great decrease in temperature could also be held responsible. Therefore JOLLOS (1937–1939) repeated his experiments in a constant temperature box under 18 mm lead. He performed these experiments in the mountains of Colorado (4000 m) where the cosmic irradiation is five times that of sea-level, and compared the results with those obtained in Madison without lead. The first attempt showed that the mutation frequency in the 'mountain' cultures was three times as high as the control experiments in Madison. JOLLOS was furthermore able to calculate, that only about 10 percent of the '*spontaneous*' mutations obtained in Madison could be ascribed to the influence of cosmic rays. Positive results were also published by RAJEWSKY, KREBS and ZICKLER (1936) with mould cultures of *Bombardia lunata*. This field of investigations is still in its early days of development but undoubtedly there are many possibilities in this direction, which may perhaps clarify an explanation for the occurrence of '*spontaneous*' mutations.

Alongside the great theoretical significance of this problem there is also a practical side, that is to say, *germ damage* as the result of *alcohol* and other *toxins*, as a social problem. It is obvious that the possibility of an effect on the genotype by such toxins has for a long time drawn the attention of those scientists, who deal with the social and practical field of genetics. An enormous amount of literature exists concerning this problem and we would like to refer for this to the reviews of P. HERTWIG (1927a, 1940), of FRETS (1930, 1931) and of SCHUBERT and PICKHAN (1938). However, there is also much controversy over this point, just as much as over the theoretical aspect of the problem. MACDOWELL and LORD (1927), on the one hand, came to the conclusion that alcoholism in mice does not have any influence on pre-natal mortality and the sex ratio. On the other hand BLUHM (1930), in an extensive study, concludes that there is positive damage by means of the influence of *alcohol*. BLUHM investigated the duration of life, capability for growth, fertility and the occurrence of deformations. With regard of the first three groups of phenomena the results were clearcut and positive, however no influence of alcohol could be established for certain deformations. Alongside these alcohol studies BLUHM (1932, 1938), obtained analogous results with *ricine* as an agent for germ damage in mice: immunization of male animals results in over-sensitivity of the offspring from this immune male animal. The descendants from a mating between a normal female and the son of a ricine-immune father are remarkably more sensitive to the toxin than are the offspring from the mating between normal males and daughters of a ricine-immune father.

From this it is clear that the father, that is to say the male individual, plays a more important rôle in the transference of the sensitivity to toxins than does the female individual, which is in accordance with the results of the alcohol investigations.

Undoubtedly these positive data are of extreme importance, however we are still left with the question of germ damage as to whether an after-effect (*Dauermodification*) or a genotypical alteration underlies the phenomena obtained. BLUHM's studies are not far enough advanced to enable us to draw a conclusion to this problem, however, it appears acceptable, that genotypical alterations have taken place but a final proof has not yet been given.

Because of this, the question became urgent (after the introduction of X-ray treatment into medicine) as to whether the results obtained in biology were valid in the case of man, i.e. if real germ damage occurs after irradiation. Although the data is still incomplete, it appears that X-ray irradiation of the reproductive system and its surrounding tissue introduces dangers with respect to the genotype, and that people who frequently work with X-ray apparatus are subject to these dangers (see HOLFELDER and VOGT, 1938, HERTWIG 1940, MULLER 1950a). The practical side of the problem became extremely important after August 1945 when large numbers of people became subject to atomic energy. Because this form of warfare can have such far reaching and serious complications, the American Committee on Atomic Casualties set up a special section composed of geneticists (see NEEL, 1947, MULLER, 1948).

As opposed to this possibility, viz., that the genotype of an individual changes as a result of a change in environment, there is a second possibility to be considered. This is that the change in the heritable predisposition in an organism changes *spontaneously* without an external cause. This autogenic change was considered highly probable around 1900, however, experimental facts to prove this hypothesis were lacking. In 1900, DE VRIES' concept of a sudden occurrence of *spontaneous mutations* gained ground. The enormous amount of data that he himself obtained from experiments with *Oenothera* helped considerably. However, after the rediscovery of the Mendelian rules (in which DE VRIES was involved) and after their recognition, BATESON (and SAUNDERS, 1902, p. 153) drew attention for the first time to the possibility that *Oenothera Lamarckiana* could not be a pure line, but was a hybrid. Undoubtedly time has shown BATESON's prophetic statement to be true, that *Oenothera Lamarckiana* was a hybrid though not a normal Mendelian hybrid (see p. 179). It is a hybrid that gives rise to two or more types of reproductive cells, which are mutually different. The first support was found in the study of HONING (1909, 1911), and the proof was completed by HERIBERT NILSSON (1915b, 1920b) and RENNER (1917, 1918a, b and c) The facts of DE VRIES were not affected by this in the least. The only thing that changed was the explanation following the facts. However the *Oenothera Lamarckiana* investigations can no longer be used as the basis for a theory concerning the problem of spontaneous change of the genotype of an organism. No evidence exists at the moment for such a change in heritable disposition. Today, the term 'mutant' for the occurrence of new forms and the term 'mutation' for the process he studied in *Oenothera* are used in a completely different sense. It is therefore necessary to describe exactly what we will mean today by these terms.

Clearly we may not consider types of plants and animals which occur suddenly in nature (and which were hitherto unknown) as mutants. It is also dangerous to classify such new forms, when they occur in laboratory cultures, as mutants without making intensive study first. DEMEREC (1931) made some interesting investigations into the occurrence of purple-speckled flowers in *Delphinium ajacis* in which he supposed '*mutable genes*' to occur. There is no evidence for accepting the occurrence of these mutable genes, since the homozygous nature of these purple-speckled individuals is doubtful, and since he did not investigate very thoroughly the internal metabolism which changes the phenotypically pink-coloured flowers into the phenotypically purple-coloured ones. In order to determine spontaneous mutation it is absolutely necessary to start with homozygous material. Phenotypic constancy over a large number of generations does not prove absolute homozygosity. We will take an example where characteristics are expressed by polymeric factors: the absolute recessive of 5 poly-meric factors is formed only once in 1024 individuals, and in the case of 10 factors it occurs once in 1048576 individuals. Absolute ho-mozygosity, with regard to the characters underlying the characteristic investigated, is extremely difficult and almost impossible to prove. Indeed though we are forced to admit that the number of genes which supposingly occur as spontaneous mutations in *Antirrhinum* and *Dro-sophila* is impressive, the proof of genotypical homozygosity of the material we started off with is yet not given and we therefore may have doubts as to whether the phenomenon is due to mutation. Therefore summaries of STUBBE (1933) and DEMEREC (1935) of the data, which they consider to be evidence for the proposition of labile genes contain, along-side very important information, a still unproven interpretation. Lack of acquaintance with the real background of genic behaviour should never be a reason for invoking the help of the panacea of *'lability'*.

However, there is a way by which the possibility of spontaneous mutation can be established. In chapter XII we saw that haploid in-dividuals sometimes occur in plant species; in the cases of these haploid individuals giving rise to offspring, we would again obtain diploid descendants. Theoretically speaking these diploids, that is to say ex-haploids, can be considered as perfect homozygotes. Therefore their genotypical purity appears to be acceptable, however we have to be aware of the possibility of occurrence of *autosyndesis* as pointed out on p. 268 (autosyndesis is the phenomenon of syndesis between chromoso-mes of one and the same haploid set) and as a consequence of this, a heterozygous pair of alleles may be formed. BLAKESLEE, MORRISON and AVERY (1927) described gene mutations which occurred in a diploid stock originating from a haploid stock of *Datura*. We may perhaps hold these cases as proven gene-mutations, if the course of the repro-duction processes was without disturbance. One may also expect that other diploid offspring originating from haploid individuals can yield important information in this respect.

The complexity of questions which underlies a change of the genotype will still hold our attention for a long time. This will be due on one hand, to the very important theoretical and practical consequences which go with the answer of this question; and on the other hand to the very difficult character of an irrefutable interpretation, since one has to take into account so many possibilities (after-effects such as Dauermodification, selective preference of particular genotypes present earlier in the genetical history of the individuals, segregation of complex-heterozygotes, true mutations). The problem of mutation is still open.

CHAPTER XX

CHANGES OF HERITABLE TENDENCIES IN PURE LINES AND POPULATIONS

The distinction JOHANNSEN made in 1903 between what he called *pure lines* (reine Linien) and *populations*, was indeed one of the sharpest theoretical distinctions ever made between concepts in genetics, in its struggle to become a more exact science. It is true that his definition of the 'pure line' concept changed somewhat in the course of the years. In 1903 (p. 9) he formulated the concept as 'individuals which descend from one single self-fertilized individual', in 1909 (p. 133; 1926, p. 178) as 'all individuals which descend from a single absolutely self-fertilized homozygous individual' and in 1915 (p. 606): 'a pure line comprises all descendants of a single absolutely self-fertilizing individual which does not have a hybrid nature of its own'. However the 1903 description of the concept only became more detailed and exact by these alterations. The individuals of a pure line are all genetically identical when compared with the parental individual. On the other hand a population is a mixture of individuals which are different in heritable tendencies. These differences can lie in the fact that some factors are homozygously present in all individuals and the remainder heterozygously present. It is obvious that such a situation can be found in a population of a district or region as well as within a taxonomical group (species or race) of plants or animals.

If one defines the concept of pure lines as sharply as JOHANNSEN did in later years, then one has to make two demands on the individual from which the pure line originates: 1st, the individual has to be an absolutely self-fertilizing individual and 2nd, the individual may not possess a hybrid nature, that is to say, it has to be homozygous in all its heritable factor-pairs. There is no doubt that these are very high demands and there is really no organism which fully meets these requirements. No plant is absolutely self-fertilizing; peas are self-fertilizing to a high extend but when they are free flowering we cannot exclude fertilization by insects and hence pollen of a plant of another genotypical tendency may be introduced, Also, beans are strong self-pollinators but still it frequently occurs that the flowers are visited by insects and in such a case pollination with other than their own pollen takes place. Wheat is a rather good self-pollinator, but due to wind cross-pollination may always occur. Broad-beans are pollinated to a high degree by pollen of other broad-beans and for maize this figure is almost 100% (see SIRKS 1923). It is true that among animals bisexual species occur, but absolute self-fertilization is very rare. Moreover the definition cannot be applied to dioecious plants neither to most of the animal species because self-fertilization is excluded in all these species.

The second demand: that the individual from which a population descended has to be homozygous in all its heritable characteristics is as difficult to realize as the preceding demand. How can we judge such a homozygosity bearing in mind the possibilities of the occurrence of *cryptomery* and of *polymery?* (see chapter VII). One may therefore not wonder at the fact, that in general, authors did not and do not take these demands fully into consideration in their genetical publications. However such a procedure is indefensible and irresponsible. Apart from the few critical contemplations such as the paper by HERIBERT NILSSON (1930a), no one really draws any attention to this point and this may be a focus for doubting the reliability of many investigations carried out in this field.

In its extreme form we may accept the concept of 'pure lines' as a theoretical one, however, one has to remember that such a concept is almost a fiction in the life of organisms and almost impossible te realize. From the point of view of genetics it is a matter of indifference whether a number of individuals, which are all homozygous in all their heritable factors and identical in their heritable tendency, descend from one single individual by means of self-fertilization or from more individuals, which are mutually identical and homozygous by means of mutual cross-fertilization. If we omit the self-fertilization as demanded by JOHANNSEN and describe the pure line 'as the total of all individuals which are homozygous in their factors and whose genotype is mutually identical' (called by LEHMANN 1914, an *isogenic unit*) then we impart to the concept 'pure lines' a more practical viability. Though such pure lines really exist the decision of their homozygosity

is still hard to make and moreover such a decision will be always somewhat doubtful if we have to take into account the possibility of
occurrence of cryptomery and of polymeric factors. In free living
organisms it is always very
doubtful if these individuals
belong to pure lines, even in the
above broader sense.

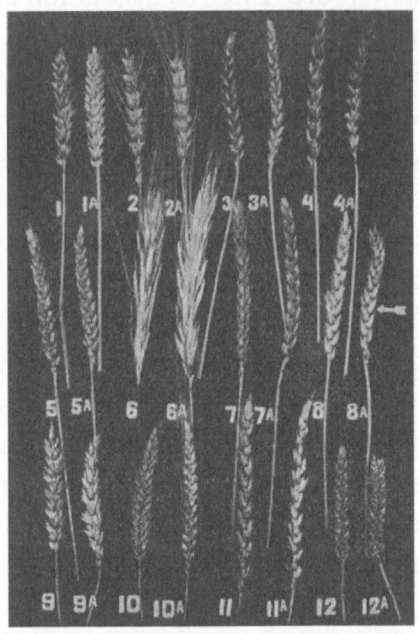

The judgment for purity, that
is to say for homozygosity can only
be carried out by means of an experiment. This experiment consists of the rearing of individual
offspring by a skilled investigator,
whose perceptive faculty is trained
with regard to these small observable differences among the material he will obtain from his breeding experiments. Therefore we
have to look for 'pure lines' among
cultivated plants and domestic
animals. It is obvious that we will
have a good chance if we look first
among the self-fertilizing plant
species, which have been handled

Fig. 197. Ears of races of wheat from
about 1850 and of the same races from
1910 (after DE VILMORIN, 1911).

by growers for a long time. A typical example of the existence of such
pure lines was given by DE VILMORIN (1911): A collection of dried
wheat ears made by LOUIS DE VILMORIN round 1850 was found by his
grandson PHILIPPE DE VILMORIN in 1910. The latter was therefore
able to compare the ears of 1850 with the ears of the same stocks as
cultivated in 1910. It turned out that not the slightest difference
existed. Figure 197 gives a picture of the ears of the same stock in 1850
and in 1910. Under number 1, for instance, an ear of the stock 'Webbis
Free trade' of 1854 and under 1A an ear from the same stock of 1910.
Number 2 is an ear from the race 'Saisette de Tarascon' cultivated
in 1853; 2A is an ear from 1908. From this example it becomes clear
that these wheat races are invariable 'pure lines' which always remain
pure in their offspring.

Is such a pure race now subject to variability or is it fixed and constant? The example of the wheat races given has already shown that nothing changed in these stocks over a period of sixty years and therefore one is permitted to conclude that the genotype of these races was really invariable. However, is alteration of the heritable tendency excluded in a pure line? The answer to this question is closely related to the answer of the question posed in the previous chapter: if one may suppose that the genotype of an organism which is homozygous in all its factors can be spontaneously altered or by environment then the same conclusion is valid with regard to the genotype of pure lines. Apart from this there is a second cause which can change the genotype of a pure line, so that these lines can be no longer considered as pure. This will be the case if generative cells with a different heritable constitution are imported from outside, that is to say cross-fertilization of one or more individuals belonging to this pure line takes place with non 'line-related' individuals. From this procedure among offspring of a pure line, individuals will arise which show a mendelian segregation etc. It is evident that under these circumstances the heritable tendency present in the pure line, is altered and with this phenomenon, the line ceases to be a pure line, that is to say the line becomes a population.

Besides these few pure lines which occur only among cultivated races and make up a small minority in the total of races, we find among the cultivated and in wild groups of organisms, in which reproduction is either by means of self-fertilization or by mutual mating. These groups can be called 'populations'. Such populations naturally comprise a number of organisms of different genotypical compositions; heterozygotes as well as homozygotes (i.e. for the same factor) can belong to a population. The simplest case is a population composed of mutually identical, but heterozygous, individuals. Nevertheless such populations show immediately the lot to which all populations are submitted, that is to say, the offspring of a number of mutually identical, but heterozygous organisms, is no longer identical to the parents but can be heterozygous and homozygous in certain characteristics. To take a simple case: we have a population of 100 rose flowering *Mirabilis* plants which are heterozygous in one pair of alleles, and whose segregation in the offspring we already know from a previous chapter. The offspring of these 100 pink *Mirabilis* plants are not all pink: among them there will be homozygous red and homozygous white plants. Hence the heritable tendency of the population is changed.

The causes of changes, which occur in every population in successive generations lie in two directions: a spontaneous change and also a change due to causes outside the population (i.e. environment), can affect the genotype of a population.

The changes, which take place spontaneously in the genotype of a population (that is to say due to the composition of this population into homozygotes and heterozygotes), are dependent on the manner of reproduction of this population. We are able to distinguish the following cases: absolute self-fertilization (*autogamy*), mating of brothers with sisters ('*sib-mating*'), mating of parents with offspring, mating of cousins in different degrees of relationship, mating of phenotypically identical individuals (that is to say AA with AA and with Aa; aa only with aa), absolute cross-fertilization (*allogamy* or *panmixy*) and autogamy mixed with allogamy from which a definite ratio can occur between the numbers of autogamous and allogamous fertilization. Further, all these possibilities can occur in populations which comprise individuals which differ in one or more characteristics. In this case these factors can still be mutually independent or more or less strongly linked. We will not discuss the details here, but only show by some examples, the methods used in these investigations. For more details we will refer to the literature dealing with this subject (DAHLBERG 1943, 1947; EAST and JONES 1919; P. HERTWIG 1927 b; HEUKELS and BONE 1915; HOGBEN 1946; HULTKRANTZ and DAHLBERG 1927; ROBBINS 1917–1918; SIRKS 1923, 1949a; WAHLUND 1928; WRIGHT 1921, 1931, 1939a).

The simplest case is that in which we deal with absolutely autogamous organisms which reproduce themselves by self-fertilization generation after generation. If we start off with a population composed exclusively of monohybrid organisms (Aa), and if we calculate the ratio of the groups AA, Aa and aa originating in the following generations, then we are able to summarize the results as follows: the percentage of heterozygotes in the whole population present, decreases during the course of the successive generations more rapidly if the organisms are mono-heterozygous and slower according as to whether more characteristics have a hybrid nature. This is represented in fig. 198; on the vertical axis the percentage of heterozygotes present in the population is indicated, on the horizontal axis the number of the generation is given; the numbers by the graphs show the number of factors which are present in a heterozygous state. One may make this clear by calculating the simple case of a monohybrid. The F_1-generation of a cross AA × aa comprises (as we saw earlier) only Aa-individuals and by self-fertilization (selfing) an F_2-generation is obtained of 25% AA,

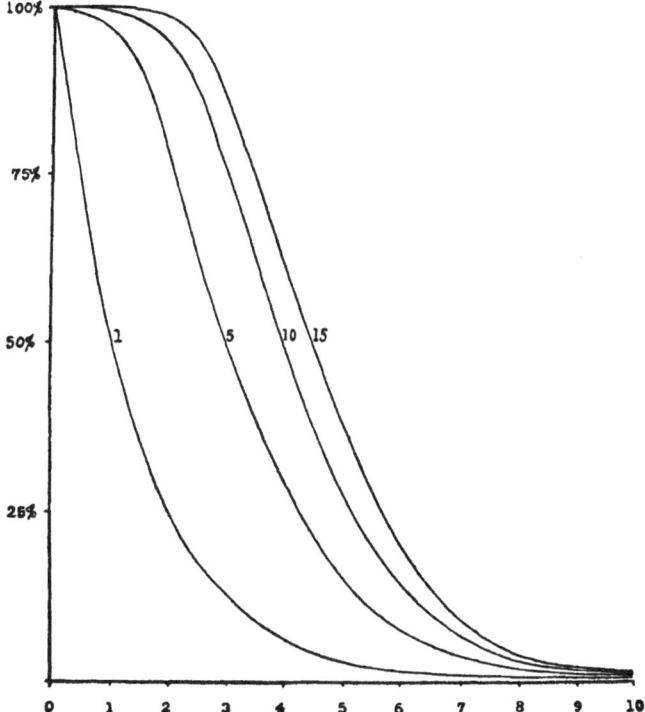

Fig. 198. Percentages of heterozygous individuals in each selfed generation when the number of allelomorphs concerned are: 1, 5, 10. 15 (after EAST and JONES, 1919).

50% Aa and 25% aa, that is to say 50% homozygotes and 50% heterozygotes.

If one supposes that each individual has 4 descendants, then the course of the following generations can be summarized as shown on next page.

The percentages of heterozygotes in F_1, F_2, F_3 etc. are successively 100; 50; 25; 12·5; 6·2; 3·1; 1·6 and subsequently decrease so much that in the 10th generation practically no heterozygotes occur. The general

formula for the number of heterozygotes $= \dfrac{1}{2^{n-1} - 1} \times 100\%$.

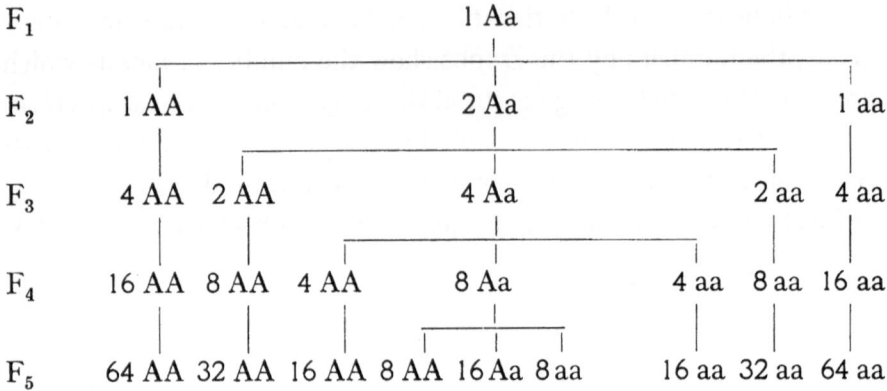

If the original F_1-individuals are heterozygous in more than one factor, then the same state will eventually be reached, however, this process takes much more time, as becomes evident from fig. 198. (One has to keep in mind, that in this figure 0 stands for the F_1-generation proper and that 1 stands for the F_2 and 2 for the F_3-generation, because EAST and JONES speak of the F_2, as the first segregating generation).

In the case when one does not apply absolute selfing, but (as in dioecious plants and animals) applies a systematic sister-brother mating (*sib-mating*) then the same result will finally be reached, but much more slowly. The general formula for the percentage of heterozygotes in the nth generation is:

$$\frac{n^{th} \text{ term of the Fibbonaci series}}{2^{n-1}} \times 100 \text{ or also}$$

$$\frac{(1 + \sqrt{5})^n - (1 - \sqrt{5})^n}{2^{22-n}\sqrt{5}} \times 100.$$

The comparative table (p. 475) in which the percentages for both methods of reproduction are given shows this very clearly.

The other extreme case we find in an absolutely free mutual mating of all individuals present (*panmixy*). Assume in this case also that the F_1 is composed of 100% Aa-individuals, hence the F_2 has a composition of 25% AA, 50% Aa and 25% aa. If each AA-individual can mate with another AA-individual as well as with Aa- and with aa-individuals, and likewise the Aa's and the aa's; the result can be summarized as follows, providing that each mating results in a progeny of 4 individuals:

	AA	Aa	Aa	aa
AA	4 AA	2 AA : 2 Aa	2 AA : 2 Aa	4 Aa
Aa	2 AA : 2 Aa	1 AA: 2Aa : 1aa	1AA : 2Aa : 1aa	2 Aa : 2 aa
Aa	2 AA : 2 Aa	1AA : 2Aa : 1aa	1AA : 2Aa : 1aa	2 Aa : 2 aa
aa	4 Aa	2 Aa : 2 aa	2 Aa : 2 aa	4 aa

If true panmixy occurs, that is to say all possible matings occur in the same ratio, then the sum of all individuals originating in the F_3-generation will be: 16 AA + 32 Aa + 16 aa or put into other words: in the F_3 the ratio 1 AA : 2 Aa : 1 aa again occurs, i.e. the same ratio we already found in the F_2-generation. By continuing panmixy this

Generation	Selfing			Brother × sister		
	AA	Aa	aa	AA	Aa	aa
1	0.00	100.0	0.00	0.00	100.00	0.00
2	25.00	50.00	25.00	25.00	50.00	25.00
3	37.50	25.00	37.50	25.00	50.00	25.00
4	43.75	12.50	43.75	31.25	37.50	31.25
5	46.88	6.25	46.88	34.38	31.24	34.38
6	48.44	3.12	48.44	37.50	25.00	37.50
7	49.22	1.56	49.22	39.85	20.30	39.85
8	49.61	0.78	49.61	41.80	16.40	41.80
9	49.80	0.40	49.80	43.36	13.28	43.36
10	49.90	0.20	49.90	44.63	10.74	44.63
11	49.95	0.10	49.95	45.65	8.70	45.65
end	50.00	0.00	50.00	50.00	0.00	50.00

ratio will never change: in all the following generations the number of the individuals AA : Aa : aa shows the characteristic ratio 1 : 2 : 1 (HARDY 1908).

With regard to this a fundamental antithesis exists between autogamy (continued selfing) and panmixy (continued mutual fertilization of all types on hand). In the first case the percentage of heterozygotes decreases constantly until it reaches nought as we already saw. Therefore the population is split into groups exclusively of homozygotes. By unrestricted mutual mating of all individuals, the characteristic F_2-ratio of the groups of genotypes of which the population is composed is maintained in the successive generations.

Between absolute autogamy and complete panmixy lie methods of reproduction, in which both are combined. Next to autogamy, allogamy (cross-fertilization) plays a rôle. HEUKELS and BONE (1915) calculated this case mathematically. We borrow from their study the case of a population in which x self-fertilizations take place against y cross-fertilizations. Their conclusion is that 'If the plants from a population which differs only in one pair of characteristics are exposed through many generations to x self-fertilizations towards y cross-fertilizations, then the number of homozygotes will finally be $\dfrac{x + y}{x + 2y}$ and the number of heterozygotes $\dfrac{y}{x + 2y}$ '(1915, p. 303). That is to say, if for instance $x = y$, (hence allogamy occurs as frequently as autogamy) then the end will be that 2/3 of the total population are homozygotes and 1/3 are heterozygotes. The following nubmers, i.e. percentages for the three groups AA, Aa and aa in the successive generations, show this very clearly:

		AA	Aa	aa
	F_1	0.00	100.00	0.00
Allogamy	F_2	25.00	50.00	25.00
occurs as	F_3	31.25	37.50	31.25
frequently as	F_4	32.81	34.38	32.81
autogamy	F_5	33.20	33.60	33.20
	F_6	33.30	33.40	33.30
	F_∞	33.33	33.33	33.33

After some generations an equilibrium is gradually reached in which 2/3 are homozygotes and 1/3 heterozygotes.

For the remaining ratios between x and y, we refer to the graphical representation of figure 199 in which the theoretical numbers of the dominant A-type (AA + Aa) in the successive generations for $y = 0\%$ $y = 10\%$ etc., are given.

Such theoretical mathematical derivations are important from the point of view of mathematics but from the point of view of biology they only form a working hypothesis (see FISHER 1949a) and may lead to experimental investigation or may give points of contact with other biological problems. In the first place we will consider the theorem, that prolonged selfing will lead to a break up of the population after a certain number of generations into a number of separated 'pure lines', that is to say, into groups of homozygotes. This will also be the case if we force plants to self-pollination, although the organisms are cross-fertilized in the normal course of nature. For instance: maize plants which are always cross pollinators (by wind) and are therefore exposed to fertilization by pollen of neighbouring plants can, by means of artificial methods, be pollinated with their own pollen and hence, change from an allogamous manner of fertilization to an autogamous one. The obvious consequence will be that after some generations the offspring loose a great part of their multiformity and break up into some lines of homozygotes. This

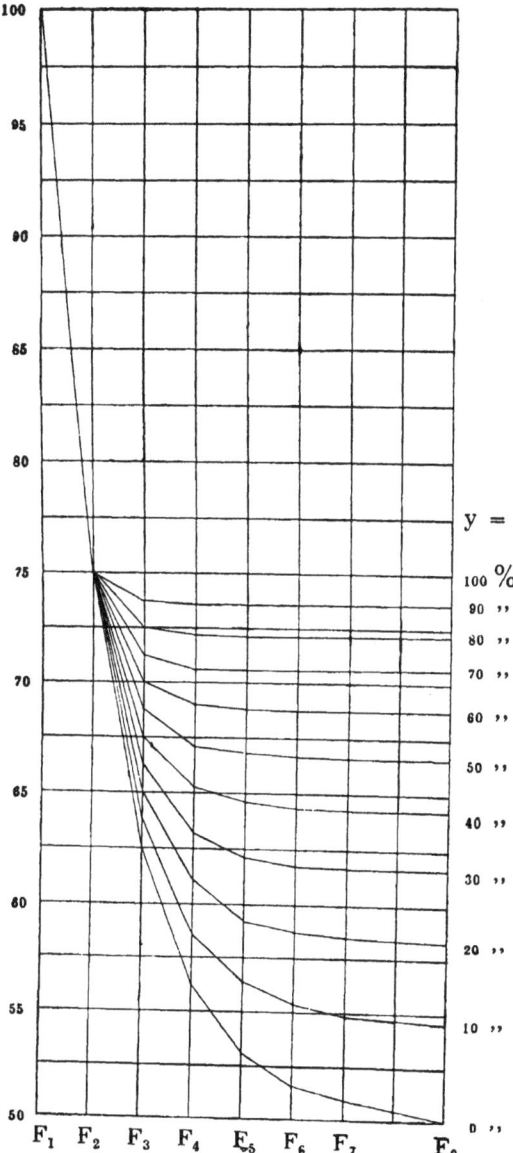

Fig. 199. Course of the percentage of dominants in the successive generations of a monohybrid by y% cross-fertilization (after SIRKS, 1923).

forced selfing of plants which are self-pollinators by nature, is a process which forms one of the most important problems from a biological point of view, that is to say the problem of harmful or non-harmful results from *inbreeding* (EAST and JONES 1919; FEDERLEY

1928). Inbreeding is the forced fertilization of the egg-cell with a male reproductive cell, which is produced by the same or by a very related organism. For centuries this was already one of the most urgent problems in applied genetics, in the practice of the breeder and grower, as well as in medicine; that is to say the problem of whether the mating of closely related individuals (*consanguineous marriages*) is harmful and objectionable to the offspring.

The important consequences of this question have been the stimulus for investigations into this question which had the purpose of finding a direct answer.

At first nearly all investigators came to the same conclusion, that prolonged inbreeding in the rat leads to retrogression in vitality and increase in sterility. However in all these experiments one can indicate sources of errors and in an extensive study of KING (1918) this problem was cleared up to a certain extent. She was able to give an answer to this question with the accuracy required by genetics today. Her answer is completely satisfactory and lacks the vague uncertain contemplations of earlier investigators.

The experiments of KING started with a litter of four almost normal albino rats (*Mus norvegicus*): two males and two females. From these females both lines A and B originate of which over 25 generations were mated constantly by mating two animals of the same litter.

To start with, almost all females of the same generation were used for breeding but after the sixth generation twenty young ♀ animals were selected out of a thousand which were used as first mothers.

In the beginning KING obtained the same results as had already been found by other investigators i.e. less viability, dwarf-forms, miscarriages, increasing sterility etc. However it was striking, that in the great rat cages, in which free pairing could occur, these phenomena also occurred and they disappeared after a radical alteration of the diet. It is possible that there are two reasons for the disappearance of these degenerations i.e., the change in diet and 2ndly elimination of strains which are genetically weak. The first cause was held as most important by KING and perhaps rightly so. The extraordinary importance of this series of experiments is that it became conclusively evident that inbreeding need not be harmful, however one cannot deny, that in many cases disadvantageous results are involved. For instance it is not true that prolonged inbreeding need not always cause a decrease in bodyweight. Figure 200 was borrowed from a preliminary communication of KING (1916); both lower rats are the first parents of an inbred family and both upper, which not yield to the others in size, were two rats originating from the twenty-fifth inbred generation of the offspring of these first parents.

None of the physiological characteristics of the first parents which were investigated, were decreased in the offspring after inbreeding; this was also the case for fertility. Figure 201 gives a representation of

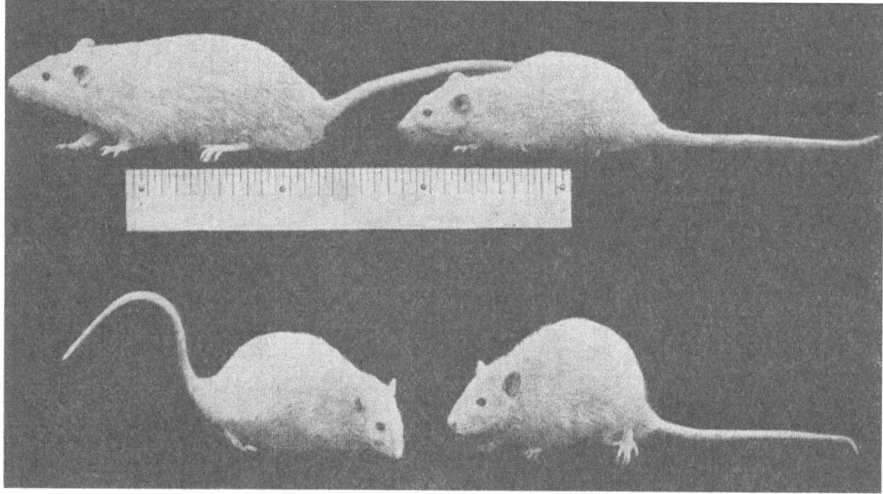

Fig. 200. Below: both ancestral parents of the rats above, which are descendants in the twenty-fifth generation obtained by means of inbreeding. The weight of these descendants is certainly not less than that of their ancestral parents (after KING, 1916).

this. In the beginning, the average number of animals per litter was 6·8; in the second generation it remained the same, in the third generation it decreased to 5·0. It would seem that we deal here with a harmful result of inbreeding, however this premature conclusion is incorrect, because after the third generation an increase of the fertility occurred. In the 4th generation the fertility amounted to an average of 6·5 animals per litter, in the 5th generation this average was 7·2 and after that generation the average remained almost constant up to and including the 25th generation (highest average 7·7 lowest 6·9). Hence there we find confirmation of the view that inbreeding does not necessarily have detrimental results.

The explanation of these results is in complete accordance with the theorem that absolute inbreeding leads to an increase of homozygosity. The material with which KING started was rats which had been reared over a long period by rather close blood related matings, which had reached a rather high degree of homozygosity. The inbreeding experiments which followed increased this purity and eliminated the small differences still existing. That there were differences in the beginning is evident from the fact that the strains A and B were not

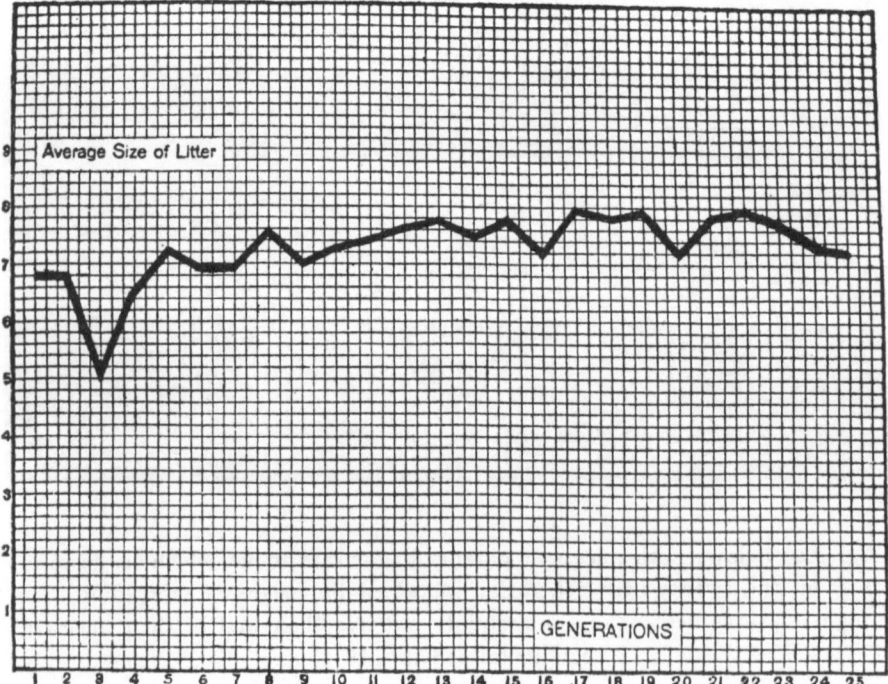

Fig. 201. Graph of the average number of the litter over twenty-five generations of inbreeding (after KING, 1918).

identical in all points: animals of strain A were slightly more fertile, somewhat earlier in puberty, and lived to a slightly older age than did the animals of strain B.

Important analogous results with guinea-pigs were obtained by WRIGHT (1922), WRIGHT and EATON (1929). Their cultures comprised of more than 34,000 and some inbred strains in their cultures remained on the same level with regard to vital properties, as did also the non-bred control animals. Other families decreased slightly with regard to these properties, but after ten generations reached a further stable 'inbred minimum'. Homozygotes therefore need not always be the weaker or less viable when compared with heterozygotes. It was observed in numerous experiments, that plants and animals which are obtained by cross-breeding, show a stronger growth than the parents, which belong to purer lines. SHULL called this 'heterosis' (cf. GOWEN, 1952). Well defined examples of such phenomenon are shown in studies of corn by JONES (1918, 1925) and also SHULL, EAST and HAYES, who originated the work in this field. We can explain the phenomenon by a single picture. In figure 202, the yields of two inbred stocks of corn are pictured from right to left,

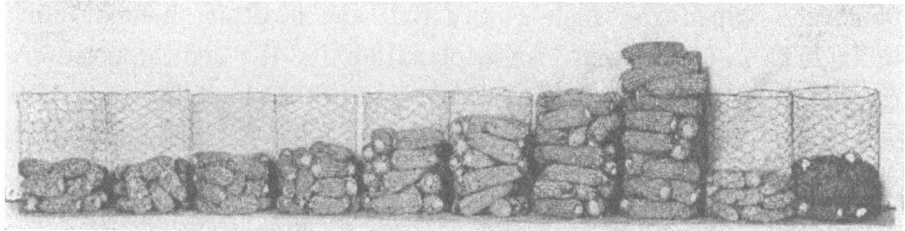

Fig. 202. Two inbred corn stocks (right), their F_1 upto and including F_8 after selfing (from the right to the left) (after JONES, 1925).

after they reached their 'inbred minimum'; then follow successively, the yield of the F_1-plants and that of the F_2 etc., up to and including F_8. It is not possible to deny that a sudden increase in yield in the F_1-generation occurs; the decrease after inbreeding this F_1 is also evident. The same phenomena took place with regard to the height of the plants and other quantitative characteristics, in general, of all growth intensities.

Several suppositions have been made to explain this striking ability for growth of hybrids. From accurate analysis it became evident, that heterosis can be brought about in some cases by a certain gene combination. KEEBLE and PELLEW (1910) had already found that by means of cooperation of several factors which originate separately from both the parents, an increase in measurements of organs can result. Regarding this, the results of experiments with *Vicia faba* (SIRKS 1931) already discussed in this book, point to the existence of cooperating genes. In the example of *Vicia Faba*, the phenomenon is caused by polymeric genes (see p. 138). However it may not be so simple in other examples. For example one may object to such an explanation in corn because if the polymeric hypothesis can be applied in this case, it would be possible to obtain homozygotes in the long run, which breed true to type for vigorous growth, however so far we are not able to show this and we may consider such a homozygosity as wellnigh impossible.

Other explanations were also given: EAST and HAYES (1912) and SHULL (1911 a) tried to solve the problem by proposing that the hybrid nature is a stimulus to a vigorous growth. Harmful genes could be outvoted by their more healthy alleles. From this point of view the disadvantages of inbreeding should be due to the fact that families, homozygous for these harmful factors segregated by means of a Mendelian scheme, in such a way that the healthier homozygous individuals did not occur. An explanation for the last supposition can hardly be found, and a case was sought in the possibility that this was due to the stimulus of heterozygosity. JONES (1917) explored this view in much more detail and put forward that the presence of linked genes, which are complementary to one another, may underlie the phenomenon. This complementary action is explained by JONES by means of three linked genes, which reach their utmost favourable effect, if all are present. One of the parents possesses AAbbCC, the other aaBBcc. The three genes A, B and C are dominant over a, b and c. Hence the hybrid will be AaBbCc and will, due to the presence of all three factors, have a vigorous development. This AaBbCc-hybrid will only give rise to gametes AbC and aBc, due to the linkage between A and b and C on one hand and of a with B and c on the other hand, and hence the progeny of the F_2 and later generations are doomed to be satisfied with either A and C or with B with the result that none of them reaches the maximum vigour, provided that no double crossing over takes place, which would give rise to a generative cell with constitution ABC. Such a new gamete would have all three growth factors and if pollinated with a cell with the same constitution, a vigorously growing homozygote i.e., AABBCC individuals could be obtained.

Undoubtedly this renewal of the theory by JONES gave a much more acceptable supposition as a basis of the heterosis-phenomenon, than did the earlier propositions of SHULL and EAST but it still remained unsatisfactory due to lack of sufficient evidence. Hence several attempts to approach the real causes of the phenomenon were made: by ASHBY, from a physiological point of view and by EAST, from a genetical point of view.

In a series of accurate physiological investigations with corn and tomato crossings, ASHBY (1937) was able to show that the dryweight of plants during seventy days after germination gives rise to a growth curve. It was found by comparing the growth curves of the hybrid and one of the parental types (i.e. the parent with the utmost growth rate) that both curves were completely parallel. Hence the growth rate seems to be based upon a heritable dominant gene; the difference between hybrid and both parents was that the weight of the embryo was already greater than that of both parents (figure 203). In itself this is very important information, but it gives no further explanation of the occurrence of heterosis as is also acknowledged by ASHBY. It shifts the observation from fullgrown plant to embryo present in the seed.

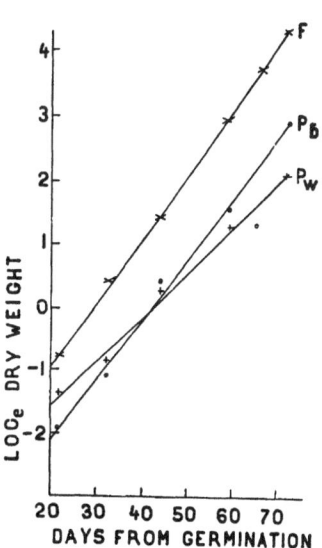

Fig.203. Increase in dry-weight of two inbred cornstocks and their hybrid (after ASHBY, 1930).

EAST (1936) tried to give an amplification of JONES' hypothesis by supposing, that inbreeding leads to exposure and elimination of harmful recessive genes, which process is completed after almost 8 generations. The result therefore should be that the original population breaks up into a number of lines of different nature. Crossing of these lines causes heterosis and it is EAST's conclusion, that the remaining favourable genes give rise to the phenomenon of heterosis by means of their mutual cooperation. RASMUSSON (1933) on the strength of observations and mathematical calculations derived the hypothesis, that for quantitative characteristics, a reciprocal cooperation of the genes involved is necessary. The effect of each gene is dependent on the other genes present; the observable effect of a certain gene is proportionately smaller when the number of genes which act phenotypically in the same direction increases. This hypothesis and its consequences blocks an explanation of the heterosis-phenomenon and therefore EAST changed this view in such a way, that he proposes differences in function between the cooperating genes together with a completely supplementary effect. However this change in JONES' hypothesis is not more than a supposition; there still remain other possibilities to be discussed.

One of these, a possible significance of the *plasm* in the occurrence of inbred degeneration and heterosis was put forward by HERIBERT NILSSON (1937a). In his studies with rye he applied different degrees of inbreeding; the isolation of a single ear (I), the combination of two plants originating from such individual isolations (isolation hybridization IK), the combination of any two selected plants (PK), the hybridization of two individuals from two selected lines (selected special stocks, EL), the progeny of free pollinated plants in selected lines (EK) and the group-hybridization (GK), in which two selected lines are also combined, but both the offspring kept separate. The EK- and GK-results did not show a practical decrease of yield but the yield of the EL receded to 90%, that of PK to 84% and that of IK to 76%. There is thus a clear-cut difference between the productivity of the several progeny. The strong depression which takes place in the very heterozygous offspring of selected free pollinated plants (EL) indicates that an explanation in the sense of JONES' is very difficult to accept, since, during the investigation by chance only plants with un unfavourable combination of genes should have been used. Therefore NILSSON tried to find the explanation in the effect of the plasm. It seems to be important that the offspring of I have only one single type of plasm at their disposal i.e., that from the mother. IK has maternal as well as a very small amount of one paternal plasm. PK-descendants are also 'diplasmic', however, they form a mixture of individuals with two types of plasm. The group of EL-plants was poly-plasmic: a high number of individual quantities of unknown parental plasm were combined with one kind of maternal plasm. Finally the EK- and GK-groups comprise mixtures of two maternal and two parental types of plasm. Hence, the occurrence of *homoplasmony*, that is to say, one type of plasm inside the group of plants leads to inbred phenomena. Alongside the possibility of explaining the inbred results on a basis of the activity of genes, HERIBERT NILSSON put forward the possibility, that inbred-degeneration may occur as a result of a plasmic effect.

WINGE and LAUSTSEN (1940) agree with such a view when they try to explain the inbreeding in *Saccharomyces* (yeast) from their observations. Diploid yeast cells can be obtained from haploid cells in three ways i.e., 1st; two different haploid spores can fuse and form a *spore-zygote*, which can be heterozygous, 2nd; a single haploid spore can multiply itself vegetatively after which, two of the daughter cells fuse into one single diploid cell (homozygous *cell-zygote*) and 3rd; after the first nuclear division in the haploid spore (that is to say before the cell division occurs), both the nuclei fuse again, and a homozygous yeast originates by means of nuclear fusion inside the same plasm. Yeast originating from spore-zygotes have a fair germination percentage, as do those originating from cell-zygotes, whereas those originating from nuclear fusion very seldom germinate. This difference can also be observed in the dry-weight production. It is obvious that we can accept an influence of the plasm here and WINGE and LAUSTSEN believed on good grounds that certain components of the plasm (the so-called chondriosomes) may be held responsible. The problem of inbred-degeneration and the problem of heterosis which is closely related to it has many aspects: the final solution seems to lie in an explanations by means of a combination of nuclear genes together with plasm-qualities.

Let us now return to the problem-proper, concerning the change in the composition of a population. It is obvious that we have to find out if the mathematical regularity to which this change is subject (according to probability) can always be observed. With regard to the nature of the experimental material as biological material it may be expected that causes will exist which influence in one or another manner this course of a population. These causes may lie in the organism itself as well as in the surrounding environment (SIRKS 1949a).

Among the numerous possibilities mentioned in chapter XVII as causes of a true-breeding hybrid nature, the conclusion of RENNER was also discussed. He found that in *Oenothera Lamarckiana* both homozygous combinations velans-velans and gaudens-gaudens succumb as young embryos. Both these homozygous types are therefore not viable (*elimination of zygotes*). MULLER (1918) obtained analogous results in *Drosophila:* in certain crossings, combinations were obtained which were only viable in a heterozygous state. MULLER ascribes these to 'lethal' genes. There are two different linked lethal genes present, which in a homozygous state cause death. In the heterozygous state they are overshadowed by non-lethal dominant allelomorphic genes. If we call these recessive genes l_1 and l_2 then the heterozygote is $L_1l_1L_2l_2$; the gametes are then L_1l_2 and l_1L_2 and the homozygotes $L_1L_1l_2l_2$ and $l_1l_1L_2L_2$ are not viable. Such pairs of homozygously lethal genes are called by MULLER 'balanced lethals'.

Beside this possibility of elimination of both the homozygous combinations as a result of *zygote-lethality*, there are a number of cases in which one of the homozygotes is viable and the other is eliminated. The most classic example is that of yellow mice. In 1905 and 1908 CUÉNOT reported that he was unable to rear homozygous yellow mice, which occurred several times in his experiments. Their progeny, by mutual mating always gave a segregation into 2 yellow: 1 different colour, so that the question arose as to whether homozygotes could not be formed or succumb as embryos. A number of investigators confirmed CUÉNOT's investigations, and other cases in mice were obtained (see CUÉNOT 1928). Some investigators also studied the uteri of the females, from which theoretically speaking homozygous yellow animals have to be born, to ascertain if really abnormal, degenerated embryos occurred. The percentages of 24·54 and 18·7 dead embryos obtained agree satisfactorily with the expectation (25% degenerated animals).

The *Drosophila* studies by MORGAN and his colleagues have also given a number of cases in which certain homozygotes were not viable. This may not surprise us after the previous discussion. In this material also, a new nature of 'lethal' genes appeared; they can occur as sex-linked genes that is to say, linked to the gene for sex. If this is so, then the male in which this lethal factor is present only once will be eliminated. On the other hand a female will be eliminated if the factor is present in a homozygous state.

For instance, let us suppose that a lethal factor (according to MORGAN) is localized in one of the X-chromosomes, which can be indicated by X_1, then an XY-male will be normal, whereas X_1Y-males are not viable; an XX_1-female, like an XX-female, will survive, only X_1X_1-females succumb at an early stage. The only possible mating, in which the lethal gene is involved will be the mating XX_1 with XY. From this mating XX_1- and XX-females will originate, both viable and beside this only the viable XY-males, i.e. 50% (the X_1Y) of the males are not viable. Such a *'zygote-lethality'* in which zygotes with a certain genotypic tendency are eliminated, is a very frequent phenomenon in the plant kingdom. For instance in the offspring of plants which are heterozygous in a gene, necessary for chlorophyll-production, we will find lethal recessive individuals. We would like to refer for summaries of this process, in animals and man, to the studies of MOHR (1926, 1931, 1939) and the book of HADORN (1955).

Gametes can also be destroyed by lethal genes. This phenomenon of *'gamete-lethality'* seems to occur quite frequently in the plant kingdom. However, it is not wholly correct to consider the succumbing of embryo-sacs or pollen-grains as 'gamete-lethality', because these reproductive cells are not true gametes, but are homologous with the spores of the lower plants. Hence, RENNER (1924b, p. 332) prefers the term *'gone-lethality'* however, the name gamete-lethality has found its way into the language, and it will be very hard to alter this term.

BELLING (1914) gave us a clear-cut story of gamete-lethality. He obtained his material by hybridization of the so-called 'Florida Velvet' (*Stizolobium deeringianum* Bort.) with three other *Stizolobium*-species, i.e., *S. niveum* ('Lyon bean'), *S. hassjoo* ('Yokohama bean') and *S. niveum* var. ? ('China bean').

The hybrids of the Velvet bean crossed with others were not as completely fertile as the parents. The F_1-individuals, however, formed macroscopically normal anthers, but by microscopical investigation, it became evident that half of the pollen in all the plants was shrunken, that is to say, the ratio of healthy to degenerated pollen was 1 : 1. The egg-cells showed the same phenomenon: 50% of the number investigated developed into normal egg-cells and the other 50% succumbed even before complete maturity was reached. The F_1-individuals of these crosses were therefore *semi-sterile*, according to the nomenclature of BELLING.

In the F_2 this generation a segregation occurred in the sense, that almost 50% of the number of the F_2-individuals again reached normal fertility, whereas the remaining 50% were not distinguishable in any way from the semi-sterile F_1-plants. The semi-sterile F_2-plants (as did those of F_1) gave rise to degenerated pollen, the remaining 50% were normal; likewise 50% of the egg-cells were sterile and 50% were fertile.

The progeny of the completely fertile F_2-individuals, cultivated as an F_3-generation, remained as fertile as the parents had been. The porgeny of the semi-sterile plants again segregated into 50% fertile and 50% semi-sterile individuals.

Data obtained in this way gave BELLING the opportunity of deriving his hypothesis: If K is a factor present in the Velvet bean but not in the other three parents (in which this factor is replaced by a factor L) and the presence of K or L is required for normal development of pollen grains and egg-cells, then the one parent is KKll and the other parent kkLL. Hence the F_1-hybrids will have the formula KkLl and as normal di-heterozygotes they ought to make $\male$ and $\female$ gametes, each in the four types: KL, Kl, kL and kl. The gametes KL are not viable because K and L inhibit each others effect, and also the gamete kl is not viable because one of the factors K or L is necessary for the normal development of the gametes. Due to this, the F_1-individuals are semisterile and therefore 50% of their egg-cells and 50% of their pollengrains succumb (i.e., KL's and kl's). The remaining 50% survive, i.e., Kl and kL. The result will be the origin of an F_2-generation composed of KKll + 2 KkLl + kkLL. The first individuals mentioned are completely fertile, because they possess only one of the fertility genes in a homozygous state, the latter group also, and the middle group, which comprises half the number of individuals, are semi-sterile, and identical to the F_1-plants. All semi-sterile individuals are therefore heterozygous in two factors, but only two out of the four possible types of gametes will develop completely, both the others degenerate.

In principle, these characters (genes) K and L can be compared to the balanced genes found by MULLER, however, their effect is only observable in gametes. The lethal effect is caused either by combination of both non-allelic dominant factors or by combination of both nonallelic recessive factors.

It seems that gamete-lethality very seldom occurs in animals; a proven case of this phenomenon is not yet known.

Both forms of lethality can be active completely independent of other circumstances but, in some cases their effect is only observable in a certain 'cytoplasmic environment' (see p. 353).

Yet one has to be very careful with the supposition that we are dealing with lethal factors as the cause of the succumbing of zygotes and gametes; the normal development of these depends on numerous physiological processes, hence a disturbance in these processes can also express a sterility. This is especially valid in crosses if the maternal organ (seed-coat) does not fit the hybrid embryo. The investigations of LAIBACH (1925, 1931) showed very clearly, that the embryos of the cross *Linum austriacum* × *L. perenne*, which succumb at an early stage inside the seed-coat, can be brought to ripeness and maturity, providing they are isolated from the seed-coat at a suitable moment and are cultivated artificially afterwards.

We can only find an explanation for the true nature of these lethal factors in a few cases. In the above mentioned case of plants lacking chlorophyll in recessive homozygotes we need no other explanation: the plant is not able to assimilate and has to stop growth as soon as reserve food-store is used up. In general we are able to trace back the effect of the lethal factors to abnormal processes in development (which will be discussed in more detail in chapter XXI). For instance STARK (1918, 1919) found sex-linked lethal factors (genes) in *Drosophila* which cause development of tumors in male larvae carrying this gene as well as in those females which are homozygous with regard to this tumor-gene.

Beside lethal genes, which unconditionally destroy the homozygotes or gametes involved, there are still sub-lethal genes, which weaken the organisms if present in a homozygous-recessive state. TAMMES (1914) for instance found in flax an exception to the usual segregational ratio of 3 : 1, that is to say, a shortage always occurred of recessives in a cross of blue-flowering (dominant) and white-flowering plants. The deficit of white flowering segregation products was due to the fact, that the blue heterozygotes yielded proportionally too few seeds, from which otherwise white-flowering plants would have originated. Moreover seeds from which white plants will originate, do not have such a favourable germination, hence a part of the homozygous white plants are eliminated.

It is evident from the examples given of lethal and sub-lethal genes (which cause the early succumbing of a homozygous individual) that if a population breaks up into homozygous lines, which is the direct result of continuous autogamy and of inbreeding over a long period, this can lead to a dangerous situation. Completely independent of environment and wholly as a necessary result of its manner of reproduction, a population may change its character over a period. The harmful results which inbreeding can bring along lie wholly in this fact, however, as was pointed out experimentally, this need not always be the case.

Up to now we have discussed the course of a population which resulted from its own force or from internal influences; no influence from the outside has been cited. If the latter occurs, then the lot which the genotypical constitution undergoes cannot always be determined, because selection in the population will take place. All kinds of complications will occur, which defeat every mathematical derivation. Only in case where we are able to regulate the environment and have the effects of environment in our hands will it be possible to determine the absolute results. For instance when we select in one particular direction, or in several directions, we will then be able to predict the possible course of the population over several generations. WENTWORTH and REMICK (1916) carried this out for three cases: firstly we select one individual from each of three groups AA, Aa and aa, and give them all the same chance of reproduction. Undoubtedly, the results will be almost the same as in the case of autogamy (1 AA : 2 Aa : 1 aa ratio), only here one promotes the decrease in the percentage of the heterozygotes and the increase in the percentage of the homozygotes, because we give only to one and not to both the Aa-individuals, the opportunity of reproduction, The second case concerns the selection and mutual fertilization of dominant (AA or Aa) on the one hand, and that of the recessive (aa) in the other hand. The final result will again be the same as in the case of autogamy, only the process is delayed, and only after the course of more generations will a separation into homozygote-dominants and homozygote-recessives be obtained. As a third case we can mention the exclusive selection of dominants (AA and Aa) which naturally leads to the dying out of the recessives (aa) and also to an accumulation of AA-types. In the following three tables the difference in these three manners of selection will become evident:

	Selection of AA, Aa and aa.			Selection of (AA or Aa) and aa			Selection of (AA or Aa).		
	AA	Aa	aa	AA	Aa	aa	AA	Aa	aa
F_1		100			100			100.0	
F_2	25.0	50.0	25.0	25.0	50.0	25.0	25.0	50.0	25.0
F_3	41.7	16.6	41.7	33.3	33.3	33.3	44.4	44.4	11.2
F_4	45.8	8.2	45.8	37.5	25.0	37.5	56.3	37.4	6.3
F_5	47.9	4.2	47.9	40.0	20.0	40.0	64.0	32.0	4.0
F_6	48.6	2.8	48.6	41.7	16.6	41.7	69.4	27.7	2.8
F_7	49.4	1.2	49.4	42.8	14.4	42.8	72.0	24.0	2.0
End	50.0	0.0	50.0	50.0	0.0	50.0	100.0	0.0	0.0

In this latter method: selection of AA and Aa only, the result will undoubtedly depend to a large extent on good luck; the mathematical calculation cannot give more than a norm; if chance co-operates, that is to say one uses more AA- than Aa-individuals for the reproduction, then purity will be soon reached. As a demonstration of this we refer to the work of MACDOWELL (1915) concerning superfluous chitine bristles in *Drosophila*, in which the homologous AA-form was obtained by the third generation in an almost pure state. It becomes evident from his work how a radical selection of individuals of a certain genotype can accelerate the process of establishing pure lines.

If we now go a step further and leave the influence on the course of the population to Nature, then the process will become so complicated, that a mathematical contemplation cannot be of use. The influence by which selective causes can affect the genotypical constitution can be grouped under four headings. Firstly: in earlier years it was generally accepted that segregation of hybrids which takes place is independent of all accidental circumstances, however, such a view was based entirely on mere suppositions, which were found to be wrong. In genetical literature there is data which states most positively that the ratio of the gametotypes can be altered by the influence of environment. We may refer here to the influence which temperature and other conditions of life wield upon the process of crossing-over (see p. 171). This influence changes the ratio of the recombining gametes (crossing-over gametes) to the non-recombinants. The composition of the progeny can also be dependent on different circumstances even in the case when the ratio of the gameto-types formed agrees with the theoretical expectations.

In numerous investigations with plants it was established that pollen-tubes in which the nuclei of different genotypes are transported can differ in their growth-rate. This results in a race between the different types if the plant is pollinated with a mixture of these pollen-grains. This race is called 'certation' by HERIBERT NILSSON (1915b), and a disturbance in the mendelian segregation ratios occurs. The simplest way to show this is by back-crossing the hybrid in both reciprocal ways to its recessive parent. CORRENS (1902a) obtained a segregation between smooth and shrunken in his corn crossings; instead of 75 smooth : 25 shrunken he observed a ratio of 84 : 16. If the hybrid Gg plants (as parents) are pollinated with pollen of a gg-individual (hence there are no mutual differences in the pollen) then the F_1R-progeny segregated into 50% Gg and 50% gg plants; the reciprocal cross Gg as father (hence with G and with g pollen) with the gg-mother plant yielded a progeny of 58% Gg and 42% gg seed. Originally CORRENS had given an other explanation to the phenomenon, however, later (1917, p. 668) he showed that in this case a certation is active as an interfering influence. In the same publication he gave a communication concerning *Melandrium* (the ragged robbin), in which he found that pollen-grains which will give rise to ♀ plants have a more rapidly growing pollen-tube than pollen which gives rise to ♂ plants. Hence, a pollination with a little pollen will give rise to a relatively higher number of male descendants, than will pollination with a large amount of pollen in which the ♀ - determining pollen-tubes will more easily oust the others. This was the fact CORRENS found: in the first case 42·96% male plants originated and in the second case 29·86%.

These differences in the growth-rate of the pollen-tubes can be based upon special certation genes which differ in their intensity (SIRKS 1926a, 1929a), however, these differences can also be due to the constitution of the chromosome configuration (BUCHOLZ and BLAKESLEE 1932); moreover they may depend on environment (e.g. temperature) (HERIBERT NILSSON 1920a). Hence it is clear that temperature of the surroundings can encroach upon the course of segregation even to such a degree that it governs the change of the genotypical composition of the population. In this case *selective fertilization* takes place.

Secondly it appears that Nature can encroach upon the genotypical constitution of a population by stimulating a mating between two definite individuals on the one hand and on the other hand by repressing matings between other individuals. This phenomena can be based primarily upon the heritable tendency of the individuals and will be expressed by sexual reproduction: *sexual* or *physiological isolation*. In a small way this can be observed in plants which are *self-sterile* (i.e., plants which cannot be selfed) and which give rise to completely normal egg-cells and pollen-grains, but from which each individual can neither be pollinated with its own pollen, nor by the pollen of certain other individuals. This phenomenon of self-sterility (*incompatibility*) is also based upon the effect of genotypical factors (see CORRENS 1912b and the summaries of EAST 1929a and LEHMANN 1928 and also the mathematical derivations of WRIGHT 1939b). This effect also finds its expression in a difference in growth-rate of the pollen-tubes. The growth-rate of the non-fertilizing pollen-tubes is, according to EAST (1926) so low that they never attain the pistil-length of the flowers in question (see fig. 204) and hence they are never able to penetrate the micropyle.

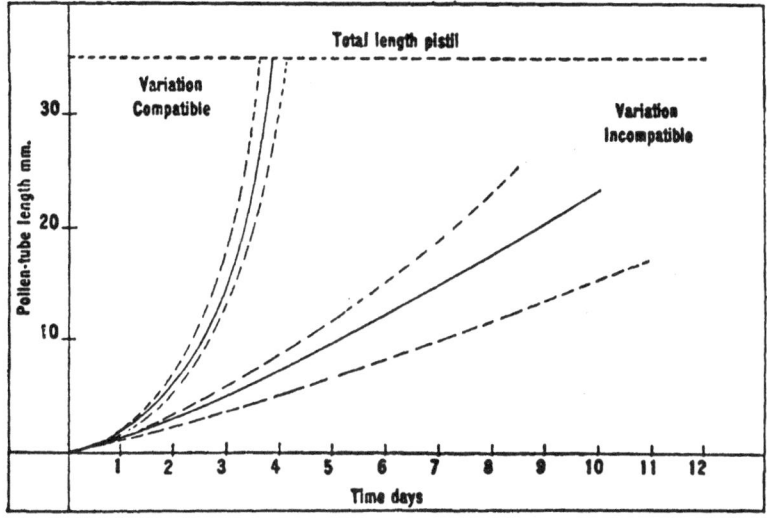

Fig. 204. Growth-rate of the pollen-tubes after compatible and after incompatible pollinations (after EAST, 1926).

Such a sexual-physiological isolation of certain genotypes can be due to several processes relating to that of fertilization. For instance in plants this may be due to differences in flowering time (which has a genotypical basis, but which is also rather strongly modified by environment) and in animals to all accidental details by which fertilization is brought about.

The third type of encroachment upon the composition of a population and of segregation of an extensive group into smaller units of a more simple constitution, will be found in the effect of a *geographical isolation* (see REINIG 1939); all boundaries, that is insurmountable natural barriers (seas, rivers, swamps, hills, mountains etc.) can lead to the breaking up of a population, which was originally equally distributed, (with regard to genotypical factors) into smaller sub-populations, which only possess a part of such a genotypical variety. These sub-populations are therefore more limited in their genotypical composition. These causes are in no way related to the organism itself, they are part of the environment: the genotype of the individuals in this process is completely passive.

The fourth manner in which smaller groups can originate from a population, as a combination of genes together with the effect of environment, lies in the difference in *vitality* which can occur between individuals of different genotype. This difference in vitality finds its extreme in the occurrence of lethal genes, which are fatal to the organism, even under the most favourable circumstances. However, next to this we find all degrees of vitality, which find expression under different circumstances and in different ways.

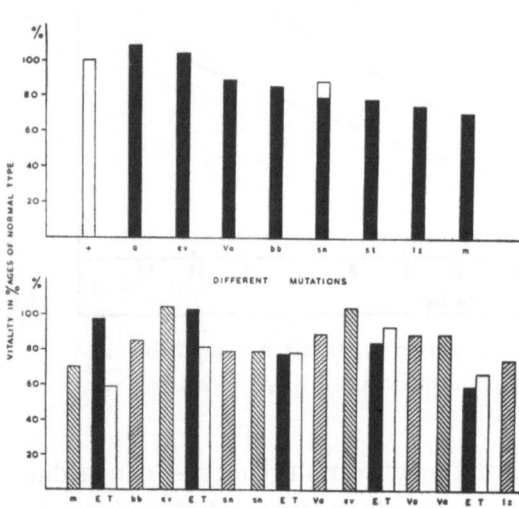

The several natural factors which can simplify a genotypical population by means of a decrease or increase in vitality belong, in general to three groups; 1stly, reciprocal influence of other genes which also belong to the genotype of the individual, 2ndly, influence of the population density and 3rdly, influence of the temperature of the environment.

Fig. 205. Influence of different genes and of their combinations upon the vitality of the *Drosophila-types* concerned (after BAUER and TIMOFÉEFF-RESSOVSKY, 1943).

In an important series of experiments by TIMOFÉEFF-RESSOVSKY (1934b) it became clear how the vitality of *Drosophila*-strains is influenced by definite genes and how the consequences of these genes behave under different environments. Presence of the gene itself has an increasing or decreasing effect upon vitality (fig. 205, upper part): the factors A (abnormal, dominant) and ev (eversae, recessive) intensify vitality to 108 and 104 (if we take the vitality of the wild strain as equal to 100). The genes Va (Venae abnormes, dominant), bb (bobbed, recessive), lz (lozenge, recessive) and m (miniature, recessive) decreased the vitality to 89, respectively

85, 74 and 70, whereas the recessive gene, singed (sn) reduced the vitality-coefficient of the females to 88 and that of the males to 79. One would now expect that the combination of two such genes in the same individual would result in an additive effect and hence the combination m-bb would give the value $\dfrac{70 \times 85}{100} = 59$.

However, for the combination m-bb, a vitality coefficient of 97 (calculated 59) was obtained, for ev-sn, this value was 103 (calculated 82), for sn-Va 77 (calculated 78), for ev-Va 84 (calculated 93) and for Va-lz 59 (calculated 66). Hence in the first two combinations, the vitality is less affected by both the factors than the summation of their influences separately. In the third, the results are almost equal to the theoretical expectations, and in the last two the decrease in vitality is greater than the additive effect would predict (fig. 205, lower part). The effect of a particular gene upon the vitality was hence highly influenced by the gene-environment.

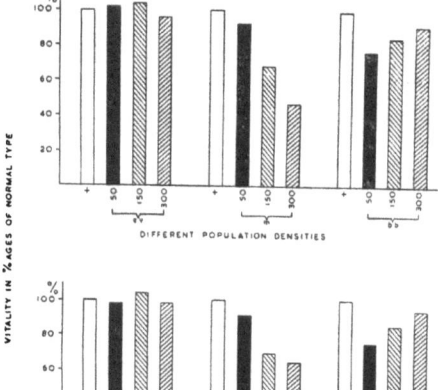

Fig. 206. Influence of population-density (above) and of temperature (below) upon the effect of the vitality-genes in *Drosophila* (after BAUER and TIMOFÉEFF-RESSOVSKY, 1943).

Environment is also important for the influence of a single gene: the density of the population in a culture tube (measured by the number of eggs per unit of medium substrate) appeared to act in different ways upon the different genes; i.e., for the gene ev an underpopulation of 50 eggs and a moderate overpopulation of 150 eggs was usually more favourable than unfavourable (102, respectively 104), whereas a highly over-populated tube, slightly decreased the vitality (96). For the genes m and bb these numbers were 93, 69 and 47, and respectively 77, 85 and 92. It appears that an overpopulation is very harmful for the gene 'miniature', however it has a mere favourable effect upon the gene 'bobbed' (fig. 206, upper part). The influence of temperature was analogous: at 15°C, 25°C, and 30°C the vitality coefficients were for ev 98, 104 and 98, for m 91, 69 and 64, and for bb 75, 85, 94, so that ev can be considered as almost insensitive to temperature, whereas m-animals can be reared better at a low temperature, b-animals on the other hand like a higher temperature (fig. 206, lower part).

Gene-environment as well as outside environment (population density and temperature) has an influence upon the vitality and an influence upon the reproductive ability of certain genotypes. Hence they encroach deeply into the course of a population to which animals belong which have the gene constitution involved.

Opposite to the simplifying effect of environment there is another process which consistently wields an influence upon the development of a population, i.e. the introduction of new genotypes into a closed population by means of the origin of individuals of a heritable tendency which has hitherto not been present in this population. Also, individuals may join a closed population and may come in contact with the genotypes already present. New genotypes (as we saw in a previous chapter) may arise by means of mutation and it does not matter if this is a genome-, chromosome- or gene-mutation. In almost every wild population one can observe the occurrence of new genotypes, which may perhaps be considered as mutations: the genus *Drosophila* yielded many examples of this phenomenon (SPENCER 1947). Joining of new genotypes to an already established hitherto closed populations is called *introgressive hybridization* (ANDERSON 1949) and this process occurs quite frequently in Nature. The contact between new genotypes or newly introduced genotypes with the already present ones can lead to the following results: competition in the struggle for life between the different genotypes and the possibility that the different genotypes unite by means of hybridization.

Competition in struggle for life, an inevitable factor in all populations in Nature, can also be shown by experiments. l'HÉRITIER and TEISSIER (1937) have obtained important data in this field with *Drosophila* cultures. The strain 'Bar' with slit eyes for which this dominant characteristic is harmful (because the vitality of homozygous as well as of heterozygous animals is decreased) endures badly the neighbourhood of animals with normal eyes. That is to say, if some couples of normal animals are introduced into a well populated culture of the abnormal strain, then after 100 days the number of Bar-animals is found to have decreased below 40% and after 500 days they have practically disappeared. However if one adds some normal grey animals to a culture of black flies (with the ebony factor, which in homozygous state decreases vitality, but which in heterozygous state has an increasing influence) then in this case also the normal grey animals get the upper hand but the percentage of ebony animals drops within certain limits, but never below 20% (fig. 207 upper part).

An analogous struggle for life occurs when a population of the species *Drosophila funebris* was enriched with some couples of *D. melanogaster;* after 100 days the *funebris*-animals had almost disappeared. They did not succeed in gaining back the lost territory but remained in only a small percentage as a part of the new mixed population (fig. 207, lower part).

In these cases an increase of the genotypical pattern of the population leads to a radical change, by which the initial enrichment can again be nullified for a great part (see DOBZHANSKY 1939).

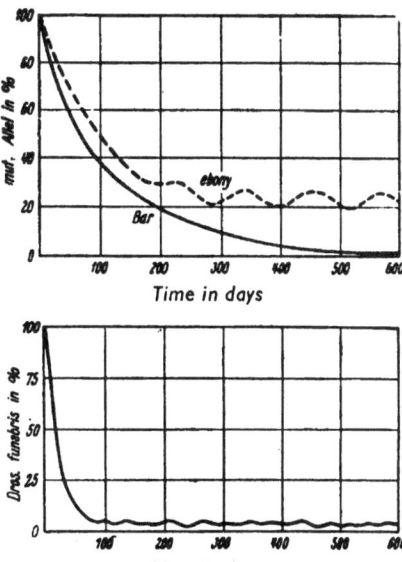

A more permanent possession of new genotypical factors is reached, if both the populations which 'live' together do not compete with one another but establish a mutual association, and if the hitherto separate populations form an association with sexual relationship, that is to say, that the individuals mutually mate and from these fertilizations (matings) fertile hybrids are born. A striking example of this was given by CLAUSEN (1933), who investigated the distribution of both the plant species *Pentstemon* laetus and *P. azureus* in California and found that the 'species' *P. neotericus* occurred in a frontier area of both these former populations (fig. 208). The latter species united numerous characteristics of the two other species. From karyological investiga-

Fig. 207. Struggle for life between different genotypes of *Drosophilia* (after l'HÉRITIER and TEISSIER, 1937, from TIMOFÉEFF-RESSOVSKY, 1939).

tions it became evident, that *P. neotericus* (with the haploid chromosome number of 32) has united both the chromosome configurations of the other species, which contains 8 in *P. laetus* and 24 chromosomes in

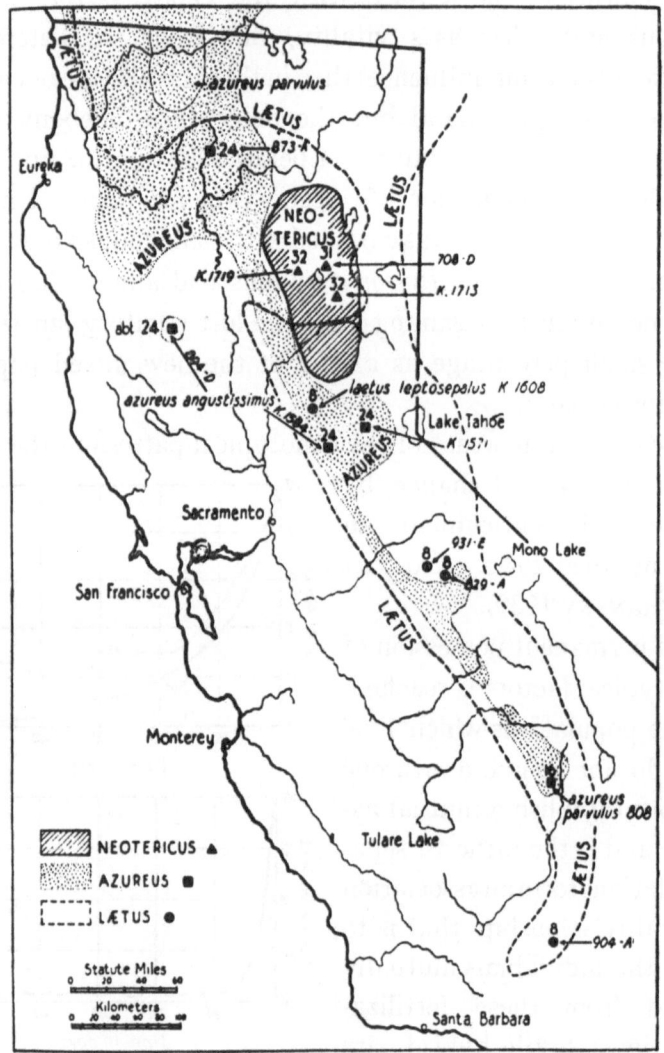

Fig. 208. Spreading of the species *Pentstemon laetus*, *P. azureus* and *P. neotericus* in California (after J. CLAUSEN, 1933).

P. azureus (fig. 209). In this way different populations can form a new unit by combining their chromosome sets. This unit will have an increased number of chromosomes and hence an enriched genotype.

The inconstancy of each group of cohabiting and mutually mating

organisms offers a continuously diversified and colourful picture. Sometimes the group gets poorer and then again becomes richer in heritable factors. This change can be influenced by environment; however it may also occur spontaneously. This extremely important

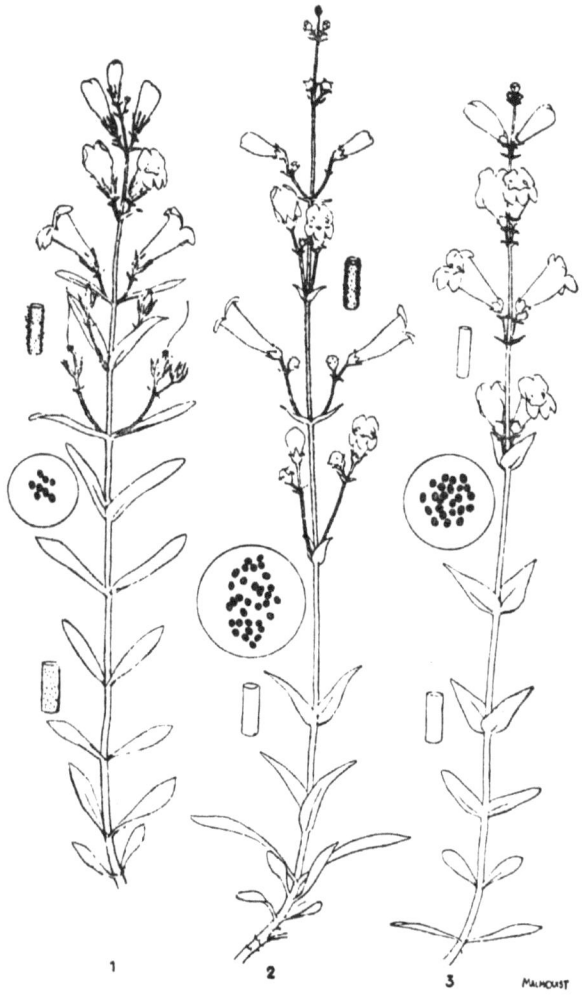

Fig. 209. The species *Pentstemon laetus* (left), *P. azureus* (right) and *P. neotericus* (middle) and their chromosome configurations (after J. CLAUSEN, 1933).

field also forms the starting point for three spheres of studies in which biology is involved: the *taxonomical concept of the species* (ALLAN 1937, 1949, ANDERSON 1937, CUÉNOT 1936, DOBROVOLSKAJA-ZAVADSKAJA 1929, HARLAND 1936, HUXLEY 1940, MAYR 1948, RENSCH 1929, 1939), the *evolution of living Nature* (DOBZHANSKY

1941, GOLDSCHMIDT 1940, HEBERER 1943, HUXLEY 1945, JEPSEN, MAYR, SIMPSON 1949, LAM 1946, MELCHERS 1939, PÄTAU 1939, SCHMALHAUSEN 1949, SIMPSON 1944, 1949, STEBBINS 1950, TIMOFÉEFF-RESSOVSKY 1939, UMBGROVE 1943, WRIGHT 1931, 1939a) and that of the quantitative and qualitative problems, which the *human population* carries with it (DAHLBERG 1948, GLASS 1954, PEARL 1939, SIRKS 1948). This is a complexity of problems to which biology and human society try to find an answer.

Population genetics has developed from a diversity of more or less related but separate problems into a well-constructed field of research (BLAIR 1953, LERNER 1950, 1954, MATHER 1953, WRIGHT 1951).

CHAPTER XXI

THE DEVELOPMENT OF THE INDIVIDUAL AND ITS HERITABLE TENDENCY.

The lapse between just-fertilized egg-cell and fully grown and mature individual, from a developmental point of view an enormous one, is fraught with all kinds of possibilities. On the one hand the gene complex (present in the fertilized nucleus) and on the other hand, the surrounding plasm. This plasm is perhaps partly independent and partly cooperating as a reaction substrate for the genes in order to establish the phenotype. Both components are influenced in their attempt to express themselves, that is to say, they are favoured or inhibited by external circumstances of life, (i.e., by the environment). By means of cooperation of these three groups of causes, thrown together in each others company, the body of the organism originates as a unit, as a living creature and a differentiated complex of cells. If genetics is to answer the problems to it, then we have firstly to exclude the effect of difference in environment, as far as is possible. In this way we are able to derive some conclusions from our experiments which may be due to interaction of genes and plasm. This however will entail very great difficulties.

The external environment can wield a great influence upon the effect of the gene, but it is also possible that it does not interfere at all (as was discussed in the introductory chapter). This environment can be controlled (and must be controlled) by precise experiments to such an extent that it does not wield a disturbing influence upon the results of that experiment. We know (especially from investigations with *Drosophila*) that environment is not limited to the external environment only; indeed it may be present in the immediate neighbourhood of the gene, that is to say, the surroundings of the gene may perhaps have an influence upon the effect of the gene proper. This *'gene milieu'* can be shown in a phenomenon known as the *'position effect*'. STURTEVANT, in his investigations dealing with the Bar-gene of *Drosophila* (see also p. 312), which causes slit eyes, brought to light the striking fact that the decrease in number of facets in the eye in homozygous Bar-animals is lower than in heterozygous doublebar-normal animals. Analogous combinations show the same phenomenon, though in both cases the same gene is doubly present; in one case it is distributed over two homologous chromosomes and in the other case united in one chromosome. This is evident from the table given on p. 314: the average number of eye facets in a normal B/B-female is 779·4, that of a heterozygous doublebar/normal female (B_2B_2/B) 45·4, and of a homozygous bar/bar (B_2/B_2) female 68·1; likewise for a heterozygous double-infrabar/normal (B_1B_1/B) female 200.2, for a homozygous infrabar (B_1/B_1) 348·4. The presence of two bar- or infra-bar-genes does not always express itself in the same way; it depends on the position of the gene in the chromosome, how deeply this gene can encroach into the development of the phenotype. Since this time numerous chromosome irregularities have become known and the number of known expressions of this position-effect has naturally increased with the possibility of dealing with translocations and inversions. We meet a very important problem here, which has special importance for the physiological aspect of genetics (see OFFERMANN 1935, DOBZHANSKY 1936b, TIMOFÉEFF-RESSOVSKY 1940c, LEWIS 1950). However it is striking, that in other organisms which have been studied karyogenetically like *Zea Mays*, and in which many cases of translocations are known, no examples of this position-effect were observed. Hitherto it had seemed likely that the position-effect has an influence upon the quantitative expression of the gene only and not upon qualitative processes.

  If all these influences of environment are eliminated then genetical experiments in this field dealing with the development of the individual, must first follow those threads which lead from fertilization-product through embryonic development to the fully-grown phenotype, and hence must lead to a concept of the development of the individual as a whole.

  As we have already seen (p. 17) there are two paths which these investigations can follow: the *cytomorphosis* (SCHAXEL 1915), which investigates processes of development in the fertilized egg-cell and hence starts at the beginning, and that of *phenogenetics* (HAECKER 1918, 1925) which traces back step by step the development of the phenotype. Both these possibilities can be approached in two ways: the *developmental-mechanical* method, which investigates the mechanical course and the *developmental-chemical* method in which the chemical processes during the development provide the field of interest.

In the case of asymmetry in the normal dextral shaped snail shell of *Crepidula* studied by CONKLIN (1903) it was possible to reduce the asymmetry to the earliest stage of development, that is to say to the first division which is a 'dexiotropic' cleavage. Both nuclei with their surrounding plasm turn clock-wise inside the cells at this stage. Hence, these studies formed the first contributions to genetics from the aspect of developmental-mechanical cytomorphosis: CONKLIN drew the conclusion that the cause of the dextral shaped shell must lie in the structure of the uncleaved egg. Therefore we will link these results to those obtained by BOYCOTT and others (1930) in the genetical field (p. 344) so that all these together form one of the most eloquent examples of cytomorphosis. The relationship between a particular part of the phenotype and its developmontal cause present in the egg-cell remains a problem for the time being, which is likely to be of a biochemical or physicochemical nature.

On the other hand the phenotype of an organism is a harmonious unit, that is to say it consists of differentiated cells, all of which originated by somatic divisions from one single fertilized egg-cell. Are these body cells mutually similar in genotypical tendency? Does their plasm structure show profound differences? To which causes one can ascribe the fact that a muscle cell is built differently and has other properties than a brain cell; that an apparently homogeneous organ such as the liver is composed of cells of very different functions; that a cell from the leaf-parenchyma of a green plant shows another phenotype, performs other functions than does a cell of the epidermis or of the sclerenchyma? These are a number of questions, which are closely related to the problem of genetics because they bear upon the genotypical constitution of the somatic cell and its development into the phenotype. These and related questions, often unasked, form the boundary between genetics and embryology in general. It is an extremely difficult area, which we investigate here; however, it raises expectations of a prolific future, expectations, from which only a small part are today realized. The borders of the field of developmental mechanics (embryology) and its relation to genetics (see HUXLEY 1932, E. E. JUST 1936, JUST and others 1937, MORGAN 1934, RAVEN 1954, SPEMANN 1936, WOODGER 1930–1931) are sometimes differently mapped. SPEMANN (1924) considers the question of localization of the heritable tendency as the principal problem, as well as the nature and the activation of the heritable factors. It is possible that by putting the problem in this way there may be a tendency to annex purely cytological problems, but in general it seems that SPEMANN's point of view is correct.

Different methods can serve the same purpose: the purely embryological investigations into the fertilized egg-cell and its further course of life (RAVEN 1954); the formation of chimerae between germ-layers of different genotypes that is to say between ectodermal, entodermal and endodermal layers (v. UBISCH 1939) and the tissue cultures under artificial conditions (botanical, BECKER 1938, GAUTHERET 1942, P. R. WHITE 1936–1946, 1943; zoological, CAMERON 1950, FISCHER 1942). For the time being we are still fumbling and searching for final conclusions with regard to the balance between genotype, plasm and environment.

How far the genotype may be independent of the environment, so that it can pursue its own nature through the whole development of the organism, and how far the environment has a decisive influence upon phenotypical external appearance of the organism and of all its parts, we do not know. SPEMANN and his co-workers have carried out a beautiful piece of work in this sphere with their admirable transplantation experiments, from which we only can mention some aspects here.

The larvae of *Triton* (salamanders) have four thread-like organs close under their mouths with which they attach themselves temporarily to waterplants; in frogs (*Rana*) these organs are disc-shaped suckers and in *Axolotl* they are wholly lacking. If a part of the abdominal epidermis (far from the mouth) is transplanted from *Triton* or *Rana* in the young embryo of the same species, to the neighbourhood of the mouth, then these cells will give rise to suckers in this place whereas, in their initial place this would not occur. Hence one may conclude, that the changed 'environment' (i.e., the conditions prevalent in the surroundings of the mouth or the influence of neighbouring cells) may have induced the formation of these suckers from material which originated from the abdominal epidermis. However, if we go a step further and we perform a transplantation of abdominal epidermis of the one species (donor) into an embryo of the other species (host), then it appears that this induction also is limited: MANGOLD (1931) found that a piece of abdominal epidermis of *Triton* transplanted to the throat area (oral field) of an *Axolotl* embryo still gave rise to thread-like suckers (in *Axolotl* they are lacking), whereas reciprocally, abdominal epidermis cells of the *Axolotl* transplanted in the throat area of a *Triton* embryo did not give rise to these organs. Even more striking is the communication of SPEMANN (1932), that abdominal epidermis cells of *Rana* as donor, when transplanted to the neighbourhood of the mouth of *Triton* as host give rise to suckers, however not to thread-like organs which would suit *Triton* embryos but to disc-shaped suckers which are common in *Rana*. In the abdominal epidermis of the frog the formation of suckers is induced by the influence of the new environment, however they develop further and their own path is determined by the *Rana* tendency in them. The influence of the new environment is limited to the development of suckers; the shape will still be determined by the nature of the donor (*Rana*). The most obvious thing to presume in this case is that a change of the plasm occurs under the influence of the environment however, the shape of the organ can be considered as genotypically established.

The statement of SCHLEIP (1927, p. 75), 'that all which can originate is preformed by the genotype and yet that which really will originate — between the first moment of differentiation in the egg-plasm and the final properties as characteristics of the full-grown body — is the result of a reaction of genes upon conditions of development, and hence of an epigenetic nature' appears to be too sharply stated. The constancy with which the fixed cleavage line determines the whorl of the full-grown snail shell of *Crepidula* and the waywardness with which the abdominal epidermis cells of *Rana* give rise to their own typical suckers in the oral field of *Triton*, proves that the expression of the genotype in the full-grown organism can be of a preformed tenor, which is not influenced in any way by the environment.

Besides the possibility of investigating the effect of the heritable tendency in genes and plasm by means of a study of the normal embryological development, there is still a second method, which tries to lead back developmental defects to peculiarities of the heritable constitution. In Chapter XII the possibility of dealing with deficiency was mentioned among the numerous irregularities, which could be observed in chromosomes, i.e., lack of a certain minute part of one of the chromosomes. Together with this missing part of the chromosomes certain genes can also disappear. These lacking genes have an influence upon the normal course of development then this normal course will also be disturbed due to this lack of genes. It was POULSON (1940, 1945) who traced the development of *Drosophila* in which the X-chromosome showed a series of deficiencies: there are several intermediate forms between complete absence of this X-chromosome ('nullo-X') and the absence of a very small single segment in the chromosome (not more than one or two 'loci' i.e., 'Notch-8-deficiency'). In the nullo-X-eggs, two hours after fertilization (the whole development of the embryo to grub takes 22 hours) the development of the embryo has already become irregular, that is to say during the blastula state which is composed of 256 cells. If less is missing then a further stage is reached, hence the Notch-8-animal has a normal development for eight hours after fertilization but, it becomes distinctly abnormal at the 12 hour stage.

McCLINTOCK ((1938, 1941, 1950) gave an analogous example in her corn investigations. She was able to show that a definite deficiency in one of the ten chromosomes is usually made up for by the presence of a small ring chromosome. However, this ring chromosome sometimes fails to appear during somatic divisions and cells originate, which do indeed possess the deficient chromosome, but not the supplementary ring chromosome. In these cells the absence of the chromosome part (and hence the absence of the gene located at this deficiency) acts as a disturbance such that the cells only divide at the utmost a few times after which they swell to great vacuolised cells.

Not only by means of cytomorphosis, but also by means of *phenogenetics* is one able to approach the mechanical course of development and in a more exact manner than before, because, in these investigations genetically known material can be used from the very beginning, and the effect of a definite gene can be followed step by step. In applying this method we will favour the term *'developmental physiology'* because the physiological nature of the gene comes especially to the foreground, as opposed to mechanism of the development. An excellent review of the data under discussion was compiled by BONNEVIE

(1940), and we will further refer to the publications of BRACHET (1945), FLORKIN (1949), GLASS (1949), GOLDSCHMIDT, (1928, 1938a), HERTWIG (1939), KOLTZOFF (1935) and WADDINGTON (1947) and for pathological developmental processes to the papers of FISCHER-WASELS (1938) and of NACHTSHEIM (1938–1940). Two examples from this literature will show some results which can be reached in this direction.

Important data can be found in mice strains, which are described by BAGG and LITTLE (1924) and which can be distinguished from normal mice by the possession of one recessive gene. As full-grown animals these 'blister mice' are characterized by all sorts of abnormalities of eyes and legs; the eye aberrations can occur on one side or on both sides and to different degrees i.e., from a small decrease of the opening between both eyelids upto eyelids completely closed by coagulated blood and even closed by skin. The leg deformations occur on one or more legs, and express themselves as so called club-feet or by incomplete fingers or toes or by the growing together of fingers and toes. Such bodily defects are of course not primary defects; they can be ascribed to disturbances in the development of these organs.

A careful, embryological investigation by BONNEVIE (1934) gave a very surprising result: as a primary cause, an abnormally heavy secretion of liquid from the fourth ventricle was observed. The fourth ventricle is located in the medulla oblongata. In normal mice this fourth ventricle is temporarily not wholly closed, an opening allows excretion of fluid so that the pressure can be controlled. However in blister-mice the amount of fluid is unprecedentedly great and in young embryos of 7 mm. this amount of fluid is piled up under the skin in the neck-region, so that a blister filled with clear fluid becomes visible. In a later stage these blisters migrate along fixed routes and form secondary blisters at definite parts of the body. This migration still continues after disturbing the normal course of development to a greater or lesser degree. Therefore these blisters move from the neck region along the head and back. On the head they move to both sides, along the shallow grooves of the vaulted mid-brain to the region of eye and nose. Along the back they are established firstly in the shoulder region, from which they reach as far as both forelegs and from this shoulder region they creep to the backside of the animal and from here they may move further down into the hind legs or the tail. The blisters are filled with clear spinal fluid, however if they remain a long time on the same place, then the increasing pressure of the fluid becomes too high for the underlying tissue so that blood is released which mixed with the clear content of the blisters.

At all places where these blisters are located or even temporarily located, the normal development of the tissues and organs is disturbed, that is to say, the development of the toes can be influenced and hence may result in a growing together or branching of the toes and further the whole frame of the leg can be thrown into disorder through these blisters. Their presence encroaches especially deeply upon the development of the eye: the first stages are absolutely normal, even in the development of the lens and retina. However, the skin is lifted up. by the presence of such blisters round the eye. In some cases when the skin again becomes normal after disappearance of the blisters the stage has been passed in which the epidermis cooperates in normal development of the cornea and eyelids. This causes many disturbances. In the case when the blister does not move and roots under the epidermis, then it will be filled with coagulated blood which will lock up the whole eye from the outside world.

Hence an apparently unrelated complex of aberrations in the structure of legs and eyes can be reduced to one and the same cause, i.e., too great a secretion of spinal fluid from the fourth ventricle of the brain and this primary cause is again based upon one genetic factor.

The second example concerns the defective development of the tail in mice, which sometimes leads to a shortening or even to a complete disappearance of the tail. In the investigations carried out by DUNN and his co-workers (1936–1939) facts were revealed about the genetical background of this striking characteristic which occurs in some types of mice. CHESLEY (1935) and GLUECKSOHN-SCHOENHEIMER (1938, 1940) have carried out the embryological aspects of this problem. Three different types were studied: the brachy-type, the D-short-type and the Shaker-short-type. The first strain, that is to say, the brachy-type, owes its characteristics to the occurrence of a series of four multiple alleles: T-t-$t^°$-t'; the second possesses a dominant S^d-factor, and the Shaker-short mice have a recessive s^t-factor, which gives rise to this phenomenon. The results which are due to the several homozygous and heterozygous combinations of genes are given in the following summary:

Series of factors T-t-t°-t'

Brachytype

TT Short tail; crinkled medulla tube, chorda dorsalis absent, no development of the hind legs, hind-part degenerates; *the embryo dies on the eleventh day.*

Tt Short tail; in the lumbar region the chorda dorsalis is branched, normal medulla tube, in the tail region a branched chorda and branched medulla tube, which causes the tail-end to drop off; the animals remain alive. Offspring 2Tt : 1 tt.

tt Normal tail: the animal remains alive.

Anury-type A

tt° Normal tail; the animal remains alive. Offspring 1 tt : 2 tt°.

t°t° Tail-less; on the fifth day still no ectoderm differentiated and abnormal cells occur in entoderm; mesoderm formation which should have occurred on the seventh day fails to appear; further differentiation fails. *The embryo dies on the seventh day of development.*

Anury-type 29

tt' Normal tail; the animal remains alive. Progeny 1 tt : 2 tt'.

t't' Tail-less; *the embryo does not fasten itself to the wall of the uteri; dies before the fifth day.*

Combination types

Tt° Tail-less, but remains alive. Offspring Tt° only because TT and t°t° die early as embryos.

Tt' Tail-less, but remains alive. Offspring only Tt', because TT and t't' die as embryos.

t°t' Normal tail; remains alive. Offspring only t°t' because t°t° and t't' die as embryos.

Factors S^d and s^d

D-short-type

S^dS^d Tail-less. Abnormalities in hind-part, i.e., lacking an anus and external genitals. *The animals are alive at birth, but die in 24 hours.*

S^ds^d Short tail stump. Irregularities in the urogenital system. Animals remain alive. Offspring 1 S^dS^d (dies in 24 hours): 2 S^ds^d : 1s^ds^d

s^ds^d Normal tail; animals remain alive.

Factors S^t and s^t

S^tS^t and S^ts^t Normal tail, animals remain alive.

s^ts^t Short tail. Tendency to turning movements (shaker mice) disturbance of the organs responsible for the equilibrium, also deaf; sterile to a higher or lesser degree. Anatomically: general anomaly of the ectoderm; narrow brain roof and narrow brain cavity; small amount of cerebrospinal fluid; slow heart beat; too high a blood pressure which causes swelling of the tail end in the embryonal stage which again results in degeneration of this part of the tail.

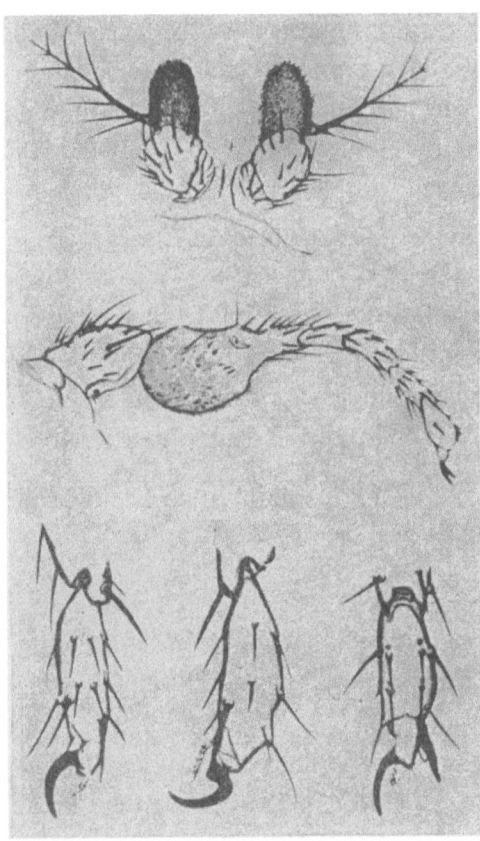

Fig. 210. Above: normal antennae of *Drosophila*. Middle: antennae of the *aristapedia*-type. Below: left two *aristapedia*-legs, right a normal leg (after BALKASCHINA, 1929).

From this data it is evident how an apparently simple characteristic such as 'tail reduction' can be accomplished in very different ways by different genes. These genes encroach at a given moment upon the development of the individual in a way characteristic for the gene involved. BALKASCHINA (1929) has given us an example of how sharply this encroachment is adjusted by the gene. In *Drosophila* one single recessive gene brings about the alteration of an organ into a homologue. By means of its influence, the top of the antenna, the so-called arista, becomes a leg (fig. 210). In normal animals, the segmentation of the antenna starts in larva $4^1/_2$ days old (in 25%) and ends in the young pupa, however, in an *aristapedia* individual this segmentation started when the larva is 2 days old, i.e., in the stage of development of the segmentation of the future legs. Due to this synchronism, the organ which is destined to become an antenna, becomes segmented like a leg and develops along this direction. Hence the action of the gene is based on the fact that the embryological process of the antenna segmentation is moved forward and shifted to a moment at which the leg segmentation is also taking place.

In the light of this genetical embryological data we can also consider the effect of lethal factors, which we discussed earlier in the book (p. 485): genes cooperate in the normal course of development of the individual, but gene and chromosome irregularities also cause a deviation from the ideal development, a disturbance in the course of things and, by means of this, damage to the individual which can become so great that the viability incurs the risk of being finally annihilated. In this case the heritable configuration causes total lethality.

Whereas these embryological investigations try to establish how far the environment on the one hand, and the genotype (or plasm structure) on the other hand play a rôle, the rôle of the genetical tendency is not fully established. The chemical background of the development of heritable tendency into phenotype is not clarified by these embryological investigations. To accomplish this, embryological investigations in the field of chemistry and biochemistry are absolutely necessary.

With this we put in practice the method of investigation already earlier mentioned (p. 327) into the nature of genes, which has as study-subject, the effect of the gene during the development of the organism.

The chemical investigations, starting off with the phenotypical characteristic established by this gene, try either to reduce this characteristic to the causes, on which it is based, or try to find the differences in the chemical constituants represented in reproductive cells of young developmental stages and relate these differences to that which will develop from it in a later stage. The first method became especially fashionable in recent years; it presumed that specific 'gene-products' or 'gene substances' are produced by genes, which find their expression into the phenotype through the medium of the protoplasm. Undoubtedly scientists have succeeded in making such a supposition very acceptable (see BEADLE 1945, HALDANE 1942, PLAGGE 1939, SIRKS 1946, SYMPOSIUM 1950b).

It is obvious, that such gene-products will firstly act *intra-cellularly*, that is to say inside the cell in the nucleus in which the genes are localized. A schematical representation of this process is given in fig. 211 which is borrowed from the study of KÜHN (1939): A and B are two genes localized in the chromosomes. From these two genes, products are secreted which by means of intermediary reaction (IRB and IRA) give rise to the corresponding phenotypical effects. Such an intracellular effect of the gene is in most instances not easy to approach for investigation because the effect inside the cell as a whole does not permit a comparison with parts where the gene is not effective. Very favourable material HÄMMERLING found (1934, 1935) in *Acetabularia*, a parasollike sea-weed, uni-cellular, and with a clear differentiation into an umbel, a stem and root-like suckers, (i.e., the rhizoids). There are two species: *A. mediterranea* with a stem of about 5 cm, an umbel which in the stage of spore formation is divided into 75 sectors and which has a diameter of about 1 cm. *A. Wettsteini* is shorter and more compact with only 15 sectors in the umbel. These

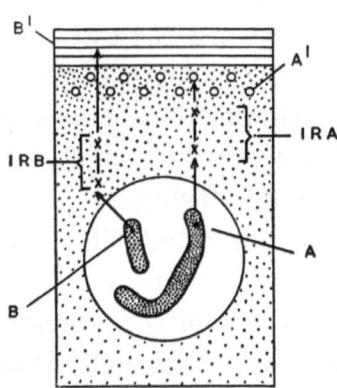

Fig. 211. Scheme of the intra-cellular effect of gene-products (after KÜHN, 1939).

algae are uni-cellular with only one nucleus, which lies in the base of the stem near the rhizoids. On completion of the umbel, the nucleus migrates to the umbel and yields, after repeated divisions the required number of nuclei for the spores. HÄMMERLING has carried out important regeneration experiments with this organism from which it became evident that a definite morphogenetical substance is necessary for the formation of the umbel, that is to say, a substance produced by the cell-nucleus. That we may speak here of a gene-product or substance is evident from the fact that after grafting of a long top-part of *A. mediterranea* upon a lower part of *A. Wettsteini* (in which the nucleus and hence the genes are located) the *mediterranea-parts* still regenerate, apparently because in this part *mediterranea*-products are accumulated which are gradually consumed, however, finally a new umbel is formed which shows a pure *Wettsteini*-type. Hence, the system was composed of mixed plasm of both species and a nucleus of *Wett-*

steini which also contains the genes and produces the gene products. In this way the occurrence of intracellularly acting gene-products was established with certainty.

Besides the direct intra-cellular effect of genes (by means of gene-products) there exists the possibility that these gene-products are transported through the body and hence, have an *inter-cellular effect*. We are indebted to STURTEVANT (1920) for the first communication in this field, he dealt with a gynandromorphic *Drosophila* in which the male part possessed only one X-chromosome per cell-nucleus and in this X-chromosome was located the recessive gene v for vermilion coloured eyes, whereas the feminine part has two X-chromosomes and the dominant gene for wild coloured eyes (i.e., V). When the eyes were situated in the male part then their colour was that of the wild type in spite of the recessive vermilion genotype present in this part of the body. Hence the gene-product for wild coloured eyes was produced in the feminine part of the body and transported to the eyes which are situated in the male part.

Analogous experiments and ones very similar with these dealing with gene-substances which have an intercellular effect were carried out by KÜHN and his co-workers CASPARI, HENKE and PLAGGE (see PLAGGE 1939) with the meal-moth *Ephestia Kühniella*. The gene a+ which is pleiotropic and dominant over a, expresses the following characteristics: black colour of the imaginal eyes (aa red), brown colour of the testes (aa colourless), pink larval skin (aa white), dark brown brain colour (aa pink), a high amount of pigment in the larval eyes (aa few), high speed of development and great vitality (in aa-animals both characteristics are much reduced). It turned out that in aa-animals, that is to say animals which do not have these colours by virtue of their own genotype, it is possible to 'colour' these organs by means of transplantation of definite tissues out a+a+-animals into the recessive individuals. These a+a+-tissues secrete a gene-substance, which in its turn now has an effect in the recessive individuals. For instance, by transplanting a+a+-testes, the imaginal eyes, the brains and the testes of the host animals also become coloured. By early transplantation (that is to say before the last but one larval stage), the larval eyes and the larval skin also become coloured. This effect takes place very quickly; if an a+-testis is transplanted into an aa-pupa in the ninth day of development and is again removed after 24 hours then a black colouration of the eyes of the aa-animal will be obtained; in 16 hours the colour has already become coffee-brown. Such a testis can be transplanted almost ten times in 10 successive host animals (aa) and it still keeps its potentiality for transference of the gene substance.

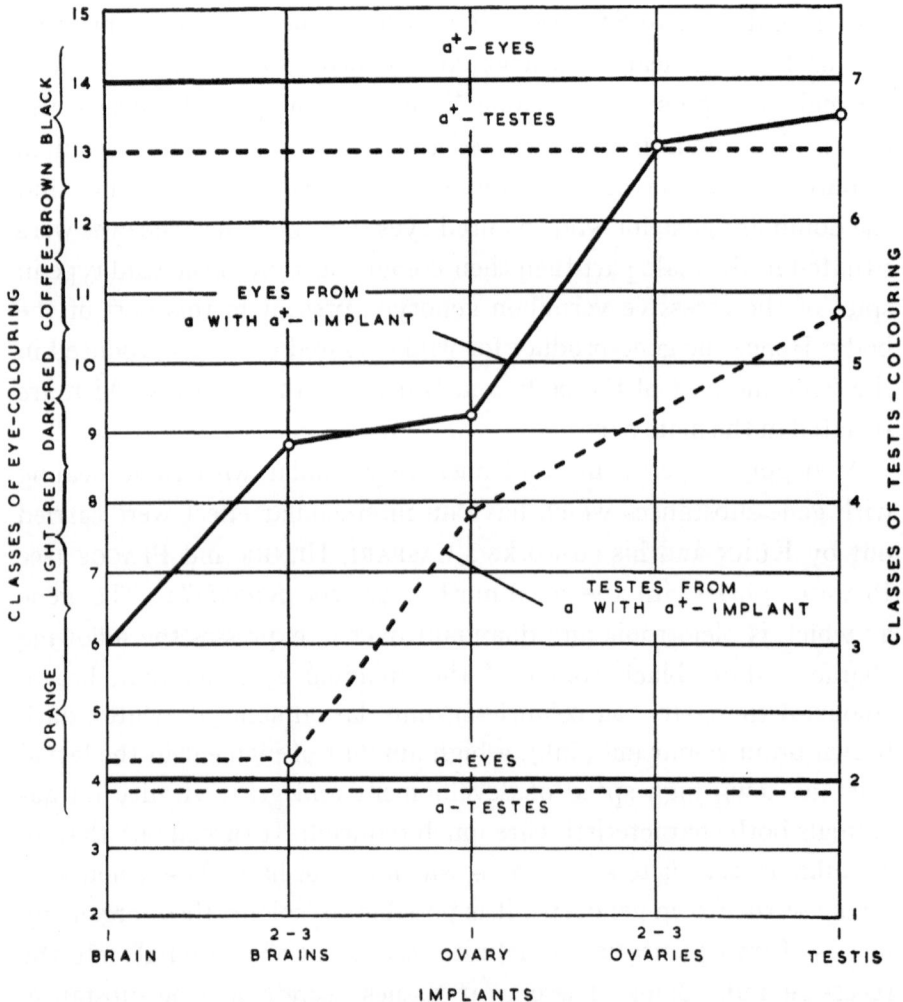

Fig. 212. Comparison of the effect of brains, ovaries, and testes with regard to the production of gene-products (after PLAGGE, 1939).

Transplantation of a+-ovaria or a+-brains also causes colouration of the imaginal eyes but not so intensively as does transplanting testes.

Two ovarian implantations reach the same effect with regard to eye-colour as does implantation of one testis, two brains almost correspond in effect with one ovarium. There is however less agreement concerning the colouration of the testes: two brains have almost no effect, and even two ovaria do not have the effect of one testis alone (this ratio is represented graphically in fig. 212). An oviduct of an a⁺-animal is absorbed after transplantation, however, it still wields a colouring effect. Hence it is obvious that this gene-substance can either accumulate in organs from which it can be used or the gene-substance (product) can be produced directly. This finds even stronger expression in the experiment of transplantation of a⁺-testes into a recessive female aa-larva. These testes give off the gene-product, which is accumulated in the ovarium of the recessive animal which does not itself possess the gene. Due to this transplantation the progeny of such an artificially impregnated female crossed with a recessive male still have coloured eyes for some time (fig. 213).

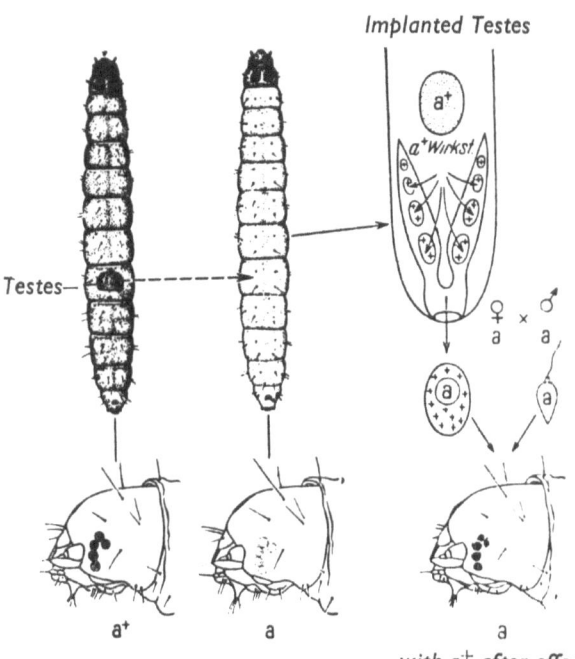

Fig. 213. Storage of a+-gene-product by the egg-cells and its effect on the pre-determination of the eyes of the offspring larvae. The a+-gene product is represented in the figure by+ (after PLAGGE, 1939).

Rather analogous with this study are the experiments of EPHRUSSI, BEADLE and CHEVAIS (1935, 1938, 1939) with regard to the gene-substances which are responsible for the eye-colour in *Drosophila*. On the grounds of transplantation experiments of the eye-disc of the fullgrown fly from the dark-eyed wild type (carrying the dominant v⁺ and c⁺ genes) into flies with much lighter coloured recessive vermilion (v) and cinnabar (cn) eyes, and reciprocally it became evident that a v-disc as well as a cn-eye transplanted in a wild coloured fly will adopt the wild colour. A v-disc transplanted on a cn-fly gave the eye colour of the wild type fly, however, cn-eyes transplanted into v-flies preserve their own cn colour. Hence the two substances which are produced for the expression v+-gene and the cn+-gene are not identical. A v-eye transplanted into a cn-host itself carries the cn+-gene and finds the v+-gene in the host which enables the development of the normal wild colour by means of interaction of v+ and cn+. Hence it appears highly acceptable that the v-type does not possess the v+- factor but the cn+-factor. On the other hand the cn-type contains the factor v+ but not cn+. Further it is evident that v+ is necessary as a substructure for cn+.

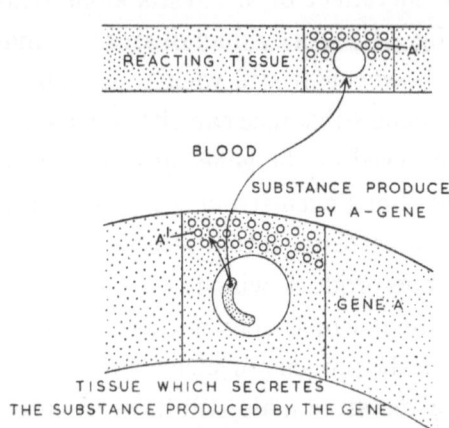

Fig. 214. Scheme of the inter-cellular action of gene substances; the A-substance is transported by the blood in the body to the reacting tissue (after KÜHN, 1939).

In both the examples discussed, that of the a+-factor in *Ephestia* as well as that of the v+- and cn+- genes in *Drosophila* the production and the transport of the geneproducts are of an intercellular nature. They are produced by the implanted organs and spread in the body of the host or are formed by the cells of the body of the host and act upon the implanted organ (fig. 214).

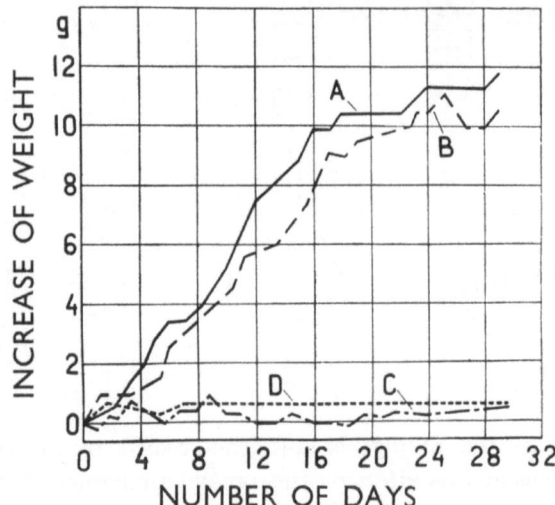

Fig. 215. Growth of normal mice (A), of dwarf mice without (D) and with implantation of normal (B) and dwarf hypophyses (C) (after SMITH and McDOWELL, 1930).

Beside these intracellular and intercellular effects of genes by means of gene-products, there still remains a third possibility, i.e., the *hormone*. This substance is produced by a definite organ, a gland with internal secretion. The formation of this gland however, is controlled by a definite gene (Symposium 1942). SMITH and McDOWELL (1930, 1931) gave the first example of this control, which was later amplified by KEMP and MARX (1936–1937). A dwarf type of mice depends on the presence of one single gene (dwarfism) which behaves towards the gene for normal growth as a recessive. The development of this dwarf type is much delayed and incomplete due to the defective condition of the frontal lobes of the hypophysis. By daily feeding of the frontal lobes of the hypophysis (pituitary body) of normal animals, dwarf mice are able to reach almost normal size and their vitality becomes equal to that of normal mice. Transplantation experiments (fig. 215) had analogous results.

The dwarf-pituitary gland appeared to differ from the normal one mainly because they do not form growth-hormones in the eosinophilic cells of the frontal lobes. Hence, the abnormal gene for dwarf

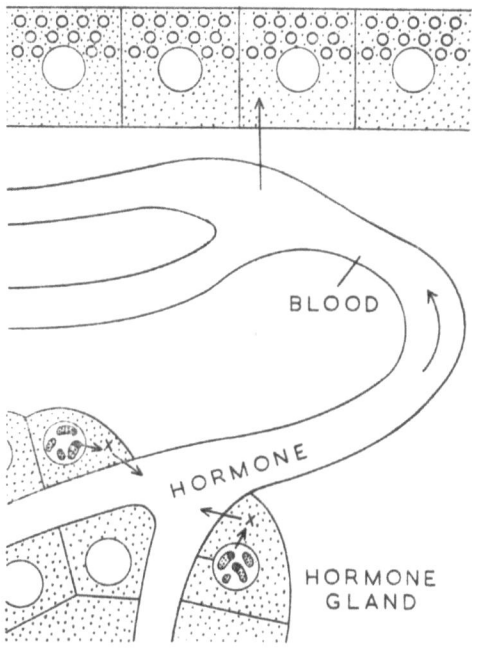

Fig. 216. Scheme of formation of a hormone in a gland of internal secretion under influence of a gene (after KÜHN, 1939).

development causes the lack of development of the pituitary gland by means of a definite gene-product. This again causes the secondary effect that no hormone is developed by the under-developed pituitary.

In the same way the *organ-forming effect of a single gene* is demonstrated in the study of HADORN and NEEL (1938) and of SCHARRER and HADORN (1938) concerning the pupation in *Drosophila* which depends on the normal operation of the pupation gland located near the larval brains (gland of Weismann, corpora allata). In a certain

stock of *Drosophila* which differs in one single recessive gene from
the normal, the pupation gland is badly developed. This again causes
the pupation to be very delayed or even completely lacking. Transplantation of a normal pupation gland into an abnormal larva
restores the normal course of pupation.

Also, from botanical aspect some information became known. VAN
OVERBEEK (1935, 1938), DE HAAN and GORTER (1936), WETTSTEIN
and PIRSCHLE (1938) possibly demonstrated gene produced growth-
substances whereas MELCHERS (1941) refers to so-called hormones
responsible for the bearing of blossoms as products from genes.

The chemistry of these gene products, or substances, form a rich
field of study. Besides the substances of *Ephestia* and *Drosophila* we
find analogous objects in the colour of the hair of rabbits and guinea
pigs, in the serum of pigeons, in the colour substances of flowers and
other plant parts (anthocyanes, flavones, carotenes), in the metabolism
processes of the mold *Neurospora* and in mans' urine-metabolism.

The biochemical approach in *Ephestia* and *Drosophila* was the first
step which was principally carried out by E. BECKER (1938, 1939, 1942).
Several important details came to light in these investigations i.e.,
that the same chain of gene substances in different animal species can
bring about the formation of pigments with very divergent and different chemical constitution (*ommines* and *ommatines*). Hence we must
conclude that these gene substances or products do not encroach upon
the last stage of the pigment formation but already much earlier upon
the basis of the chain of processes which later differentiates in several
directions. Such a point of view becomes especially evident from the
investigations of EPHRUSSI (see Symposium 1942) and is pictured
in fig. 217. The so-called pupation gland in *Drosophila* contributes towards the development of an eye-disc, in which tryptophan originates
by means of the activity of the pupation hormone. The genotype vv
is not able to bring about further conversion of this substance and
leaves the eye vermilion; gene v$^+$ converts tryptophan via α -oxy-
tryptophan into kynurenine, that is to say, the v$^+$-substance, which is
thereupon altered into the cinnabar-substance (a chromogene). On the
other hand a substrate is formed in the eye-disc which develops under
the influence of st$^+$-gene (the scarlet-gene) into an element (basic substance) for brown colour. Together with the cn$^+$-substance this element
for brown colour gives rise to the brown colour (ommatine). The same

substrate from the eye-disc can be transformed by the bw+-gene (brown) into a basic substance for the development of red colour of which nothing further is known for the time being. In the last resort, the brown colour and red colour come to expression together in the

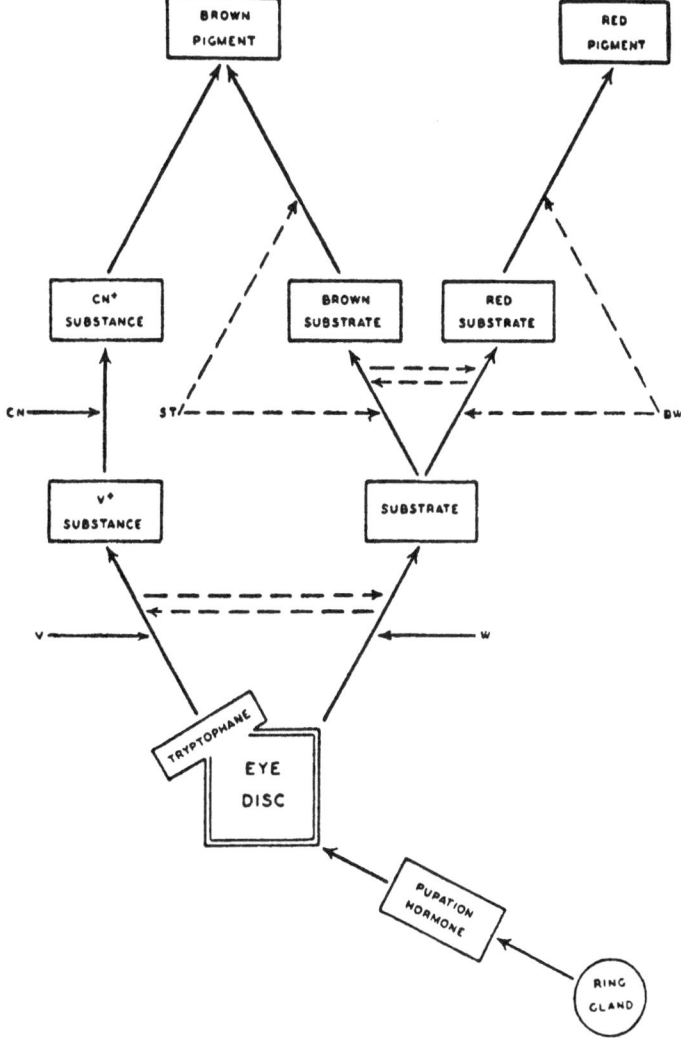

Fig. 217. Scheme of gene-action in the origin of the wild colour in the eye of *Drosophila* (after EPHRUSSI, 1942).

full-grown eye which owes its wild colour to this cooperation of brown and red colour.

The *melanins* in animals anyway seem to be a promising object in this field of genetics (SERRA 1943). The so-called Himalya rabbit is wholly white at birth, but it shows a typical acromelanism as a full-grown animal i.e., black nose, black earpoints and a black tail-point. This typical colouration depends on one single gene c^n which is recessive towards the gene C for total colouration and dominant towards the gene c for complete albinism. SCHULTZ (1915) showed, that other parts of the body can also form black colour, that is to say, if the animal is sheared locally and the new hair-growth takes place at reduced temperature levels. The critical temperature below which colour can be formed is about 33°C. After birth, cooling down of the extremities occurs which causes acromelanism. If new-born animals are exposed to higher temperatures, then they remain wholly white.

DANNEEL (1941) and his co-workers have investigated this problem in more detail. The result is that cooling down occurs during a sensible period of hair formation. The duration of such a period may be limited to five minutes. The process of colour formation is composed of two parts. The first period comprises the formation of a precursor, which is temperature sensitive and which cannot be inhibited by cyanide. The formation of this precursor is an anoxydative process. Hence the conclusion may be drawn that here an oxydative enzyme is formed from its pro-enzyme. In the second period oxygen is necessary and also, the process can be inhibited by cyanide, but on the other hand the process is not temperature sensitive. Investigations based on an extract method are still necessary; a complete picture cannot be drawn for the time being.

A favourable object for such investigations is the guinea pig, which was studied by WRIGHT (1945 a and b). Here two series of hair colours are available: the xanthines (red or yellow) and the melanins (black, sepia, brown); three genes (A.B.E.) prepare the hair follicles for uptake of a definite colour and hence they determine the type to which the hair will belong i.e., E determines brown, B sepia, whereas A (agouti) inhibits the formation of melanins in specific parts of the hair, so that only red or yellow show up instead of the anticipated brown and black colours. Besides these genes there are other ones such as C (a series of multiple alleles), F and P which all have a definite chemical effect because they influence enzymatically the formation of yellow, brown or sepia melanins from their mutual precursor for melanins (3.4-dihydroxyphenylalanine). Figure 218 gives a schematical representation of these processes.

The differences which occur between the antigenes in doves are more of a physiological chemical nature: the pearl-necked dove (*Streptopelia* chinensis) differs at least in ten specific so-called antigenes from the ring-dove (S. risoria) and these differences are again based upon different genes in the nucleus. The method applied by CUMLEY and IRWIN (1942) is rather simple (fig. 219). By means of injection of the blood of S. chinensis (pearl-neck dove) into a rabbit, anti-bodies are formed which go with the pearl-neck antigenes. The ability of agglutination of this rabbit serum with regard to the red blood bodies of the ring-dove is now studied and by means of agglutination the differences between the types of antigenes present in both pigeons can be established. It is

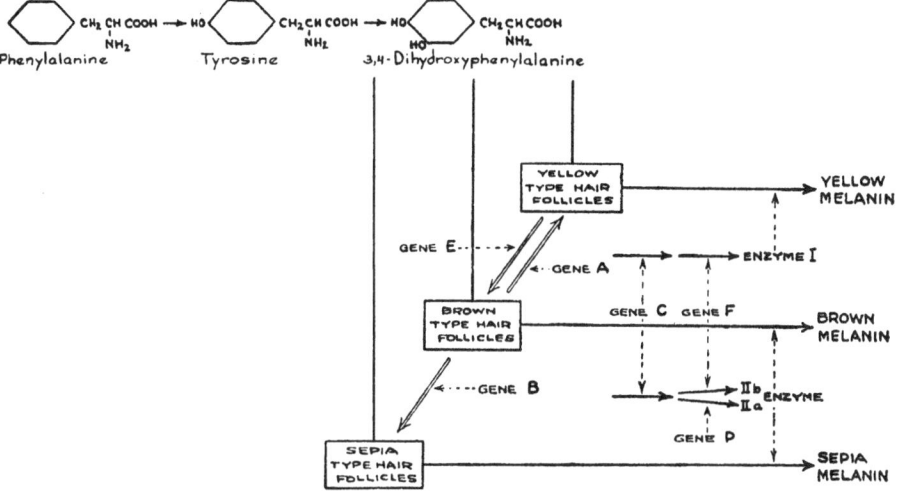

Fig. 218. Scheme of the formation of melanins in *Cavias* (after WRIGHT from BEADLE, 1945).

even possible to develop a rabbit serum which responds to one definite antigene and by means of this method it is possible to determine the presence of definite antigenes for each individual originating from the cross pearl-neck dove × ring dove in the F_2- or following generations. Hence, from this kind of investigation it is also possible to draw a conclusion with regard to the effect of the underlying genotype (see IRWIN 1946, 1947).

Good results are also obtained by comparing genetical investigations into flower-colours with those obtained in biochemical investigations. Although started rather long ago (WHELDALE 1914) this study made great progress only after 1930 (SCOTT MONCRIEFF 1936, LAWRENCE and PRICE 1940, HALDANE 1942). The basic formula of the anthocyanes can be found in pelargonidin (see formula p. 523), which is capable of

several alterations. Oxidation for instance of position 3′ will give rise
to cyanidine, oxidation of 3′ and 5′ alters pelargonidin into delphinidin. These three basic types are easy recognizable by their chromato-

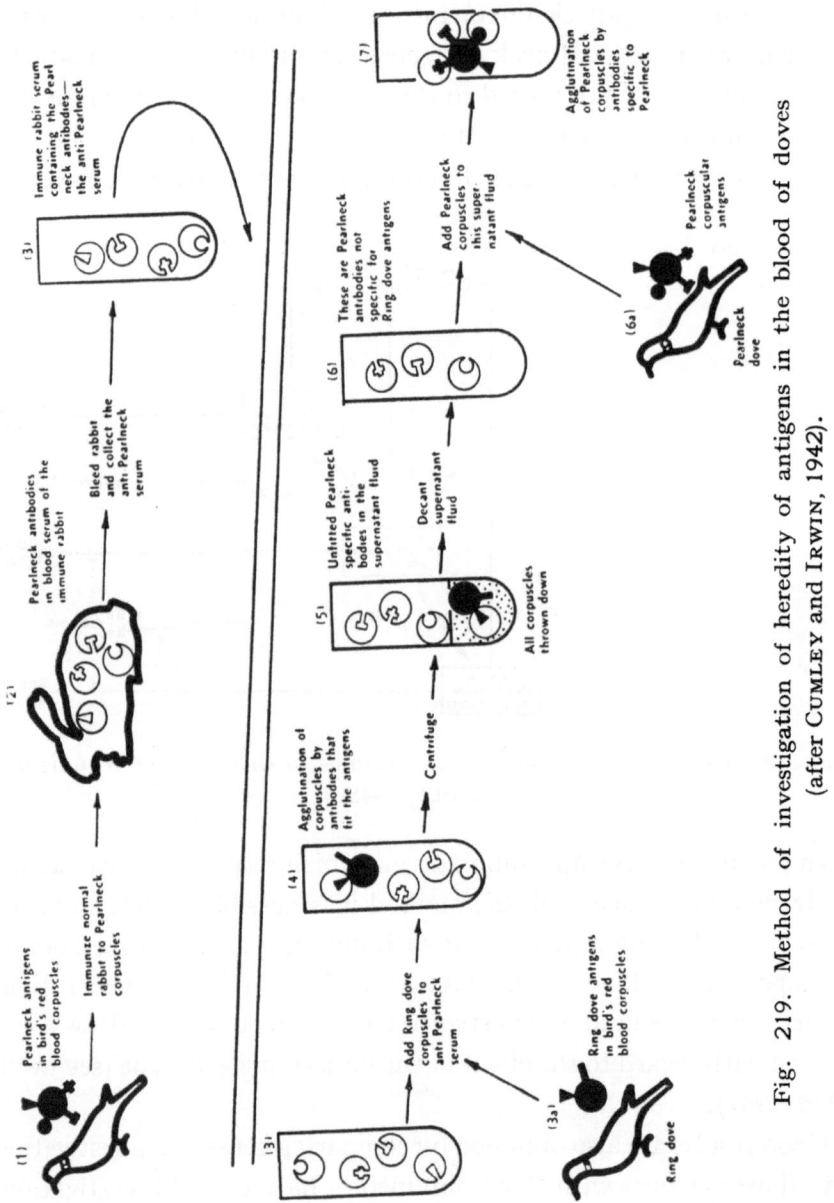

Fig. 219. Method of investigation of heredity of antigens in the blood of doves (after Cumley and Irwin, 1942).

grams. Methylation of 3′ of cyanidin yields paeonidin, methylation
of delphinidin of the 3′ position gives the petunidin, and methylation
of 3′ and 5′ malvidin and of 3′, 5′ and 7′ gives rise to hirsutidin.

Hence the colour depends upon methylation and also upon several cooperating factors: increase of pH of the cell-sap and increase of co-pigments (anthoxanthines and tannins) cause a more intensively blue colour. Methylation gives rise to a redder colour. The structure of the seven main anthocyanidins can be derived all from the following formula of pelargonidin (-chloride) i.e., the 3.5.7.4'-tetra-hydroxy-2 phenylbenzo-pyryliumchloride:

With OH at position 3': cyanidin; with OH at position 3' and 5': delphinidin; with OCH_3 at position 3': paeonidin; with OCH_3 at position 3' and OH at position 5': petunidin; with OCH_3 at position 3' and 5': malvidin; with OCH_3 at position 3', 5' and 7': hirsutidin and coupled with monosides or biosides at position 3' or/and 5' all sorts of mono- and diglucosides (anthocyanes), for instance pelargonin-chloride $C_{27}H_{31}O_{15}Cl + 2\ H_2O =$ pelargonidin chloride $C_{15}H_{11}O_5Cl + $ glucose $2\ C_6H_{12}O_6$, can be obtained.

In *Lathyrus odoratus* and *Primula sinensis* especially, as also in a number of other species, genes were discovered which each have a very specific effect: 1) oxidation at 3'; 2) oxidation at 5' as 3' is already oxidized; 3) simultaneous oxidation at 3' and 5'; 4) methylation of the hydroxyl group at 3'; 5) methylation of the hydroxyl groups at 3' and 5'; 6) substitution of a sugar rest at 3'; 7) substitution of a sugar at 3' and 5'; 8) coupling of an acid rest at an unknown position; 9) change of the pH and 10) addition of a co-pigment. The great variety in red and blue colours, which we will find in so many plants can be genetically determined in this way and can be explained by biochemical investigation.

The same method can be applied for the recognition of yellow cell-sap colour like anthoxanthines and flavones which are all related to the anthocyanes and also for the recognition of green, orange and yellow plastid colours i.e., chlorophylls (v. EULER 1929–1935), xanthophylls and carotenes (DOUWES 1943).

In the early days of genetics the expression of a definite genotype was usually observed by means of morphological characteristics. In cases of a simple Mendelian segregation, dealing with colour, forms of organs etc., we are able to distinguish different genotypical types in an offspring. However as soon as we deal with a complex system or with a system in which interaction of genes occurs upon the phenotype we will find that morphological characteristics can show transitions from one type to another and a final conclusion is hard to derive.

As more became known in the field of genetics, more attention was drawn to possibilities of investigation in which no interaction takes place or at the utmost, to an understandable degree. In plants and animals the morphology is a result of almost balanced processes, that is to say, many metabolic and developmental processes find their joint expression in one characteristic. Therefore one may not wonder that genetics tried to obtain a more overall picture of their field by studying these elementary systems. To-day we will find in every branch of genetics a rich literature concerning investigations into fundamental processes. In human genetics there is the case of *phenyl-ketonuria* and the metabolism of *phenylalanine* and *tyrosine*.

Already in 1909 GARROD in his book 'Inborn Errors of Metabolism' gave genetical references to the disease *alcaptonuria* in man. We in our modern times will see in GARROD a pioneer in a field of study, which was very fruitful for genetics, however, in his own time nobody appreciated the profundity and significance of his thoughts.

We already saw that phenylalanine is transformed via tyrosine into 3.4-dihydroxyphenylalanine, which again gives melanines under the influence of a dominant gene; the recessive allelomorph interrupts this process which in its turn causes *albinism*. On the other hand this phenylalanine can be broken down in the following steps: 1) phenyl-pyruvinic acid, 2) p-hydroxyphenyl-pyruvinic acid, 3) 2.5 dihydroxy-phenylpyruvinic acid, 4) homogentisinic acid (alcapton), 5) aceto-acetic acid, 6) carbon-dioxyde and water. Step 1 - 2 is influenced by a dominant gene; if this is a recessive gene the heritable aberration occurs which is known by the name of *phenylketonuria*. Phenylketo-nuria causes a high amount of phenylpyruvinic acid to be accumulat-ed in the urine. Between step 2 and 3 another gene is the cause of a chemical reaction. This gene as a recessive allele gives rise to *tyrosinosis*. The step between 4 and 5 is controlled by a dominant gene; its recessive allele causes an accumulation of alcapton in the urine, known as *alcaptonuria*.

It is clear that in this study, by means of chemical investigation parallel with the genetical investigation, the pathway of these aberrations was discovered, that is to say the rôle which some dominant

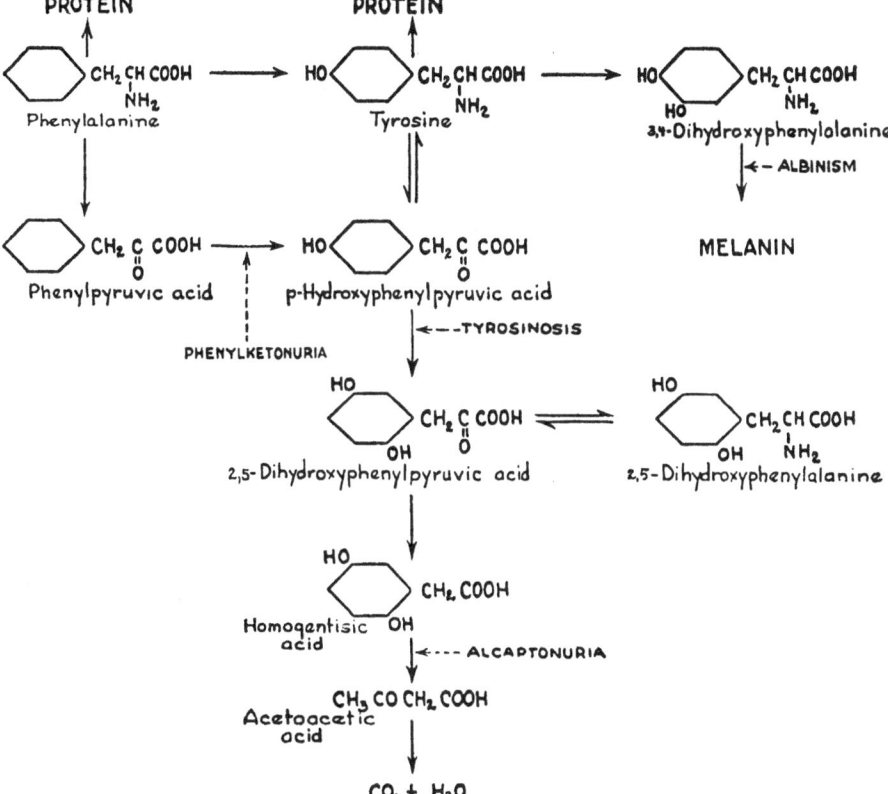

Fig. 220. Gene action in urine metabolism in man (after HALDANE, 1942).

genes can play in the chemical processes involved in the formation of urine in man (see HALDANE 1942, PENROSE and RIMINGTON in Symposium 1950b; also fig. 220).

Undoubtedly these preliminary investigations of gene action and biochemistry were the first steps in a direction where genetics made considerable progress in the field of fundamental research. Yet the significance of this fundamental research came to light not earlier than about 1940. In their almost classical investigations BEADLE, BONNER, HOROWITZ, MITCHELL and TATUM (1945) showed for the first time a mechanism of tryptophan synthesis in living organisms.

In a long series of papers the school of BEADLE developed a new branch of genetics that is to say the genetics of *Neurospora crassa*. From their investigations it became known that the synthesis of tryptophan and of niacin and many other compounds was genetically controlled. Therefore substantial contributions of bio-chemical and genetic investigations in mutant strains of micro-organisms into the biosynthetic pathways involved in the formation of many amino acids and vitamins have been made. From the studies of BONNER (1951) we discussed already (p. 324) the pathway of niacin which lead to such remarkable modern concepts in the field of genetics, more remarkable since these contributions were made with help of microbiology and biochemistry, side-branches of biology, which play such an important role in the allover picture of genetics to-day.

From *Neurospora*, as well as other microorganisms, there can be isolated mutant strains which have lost the ability to form such vital cellular constituents as vitamins and amino acids. In general bio-chemical mutants have been obtained by treating conidia with muta-genic agents as ultra-violet, or X-ray radiation or a variety of chemical substances. Such mutant strains which have lost the ability to per-form a single biochemical reaction differ from the parental strain by alteration of a single gene, i.e., invariable 1 : 1 segregation of the nutritional requirement has been obtained when such strains are crossed with the parental strain. These isolated strains in general differ from the parental strain by requiring an exogenous source of a known vitamin, amino-acid or similar metabolically vital compound. The requirement for an end-product has been traced in certain instan-ces to an inability to perform a specific biochemical reaction involved in the formation of the end-product. The identification of this specific reaction is premised primarily on the ability of a mutant strain to use a specific compound for growth, but not the immediate precursor of this compound. Mutant strains requiring the same end-product may or may not be allelic. However allelic strains are invariably blocked in the same biochemical step, while non-allelic strains are invariably blocked in separate steps in the series of specific biochemical reactions involved in the formation of the end product. For instance biochemical studies of genetically different mutant strains of *Neurospora* have shown the action of definite genes in the ornithine-citrulline-arginine chain in the metabolism of proteins (fig. 221).

Other studies concerned strains which require niacin for growth, as has been discussed already in chapter XIV (p. 324). The work of several investigators has led to the conclusion that niacin is formed from tryptophan through the intermediate compounds formylkynurenine, kynurenine, 3-hydroxy-kyrurenine, 3-hydroxy-anthranilic acid, and an aliphatic compound of undetermined structure.

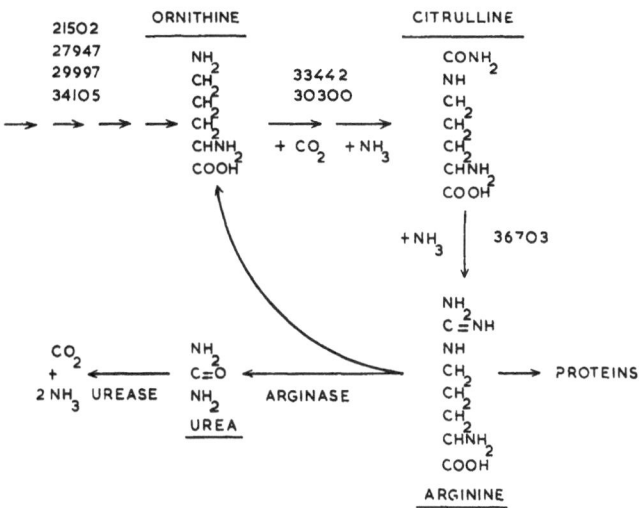

Fig. 221. Scheme of formation of arginine under influence of specific genes in *Neurospora* (from HOROWITZ a.o., 1945).

Similarly tryptophan has been found to be formed from anthranilic acid, though there is not complete agreement at the present time concerning the role of anthranilic acid as a normal intermediate. Direct precursors of tryptophan are indole and serine, which together are condensed to tryptophan with help of the enzyme tryptophan desmolase. Gene mutations which result in the loss of the ability of mutant strains to carry out particular reactions shown in fig. 133 (p. 324) are known in *Neurospora*.

Some mutations lead to requirement for tryptophan and to an inability to convert indole to tryptophan, while others lead to an inability to convert kynurenine to 3 OH-kynurenine. The striking fact about such mutations, however, is that all strains which require niacin for growth and which are unable to convert kynurenine to 3 OH-kynurenine, for instance, are genetically identical or are representative of changes of very closely linked genes (see p. 324). Therefore alteration of the same locus always results in a loss of ability to perform the same biochemical reaction. Hence as most biochemical reactions are enzyme catalyzed, the fact that genetic change leads to loss of ability to perform a certain reaction implies that genetic change results in loss of a functional enzyme. In general, loss of activity of a particular enzyme is associated only with the alteration of a single gene, however, cases are known which show a multiple genic control of a single enzyme.

The simplest theory as put forward by HOROWITZ (1951) is that single enzymes are controlled in their specificity by single genes. In this theory enzymes are essentially gene-products and genes serve in their primary rôle as a template for enzyme formation. The gene, in its rôle as a template, might use non-specific polypeptids and confer specificity on them, or it might assemble the enzyme starting with amino acids and serve as a template from the very beginning of the entire process of enzyme synthesis. If this postulated relationship were the actual relationship of gene to enzyme, enzyme formation would occur in the nucleus. A slight modification of this relationship, however, would permit enzyme formation to occur in the cytoplasm. For if the gene were to form a product closely resembling itself, this product could move to the cytoplasm and there serve as a template for enzyme formation.

However, if the gene serves primarily or secondarily as a template for enzyme formation, one might expect that genetic alteration should result either in complete loss of enzymatic activity or in an enzyme with altered properties. In order to assess such a hypothesis, the types of enzyme-change that are found associated with genetic change must first be determined. It is possible to determine experimentally whether a genetic change has resulted in complete loss of ability to perform a given reaction. If genetic alterations do lead to a complete loss of ability to perform particular biochemical reactions, tracer atoms incorporated into intermediates after the point of block would be expected to be found in equal concentrations in the end products. Therefore the completeness of a genetic block can be judged by the dilution of tracer in an end product, provided no exchange of tracer occurs and no alternative paths of end product are functional (BONNER, YANOFSKY and PARTRIDGE, 1951). BONNER was able to show that in the case of tryptophan formation a genetic block characteristic of a strain (T-3) does not lead to complete loss of ability to form tryptophan. The completeness of genetic blocks has also been examined experimentally by another technique. It is a well known fact that meny strains of *Neurospora* accumulate characteristic compounds. The basis of the use of double mutants can be most easily shown diagrammatically

$$C \xrightarrow{\text{gene 3}} B \xrightarrow{\text{gene 2}} A \xrightarrow{\text{gene 1}} EP$$

Assume that a given end product (EP) is formed through a sequence of reactions involving three other intermediate compounds. Gene 1 controls the conversion of A to EP and, if defective, causes the accumulation of A. Gene 2 controls the conversion of B to A. The double mutant having defects in both gene 1 and 2 should not accumulate A, provided the alteration of gene 2 has resulted in a complete block of the ability to convert B to A. Gene 3 is concerned with the conversion of C to B. The double mutant having defects in genes 1 and 3 likewise should not accumulate A if grown in the absence of B.

However, if B is present, A should be accumulated and, in turn, the extent of accumulation of A should be a function of the amount of B added. This method is only valid if there are no alternative pathways of formation of A. Investigations carried out by the school of BONNER showed that there are strains which are able to carry out their presumed blocked reactions under growing conditions. Gene-change in these cases frequently leads to a quantitative alteration in ability to perform a certain reaction and only infrequently to a complete loss of ability to perform it. It still remains at this moment a question if a change which results in changed ability to perform a given reaction is due to a change in the specificity of the enzyme in question, or has genetic change resulted in a change in the rate or time of enzyme formation, due to some change in the formation of a necessary precursor or some change in the enzyme-forming system (BONNER 1952). Only a few superficial investigations into this problem are known. The tryptophan desmolase mutant strain T-1 cannot convert indole to tryptophan. Examination of extracts of this strain have failed to yield any evidence of the presence of tryptophan desmolase, the enzyme which couples indole and serine (YANOFSKY 1952). A suppressor mutation at a second locus, however, partially restores the ability of this mutant strain to form tryptophan, and extracts of strain T-1 bearing the suppressor gene are found to contain tryptophan desmolase. The tryptophan desmolase of this strain has been carefully compared with that of the normal strain and has been found to be kinetically identical. Therefore a strain carrying the defective gene which originally resulted in loss of ability to form tryptophan desmolase, can now form tryptophan desmolase, and the tryptophan desmolase formed is normal.

The tryptophan desmolase activity in the suppressor strain is far lower than in the parental strain, yet far exceeds that in the mutant strain. In conclusion BONNER (1951) abandoned the almost classical hypothesis of one gene- one enzyme relationship as put forward by HOROWITZ (1951): 'These observations, I feel, are particularly noteworthy since in this instance alternative pathways of tryptophan formation from indole have been substantially ruled out. Either two independent genes are then capable of forming the same enzyme, or two different genes can affect its formation' (1951, p. 149) and further 'The suppressor mutation, however, results in formation of apparently normal enzyme. This could imply that two independent genes are both capable of mediating the formation of the same enzyme. However since it is known that genetic alterations lead to alterations in time and rate of lactase formation, it is equally plausible to believe that the original mutation in strain S-1952 (suppressed) did not directly affect the specificity of tryptophan desmolase, but rather altered its formation in point of time and rate. The suppressor mutation in turn might then render a sufficient biochemical change to permit earlier or faster formation' (BONNER 1951, p. 150).

The simple picture of GOLDSCHMIDT (see p. 309) considering a gene as an enzyme may be wrong from a strictly biochemical point of view, but on the other hand we have to admit that this view has close connections with our modern interpretation of enzyme formation as being genic controlled, even in the case when enzyme formation occurs in cytoplasmic particles (as put forward by BONNER); whereas the gene controls the ability of cytoplasmic particles to form replicas in active form.

From a genetical point of view such a conclusion is of immense importance. Perhaps we may not completely understand the complexity of gene-enzyme relationship, however, we get the thread which runs through this problem. To obtain an overall picture of gene activity we undoubtedly need more data.

Genetical and cytological investigations are especially presumed to carry us over the barrier which now exists. At this moment there are already signs that the close alliance between biochemistry and genetics meets controversies at several points. One may account for this as being due to limitations in the successive fields, however, the whole concept of genetics is that of a dynamic science, that is to say a science aware of the possibility that living Nature is not always caught easily in a test tube.

It appears that too strict a mechanism (one gene-one enzyme) as postulated by HOROWITZ cannot be upheld. With regard to BONNER's point of view that more genes might be involved in the synthesis of an enzyme; this is not yet proven from a general point of view. Final proof is difficult to obtain; in this respect we will think particularly of modifying genes, such as common in *Drosophila*. These modifying genes are so similar in their effects and there are so many that they often cannot be individually identified, symbolized, or located on chromosomes. Some evidence in this direction seems to be produced, however, not enough data is available at this moment to put such a conclusion forward.

Whereas these genetico-chemical studies give on one hand a support by trying to interpret the phenotypical differences in the full-grown individual, there are, on the other hand, some studies which tried to discover differences at as early a stage as possible i.e., these studies try to approach the genotype directly.

Pollen-grains yield very favourable material for such investigations. Pollengrains have the reduced number of chromosomes and therefore belong to two or more types when originating from heterozygous plants. We already mentioned in this book investigations into the qualitative differences between pollen-grains (p. 368); BRINK (1929) gives more examples of this kind of study.

If we say that developmental-mechanical (embryological) and developmental-chemical (chemical embryological) investigations are making progress by following step by step the relationship between gene and phenotypical characteristic or property, then such a statement will only be valid when applied to qualitative characteristics, i.e., in so far as the phenotype is characterized by quantitative differences this method of investigation does not offer an opportunity of gaining information necessary for a better understandng of to the process of development. It is obvious that the study of these quantitative characteristics and their development will in general be more difficult and this can be mainly ascribed to the strong modifiability under influence of environment. However, there are other methods which may give us some information; in the first place we will mention a comparative investigation between the *autopolyploids* (see Chapter XII) and the original races, from which these polyploids arose. Through duplication or accumulation of chromosome sets, (i.e., the genome), a different proportion between the quantity of chromosomes and the plasm on the one hand can be achieved and on the other hand it is possible to preserve different genomes in different combinations.

Again we will mention in this respect WETTSTEIN's (1924–1928) interesting studies with mosses which have also yielded important data in this field. We saw earlier (p. 251) that considerable quantitative differences in cell size can be observed between haploid, diploid, triploid and tetraploid individuals of *Funaria hygrometrica*. These differences are exclusively caused by a multiple of the same genome, whereas the plasm-quantity remains the same. We thus meet here the old problem of the nucleus-plasm relationship. It was believed that the proportion in cell-size between haploids and polyploids should keep step with the number of genomes. However it became evident from WETTSTEIN's investigation that this is not exactly the case. In the first place WETTSTEIN found that this nucleoplasm relationship depends upon environment. The ratios differ considerably according to the medium available. For instance in *Amblystegium* WETTSTEIN found that the proportion in cell-size between n and 2n on common garden soil was 1 : 1·88, whereas on garden soil enriched with Knops solution a ratio of 1 : 3·14 was obtained. Other soils used as a growth-medium yielded intermediate ratios. Each soil has therefore a specific '*reaction constant*' which we can call s.

Next to this there exists a *race-constant;* comparison of different species and races grown on the same media led to the conclusion that very different race-constants occur. *Bryum caespiticium* and *Physcomitrella patens* yielded the lowest and the highest constants which are respectively 1 : 1·45 and 1 : 3·94. By means of accurate measurements carried out in *Funaria hygrometrica*, in which the ratio was 1 : 1·82, the general rule could be postulated that cell volume of an n-ploid form is proportional to cell-volume of the haploid and that this ratio is expressed in the formula $V = V_1 K^{n-1}$, in which V stands for the cell-volume, K for a specific constant and n the coefficient of polyploidy (haploid n=1, diploid n=2, triploid n=3 etc.). Figure 222 represents graphically the cell volumes for polyploids of the species *Physcomitrella patens* (Ph), *Funaria hygrometrica* (Hy) and *Physcomitrium piriforme* (Pi). It is obvious, that this method of investigation can be extended to a study on *allopolyploids* in which one or more genomes of different species are combined. These experiments led to the general result and conclusion that the K-values are much lower in these alloploids than in the autopolyploids. Figure 222 also

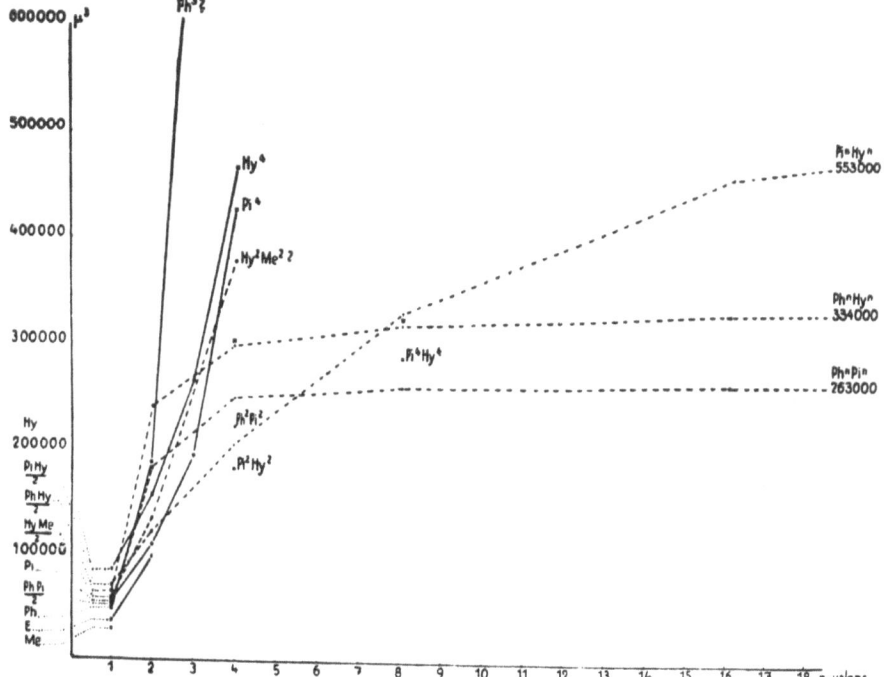

Fig. 222. Cell-volume in polyploid mosses (after WETTSTEIN, 1928).

represents graphically the ratios of cell volumes in the combinations HyMe (Me = *Funaria mediterranea*), PiHy, PhHy and PhPi. From this part of the figure it is evident that the graphs corresponding with these combinations do not show such an enormous increase as those corresponding to autoploids.

The existence of both these factors made the assumption of the occurrence of a fixed nucleus-plasm relationship dubious and together with other information obtained later by WETTSTEIN (1937–1942) it seems that the ratio between the amount of plasm (expressed in cell-size and organ-size) and number of chromosomes is highly variable. Regeneration experiments with the moss *Bryum caespiticium* led to the development of a diploid-tetraploid form with a cell-size of 34,000. This new form was derived from a normal haploid-diploid plant with a cell-size of 17,000 cubic micron. However, this giant form lasted only temporarily; over the course of the years this character (*gigas*) decreased more and more, so that the plant in 1939 at an age of twelve years showed an average cell-size of 18,000, that is to say very nearly the same measurements as the original haploid-diploid ancestor. However the number of chromosomes had not decreased over twelve years of time and hence the plant was still sticking to a diploid-tetraploid lifecycle. Comparison with diploid-tetraploid specimens of the same species which occurred in free Nature showed that in these plants also the actual sizes did not differ greatly from those of haploid-diploid plants.

These investigations and their conclusions were finally expressed in the rule: that artificial duplication of the number of chromosomes leads to a considerable increase in cell-size, however, no indissoluble relationship exists between cell-size and number of chromosomes. This rule was confirmed in studies with the phanerogamous plant species *Arenaria serpyllifolia* (WOESS 1941); in this plant also the cell-size increased twofold in a tetraploid which was obtained by colchicine treatment, whereas the tetraploid cells of plants which occur in free Nature are much smaller and almost the same size as those of diploids.

That the last word concerning the significance of the chromosome number has still not been spoken is clear from the very divergent results which were obtained in numerous studies with autotetraploids obtained by colchicine treatment.

In some experiments a considerable enlargement of the cells was obtained whereas in other races, even closely related races, such a result could not be observed (GYÖRFFI 1941, NEBEL et al. 1941, PIRSCHLE 1942, STRAUB 1941).

An analogous quantitative effect (by means of genome-duplication) was observed and chemically studied by LINDSTROM and GERHARDT (1926, 1927) in *Zea*. It is a well-known fact, that in corn the embryo originates from a single fertilization process of a haploid cell-nucleus with a haploid nucleus from the pollen-tube. Hence the embryo inherits equally from both parental sides. However, the *endosperm* (reserve food) arises from the fusion of two haploid embryo-sac nuclei with one haploid pollen-tube nucleus. The endosperm is influenced twice as strongly by the mother as it is by the father plant. The presence of starch in the endosperm is based on one single geno-typical gene S. Therefore it is possible to compare endosperms orig-

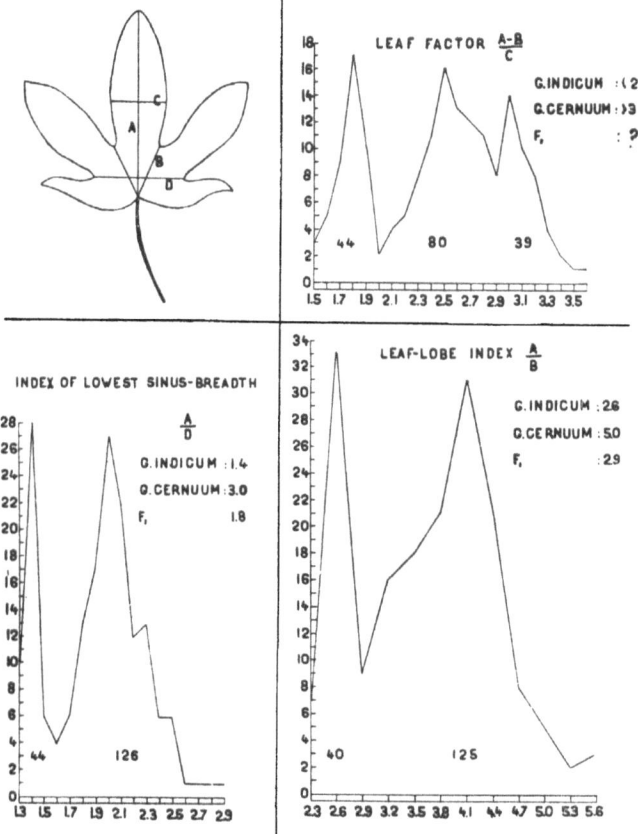

Fig. 223. Correlation between the leaf-measurements caused by one factor in *Gossypium* (after AFZAL, 1930).

inating from different pollinations SSS (= the pure starchy stock), SSs (starch as mother and sugar as father), Sss (sugar as mother and starch as father) and sss (the pure sugar corn). For instance: a pure starchy corn contains 58·3% starch and 5% sugar in the endosperm; a pure sugar corn 18·1% starch and 44 1% sugar.

Though the character for starch is dominant, it is still possible to distinguish chemically the four types SSS, SSs, Sss and sss, because the carbohydrate indices (% sugar + dextrines : % starch) amounted in SSS to 0˙09, in SSs to 0˙11, in Sss to 0˙15 and in sss to 2˙00. The influence of the presence of three, two one or none S-genes could be recorded in figures; the phenotype was almost proportional to the quantitaties of the S-gene.

One would expect that the results of crossings which are usually explained with the help of possible existence of polymeric genes are of importance for a study concerning the relationship between the multiple of one and the same gene and the phenotypical expression of these multiple genes (p. 130). However, the grounds for such a pos- tulate would be very weak, since *polymeric genes* were never analys- ed sufficiently and since we are not able to draw any conclusion with regard to equality (*isomery*) or inequality (*anisomery*) of the cooper- ating genes. If the presence of isomeric genes is acceptable we never know the exact number in which they are present. This is also the case when the heredity of certain quantitative properties appears to be based upon polymeric genes: the material is not suitable for an investigation into the development of these quantitative properties as a result of multiple action.

The future of a study of quantitative characteristics to ascertain if these are due to definite factors looks more promising (SINNOTT and DUNN 1935). We have already seen (p. 137) that data was obtain- ed, which shows clearly, that different quantitative properties are based upon one single pair of alleles. A very simple case of such a monohybrid nature of a quantitative character was found by AFZAL (1930), who studied the measurements of leaves in a cross of *Gossypium indicum* and *G. cernuum*. AFZAL took four fixed measurements of a quinary palmate leaf of a cotton plant (see fig. 223): A represents the length of the middle leaf lobe from leaf-base to top, B stands for the measurement from leaf-base to lowest point of the middle serations, C for the largest width of the middle leaf lobe and D, the greatest distance between both serations. From these measurements he calculated three standard leaf-measurements: the

leaf-factor $\dfrac{A-B}{C}$, index of lowest sinus-breadth $\dfrac{A}{D}$, and the leaf-lobe

index $\dfrac{A}{B}$. In his F_2-generations he found, as the graphs show (Figure

223), a clear-cut monohybrid segregation; the ratios obtained were as follows 44 : 80 : 39, 126 : 44, and 125 : 40. Hence, we can assign a specific quantity of phenotypical expression to each of these isolated genes.

In an almost analogous manner we can ascribe the measurements of leaves, fruits and seeds of *Vicia Faba* to very definite genes (see p.138).

These genes are polymeric genes and anisomeric, although

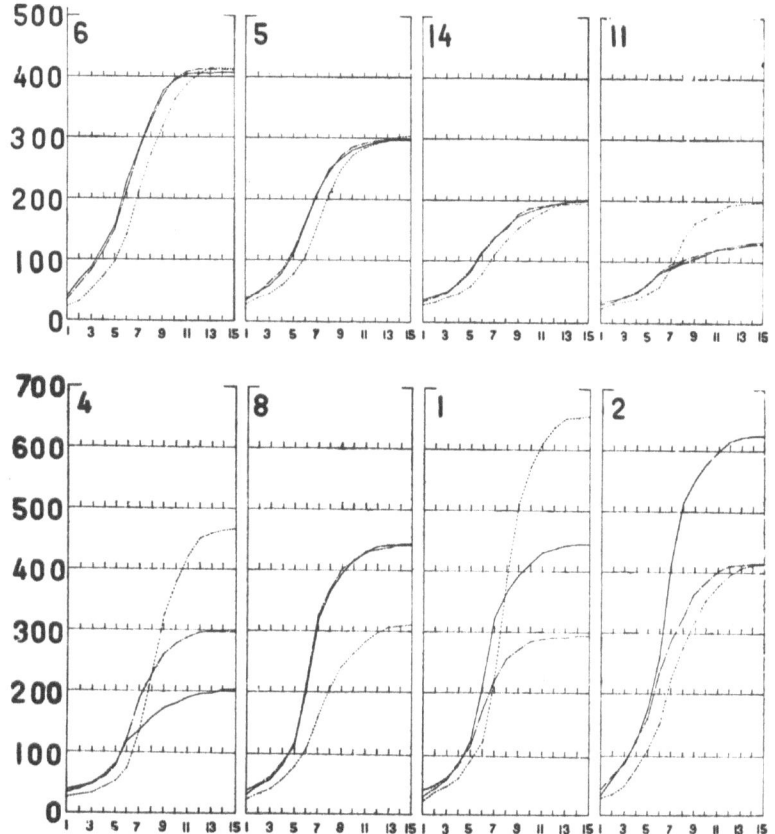

Fig. 224. Growth-rate of leaf-measurements in eight types of
Vicia Faba (after SIRKS, 1931).

some are represented by a series of multiple alleles. For instance, general growth factors are found to be due to a series of four multiple alleles G_4-G_1, one pair of additional factors for the leaf-breadth W_1-w, one pair for basic length of the leaf B_1-b and three multiple genes for the terminal leaf-length T_2-T_1-t. For each of the eight 'pure lines' investigated, the development of the three leaf measurements could be established graphically (figure 224). It became evident that the growth of the leaf was completed in almost fifteen days and that the growth of the terminal leaf length took place slightly later than the development of the breadth of the leaf and of the length at the base of the leaf (which keeps step). For the three pure lines 6, 5 and 14, which possess only G-genes and the other genes in their recessive form (respectively G_4wbt, G_3wbt and G_2wbt) the three growth-curves are similar whereas for the remaining pure lines it turned out that one of the measurements showed a more rapid growth than both the other ones, and also all three measurements could differ.

A close comparison of both figures with the genotypical formulae will make this clear. The unbroken line represents the leaf breadth; the striped- dotted line represents the length at the base of the leaf and the dotted line represents the terminal length. In pure line 11 (formula G_1wbT_1) both growth curves for leaf-breadth and for length at the base of the leaf merge together; the growth curve for the terminal length rises more rapidly and reaches, in the same time of 15 days, a greater end-value than do the other two curves.

In the line 4 ($G_2wB_1T_2$) all three growth curves differ from one another: the leaf-breadth is the least steep, the leaf-length at the base rises more rapidly, whereas terminal leaf-length rises very rapidly in this example. This is in complete agreement with the genotypical formula. Likewise the three remaining curves show growth-speeds, which can easily be derived from their genotypical formulae (8, $G_3W_1B_1t$; 1, $G_3W_1bT_2$; 2, G_4W_1bt). The most important conclusion which can be drawn is that growth-speed is influenced or caused by these genes. The curves were identical for the leaf measurements investigated, in the case when the genotype of a specific leaf-measurement was also identical. Hence: the factors involved appear to form the genotypical basis for substances which determine the growth-speed and in this way both development and final measurements of an organ are determined by the same gene.

Hence it seems to be proven that quantitative properties of the phenotype can be based upon one single pair of alleles, and that in some instances these alleles form a series of multiple alleles. We put forward the question as to how far the phenotypical quantity is based upon a possible quantitative nature of the genes. One may say the phenotypical differences in measurements which are due to the occurrence of the genes G_4, G_3, G_2 and G_1, are also the result of quantitative differences between the genes which cause these differences? This question is closely related to the problem of the nature of the genes, (as was discussed in Chapter XIV). As already mentioned in that chapter, GOLDSCHMIDT believes that we are allowed to see in every gene, a specific amount of enzyme and following the track of this concept one can consider the nature of differences in quantity as being differences between the members of such a series. Such a relationship between a quantitative nature of the genes involved and the quantitative differences in phenotype seems to be highly possible. For instance, if we take the example of a series of *multiple alleles* which influences the degrees of wing reductions in *Drosophila*: different degrees of wing reduction are expressed by the eight alleles 'vg' (vestigial). In figure 225 four of these are represented which are successively caused by the genes vg^a (antlered), vg^s (strap), vg (vestigial) and vg^{no} (no wings). Hence it appears that the step which we must make in order to conclude that these gradual wing reductions are due to quantitative differences between the members of the series of alleles is not too risky.

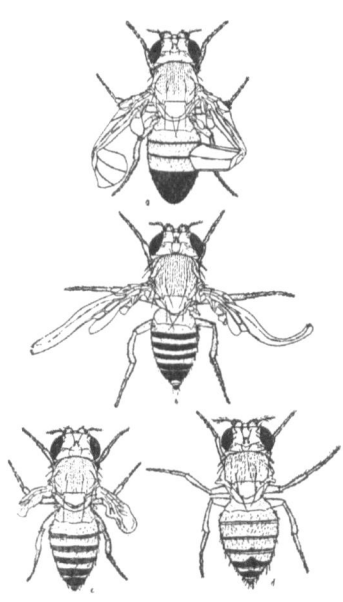

Fig. 225. Multiple allelomorphism of the wing shape in *Drosophila* (after STERN, 1930 b).

The relationship between quantitative differences of allelomorphs and their phenotypical expression is shown in an investigation carried out by STERN (1929c) concerning the length of the bristles in *Drosophila*. This length is influenced by a series of five alleles, which are known by the term 'bobbed'. Three of these (the normal, dominant +; a type bb, and a lethal bb^1, which is lethal in the absence of the other members of the series) are located in the X-chromosome. Two others (bb' and bb'') are probably situated in the Y-chromosome. Using crosses, in which the male flies have none, one, two or three Y-chromosomes respectively next to one X-chromosome, and female flies with none, one or two Y-chromosomes next to two X-chromosomes, STERN could establish a great number of gene configurations by the length of the bristles. Figure 226 gives a schematic drawing in which relative quantities derived from numerous measurements of the four allelomorphs bb′, bb, bb″ and bb¹ are represented. It represents the differences in length of the bristles caused by different combinations of alleles from the series mentioned: a and b represent female combinations (XX) of which the homozygous bb¹bb¹ is a lethal form, c and d are male individuals (X) among which bb¹ is not viable. On grounds of this data

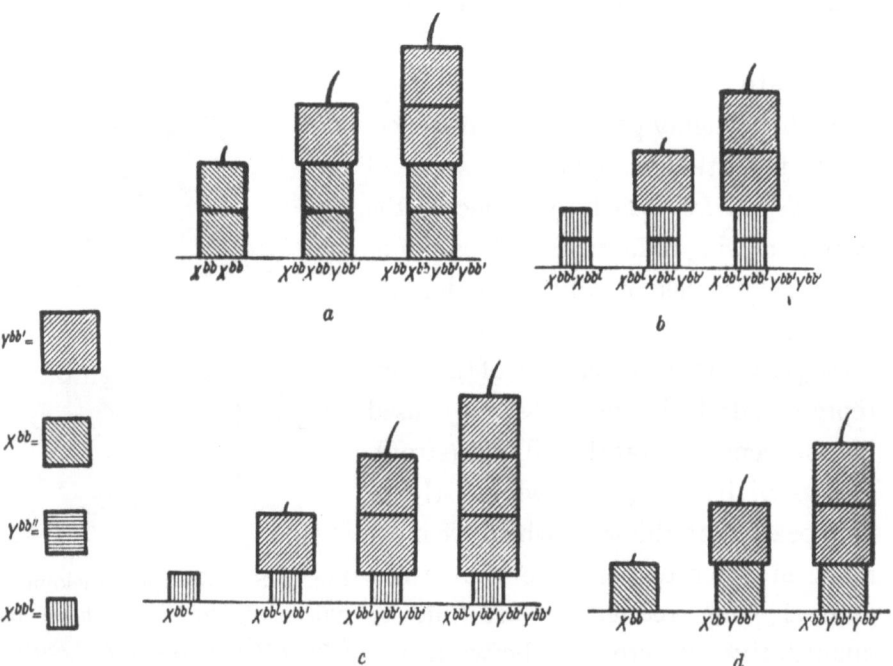

Fig. 226. Addition of multiple allelomorphs in bristle hairs of *Drosophila* after STERN, 1930 b).

STERN derived the theory that the quantities of multiple alleles can be added together, so that heterozygous combinations will usually be intermediate between both homozygotes. The one homozygote will always excel the heterozygote, the other homozygote is lower. Undoubtedly this situation is frequently found in multiple allelomorphs with the exception of the 'normal' gene, which according to the theory would have the greatest quantity and should be completely dominant. STERN tries to give an explanation to this irregularity in the sense of a physiological limitation in which the homozygous 'top'-form in the series of multiple alleles is prevented from expressing itself quantita-

vely above that of its heterozygous combination. The view of STERN
may be acceptable in many instances but several questions still remain
unanswered. One example: how to explain the phenomena which occurs
in a series of multiple alleles in which the higher series is always
completely or almost completely dominant over the lower one. How far
the quantitative differences in the phenotype are also based upon the
quantitative differences of the allelomorphs in this case is still not
known.

It has been mentioned several times in this book that the view that
the gene is a specific
amount or quantity,
originated from GOLD-
SCHMIDT, who developed
this idea (see GOLD-
SCHMIDT 1927, 1928b).
A single example will
make it clear how one
is able to explain a
number of phenotypic-
al phenomena on the
basis of this theory.
We will first discuss

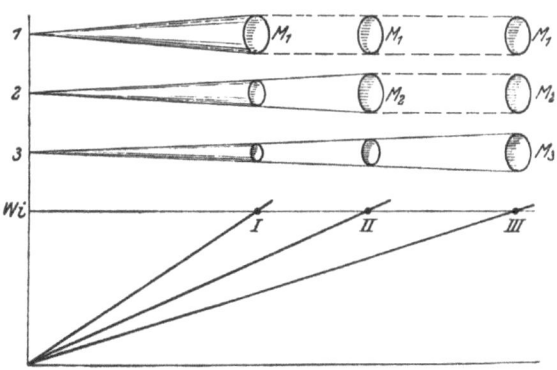

Fig. 227. Scheme of development of colour pattern
in butterfly wings (after GOLDSCHMIDT, 1927).

the phenotypical differentiation which brings about the *colour-
pattern* in the wings of a butterfly, which can very easily be studied.
Figure 227 represents the process which can explain this very
complex result schematically. A specific chemical process which
is brought about by only one gene, causes the epidermis-cells to
become adapted to take up pigment. This occurs with different speeds
during development of different parts of the wing (the diameter
of the cones 1, 2, and 3 stand for the speed involved). Besides this,
three reactions take place with different speeds for pigments I,
II and III. These pigments are all taken up by the scales of the wing.
At the moment that pigment reaction I is finished, only the scales
with a speed $1(M_1)$ have attained the ability to take up this pigment
and hence the pigment is deposited only in these scales. If pigment II
is ready, then the scales of the two regions M_1 and M_2 are already
completed, whereas the scales M_3 can only take up the slow-developing
pigment III. The chief principle of this hypothesis is, that the element

'time' is taken into account as an important factor in the result of these processes and that the colour pattern of butterfly wings is reduced to a cooperation of two groups of developmental reactions which take place at different speeds.

Finally, a word about the phenomenon of *dominance* (see FISHER 1941 and HALDANE 1939) which is also important from the point of view of the developmental physiology. GOLDSCHMIDT gave an explanation on grounds of his contemplation which supposes two different possibilities: 1) whether it is possible that complete dominance is

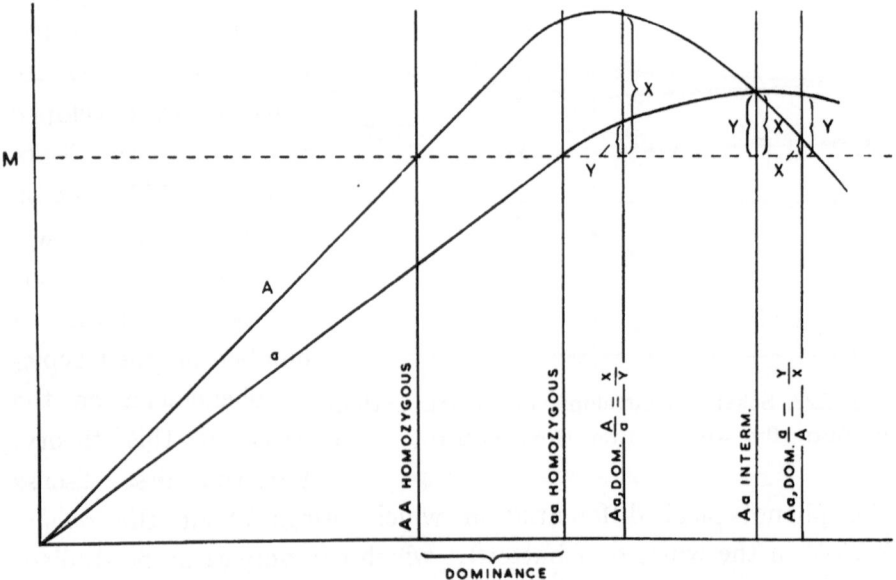

Fig. 228. Scheme of dominance (after GOLDSCHMIDT, 1927).

caused if, from two developmental reactions, which run parallel one to another, the fastest reaction (controlled by the greatest quantity of 'formative' substance) wins the competition, and 2) whether it is possible that the effect of these 'formatives' is an additive effect instead of an alternative effect. In the case of an additive effect, the final result is proportional to the quantities of the substance available. For instance, assume that the reaction of both the alleles A and a take place according the curves given in figure 228, then each reaction must reach a definite 'reaction minimum' before its effect will become visible. This minimum is represented by the horizontal line M. If the surpluses above the minimum produced by both reactions are equal, then the heterozygote is intermediate in its phenotype. If the surplus of A is greater than that of a (x greater than y), then the characteristic is caused by A (dominant) and in the reverse case (y larger than x) a is dominant.

Undoubtedly GOLDSCHMIDT succeeded in deriving a series of interesting views of developmental processes from his basic hypothesis: 'the gene as a specific quantity'. However, KÖHLER and FELDOTTO (1937) believed on the grounds of their investigations, that no difference exists in speed of development between the scales in the different regions of the wing. Hence, a more detailed study is necessary to determine how far these views and perhaps their basis also must be altered in some respects to agree with the experimental data. However, his expositions form an attempt to bridge the great distance between gene and phenotype by considering the gene as a source of reaction processes which through developmental chemistry and developmental physiology lead to a final form which we can observe and which for us remains as the expression of its heritable basis, i.e., the phenotype.

CHAPTER XXII

THE BOUNDARY OF GENETICS AND ITS APPLICATIONS

From the examples with which the theorems and theses of genetics are usually supported, we could perhaps draw the conclusion, that genetical rules can be applied to properties of an anatomical nature only i.e., shape, structure and colour, but that on the other hand, properties of a physiological nature and especially those in the realm of psychology would not fit into these deductions. Hence it could appear that genetics, and especially its applications, would be of an almost theoretical value. This conclusion was made more than once and it seems obvious, because the first studies, which led us to the basic rules of genetics dealt exclusively with such anatomical properties and the mechanism underlying these. However, this conclusion is wrong and it is very fortunate for genetics that it is so, because this greatly increases its influence in practical directions, especially in those fields which try to explain several *physiological and psychic characteristics*.

It is obvious that the recognition of physiological characteristics as 'mendelian' characters took more time especially if we bear in mind four facts: 1) physiological characters are more sensitive to environment than morphological or anatomical characteristics; 2) their genotypical basis is usually polymeric; 3) physiological characters are more strongly pleiotropic, that is to say a correlation exists between them; 4) a number of physiological characteristics can only find expression under specific circumstances, which cannot always be realized in the experiment (winter-hardy cultivated plants, disease-resistance of plants, animals, etc.).

Yet in this field of genetics also, progress went on and many interesting items concerning the inheritance of physiological characteristics became known. We will refer first to the data which was mentioned earlier concerning chemo-genetics and which were already of a partly physiological nature. Further we can point to the results of investigations like those of TRIMBLE and KEELER (1938) with dogs: the Dalmatian bulldog is characterized by a high amount of *uric acid* in the urine (compared with other races of dogs in which this uric acid is replaced by *allantoïn*). A pedigree of Dalmatians (fig. 229A) showed that all the 21 dogs studied exhibited this characteristic; some of these dogs were mated with Collies and all eleven F_1-animals produced the same low uric acid percentage as found in the Collies. Back-crossing of F_1-individuals with Dalmatians yielded litters with the following compositions: 2 low + 2 high, 3 1 + 1 h, 2 1 + 2 h, and 1 low + 2 h, hence a total of 8 low against 7 high or an almost 1 : 1. Three litters from heterozygotes were composed of 16 animals of which 14 had a low percentage and 2 had a high percentage (6 1 : 0 h, 3 1 : 1 h, and 5 1 : 1 h). Therefore it is obvious that low production of uric acid and high production of allantoin in Collies is based on one single dominant gene, which when present in Dalmatians as a pair of recessive allelomorphs is responsible for a high uric acid percentage. Figure 229 represents some mating results. Whereas this chemical difference in metabolism does not influence the health of dogs or their constitution we find in the so-called *haemophilia* an example of a difference (as yet chemically undetermined) which, is of a much more serious physiological

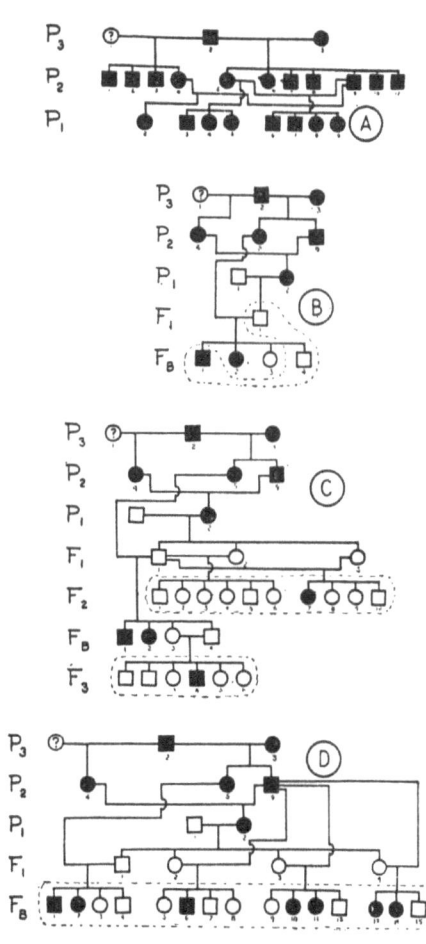

Fig. 229. Some results of the cross Dalmatian dog × Collie (black = high amount uric acid; white = small amount uric acid; P_3, P_2 and P_1 great-grand-parents, grand-parents, parents; F_B back-cross). (after TRIMBLE and KEELER, 1938).

nature. This disease reveals itself in men but is handed over by apparently normal women to their sons.

The characteristic of clotting of normal blood exposed to air is greatly decreased in patients with haemophilia, so much so that very insignificant wounds remain bleeding and may become dangerous over a period of time because of loss of blood. Moreover, patients very often suffer from haematomes in joints such as the knee and hip. ZAHN's novel: „Die Frauen von Tännö" shows how common such a heriditary disease can become in regions where many

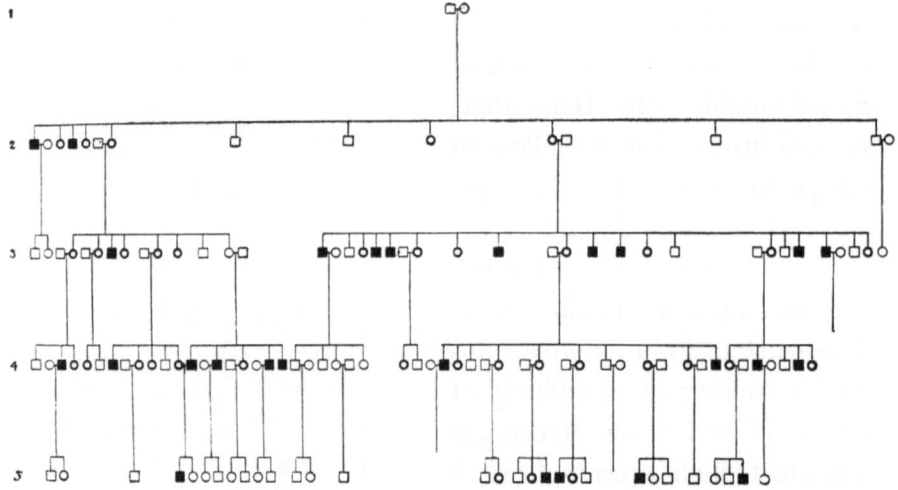

Fig. 230. Pedigree of a haemophilic family in 5 generations; squares males; circles females. All patients are males (after LOSSEN, 1905).

consanguineous marriages take place. In this novel the serious significance of this evil for a whole district is pictured. From LOSSEN's study (1905) it becomes evident, that only men are affected (black squares in the pedigree of figure 230); females can be apparently normal and yet transfer the tendency to their sons. This is clear if we consider for a moment the inheritance of sex: man is heterozygous XY, the female homozygous XX. In the case when the factor for haemophilia is linked with the factor for sex we can indicate this 'diseased' sex-factor as X_1. An XY-man is therefore normal; an X_1Y-man will show haemophilia. Contrary to this a woman XX will be completely normal and an X_1X apparently normal. From a marriage of the latter with a normal man, two kinds of sons will originate, i.e., XY (normal) and X_1Y (haemophilic) and two kinds of daughters XX (completely

normal) and X_1X (apparently normal). According to this view haemophilic females would be X_1X_1 and these can only originate from a
marriage of a haemophilic man with an apparently normal woman.
However, such a marriage was never observed in statistical investigations and hence it is hard to say if haemophilic women are viable.

As a botanical analogy of this disease we will refer to the *Sordago-
disease* of *Mirabilis Jalapa* (CORRENS 1915). The Sordago-disease is a
disease of the palisade tissue in the leaf, which shows anatomically
observable symptoms due to this disease, that is to say leaves become
brown and roll up, and at the same time this progress interferes with
the physiology of the leaf which manifests itself in a disturbance of the
metabolism. Also, Sordago-disease is handed over to the progeny:
diseased plants originate in the progeny of normal plants and these only
yield diseased plants by selfing. Hence the Sordago-characteristic is
recessive towards the normal outward appearance of the plant and
responsive to the rules of genetics and differs from the normal one in
one single heritable factor. Nineteen heterozygous normal plants
yielded 519 normal plants: 185 Sordago-diseased plants (D : m = 0·81).
This ratio agrees sufficiently with the theoretical 3 : 1 segregation to
allow the Sordago-disease to be described as being due to a recessive
factor which shows a Mendelian segregation.

A completely different picture appears from the effect of one, prevalent gene in fowl; the so-called *frizzle-fowl* differ from normal not
only in this typically morphological characteristic but also by a
series of other morphological and physiological characteristics, which
are all due (to a higher or lesser degree) to the physiological consequence
of the frizzle characteristic viz. too great a loss of body-heat. Hence
much extra heat has to be produced to compensate this loss: the
basal metabolism is very much increased (at 17°C it is increased
more than 100%); the heart-beat is considerably accelerated (at room
temperature by 27% in hens and 68% in cocks) and is very labile: the
heart shows hypertrophy and it circulates less blood in the same
period of time; at low temperatures heavy lymphocytosis occurs,
leucopeny and haemoglobin values are below normal; rapid destruction of erythrocytes; respiration is reduced; and less fat production in
younger animals (this production increases in older fowl); high death
rate of chickens; late maturity and beside this a number of anatomical
aberrations (crop, stomach, pancreas, kidneys, spleen are all enlarged,

ovaria are smaller and testes larger than normal). Important physiological changes are found, due to anatomical changes of the thyroid
gland and adrenals. This whole syndrome of physiological and anatomically different characteristics as compared with a normal

Fig. 231. Experimental field of different races of
Cowpea (*Vigna sinensis*) which are all infected at
the same time with a harmful bacteria (after ORTON,
1911).

fowl, can be ascribed to one single gene, which affects the feather
structure so that it develops in completely different lines and hence
gives rise to all sorts of complications (BENEDICT, LANDAUER and
FOX 1932; LANDAUER 1937).

These heritable aberrations and diseases, which are transferred
directly from parents to their progeny by a gene causing anatomical
or physiological deviations of the organism must be sharply distinguished from another group of heritable characteristics which
have significance for the epidemiology of diseases, that is to say, the inheritance of a power of resistance or a susceptibility towards one or
another parasitic infections. It is a well-known fact that a great difference exists between different species with regard to resistance, also it
is known that races and even individuals of the same species are
characterized by differences of resistance. This is shown in figure 231,
in which pictures are given of two types of cow-peas (*Vigna sinensis*)
(a very common American crop). Both types (races) were infected by

ORTON (1911) in the same manner and the picture clearly shows that one race remained completely healthy whereas the other race of *Vigna sinensis* was very much affected. Individuals can also differ in resistance against diseases. The occurrence of germ-carriers is a well-known phenomenon in man. These individuals are infected with some germ or another, they do not however suffer from the disease, that they undoubtedly carry. For this reason these people form an often unknown source for contamination.

For a long time scientists held the view that heritable factors play a rôle in causing this *congenital resistance*. However, in most cases the course of such an inheritance is not as easy to discover as in the study of ORTON, that is to say not much is known with any certainty with regard to such heritable characteristics. In most cases immunity is dominant; there are however, instances in which the F_1 was composed of different types so that it appeared that the parents in question were not absolutely homozygous for these characteristics.

An immunity against parasites was clearly shown in the hybridization experiments of HAGEDOORN (1919). HAGEDOORN worked with mice, which he investigated with other purposes in mind, but which yielded results which are of significance for our knowledge of the immunity question as a problem in genetics. HAGEDOORN made crosses between the small Japanese shaker mice and big white mice in order to investigate the inheritance of body-weight characteristic in both races of mice. In January 1919, cultures were made of the parental types, F_1-animals, a great number of F_2-individuals and backcrossings of F_1-animals with the Japanese shaker type and with the big white mice. Suddenly a mass *staphylococcus infection* occurred, which caused the death of all Japanese shaker mice, but the big white mice remained completely healthy in spite of the fact that they were kept in the same cages. The whole process of infection, illness and death of the Japanese shaker mice took place within six weeks. It is interesting to trace what happened to the F_1-generation, the F_2's and the several backcross products. The hybrids did not die and hence the resistance of the white mice was dominant. Out of 125 F_2-individuals (31 litters) 34 died, so that 91 were left alive. The back-crosses of the F_1's with big white mice all (50) remained healthy and of the backcrosses of F_1's with Japanese shaker mice 32 out of 57 died. Summarizing, the numbers agree with the supposition that the resistance of the big white mice depends on one single dominant factor. Hence the whites are AA (resistant) ; the Japanese aa (susceptible); the F_1-generation (Aa) was resistant and yielded 75% AA and Aa in the F_2 (91 found resistant) and 25% aa (34 found sensitive). The backcrosses $F_1 \times$ white (Aa × AA) may yield exclusively resistant animals (AA and Aa); this result was also obtained; there were only 50 healthy animals. The other back-cross $F_1 \times$ Japanese (Aa × aa) must theoretically segregate into 50% Aa (25 obtained resistant) and 50% aa (32 obtained sensitive).

This 'power of resistance' against infectious diseases, which has a genic basis occurs frequently in plants as well as in animals, however, it is often very hard to determine and often it is very seldom that we can hold one gene responsible for this phenomenon. Summaries con-cerning genetical disease resistance in plants were given by HANSEN (1934), ROEMER, FUCHS and ISENBECK (1938), KOEHLER (1940) and SCHARNAGEL (1943), in animals this was carried out by GOWEN (1933, 1937) and LAMBERT (1933).

We also find such a genic based resistance in plants against all sorts of unfavourable environmental circumstances (low temperature, drought, etcetera). In these instances we deal with a physiological resistance, which is controlled by a genetical mechanism ((Bibliogra-phy 1939; FUCHS and VON ROSENSTIEL 1941).

Since 1900 NILSSON-EHLE (1913) dealt with the problem of frost resistance in his wheat investigations. Detailed comparative studies showed him that, with regard to frost resistance, a great difference occurs between types of wheat which were cultivated over a long period in Sweden. The conclusion drawn from his careful investiga-tions which had been continued over many years was this: The resist-ance against frost is a complex property which may be based upon the presence of several genes, so that its inheritance is complicated.

The polymeric nature of these genes can be established by three ex-perimental facts: 1st. the F_2 of a cross between a very resistant and a very sensitive type shows a number of intermediates, which altogether form an uninterrupted chain between both of the parent types; 2nd. individuals which equal either parent in their resistance or in suscep-tibility are very rare in these F_2-generations; 3rd. from crosses be-tween rather resistant plants an F_2-can occur, in which there are plants which excel in their resistance and also those which lag behind com-pared with both parents and these degrees of resistance are inherited

by the descendants as new combinations of characters. The segregation, after crossing two fairly frost resistant individuals clearly shows
that these degrees are again combinations of characters which can
group themselves independently one from another. A polymeric genic
basis for the characteristic of '*frost resistance*' appears highly acceptable. However, it is impossible to draw any conclusion about the
system of these polymeric characters and hence further conclusions
cannot be made.

A clear-cut case of *sensitivity to cold*, without freezing being involved
was found by CORRENS (1913) in a stock of *Mirabilis jalapa* (forma
delicata). These plants had been already very much damaged by a
temperature decrease down to $+ 5°0°C$. The offspring of the cross
delicata × normal are not sensitive and the F_2 segregated into 75% non-
sensitive and 25% sensitive plants. The difference between the cold
sensitivity of NILSSON-EHLE's wheat plants and of CORRENS *Mirabilis*
consists of two facts: the first were all sensitive to frost but it is probable that they are not sensitive to frost-free, low temperatures and
their resistance was based upon a complex of polymeric genes; the
latter plant was already sensitive as form *delicata* to a temperature of
almost 5°C, whereas the normal type endures this temperature
well. The difference between the normal type and the *delicata*-type is
based upon one dominant gene present in the normal type only.

Psychic properties are even more difficult to approach experimentally with regard to the heritable mechanism involved than was the
case with physiological properties. The main difficulty is that, for the
time being we do not have enough methods available to measure these
properties. Not much is known about the boundaries between the
effect of inheritance and the effect of environment. It is possible
that the genetical basis of these properties are much more complex
than we think, because they are often partly founded upon other
somatic (anatomical-histological, as well as physiological) characteristics and an analysis of these separate elements has not yet been
carried out.

There is also the other reason: that it sometimes seems to be difficult for scientists to see the results of their investigations into the
heritable basis (carried out as exactly as possible) in an objective
light and to make an interpretation free from own pre-conceived
opinions. The data obtained till now is neither absolutely indispu-

table nor mathematically certain. The incompleteness of the data sometimes provides grounds for a discussion of deductions to those who do not wish to recognize the reliability of the results. It is like a controversy between legal and illegal in a lawsuit: a skilful lawyer will always discredit evidence even when an unprejudiced judge of the data must admit its convincing power.

Four methods of investigations can serve us, i.e., purely statistical, genealogical, study of twins and experiments with animals (SIRKS, 1948).

For instance, purely *statistical data* usually leads to the conclusion that children of a father with an intellect above average, also show the tendency for a positive intellect. DUFF and THOMSON (1923) found that children of academically educated fathers have, in general, an intelligence-quotient (I.Q.) of 112, whereas the I.Q. for children of factory workers is almost 100, and that for children of unskilled workers is 96. Such figures indicate clearly that inheritance of intellect may be involved, but on the other hand, this data is contestable because the possibility of the environment being involved in the determination of the I.Q. has not been excluded. However, TERMAN (1925) showed that when he classifies gifted children according to the intelligence degree the profession of their fathers demand; and when he compares these percentages with the percentages of the fathers of a particular intelligence class in the total population, then the result was as follows:

Degree of intelligence, which profession of father demands	Percentage of fathers with gifted children	Percentage in wich the intelligence classes of the fathers occur
15 and over	26·8%	2·2%
12–15	26·8%	4·5%
9–12	36·1%	37·0%
6–9	8·9%	13·4%
3–6	1·3%	42·9%

The environment (schools) for all the children was similar and hence the conclusion of TERMAN is that we deal with a congenital heritable giftedness. Undoubtedly this view can be easily upheld and can be defended on the grounds of the results obtained.

A great amount of *genealogical data* is available since the classic work of GALTON (1869, 1883, 1889). Many of these studies deal with all sorts of giftedness and they form a sizeable piece of evidence

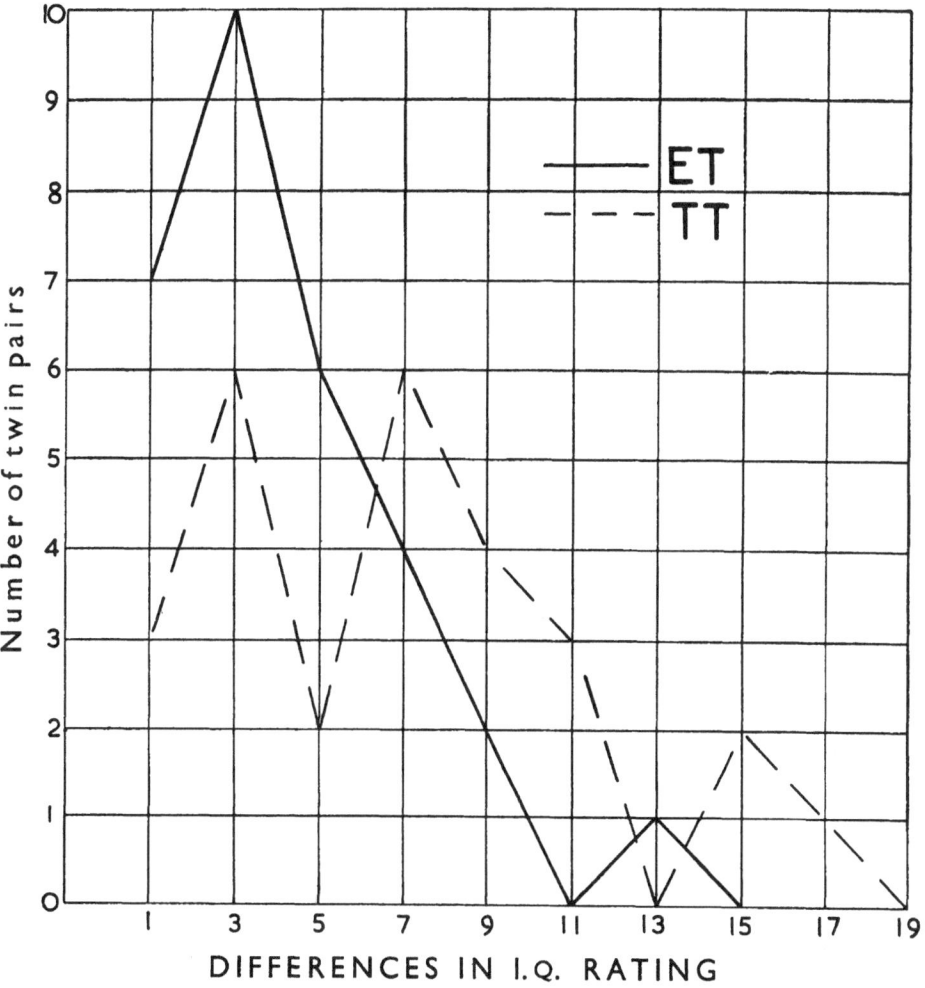

Fig. 232. Difference in intelligence quotients of the partners of identical (ET) and non-identical (TT) twin pairs (after VON VERSCHUER, 1930).

for the unprejudiced reader. These studies donot deal exclusively with extreme cases of giftedness but also with values of ability in the middle of the spectrum of intelligence. One cannot deny that in these instances also the family milieu, and family tradition play a rôle. However, all this data still points to the fact that an important heritable element is involved.

We will also mention the third method, that is *studies on twins*. This method is more convincing since the influence of environment can be eliminated. VERSCHUER (1930) gives results of studies with one-egg twins (identical, monozygous) and fraternal or dizygous twins orginating from two fertilized eggs, which were delivered simultaneously. He clearly showed (fig. 232), that the points of difference between the I.Q. 's of the partners of monozygous twins are fewer, than those in dizygous twins, educated under similar circumstances. This great accordance between the two partners of one homozygous twin was also ob-

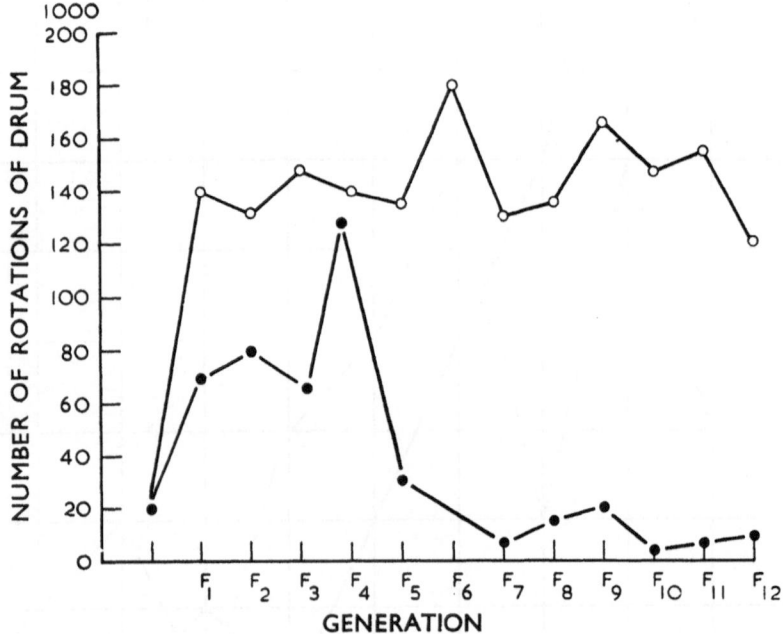

Fig. 233. Average of rotation numbers of the so-called activity drum in experiments with active (above) and non-active (below) rats (sib-mating was applied after the F_4, by which 'pure lines' were reared). (after RUNDQUIST, 1933).

served by VERSCHUER in his attempt to answer several special questions which arose by applying the so-called Rorschach-method. His results could be confirmed by many other investigators.

In many fields, this study in man can be amplified by data obtained from *experiments with animals*. The great advantage of these studies is, that the parents can be selected on the grounds of their psychical achievement so that only those animals are used for reproduction which show a typical characteristic.

In this way it became possible (RUNDQUIST 1933) to rear two stocks of rats from a group which differ very much in 'activity'. With the help of a so-called 'activity-drum' a rotation number could be calculated, by which this characteristic of temperament can be expressed. By means of selection of very active and rather inactive rats and by mating them to rats with the same activity, two lines originated, which differed very much in their respective activities. However, these lines still showed irregularities up to and including the third generation. In the fourth generation the progenies became more uniform, due to the fact that instead of free mating in the group, sibmating was applied. Hence, the picture changed: By the 5th generation the inactive rats yielded inactive litters, the active ones yielded active litters and the variability within the lines which was considerable until the fourth generation, now disappeared (fig. 233). All influence of environment was excluded and the conclusion that these rats possess a clear cut heritable tendency for this temperament characteristic, appears acceptable.

In general, one is able to say that with regard to psychic properties the final level (for instance the I.Q.) is not inherited, but rather that the boundaries between which the psychic properties can fluctuate by different influences of the environment (education etc.) are controlled by the individual genotype.

Much work has yet to be done in the field of inheritance of psychic characteristics before we can draw a final picture. The data now available justifies the supposition that a heritable basis underlies these characteristics.

The recognition of the significance which the study of genetics has for morphological-anatomical as well as for physiological and psychological characteristics, is in many respects a rather vital question. Undoubtedly its consequences penetrate and influence not only natural sciences, but social sciences as well and even the field of philology which at first sight appears to be completely unrelated with natural sciences.

The *physiologist* who wishes to reach reliable conclusions, must be advised that he should control his material first for mutual heritable purity. It has become known that among several animal and plant species there may occur heritable differences in susceptibility to infectious diseases, in the ability to reproduce, in vitality, in capacity of growth, in ability to assimilate definite nutritious matter and so on. It is obvious that heritable differences between individuals and heritable impurity of the living material can bring about important disturbances with regard to the results obtained.

Hence, for genetics it is a great satisfaction that from a medical point of view the importance of homozygous animals is recognized as a material for standardization of organ preparations. Therefore the *pharmacologist* must also take the conclusions of genetics into account.

Studies of *development*, (in which the epigenetic view reigned supreme for a long time, and pre-formative views had only slight consideration), came more and more to recognize, that *embryology* must also consider genetics as an important source of information.

Next to this comes *taxonomy*, which in former days limited itself to study of herbarium- and museum material and which succeeded in classifying the chaos of forms, which exists in living Nature into a very desirable clear system. Up to now taxonomy as a descriptive science mainly used the morphological method of investigation in its attempt to classify the rich diversity of living plants and animals. However, a change in the method of approach to its problems took place recently which is due to the results of modern genetics. Everyone, who has taken the trouble to cultivate wild flowers and has tried to study the inheritance within the species has been amazed over the great differences which become visible between the several lines of the pedigree. The 'species' as accepted by taxonomy as the foundation of its system of classification, is not a unity; but rather a heterogeneous mixture of different heritable forms.

The view was repeatedly put forward by circles of taxonomists that the classic concept of the species (the so-called *Linnean species*) can no longer serve as a basis for modern taxonomy. In its place the *Jordanian species* (more or less equivalent to the sub-species) was proposed. However, in following this principle to extremes it would appear unattainable. The only exact basis would be a group of individuals whose genotypical constitution is mutually identical. Such a unit was discussed by LEHMANN (1914), who called it an *isogenic unit* and *in abstracto* one cannot deny such an exact basis for taxonomy. However, in practice it will not yield any advantages, because an analysis of the Linnean species into isogenic units would lead us to the recognition that each individual is a unit in itself.

The species *Homo sapiens* is from this point of view a very favourable and characteristic example: there are only a very few individuals which are mutually genotypically identical (perhaps only monozygous twins) and hence the species must contain as many isogenic units as there are individuals. This leads to taxonomy *ad absurdum*, because from this aspect no system exists. Hence taxonomists are forced to take another road and to take other views as a basis for their field of interest. Undoubtedly an exact definition of the '*concept of the species*' has still to be formulated. Historical and recent attempts to put forward a definition clearly show the difficulties, which stand in the way of establishing such a concept (SIRKS 1949b). For discussion of this problem we will refer to the publications of DANSER (1929), DURIETZ (1930), KLEINSCHMIDT (1926), LAM (1943), RENSCH (1929), ROBSON (1928), TURESSON (1925, 1929) and especially to the recent book of HUXLEY and others (1940).

Genetics has worked in three directions to enlighten taxonomical views and problems. In the first place it was generally accepted (before modern genetics started to wield its influence), that the species should be qualified by characteristics of a different order than the variety. From the pre-mendelian period we almost inherited the dogma that *interspecific hybrids* (originating from hybridizations between species) should breed true to type, whereas *intraspecific hybrids* (originating from hybridizations between varieties) should segregate according the Mendelian scheme. This view was upheld till long after 1900. HUGO DE VRIES who certainly knew how to value the importance of MENDEL's experiments believed (1903. p. 643) that it was possible to make a difference between *species-characteristics*, which are not controlled by MENDEL's laws and *variety characteristics* which show a Mendelian segregation; a view which is completely in agreement with the ancient dogma concerning the species concept. Species characteristics should originate by so-called *progressive mutation*, that is to say they should be caused by a completely new gene, which does not find another allelomorph in the other parent by hybridization. Hence this new characteristic would remain true-breeding in the offspring but not to such a degree as in the original parent. Variety characteristics should owe their origin to *retrogressive mutation*, that is to say, to a change of a certain gene; the new gene now finds an allelomorph and is therefore able to show up according to a Mendelian scheme when segregating. For instance, between *Lychnis diurna* and *L. vespertina* (both ragged robbins) the flower colour should form a variety characteristic, which follows a Mendelian scheme in its segregation according to a 3 : 1 ratio, since red is dominant over white, however, the leaf size and the length of the stalk of the flower do not follow a segregation scheme and are therefore species characteristics. The possibility that polymery may occur is completely neglected in this hypothesis.

It was later shown by experiments that this view was incorrect and species characteristics as well as variety characteristics can follow a Mendelian scheme. For instance in the genus *Streptocarpus*, the habitus stands as a characteristic of the taxonomical section. Even the characteristic: unifoliate habitus with one single leaf and inflorescence rich in flowers in the axil appears to be recessive to the rosulate-type with more leaves forming the rosette and a poor flowering inflorescence. The segregation observed by OEHLKERS (1938–1941, 1942) in the cross *Str. Rexii* (rosulate) × *S. grandis* (uni-foliate) looked like a 3 : 1-segregation six months after sowing, however, three months later this changed into a 15 : 1 ratio. It appears most likely that in this experiment a particular genotype for the dominant rosulate form is slower in its phenotypical development than the recessive uni-foliate. In every case, evidence is given that a characteristic which is considered as typical for a species does not differ in its genetical behaviour from variety characteristics.

The second manner in which taxonomy is influenced by genetics is based upon the theorem that the chromosome configuration as the bearer of the whole genotypical property of the individual, must be typical for all those individuals, which are genotypical similar to a high degree and hence form a species. The links between karyology and taxonomy have been considerably enforced in recent years. After the pioneer work of KIHARA and ONO (1926) with the very difficult genus *Rumex*, the principle of a genome-analysis, on the basis of chromosome configurations was applied to the genera *Triticum*, *Aegilops* and its relatives (KIHARA (1930–1949) see also SCHIEMANN 1943 and SEARS 1948). We will mention a classic monographic study of the genus *Crep is* by BABCOCK (1947a, 1947b). The value of the '*chromosome-portrait*' in the solution of taxonomical problems finds more and more recognition in its zoological aspects (see WHITE 1945). Two examples will show the justification of criticism of taxonomical problems by means of karyology. In the group of the *Boraginaceae*, the place of *Brunnera macrophylla* was the subject of much divergence of opinion. Some investigators considered this plant to be representative of the separate genus *Brunnera*, others believed that it belonged to the genus *Anchusa*. Karyological investigations of SMITH (1932) proved, that the chromosome-portrait of this species differs greatly from the chromosome-portrait of all *Anchusa*-species, that is to say, in number

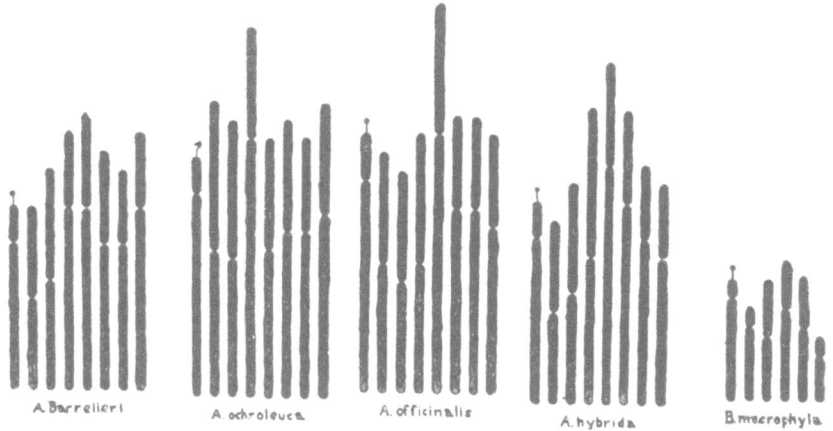

A.Barrelieri   A. ochroleuca   A. officinalis   A.hybrida   B.macrophylla

Fig. 234. Chromosome portraits of four species of *Anchusa* and of *Brunnera macrophylla* (after data of SMITH, from ANDERSON, 1937).

(6 in *Brunnera*, 8 in *Anchusa*), in size (much smaller in *Brunnera*), in form, and also by the presence of two chromosomes with sub-terminal primary constrictions which do not occur in *Anchusa*. The drawings of the chromosome constitutions in figure 234 are very striking and lead one to the opinion that *Brunnera* forms a completely separate genus which may not even be closely related to the genus *Anchusa*.

The second example of a karyological piece of evidence dealing with a taxonomical problem can be found in the investigations of McKELVEY and SAX (1933), in which the taxonomists involved were rather startled: *Yucca* is usually classified among the *Liliaceae* and *Agave* under the *Amaryllidaceae* on the grounds of the position of its ovary. However, the chromosome-portraits of both genera show an unmistakable similarity (fig. 235). Due to this similarity both genera are placed in a collective family, the *Agavaceae*, and it would seem that the whole taxonomy of the *Monocotyles* should be revised because of this.

Finally, taxonomy can be influenced by genetics by means of the study of the geographical distribution of species and their genotypic-

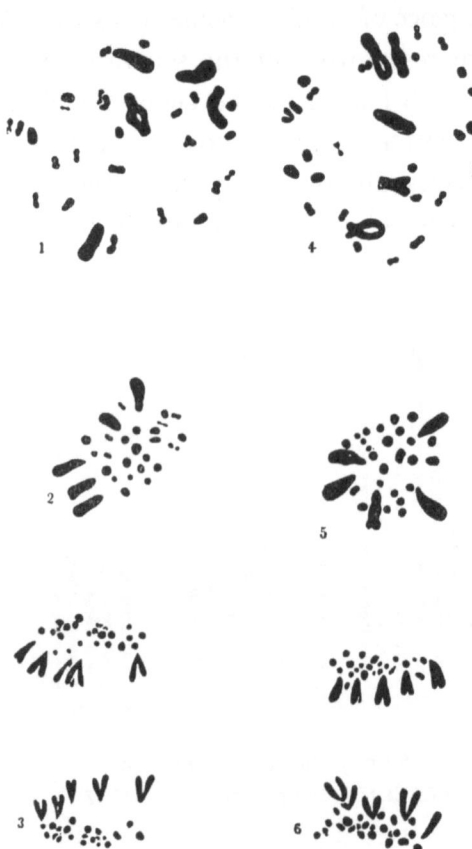

ally different varieties (COLE and others 1940, LAM 1943). This subject lies within the scope of biogeography and is extremely rich in data for genetical taxonomy. Any investigation in this field can yield important results.

For instance the study of MAYR (1944, 1948) concerning the occurrence of particular genotypically characterized groups of birds, i.e., *Pachycephala pectoralis* in the Pacific (fig. 236) clearly showed that the gene-configuration of number of pairs of alleles is characteristic for each island. Many analogous investigations concerning other organisms have now been

Fig. 235. Chromosome portraits of *Yucca* published; it is sufficient to *flaccida* (1, 2, 3) and of *Agave virginica* refer to the publications of (4, 5, 6): 1,4 diakinesis, 2,5 metaphase, 3,6 EISENTRAUT (1950) concerning telophase. (after McKELVEY and SAX, 1933). lizards, GORDON (1947) concerning fish and MANTON (1950)

concerning Pteridophytae. By means of these studies for each species a *'gene-geography'* is formed for each species which, on the grounds of a genetical analysis of the species, gives a view over the geographical variability. Moreover these studies lead us to the problem called (in American literature) by the name *'speciation'*, that is to say, the problem dealing with the origin of new species in Nature. An answer to this problem can only be given if such an answer has an experimental basis. We will mention the problem of the origin of one or more individuals in a population which differ from the general gene-configuration present in the population before these new genotypes occurred. The answer to this question can be given from two viewpoints (see chapter XIX): the possibility of mutation as *gene-, chromosome-,* or *genome- mutation* and the possibility of import of new genes into a population by means of hybridization. The second question concerns the nature of the processes by which a group of organisms can arise from genetically different individuals, hence the problem of how a new and so far unknown population may arise (see chapter XX). The

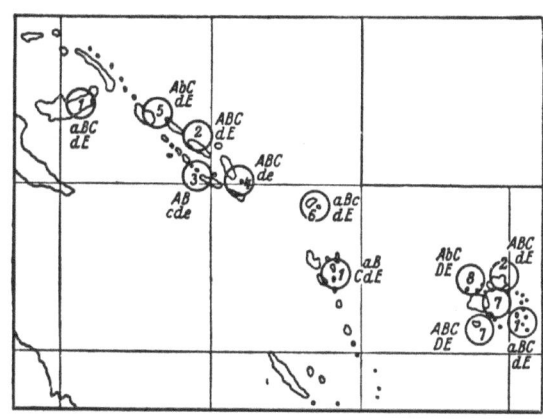

Fig. 236. Spreading of certain genotypes of *Pachycephala pectoralis* over the South Sea Islands (after MAYR from TIMOFÉEFF-RESSOVSKY, 1939).

A = yellow neck; a = white neck.
B = breast stripe; b = no breast stripe.
C = olive coloured back; c = black back.
D = yellow forehead; d = black forehead.
E = coloured wing; e = black wing.

third group of problems finally deals with the environmental adaptation of the species as a result of selection was investigated by CLAUSEN and his co-workers (1940–1948a, also SYMPOSIA 1949b, 1949c). In dealing with these questions we touch a complexity of problems which form the field of evolution.

In this field the precise data of genetics can also serve as a basis for a stronger construction of views. However the complex of evolution comprises more than the explanation of heritable variability only (SIRKS 1951). The break up of very variable groups into smaller and more stable units, that is to say, the process of the origin of species proper, belongs to its field. Here also genetics has something to say; the complex of evolution problems recently came under the influence of the results of descriptive and experimental genetics (DOBZHANSKY 1941, GOLDSCHMIDT 1940, HEBERER 1943, HUXLEY 1945, LAM 1946, MELCHERS 1939, TIMOFÉEFF-RESSOVSKY

1939) after being completely in hands of palaeontology, comparative anatomy and comparative embryology, biogeography etc., for a long period of time.

Next to *macro-evolution*, that is to say the study dealing with the complex of problems comprising the great lines of development of life and living organisms on earth, a *micro-evolution* now comes to the

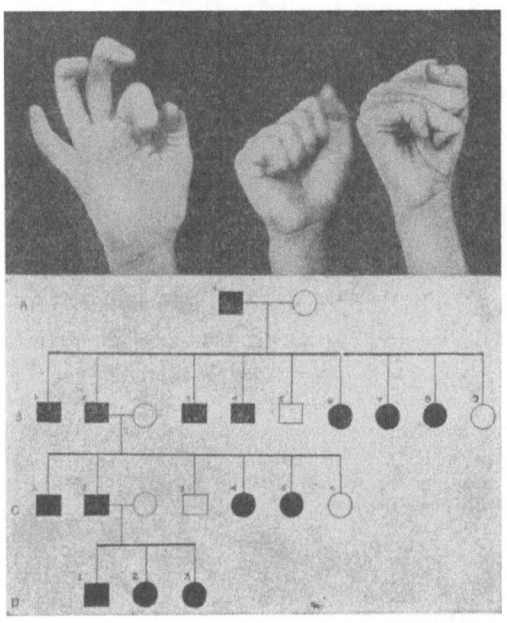

Fig. 237. Above: left and right orthodactylic hands, middle normal. Below: pedigree of a family, in which the black individuals (square ♂, circle ♀) represent orthodactylic individuals (after DUNCAN, 1917).

foreground which deals analogously with phenomena in populations of cultivated plants and reared animals; the division of evolution into these two separate fields (which are nevertheless in continuous interaction) was proposed by PHILIPTSCHENKO (1927, p. 93) and gives genetics its due place as a fundamental science in the theory of evolution. Hence an unexpected rapprochement occurred between karyology, genetics and palaeontology (see SYMPOSIUM 1941a and JEPSEN, MAYR, SIMPSON 1949).

If we now look further, even beyond the boundaries of natural

sciences, we will see how the influence of modern genetics has penetrated into all areas and observe gratefully, how the interest in genetical phenomena in man and their scientific interpretation has of late
increased considerably.

From this a very extensive field developed: a branch of investigation
which GALTON (1883) called 'eugenics'. A series of characteristics in
man which appear to have a heritable basis comprise a sequence of
harmless bodily peculiarities; for instance, orthodactyly. However,
more serious and even lethal aberrations of the normal build and
normal metabolism occur as well as characteristics of intelligence,
mind, and personality.

For instance orthodactyly as shown in figure 237 is a peculiar aberration of the hand, in which three fingers (middle-finger, ring-finger
and little finger) of the hand cannot be completely bent. In normal
man no difference exists between these fingers and the others.
There is no doubt that this aberration is heritable and that it is a
dominant character. Figure 237 gives a pedigree (borrowed from
DUNCAN 1917) which clearly shows, that 78 out of 150 children born
of marriages between an orthodactylic person and a normal person
were orthodactylic and 72 children were normal. This is wholly in
agreement with the view that all orthodactylic persons are heterozygously dominant (Aa) and all normal ones homozygously recessive
(aa). All marriages Aa × aa therefore give equal numbers of Aa and
aa descendants, that is to say 50% of the children are orthodactylic
and 50% are normal.

In the case when the aberration is of a harmless nature for the body
(as in the example given above) we can hardly speak of a disease. However if the aberration encroaches deeper in the structure of an organ
or in its function (that is to say influences the processes of life more
strongly) it may disturb the normal development and the normal
physiology of the organism. Hence one of the most urgent problems
in man is the question as to whether or not there is a genetical basis
for cancer formation (HESTON 1948, SYMPOSIUM 1945b). Physiological
processes, as we saw earlier, as well as morphological-anatomical ones
can be heritable, however, their study is in general more difficult and
their nature is genetically not so clear. As we mentioned already,
genetics wielded an influence in the study of psychic characteristics (p.
551). In this field it is not easy to determine the existence of a genetical

basis. However, we have to admit that the convincing power of the available data is becoming more convincing and it also becomes very difficult to ignore the problem itself, as to whether or not psychic characteristics are determined by a heritable tendency.

Summarizing it appears that more and more evidence is being compiled which shows that genetics also plays an important rôle in the life of man, and the importance of this rôle increases continuously. However, it is not our intention to go into this subject in more detail. We would like to refer to some works: for genetics of man to the extensive handbook of JUST and others (1939–1940) and to the excellent study of STERN (1949); for human pathology to the books of BAUR-FISCHER-LENZ (1932–1940), FRASER ROBERTS (1940), GÜTT and others (1937–1940), MOHR (1934), MULLER, LITTLE, SNYDER (1947), SIRKS and WAARDENBURG (1949) and VON VERSCHUER (1937); for special fields, for example research for the eye to WAARDENBURG (1932) and SORSBY (1951) or for the ear to VAN GILSE, HINNEN and NIEUWENHUYSE (1942), for dentistry to BROEKMAN (1950). To this list we could still add several bibliographies and a number of periodicals and journals devoted to this field of *anthropogenetics*.

There are a number of questions closely related to this complex of investigations, which the bodily and mental characteristics of man offer us (SIRKS 1948): in the first place the relationship between the heritable tendency of man as an individual and environment in which he grows up form an important group of problems (see WAARDENBURG 1927, G. JUST and others 1930); further, the problem of man as a member of a society with all the social problems related to it, such as the significance of *a-socials* (RITTER 1941), *criminals* (STUMPFL 1941, WAARDENBURG 1940) and of *mentally deficient people* (FRETS 1941, KRANZ 1938, 1940) as heritably inferior individuals of the community. Also the possibility of a lasting improvement of human society by *social measures* (SCHRIJVER 1941, VON VERSCHUER 1941) and finally the study of the *human races* and their interracial relations may be mentioned here as problems. These studies need to be on a strong scientific basis and objective in their deductions. However to maintain its scientific character is very difficult. As soon as *Eugenics* leaves the path of unprejudiced establishment of facts and hails the personal subjective element in the interpretation of the phenomena, the scientific nature of the complex of problems will be denied and the consequences of genetics are in danger of being disfigured. For every-one dealing with eugenics it is an imperative duty to safeguard conclusions made in this field against subjective bias.

Genetics finds further practical application in the improvement of *agriculture, horticulture* and *cattle-breeding*. However, in some of these circles genetical analysis does not reap the appreciation it deserves. Nevertheless, we must bear in mind that MENDEL's work played a great part in the improvement of the basis of selection- and hybridization work. For the time being, the significance of genetical analysis of crop plants is limited, because morphological characteristics are not of great importance for agriculture. Agriculture primarily needs precise descriptions (in the physiological sense) of the numerous varieties which are on the market today. We have to admit however, that genotypical characteristics underly the physiological differences of varieties and hence we must again return to genetics and its results.

It it is undeniable that the concepts, originating from genetics, have caused an important refining of the methods of plant and live-stock improvement. Recently, numerous standard works in this field were published. For plant improvement we will refer to CRANE and LAWRENCE 1938, HAYES and IMMER 1942, HUDSON 1937, HUNTER and LEAKE 1933, KAPPERT 1948, KUCKUCK and MUDRA 1950, ROEMER and RUDORF 1941- . . . , VON SENGBUSCH 1937, WELLENSIEK 1943, YEARBOOK U.S. Dept Agr. 1936, 1937 and others, and to the bibliographies of MATSUURA 1929, WARNER, SHERMAN and COLVIN 1934 and several of the COMMONWEALTH BUREAU OF PLANT GENETICS. The literature concerning the origin of economical plants is summarized by SCHIEMANN 1943 and BERTSCH 1947.

The work which the famous Institute in Svalöf has carried out during the last sixty years will serve as an example of the importance of genetics in the field of crep improvement (AAKERMAN, TEDIN and FRÖIER 1948). As an example that a single crop can give rise to extensive karyological and genetical studies we will name wheat (AASE 1935, CLARK 1936, KIHARA 1930–1949, KOSTOFF 1942), cotton (HARLAND 1939, HUTCHINSON and others 1947, STEPHENS 1947, WOUTERS 1948), corn (CRABB 1947, MANGELSDORF, 1947, RICHEY 1950), fodder plants (ATWOOD 1947). As reviews of the field of animal breeding the following works can be mentioned:

CREW 1925, HAGEDOORN 1946, KELLEY 1946, LERNER 1950, LUSH 1938, REINÖHL 1938, RICE 1934; in special fields such as cattle, SHRODE and LUSH 1947, for pigs BUCHANAN SMITH, ROBINSON and BRYANT 1938, for poultry JULL 1946 and HUTT 1949. For milk production GOWEN 1927 and BUCHANAN SMITH and ROBINSON 1933. Concerning the origin of domestical animals, KLATT 1927 and NACHTSHEIM 1949 can be named. Finally the study of KOBOZIEFF (1943) concerns the application of genetics in veterinary science.

Jurisprudence is also forced to accept the results of genetics as one of its aids. As a first example of this a contribution of MOHR (1921) will serve. In a case concerning investigation into paternity it turned out that the child in question showed unusual aberrations on the hands and the feet. This aberration, known as *brachyphalangy* or *brachydactyly* (fig. 238) concerns the length of the fingers and toes. When the phenomenon is present only to a small degree then the second phalanx of the three

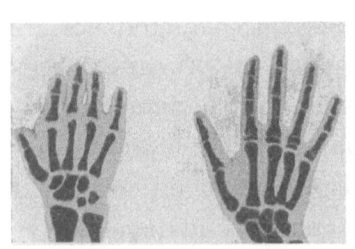

Fig. 238. Right: normal hand; left: strong brachydactylic hand X-ray photograph (after DRINKWATER, 1908).

(present in all the fingers except in the thumb) is shortened. In extreme cases the phalanx is more and more reduced and can even be missing. Brachyphalangy was found to be a dominant-heritable aberration of the normal hand by DRINKWATER (1908) and also by MOHR and WRIEDT (1919). That is to say every individual who is brachydactylic inherits this characteristic from one of the parents, in whom this characteristic has been also visible. A brachydactylic parent transfers the character to half of the offspring according to the scheme DR × RR = DR + RR. Hence if one of the parents is brachydactylic then half of the children also carry this aberration. In the family of the mother in question brachyphalangy was completely unknown; neither the mother nor the other relations showed the slightest tendency for this aberrant form of the finger. Hence the father of the child must have been brachydactylic. The person who was named by the mother as probable father showed brachydactyly to a quite high degree. More detailed study showed that this man must also have inherited this characteristic from his father, since the

aberration did not occur in the family of his mother. The brachydactyly of the probable father (a tramp, originating from another part of
the country) together with the fact that the aberration was unknown
in the region where the mother lived, was sufficient proof for the judge
to consider him as the real father and to sentence him to the partial
support of the child. For a survey of the significance of genetics in
paternity investigations we refer to the publication of BAUERMEISTER
and KUPER (1942).

The existence of blood groups found in mankind, is significant.
The starting point for the discovery of these blood groups was
the fatal consequences observed in some blood transfusions.
The investigations as to whether or not these blood groups were
genotypically based, led to important discoveries, which have an influence in jurisdiction, especially in the investigation into paternity.
It started in 1900 with the system A-B-O, later in 1911 this was extended to A_1-A_2-B-O. In 1928 the M-N-system followed and since that
time many other blood groups were discovered, of which the Rhesus-
system (1940, see Symposium 1948b) is the most important one. However, other systems have been put forward, e.g. S, P, Lutheran, Lewis
and Kell. RACE calculated that 23616 different serological types can
be distinguished by means of the existence of these bloodgroups. The
same significance can be attached to the structure of finger-prints, the
structure of the external part of the ear and also in striking cases to the
colour of the skin, hair and eyes, to the form of the hair and the combined features of the face. Usually such investigations do not yield a
completely positive result for paternity, but prove absolutely
negative findings (see ESSEN-MÖLLER and GEYER 1938). There are
even cases of the mistake of two children in a maternity home, in which,
by means of genetical advice a decision with regard to maternity
could be taken (see HIRSZFELD 1936).

Finally a *philologist* (VAN GINNEKEN 1926) used the results of genetics
to draw up, a completely new theory concerning the origin and the
change of languages. Supposing that interracial marriages and the
following mendelian segregation phenomena are primary and express
themselves in differences of the organs of speech VAN GINNEKEN
considers the changes of a language as a secondary result. Hence, he
relates the two sciences philology and genetics which are apparently
very far from one another. How far these contemplations are exact is

hard to predict for the time being. However the phenomenon in itself viz. that genetics is called in as an aid in the explanation of philological problems, is very interesting.

What the future will bring us, we do not know; however the wide-spread influence of genetics upon all phenomena of living creatures which can be investigated scientifically, including man, charges genetics with the duty to make her foundations as strong as possible. To fulfil this task is the aim of the geneticist.

BIBLIOGRAPHY

AAKERMAN, A., 1920. Speltlike bud-sports in common wheat (Hereditas, 1 : 116-127).

AAKERMAN, A., O. TEDIN and K. FRÖIER, 1948. Svalöf 1886—1946. History and present problems (Lund C. Blom, 1948: 389 pp.).

AASE, H. C., 1935. Cytology of cereals (Bot. Rev., 1: 467-496).

ACKERMAN, K., 1897—98. Thierbastarde. Zusammenstellung der bisherigen Beobachtungen über Bastardierung im Thierreich nebst Litteraturnachweise. I. Wirbellose Thiere. II. Wirbelthiere (Abh. u. Ber. Ver. f. Naturk. Kassel, 42: 103-121. 43: 1-79).

AEBLY, I., 1924. Ueber die Möglichkeit einer chemischen Deutung der Bastardbildung und Mendelspaltung (Vierteljahrsschr. Naturf. Ges. Zürich, 69: 39-51).

AFZAL, M., 1930. Studies in inheritance in cotton (Mem. Deptm. Agr. India. Botan. Ser., 17: 75-115).

AFZELIUS, K., 1924. Embryologische und cytologische Studien in Senecio und verwandten Gattungen (Acta Horti Bergiani, 8: 123-219).

AGOL, I. J., 1931. Stepallelomorphism in *Drosophila melanogaster* (Genetics, 16: 254-266).

AGOL, I. J., 1929. Treppenallelomorphismus bei *Drosophila melanogaster* (Zurn. Eksper. Biol., 5: 86—101).

ALAM, M., 1929. The calculation of linkage values. A comparison of various methods (Mem. Deptm. Agr. India, 18: 1-56).

ALLAN, H. H., 1937—1949. Wild specieshybrids in the Phanerogams I-II (Bot. Rev., 3: 593-615, 15: 77-105).

ALLEN, C. E., 1917. A chromosome difference correlated with sex-difference in *Sphaerocarpus* (Science, N. S. 47: 466-467).

ALLEN, C. E., 1919. The basis of sex-inheritance in *Sphaerocarpus* (Proc. Amer. Philos. Soc., 58: 289-316).

ALLEN, C. E., 1924. Inheritance by tetrad sibs in *Sphaerocarpus* (Proc. Amer. Phil. Soc., 63: 222-235).

ALLEN, C. E., 1935. The genetics of Bryophytes (Bot.Rev., 1: 269-291).

ALLEN, C. E., 1940. The genotypic basis of sex-expression in Angiosperms (Bot.Rev., 6: 227-300).

ALLEN, E., a.o., 1939. Sex and internal secretions (Sec.ed., London, Baillière, Tindall and Cox, 1939: 1346 pp.).

ANDERSON, E., 1936. The species problem in *Iris*. III. The phylogenetic relationship of *Iris versicolor* and *Iris virginia* (Ann. Miss. Bot. Garden, 23: 471-483).

ANDERSON, E., 1937. Cytology in its relation to taxonomy (Bot.Rev., 3: 335-350).

ANDERSON, E., 1949. Introgressive hybridization (New York, Wiley, 1949: 109).

ANDERSON, E.G., 1936. Induced chromosomal alterations in maize (In: Duggar, Biological effects of radiation, 1936: 1297-1310).

ANKEL, W. E. 1927—1929. Neuere Arbeiten zur Zytologie der natürlichen Parthenogenese der Tiere (Zschr. ind. Abst. Vererb. Lehre, 45: 232-278, 52 : 318-370).

ARKELL, T. R. and C. B. DAVENPORT, 1912. The nature of the inheritance of horns in sheep (Science, N.S. 35: 927-930).

ARMBRUSTER, L., H. NACHTSHEIM und TH. ROEMER, 1917. Die Hymenopteren als Studienobjekt azygoter Vererbungerscheinungen. Experimentum crucis theoriae mendelianae (Zschr. ind. Abst. Vererb. Lehre, 17: 273-355).

ARNASON, T. J., 1936. Cytogenetics of hybrids between *Zea Mays* and *Euchlaena mexicana* (Genetics, 21: 40-60).

ASHBY, E., 1930—1937. Studies in the inheritance of physiological characters I-III (Ann. of Bot., 44: 457-467, 46: 1007-1032, N.S. 1: 11-42).

ASTBURY, W. T., 1939. X-ray studies of the structure of compounds of biological interest (Ann.Rev.Biochem., 8: 113-132).

ATWOOD, S. S., 1947. Cytogenetics and breeding of forage crops (Adv. Genetics, 1: 1-68).

AVERY, O. T., C. M. MACLEOD and M. MCCARTY, 1944. Studies on the chemical value of the substance inducing transformation of pneumococcal types. Induction of transformation by a desoxyribose nuclear acid fraction isolated from *Pneumococcus* type III (Journ. exp. Med., 79: 137-158).

BABCOCK, E. B., 1947a. The genus *Crepis*. I. The taxonomy, phylogeny distribution and evolution of *Crepis*. II. Systematic treatment (Univ. Calif. Publ. Botany, 21-22: 1-198, 199-1030).

BABCOCK, E. B., 1947b. Cytogenetics and speciation in *Crepsis* (Adv. Genetics, 1: 69-94).

BABCOCK, E. J., and J. L. COLLINS, 1929. Natural ionizing radiation and the rate of mutation (Nature, 124: 227-228).

BABCOCK, E. B. and M. NAVASHIN, 1930. The genus *Crepis* (Bibliogr. genet., 6: 1-90).

BAGG, H. J. and C. C. LITTLE, 1924. Hereditary structural defects in the descendants of mice exposed to Roentgen ray irradiation (Am. Journ. Anat., 33: 119-140).

BALBIANI, E. G., 1881. Sur la structure du noyau des cellules salivaires chez les larves du *Chironomus* (Zool. Anz., 4: 637-641, 662-666).

BALKASCHINA, E. I., 1929. Ein Fall der Erbhomoösis (die Genovariation ,,Aristapedia'') bei *Drosophila*

melanogaster (Arch. Entw. Mech., 115: 448-463).

BALTZER, F., 1914. Die Bestimmung des Geschlechts nebst einer Analyse des Geschlechtsdimorphismus bei *Bonellia* (Mitt. Zool. Stat. Neapel, 22 1-44).

BALTZER, F., 1925. Untersuchungen über die Entwicklung und Geschlechtsbestmmung der *Bonellia* (Pubbl. Staz. Zool. Neapoli, 6: 223-286).

BALTZER, F., 1928. Ueber metagame Geschlechtsbestimmung und ihre Beziehung zu einigen Problemen der Entwicklungsmechanik und Vererbung (auf Grund von Versuchen an *Bonellia*) (Zool. Anz. Suppl., 3: 273--325).

BALTZER, F., 1937. Analyse des Goldschmidtschen Zeitgesetzes der Intersexualität auf Grund eines Vergleiches der Entwicklung der *Bonellia*- und *Lymantria*-Intersexe. Zeitlich gestaffelte Wirkung der Geschlechtsfaktoren (Zeitgesetz) oder Faktorengleichzeitigkeit (Gen-Gleichgewicht) (Arch. Entw. Mech., 136: 1-43).

BALTZER, F., 1940. Ueber erbliche letale Entwicklung und Austauschbarkeit artverschiedener Kerne bei Bastarden (Naturwiss., 28: 177-187, 196-206).

BALTZER, F., und K. ZURBUCHEN, 1931—1937. Entwicklungsmechanische Untersuchungen an *Bonellia viridis* I-III (Rev. Suisse de Zool., 38: 361-371, Pubbl. Staz. zool. Napoli, 16: 28-79, 89-159).

BAMBER, R. C., 1927. Genetics of domestic cats (Bibliographia genetica, 3: 1-86).

BARRATT, R. W. and L. G. GARNJOBST, 1949. Genetics of a colonial microconidial strain of *Neurospora crassa* (Genetics, 34: 351-369).

BARTLETT, M. S. and J. S. B. HALDANE, 1934. The theory of inbreeding in autotetraploids (Journ.Genet., 29: 175-180).

BATESON, W., 1894. Materials for the study of variation (London, Macmillan Co, 1894: 598 pp.).

BATESON, W., 1907. Facts limiting the theory of heredity (Science, N.S. 26: 649-660).

BATESON, W., 1909. Mendel's principles of heredity (Cambridge, University Press, 1909: 413 pp.).

BATESON, W., 1926. Segregation (Journ. Genet., 16: 201-234).

BATESON, W. and A. E. GAIRDNER, 1921. Male-sterility in flax subject to two types of segregation (Journ. Genet., 11: 269-276).

BATESON, W. and R. C. PUNNETT, 1911. On gametic series involving reduplication of certain terms (Journ. Genet., 1: 293-302).

BATESON, W. and E. R. SAUNDERS, 1902. Experimental studies in the physiology of heredity. I (Reports Evol.Comm. Royal Soc., 1: 160 pp.).

BATESON, W., E. R. SAUNDERS and R. C. PUNNETT, 1905. Ditto II. Sweet Peas (Reports Evol. Comm. Roy. Soc., 2: 80-99).

BATESON, W., E. R. SAUNDERS and R. C. PUNNETT, 1906. Ditto III (Reports Evol. Comm. Roy. Soc., 3: 53 pp.).

BATESON, W., E. R. SAUNDERS and R. C. PUNNETT, 1908. Ditto IV (Reports Evol. Comm. Roy. Soc., 4: 1-40).

BAUER, H., 1935. Der Aufbau der Chromosomen aus den Speicheldrüsen von Chironomus Thummi Kiefer (Zschr. f. Zellf., 23: 280-313).

BAUER, H. und N. W. TIMOFÉEFF-RESSOVSKY, 1943. Genetik und Evolutionsforschung bei Tieren (In: Heberer, Die Evolution der Organismen, Jena, Fischer, 1943: 335-429).

BAUERMEISTER, W. und M. KUPER, 1942. Die erbbiologische Abstammungsprüfung (Fortschr. Erbpath. usw., 6: 144-212).

BAUR, E., 1909a. Pfropfbastarde, Periklinalchimaeren und Hyperchimaeren (Ber. Deutsch. Botan. Ges., 27: 603-605).

BAUR, E., 1909b. Das Wesen und die Erblichkeitsverhältnisse der ,,Varietates albomarginatae hort'' von Pelargonium zonale (Zschr. ind. Abst. Vererb. Lehre., 1: 330-351).

BAUR, E., 1910. Vererbungs- und Bastardierungsversuche mit Antirrhinum (Zschr. ind. Abst. Vererb. Lehre, 3: 34-98).

BAUR, E., 1911. Untersuchungen über die Vererbung von Chromatophorenmerkmalen bei Melandrium, Antirrhinum und Aquilegia (Zschr. ind. Abst. Vererb. Lehre, 4: 81-102).

BAUR, E., 1911, 1930. Einführung in die (experimentelle) Vererbungslehre (Berlin, Borntraeger. 1e Aufl. 1911: 293 pp.; 7—11 Aufl. 1930: 436 pp.).

BAUR, E., 1912. Ein Fall von geschlechtsbegrenzter Vererbung bei Melandrium album (Zschr. ind. Abst. Vererb. Lehre, 8: 335-336).

BAUR, E., E. FISCHER und F. LENZ, 1932-1940. Menschliche Erblehre und Rassenhygiene (4. Aufl., München, Lehmann, 1: 796 pp.; 2: 593 pp.; 5. Aufl. 1.2 : 511 pp.).

BAWDEN, F. C., 1950. Plant viruses and virus diseases (3rd revised ed. Waltham, Mass., Chronica botanica, 1950: 335 pp.).

BEADLE, G. W., 1932a. Studies of Euchlaena and its hybrids with Zea. I. Chromosome behavior in Euchlaena mexicana and its hybrids with Zea mays (Zschr. ind. Abst. Vererb. Lehre, 62: 291-304).

BEADLE, G. W., 1932b. The relation of crossing over to chromosome association in Zea-Euchlaena hybrids Genetics, 17: 481-502).

BEADLE, G. W., 1945. Biochemical genetics (Chem. Rev., 37: 15-96).

BEADLE, G. W., 1949. Genes and biological enigmas (Science in Progress, 6th Series: 184-248).

BEADLE, G. W. and E. L. TATUM, 1941 Genetic control of biochemical reactions in Neurospora (Proc. Nat. Ac. Sc. Wash., 27: 499-506).

BEADLE, G. W. and E. L. TATUM, 1945. Neurospora: II. Methods of producing and detecting mutations concerned with nutritional requirements (Am. J. Bot., 32: 678-686).

BEALE, G. H., 1954. The genetics of Paramaecium aurelia (Cambridge, Univ. Press, 1954: 179 pp.).

BECKER, E., 1938. Das Gen-Wirkstoff-System der Augenausfärbung bei Insekten (Naturwiss., 26: 433-441).

BECKER, E., 1939. Ueber die Natur des Augenpigments von Ephestia kühniella und seinem Vergleich mit

den Augenpigmenten anderer Insekten (Biol. Zbl., 59: 597-627).

BECKER, E., 1942. Ueber Eigenschaften, Verbreitung und die genetisch-
entwicklungsphysiologische Bedeutung der Pigmente der Ommatin-
und Ommingruppe (Ommochrome)
bei den Arthropoden (Zschr. ind.
Abst. Vererb. Lehre, 80: 157-
204).

BECKER, W. A. 1938. Recent investigations in vivo on the division of
plant cells (Bot-Rev., 4: 446-472).

BĚLAŘ, K., 1925. Der Chromosomenbestand der *Melandrium*-Zwitter
(Zschr. ind. Abst. Vererb. Lehre,
39: 184-190).

BĚLAŘ, K., 1928. Die cytologischen
Grundlagen der Vererbung. (Handb.
Vererb. wiss. I B: 412 pp.).

BELLING, J., 1914. The mode of inheritance of semi-sterility in the
offspring of certain hybrid plants
(Zschr. ind. Abst. Vererb. Lehre,
12: 303-342).

BELLING, J., 1928. The ultimate
chromomeres of *Lilium* and *Aloë*
with regard to the number of genes
(Univ. Calif. Publ. Bot., 14: 307
-318).

BELLING, J., 1933. Crossing-over and
gene rearrangement in flowering
plants (Genetics, 18: 388-412).

BEMMELEN, J. F. VAN, 1939. Over het
begrip parenteel (Afkomst en Toekomst, 5: 25-37).

BENEDEN, E. VAN, 1883. Recherches
sur la maturation de l'oeuf et la fécondation. *Ascaris megalocephala*
(Arch. de Biol., 4: 265-640).

BENEDICT, F. G., W. LANDAUER and
E. L. Fox, 1932. The physiology
of normal and frizzle fowl with
special reference to the basal metabolism (Storrs Agr. Exp. Stat.
Bull. 177: 13-101).

BERNSTEIN, F., 1929. Variations- und
Erblichkeitsstatistik (Handb. Verer.
wiss. I. C. : 96 pp.).

BERTSCH, K. und F., 1947. Geschichte
unserer Kulturpflanzen (Stuttgart,
Wiss. Verlagsges., 1947: 268 pp.).

BEYER, J. J., 1932. Ueber die Knos-
penvariationen der *Coleus hybridus*
Genetica, 14: 279-318).

BEYERINCK, M. W., 1917. De enzymtheorie van de erfelijkheid (Versl.

gew. Verg. K. A. W. Amsterdam.
Wis- en Natuurk., 25: 1231-1245).

BIBLIOGRAPHY, 1939. Bibliography on
cold resistance in plants (Cambridge, Imp. Bur. Plant Genet.,
1939: 22 pp.).

BISSET, K. A., 1950. The cytology and
life-history of bacteria (Edinburgh,
Livingstone, 1950: 136 pp.).

BLACKBURN, K. B., 1923. Sex chromosomes in plants (Nature 112: 687
-688).

BLACKBURN, K.B., 1928. Chromosome number in *Silene* and the neighbouring genera (Verhandl. V. Int.
Kongr. Vererb. wiss., 439-446).

BLACKBURN, K. and J. W. H. HARRI-
SON, 1924. Genetical and cytological
studies in hybrid roses. I. The origin of a fertile hexaploid form in
the *Pimpinellifolia - vilosa* - crosses
(Brit. Journ. Exp. Biol., 1: 557-
570).

BLAIR, W. F., 1953. Population dy-
namics of rodents and other small
mammals (Adv. Genet., 5: 2-42).

BLAKESLEE, A. F., 1904. Sexual reproduction in *Mucorineae* (Proc.
Amer. Acad. Arts Sc., 40: 200-
319)..

BLAKESLEE, A. F., 1914. Corn and men
(Journ. Hered., 5: 511-518).

BLAKESLEE, A. F., 1915. Sexual reactions between hermaphroditic and
dioecious *Mucors* (Biol. Bull., 29:
87-102).

BLAKESLEE, A. F., 1928. Genetics of
Datura (Verhandl. V. Intern. Kongr. Vererb. wiss., 1: 117-130).

BLAKESLEE, A. F., 1931. Extra chromosomes a source of variations in
the Jimson weed (Smithsonian Report for 1930: 431-450).

BLAKESLEE, A. F., 1937. Dédouble-
ment du nombre des chromosomes
chez les plantes par traitement chimique (C.R.Ac. Sc. Paris, 205:
476-497).

BLAKESLEE, A. F. and A. G. AVERY,
1937. Methods of inducing chromosome doubling in plants by treatment with colchicine (Science N.S.,
86: 408).

BLAKESLEE, A. F. and A. G. AVERY,
1938. Fifteen year breeding records
of 2n + 1 types in *Datura stramonium* (Cooperation in research

Carnegie Inst. Wash.Publ., 501: 315-351).

BLAKESLEE, A. F., J. BELLING, M. E. FARNHAM and A. D. BERGNER, 1922. A haploid mutant in the Jimson weed, *Datura stramonium* (Science, N.S. 55: 646-647).

BLAKESLEE, A. F. and M. E. FARNHAM, 1923. Trisomic inheritance in the poinsettia mutant of *Datura* (Am. Nat. 57: 481-495).

BLAKESLEE, A. F., G. MORRISON and A. G. AVERY, 1927. Mutations in a haploid *Datura* and their bearing on the hybrid-origin theory of mutants (Journ. Hered., 18: 193-200).

BLAU, M. und K. ALTENBURGER, 1922. Ueber einige Wirkungen von Strahlen II. (Zshr.f.Physik, 12: 315-329).

BLEIER, H., 1928a. Genetik und Cytologie teilweise und ganz steriler Getreidebastarde (Bibliogr. Genet., 4: 321-400).

BLEIER, H., 1928b. Zytologische Untersuchungen an seltenen Getreide- und Rübenbastarden (Verh. V. Int. Kongr. Vererb. wiss., 1: 447-452).

BLEIER, H., 1934. Bastardkaryologie (Bibliogr. Genet. 11: 393—489).

BLUHM, A., 1930. Zum Problem ,,Alkohol und Nachkommenschaft'' (Arch. Rassen- u. Ges. Biol., 24: 12-82).

BLUHM, A., 1932. Gibt es eine erworbene, auf die Nachkommenschaft übertragbare, spezifische Giftüberempfindlichkeit? (Biol. Zbl., 52: 667-673).

BLUHM, A., 1938. Ueber erworbene Immunität, Giftempfindlichkeit und Vererbung. Ein Beitrag zur Frage der Dauermodifikationen (Arch. Rassen- u. Ges. Biol., 32: 97-110).

BONNER, D. M., 1950. Genes and biochemical reaction (Colloid Chemistr. Ed. J. Alexander, Vol. 8, N.Y., Reinhold).

BONNER, D. M., 1950. The Q locus of *Neurospora* (Genetics, 35: 89-90).

BONNER, D. M., 1951. Gene-enzyme relationships in *Neurospora* (Cold Spring Harbor Symposia Quant. Biol., 16: 143-157).

BONNER, D. M., C. YANOFSKY and C. W. H. PARTRIDGE, 1951. Incomplete genetic blocks in biochemical mutants of *Neurospora* (Proc. Nat. Ac. Sc., 38: 25-34).

BONNEVIE, K., 1934. Embryological analysis of gene manifestation in Little and Bagg's abnormal mouse tribe (Journ. exp. Zool., 67: 443-520).

BONNEVIE, K., 1940. Tatsachen der genetischen Entwicklungsphysiologie (Handb. Erbb. d. Menschen, 1: 73-180).

BONNIER, G., 1895. Recherches expérimentales sur l'adaptation des plantes au climat alpin (Ann. Sc. natur. Botanique, Série 7, 2: 217-360).

BONNIER, G. and A. HANSSON, 1948. Identical twingenetics in cattle (Heredity 2: 1-27).

BOVERI, Th., 1895. Ueber die Befruchtungs- und Entwicklungsfähigkeit kernloser Seeigeleier und die Möglichkeit ihrer Bastardierung (Arch. Entwickl. Mech., 2: 394-443).

BOVERI, Th., 1904. Ergebnisse über die Konstitution der chromatischen Substanz des Zellkerns (Jena, Fischer, 1904: 130 pp).

BOVERI, TH., 1915. Ueber die Entstehung der Eugsterschen Zwitterbienen (Arch. Entw. Mech., 41: 264-311).

BOVERI, TH., 1918. Zwei Fehlerquellen bei Merogonieversuchen und die Entwicklungsfähigkeit merogonischer und partiell-merogonischer Seeigelbastarde (Arch. Entw. Mech., 44: 417-471).

BOYCOTT, A. E. and C. DIVER, 1923. On the inheritance of sinistrality in *Limnaea peregra* (Proc. Roy. Soc. London B., 85: 207-213).

BOYCOTT, A. E., C. DIVER, S. L. GARSTANG and F. M. TURNER, 1930. The inheritance of sinistrality in *Limnaea peregra* (Mollusca, Pulmonata) (Trans. Roy. Soc. London B., 219: 51-131).

BRACHET, J., 1945. Embryologie chimique (Paris, Masson-Liège, Desoer, 1945: 535 pp.).

BRADLEY, R., 1719. New improvements of planting and gardening (Sixth Ed., London, 1731: 608 pp.).

BREMER, G., 1922-1931. De cytologie van het suikerriet. I—VI (Arch.

v. d. Suikerind. Nederl. Indië 1922:
1-111; 1924: 151-180, 477-508;
1928: 565-696; 1931: 583-607,
1349-1391).

BREMER, G., 1949. Increase of chromo-
some number in species hybrids of
Saccharum (Proc. VIII Intern.
Congr. Genetics, Stockholm, 1948;
Suppl. vol. Hereditas, 1949: 541-
542).

BREZLAU, E. und HARNISCH, 1927.
Zahl der Chromosomen bei den
Tieren (Tabulae biol., 4: 83-113).

BRIDGES, C. B., 1913. Non-disjunction
of the sex-chromosomes of *Drosophila* (Journ. Exper. Zool., 15:
318-351).

BRIDGES, C. B., 1914. Direct proof
through non-disjunction that the
sexlinked genes of *Drosophila* are
borne by the X-chromosome (Science
N.S. 40: 107-109).

BRIDGES, C. B., 1915. A linkage variation in *Drosophila* (Journ. exp.
Zool., 19: 1-21).

BRIDGES, C. B., 1916. Non-disjunction as proof of the chromosome
theory of heredity (Genetics, 1:
1-53).

BRIDGES, C. B., 1917. Deficiency
(Genetics, 2: 445-465).

BRIDGES, C. B., 1919 Duplication
(Anat. Rec., 15: 357-358).

BRIDGES, C. B., 1921. Genetical and
cytological proof of non-disjunction
of the fourth chromosome of *Drosophila melanogaster* (Proc. nat. Ac.
Sc., 7: 186-192).

BRIDGES, C. B., 1922. The origin of
variations in sexual and sex-limited
characters (Am. Nat., 56: 51-63).

BRIDGES, C. B., 1923a. The translocation of a section of chromosome
II upon chromosome III in *Drosophila* (Anat. Rec., 24: 426-427).

BRIDGES, C. B., 1923b. Aberrations
in chromosomal materials (Sc. Pap.
Second Intern. Congr. Eugenics,
1:76-81).

BRIDGES, C. B., 1925a. Haploidy in
Drosophila melanogaster (Proc. nat.
Ac. Sc., II: 706-710).

BRIDGES, C. B., 1925b. Sex in relation
to chromosomes and genes (Am.
Nat., 59: 127-137).

BRIDGES, C. B., 1929. Variations in
crossing-over in relation to age of

female in *Drosophila melanogaster*
(Carnegie Inst. Wash. Publ. 399:
63-89).

BRIDGES, C. B., 1930. Haploid *Drosophila* and the theory of genic balance (Science N.S. 72: 405-406).

BRIDGES, C. B., 1935. Salivary chromosome maps with a key to the
banding of the chromosomes of *Drosophila melanogaster* (Journ. Hered., 26: 60-64).

BRIDGES, C. B., 1936. The bar 'gene'
a duplication (Science N.S. 83:
210-211).

BRIDGES, C. B., 1939. Cytological and
genetical basis of sex (In: E. AL-
LEN, 1939. Sex and internal secretions 2nd ed., London, Baillière,
Tindall & Cox, 1939, Chapt. II,
p. 15-63).

BRIEGER, F., 1933. Die genaue Bestimmung des Zeitpunktes der Mendelspaltung (Züchter, 5: 34-44).

BRINK, R. A., 1929. Studies on the
physiology of a gene (Quart.Rev.
Biol., 4: 520-543).

BRINK, R. A., and J. H. MACGILLI-
VRAY, 1924. Segregation for the
waxy character in maize pollen and
differential development of the male
gametophyte (Am. Journ. Bot.,
11: 465-469).

BRINK, R. A., J. H. MACGILLIVRAY
and M. DEMEREC, 1926. Effect of
the waxy gene in maizepollen, — a
reply to a criticism (Genetics, 11:
38-40).

BROEKMAN, R. W., 1950. De betekenis
van de erfelijkheid voor de tand-
heelkunde (2 ed,, Utrecht, Tholen,
1950: 153 pp.).

BRUNSWIK, H., 1924. Untersuchungen
über die Geschlechts- und Kernver-
hältnisse bei der Hymenomyzeten-
gattung *Coprinus* (Bot. Abh. 5:
152 pp.).

BRUNSWIK, H., 1926. Die Reduktionsteilung bei den Basidiomyzeten
(Zschr. Bot. 18: 481—498).

BUCHANAN SMITH, A. D. and O. J.
ROBISON, 1933. The genetics of
cattle. I. A survey of the literature
upon the inheritance of milking
capacity (Bibliogr. Genet., 10: 1-
104).

BUCHANAN SMITH, A. D., O. J. ROBI-
SON and D. M. BRYANT, 1938. The

genetics of the pig (Bibliogr. Genet., 12: 1-60).

BUCHHOLZ, J. T., and A. F. BLAKESLEE 1932. Pollen-tube growth in primary and secondary 2n + 1 *Daturas* (Am. Journ. Bot., 19: 604-626).

BUCHNER, P., 1921. Tier und Pflanze in intrazellulärer Symbiose (Berlin, Borntraeger, 1921: 469 pp.).

BUDER, J., 1910. Studien an *Laburnum Adami* (Ber. Deutsch. Bot. Ges. 28: 188-192).

BUDER, J., 1911. Studien an *Laburnum Adami* II. Allgemeine anatomische Analyse des Mischlings und seiner Stammpflanzen (Zschr. ind. Abst. Vererb. Lehre, 5: 209-284).

BUETSCHLI, O., 1876. Ueber die Bedeutung der Entwicklungsgeschichte für die Stamsgeschichte der Thiere (Jahresber. d. Senckenb. Ges. Frankfurt a. Main, 1876: 66).

BURNS, R. K., 1949. Hormones and the differentiation of sex (Survey of biological progress, 1: 233-266).

CAMERARIUS, R. J., 1694. De sexu plantarume pistola (Tübingen, Romme, 1694: 110 pp. Again edited by M. B. Valentin in Frankfurt a. d. Main 1700 and 1701, by J. G. Gmelin in Tübingen 1749 and translated by M. Möbius in Ostwalds Klassiker d. exakt. Wiss. Nr. 105. Leipzig, Engelmann, 1899).

CAMERON, G., 1950. Tissue culture technique. (Sec. ed. New York Acad. Press, 1950: 191 pp.).

CAROTHERS, E. E., 1931. The maturation divisions and segregations of heteromorphic homologous chromosomes in *Acrididae*. (Orthoptera) (Biol. Bull., 61: 324-349).

CASPARI, E., 1948. Cytoplasmic inheritance (Adv. Gen., 2: 1-68).

CASPARI, E., 1950. Visible plasmagenes (Evolution 4: 362-363).

CASPERSSON, T., 1936. Ueber den chemischen Aufbau der Strukturen des Zellkerns (Skand. Arch. f. Physiol., 73: Suppl. 8: 1-152).

CASPERSSON, T., 1940a. Die Eiweissverteilung in den Strukturen des Zellkerns (Chromosoma, 1 : 562-604).

CASPERSSON, T., 1940b. Nucleïnsäure-

ketten und Genvermehrung (Chromosoma, 1: 605-619).

CATCHESIDE, D. G., 1948. Genetic effects of radiation (Adv. Genetics, 2: 271-358).

CHARGAFF, E. and J. N. DAVIDSON, editors, 1955. The nucleic acids. Chemistry and biology (New York, Academic Press, 1955, 2 vols.).

CHARLIER, C. V., 1920. Vorlesungen über die Grundzüge der mathematischen Statistik (Lund-Hamburg, 1920: 128 pp.).

CHESLEY, P., 1935. Development of the short tailed mutant in the house mouse (Journ. exp. Zool., 70 :429-459).

CHITTENDEN, R. J., 1927a. Cytoplasmic inheritance in flax (Journ. Hered., 18: 337-343).

CHITTENDEN, R. J., 1927b. Vegetative segregation (Bibliogr. genet., 3: 357-442).

CHITTENDEN, R. J. and C. PELLEW, 1927. A suggested interpretation of certain cases of anisogeny (Nature, 119: 10-11).

CHODAT, F., 1943. Problèmes du déterminisme phénotypique du sexe chez les végétaux (Arch. Jul. Klaus Stift., 17: 496-512).

CLARK, J. A., 1936. Improvement in wheat (Yearb. U.S. Dep. Agric., 1936: 207-302).

CLAUSEN, J., 1926. Genetical and cytological investigations on *Viola tricolor* L. and *V. arvensis* Murr. (Hereditas, 8: 1-156).

CLAUSEN, J., 1933. Cytological evidence for the hybrid origin of *Pentstemon neotericus* Keck (Hereditas, 18: 65-76).

CLAUSEN, J., D. D. KECK and W. M. HIESEY, 1940-1948. Experimental studies on the nature of species I-III (Carnegie Inst. Wash., Publ. 520: 452 pp., 564: 174 pp., 581: 129 pp.).

CLAUSEN, R. E. and T. H. GOODSPEED 1923. Inheritance in *Nicotiana Tabacum*. III. The occurrence of two natural periclinal chimaeras (Genetics, 8: 97-105).

CLAUSEN, R. E. and T. H. GOODSPEED, 1925. Interspecific hybridization in *Nicotiana*. II. A tetraploid *glutinosa-tabacum* hybrid, an experimental verification of Winge's hypothesis (Genetics, 10: 278-284).

CLAUSEN, R. E. and W. E. LAMMERTS, 1929. Interspecific hybridisation in *Nicotiana*. X. Haploid and diploid merogony (Am. Nat., 63: 279-282).

CLELAND, R. E., 1923. Chromosome arrangements during meiosis in certain *Oenotheras* (Am. Nat. 57: 562-566).

CLELAND, R. E., 1925. Chromosome behavior during meiosis in the pollen mother cells of certain *Oenotheras* (Am. Nat., 59: 475-479).

CLELAND, R. E., 1928. The genetics of *Oenothera* in relation to chromosome behavior, with special reference to certain hybrids (Verhandl. V. Int. Kongr. Vererb. wiss. 1: 554-567).

CLELAND, R. E., 1936. Some aspects of the cytogenetics of *Oenothera* (Bot.Rev., 2: 316-348).

CLELAND, R. E. and A. F. BLAKESLEE, 1930. Interaction between complexes as evidence for segmental interchange in *Oenothera* (Proc. nat. Ac. Sc., 16: 183-189).

CLELAND, R. E. und F. OEHLKERS, 1930. Erblichkeit und Zytologie verschiedener *Oenotheren* und ihrer Kreuzungen (Jahrb. wiss. Bot., 73: 1-124).

COCKAYNE, E. A., 1938. The genetics of sex in Lepidoptera (Biol.Rev., 13: 107-132).

COLE, L. J., a.o., 1940. Symposium on the relation of genetics to geographical distribution and speciation (Am. Nat., 74: 193-278; 289-322).

COLLINS, G. N. and J. H. KEMPTON, 1911. Inheritance of waxy endosperm in hybrids of chinese maize (C. R. IV Conf. intern. génétique, Paris, 1911: 347-357).

COLLINS, G. N. and J. H. KEMPTON, 1926. Variability in the linkage values of two seed characters in maize (Bull. U.S. Deptm. Agr., 1468: 64 pp.).

CONKLIN, E. G., 1903. The cause of inverse symmetry (Anat. Anz., 23: 577-588).

CORRENS, C., 1899. Untersuchungen über die Xenien bei *Zea Mays* (Ber. Deutsch. Bot. Ges., 17: 410-417).

CORRENS, C., 1900a. G. Mendel's Regel über das Verhalten der Nachkommenschaft der Rassenbastarde (Ber. Deutsch. Bot. Ges., 18: 158—168).

CORRENS, C., 1900b. Ueber Leukojenbastarde. Zur Kenntnis der Grenzen der Mendel'schen Regeln (Bot. Centralb., 84: 97-113).

CORRENS, C., 1902a. Scheinbare Ausnahmen von der Mendelschen Spaltungsregel für Bastarde (Berlin Deutsch. Bot. Ges., 20: 159-172).

CORRENS, C., 1902b. Ueber Bastardirungsversuche mit *Mirabilis*-Sippen (Ber. Deutsch Bot. Ges., 20: 594 -608).

CORRENS, C., 1902c. Ueber den Modus und den Zeitpunkt der Spaltung der Anlagen bei den Bastarden vom Erbsen-Typus (Bot. Ztg., 60,2: 65 -82).

CORRENS, C., 1905. Ueber Vererbungsgesetze (Berlin, Borntraeger, 1905: 43 pp.).

CORRENS, C., 1907a. Die Bestimmung und Vererbung des Geschlechtes nach neuen Versuchen mit höheren Pflanzen (Berlin, Borntraeger, 1907: 81 pp.).

CORRENS, C., 1907b. Zur Kenntnis der Geschlechtsformen polygamer Blütenpflanzen und ihrer Beeinflussbarkeit (Jahrb. wiss. Bot., 44: 124-173).

CORRENS, C., 1909a. Vererbungsversuche mit blass (gelb) grünen und buntblättrigen Sippen bei *Mirabilis Jalapa*, *Urtica pilulifera* und *Lunaria annua* (Zschr. ind. Abst. Vererb. Lehre, 1: 291-329).

CORRENS, C., 1909b. Zur kenntnis der Rolle von Kern und Plasma bei der Vererbung (Zschr. ind. Abst. Vererb. Lehre, 2: 331-340).

CORRENS, C., 1912a. Die neuen Vererbungsgesetze. Zugleich zweite Auflage der Vererbungsgesetze (Berlin, Borntraeger, 1912: 75 pp.).

CORRENS, C., 1912b. Selbststerilität und Individualstoffe (Festschr. z. 84 Versamml. Deutsch. Naturf. u. Aerzte. Also in: Biol. Centralbl., 33: 389-423).

CORRENS, C., 1913. Eine mendelnde, kälteempfindliche Sippe (f. *delicata* der *Mirabilis Jalapa* (Zschr. ind.

Abst. Vererb. Lehre, 10: 130-135).

CORRENS, C., 1915. Ueber eine nach den Mendelschen Gesetzen vererbte Blattkrankheit (Sordago) der *Mirabilis Jalapa* (Jahrb. wiss. Bot., 56: 585-616).

CORRENS, C., .1917. Ein Fall experimenteller Verschiebung des Geschlechtsverhältnisses (Sitzungsber. Kgl. preuss. Ak. Wiss. 1917: 685-717).

CORRENS, C., 1918a. Zur Kenntnis einfach mendelnder Bastarde. I. Die Unterscheidung der *pilulifera*-Homozygoten und der Heterozygoten des Bastardes *Urtica pilulifera* × *Dodartii* (Sitzungsber. Kgl. preuss. Akad. Wiss. 1918: 221-231).

CORRENS, C., 1918b. Fortsetzung der Versuche zur experimentellen Verschiebung des Geschlechtsverhältnisses (Sitzungsber. Kgl. preuss. Ak. Wiss. 1918: 1175-1200).

CORRENS, C., 1919. Vererbungsversuche mit buntblättrigen Sippen. I. *Capsella Bursa pastoris albovariabilis* und *chlorina* (Sitzungsber. preuss. Ak. Wiss. 1919: 585-610).

CORRENS, C., 1921. Versuche bei Pflanzen das Geschlechtsverhältnis zu verschieben (Hereditas 2: 1-24).

CORRENS, C., 1928a. Bestimmung, Vererbung und Verteilung des Geschlechtes bei den höheren Pflanzen (Handb. Vererb. wiss. II. C: 138 pp.).

CORRENS, C., 1928b. Ueber nichtmendelnde Vererbung (Verhandl. V. Int. Kongr. Vererb. wiss. 1: 131-168).

CORRENS, C., 1937. Nicht mendelnde Vererbung (Handb. Vererb. wiss. II H: 159 pp.).

CORRENS, C. und R. GOLDSCHMIDT, 1913. Die Vererbung und Bestimmung des Geschlechtes (Berlin, Borntraeger, 1913: 149 pp.).

CRABB, A. R., 1947. The hybrid cornmakers. (New Brunswick, Rutgers Univ.Press, 1947: 331 pp.).

CRAMER, P. J. S., 1907. Kritische Uebersicht der bekannten Fälle von Knospenvariation (Natuurk. Verh. Holl. Mij. v. Wetensch., III Verz. 6,3: 474 pp.).

CRAMER, P. J. S., 1954. Chimeras (Bibliogr. genet., 16: 193-381).

CRANE, M. B. and W. J. C. LAWRENCE, 1938. The genetics of garden plants (2nd.ed., London, Macmillan, 1938: 287 pp.).

CREIGHTON, H. B. and B. Mc. CLINTOCK, 1931. A correlation of cytological and genetical crossing-over in *Zea mays* (Proc. nat. Ac. Sc., 17: 492-497).

CREW, F. A. E., 1925. Animal genetics, an introduction to the science of animal breeding (Edinburgh, Oliver and Boyd, 1925: 420 pp.).

CREW, F. A. E., 1926-1927. Abnormal sexuality in animals. I. Genotypical; II. Physiological; III. Sex reversal (Quart. Rev. Biol., 1: 315-359, 2:249-266, 427-441).

CREW, F. A. E., 1927. The genetics of sexuality in animals (Cambridge, Univ. Press., 1927: 188 pp)..

CREW, F. A. E., 1933. Sex determination (London, Methuen, 1933: 138 pp.).

CREW, F. A. E., 1937. The sex ratio (Am. Nat., 71: 529-559).

CUÉNOT, L., 1902-1911. La loi de Mendel et l'hérédité de la pigmentation chez les souris. Note 1—7 (Arch. Zool. expérim., (3) 10: 27-30; (4) 1: 33-39; (4) 2: 45-56; (4) 3: 123-132; (4) 6: 1-13; (4) 9: 7-15; (5) 8: 40-56).

CUÉNOT, L., 1928. Génétique des souris (Bibliogr. genet., 4: 179-242).

CUÉNOT, L., 1936. L'espèce (Paris, G. Doin, 1936: 310 pp.).

CUMLEY, R. W. and M. R. IRWIN, 1942. Genetic segregation of antigens (Journ. Hered., 33: 357-365).

CUSHING, H., 1916. Hereditary anchylosis of the proximal phalangeal joints (symphalangism) (Genetics, 1: 90-106).

DAHLBERG, G., 1926. Twin births and twins from a hereditary point of view (Stockholm, A. B. Tiden, 1926: 396 pp.).

DAHLBERG, G., 1943. Mathematische Erblichkeitsanalyse von Populationen (Acta med. Scand. Suppl., 148: 219 pp.).

DAHLBERG, G., 1947. Mathematical methods for population genetics (Basle and New York, Karger, 1947: 182 pp.).

DAHLBERG, G., 1948. Genetics of human populations (Adv. Genetics, 2: 69-98).

DALCQ, A., 1928. Les données expérimentales relatives au mécanisme de la division cellulaire (Biol. Rev., 3: 179-207).

DANFORTH, C. H., 1939. Relation of genic and endocrine factors in sex (In: E. ALLEN, 1939. Sex and internal secretions, 2nd ed., London Baillière, Tindall & Cox, 1939, Chapt. VI, 328-350).

DANNEEL, R., 1941. Phänogenetik der Kaninchenfärbung (Ergebn. Biol., 18: 55-87).

DANSER, B. H., 1929. Ueber die Begriffe Komparium, Kommiskuum und Konvivium und über die Entstehungsweise der Konvivien (Genetica, 11: 399-450).

DANTCHAKOF, V., 1938. Das Hormon im Aufbau der Geschlechter (Biol. Zbl., 58: 302-328).

DANTCHAKOFF, V., 1941. Der Aufbau des Geschlechtes bein höheren Wirbeltier (Jena, Fischer, 1941: 446 pp.).

DARBISHIRE, A. D., 1908. On the result of crossing round with wrinkled peas with special reference to their starchgrains (Proc. Roy. Soc. London, B. 80: 122-135).

DARBISHIRE, A. D., 1911. Breeding and the mendelian discovery (London, Cassel Co., 1911: 282 pp.).

DARLINGTON, C. D., 1931. Meiosis (Biol. Rev., 6: 221-264).

DARLINGTON, C. D., 1932, 1937. Recent advances in cytology)London, Churchill 1932: 560 pp., Sec.ed. 1937: 671 pp.).

DARLINGTON, C. D., 1935a. The internal mechanics of the chromosomes (Proc. Roy. Soc. London, B. 118: 33-96).

DARLINGTON, C. D., 1935b. The time, place and action of crossing-over (Journ. Genet., 31: 185-212).

DARLINGTON, C. D., 1936. The external mechanics of crossing-over (Proc. Roy. Soc. London, B. 121: 264-319).

DARLINGTON, C. D., 1944. Heredity, Development and Infection (Nature 154, p. 164-169).

DARLINGTON, C. D., 1949. The working units of heredity (Proc. 8th.

Intern. Congr. Genet., Stockholm, 189-200).

DARLINGTON, C. D., 1953. Polyploidy in animals (Nature, 171: 191-194).

DARLINGTON, C. D., and E. K. JANAKI AMMAL, 1945. Chromosome atlas of cultivated plants (London, Allen and Unwin, 1945: 397 pp.).

DARWIN, C., 1859, 1860. On the origin of species by means of natural selection (London, Murray, 1859: 502 pp., Sec. ed. 1860: 502 pp.).

DARWIN, C., 1868, 1888. The variation of animals and plants under domestication (London, Murray, 1868: 411 + 486 pp., Sec. ed. 1888: 473 + 495 pp.).

DAVENPORT, C. B., 1910a. The imperfection of dominance and some of its consequenses (Am. Nat., 44: 129-135).

DAVENPORT, C. B., 1910b. Inheritance of characteristics in domestic fowl (Publ. Carnegie Instit. Washington, 121: 100 pp.).

DAVENPORT, C. B. and M. P. EKAS, 1936. Statistical methods in biology, medicine and psychology (4th ed. New York, J. Wiley; London, Chapman & Hall, 1936: 216 pp.).

DAVIDSON, J. N., 1953. The biochemistry of the nucleic acids (Sec. ed., London, Methuen Co, 1953: 200 pp.).

DAVIS, B. M., 1914. Genetical studies on *Oenothera*. V (Zschr. ind. Abst. Vereb. Lehre, 12: 169-205).

DEHLINGER, U., 1937. Ueber die Morphologie des Gens und den Mechanismus der Mutation (Naturwiss. 25: 138).

DEHLINGER, U. und E. WERTZ, 1942. Biologische Grundfragen in physikalischer Betrachtung (Naturwiss. 30: 250-253).

DELAGE, Y., 1899a. Etudes sur la mérogonie (Arch. Zool. expérim. (3) 7: 383-417).

DELAGE, Y., 1899b. Sur l'interprétation de la fécondation mérogonique (Arch. Zool. expérim. (3) 7: 511 —527).

DELAPORTE, B., 1950. Observations on the cytology of bacteria. (Adv. Genet., 3:1-32).

DELLINGSHAUSEN, M. v. 1935, 1936. (cf. MICHAELIS 1931-1936 V and VIII).

DEMEREC, M., 1926. Notes on linkages in maize (Am. Nat., 60: 172-176).

DEMEREC, M., 1931. Behavior of two mutable genes in *Delphinium ajacis* (Journ. Genet., 24: 179-193).

DEMEREC, M., 1935. Unstable genes (Bot. Rev., 1: 233-248).

DESSAUER, F., 1922. Ueber einige Wirkungen von Strahlen. I. (Zschr. f. Physik, 12: 38-47).

DESSAUER, F., 1933. Quantenphysik der biologischen Röntgenstrahlenwirkungen (Zschr. f. Physik, 84: 218-221).

DESSAUER, F., 1954. Quantenbiologie (Berlin, Springer, 1954: 178 pp.).

DIGBY, L., 1912. The cytology of *Primula kewensis* and of other related *Primula* hyrbids (Ann. Bot., 26: 357-388).

DIVER, C. and I. ANDERSSON-KOTTÖ, 1938. Sinistrality in *Limnaea peregra* (Mollusca Pulmonata): The problem of mixed brood (Journ. Genet., 25: 447-525).

DIVER, C., A. E. BOYCOTT and S. L. GARSTANG, 1925. The inheritance of inverse symmetry in *Limnaea peregra* (Journ. Genet., 15: 113-200).

DOBROVOLSKAJA-ZAVADSKAJA,N.,1929 The problem of species in view of the origin of some new forms in mice (Biol. Rev., 4: 327-351).

DOBZHANSKY, TH., 1928. Studies on the manifold effects of certain genes in *Drosophila melanogaster* (Zschr. ind. Abst. Vererb. Lehre, 43: 330 -388).

DOBZHANSKY, TH., 1929. Genetical and cytological proof of translocations involving the third and the fourth chromosomes of *Drosophila melanogaster* (Biol. Zbl., 49: 408-419).

DOBZHANSKY, TH., 1936a. Induced chromosomal aberrations in animals (In: Duggar, Biological effects of radiation, 1936: 1167-1208).

DOBZHANSKY, TH., 1936b. Position effects on genes (Biol. Rev., 11: 364-384).

DOBZHANSKY, TH., 1939. Experimental studies on genetics of free-living populations of *Drosophila* (Biol. Rev., 14: 339-368).

DOBZHANSKY, TH., 1941. Genetics and the origin of species (Revised edition, New York, Columbia Univ. Press, 1941: 446 pp.).

DODGE, B. O., 1931. Inheritance of the albinistic non-conidial character in interspecific hybrids in *Neurospora* (Mycologia, 23: 1-50).

DOERRIES-RUEGER, K. a.o. 1929-1932. Experimentelle Analyse der Genom- und Plasmonwirkungen bei Moosen. I-V (Zschr. ind. Abst. Vererb. Lehre, 52: 390-405; 57: 306-432; 60: 17-38, 39-62; 62: 232-248).

DOMM, L. V., 1939. Modifications in sex and secondary sexual characters in birds (In: E. ALLEN, 1939. Sex and internal secretions, 2nd ed., London, Baillière, Tindall & Cox, 1939: Chapt. V, 227-327).

DONCASTER, L., 1908. Sex inheritance in the moth *Abraxas grossulariata* and its var. *lacticolor* (Rep. Evol. Comm. Roy. Soc., 4:53-57).

DONCASTER, L., 1914a. On sex-limited inheritance in cats, and its bearing on the sexlimited transmission of certain human abnormalities (Journ. Genet., 3: 11-23).

DONCASTER, L., 1914b. On the relation between chromosomes, sex-limited transmission and sex-determination in *Abraxas grossulariata* (Journ. Genet., 4: 1-22).

DONCASTER, L., 1915. The relation between chromosomes and sex-determination in *Abraxas* (Nature, 95: 311).

DONCASTER, L. and. G. H. RAYNOR, 1906. Breeding experiments with Lepidoptera (Proc. Zool. Soc., 1906: 125-132).

DOUWES, H., 1943. Een genetisch-chemisch onderzoek van *Eschscholtzia californica* Cham. (Genetica, 23: 353-464).

DRINKWATER, H., 1908. An account of a brachydactylous family (Proc. Roy. Soc. Edinburgh, 28: 35-57).

DUBININ, N. P., 1929. Allelomorphentreppen bei *Drosophila melanogaster* (Biol. Zbl., 49: 328-339).

DUBININ, N., 1932. Step allelomorphism in *Drosophila melanogaster*, The allelomorphs Achaete²-Scute¹⁰. Achaete¹-Scute¹¹ and Achaete³-Scute¹³ (Journ. Genet., 25: 163—181).

DUBOS, R. J., 1946. The bacterial cell

in its relation to problems of virulence, immunity and chemotherapy (Cambridge, Mass., U.S.A., Harwar Univ. Press. 1946: 460 pp.).

DUCHARTRE, 1863. Rapport sur la question de l'hybridité dans les végétaux (Ann. Sc. nat. Botan., (4) 19: 125-134)

DUFF, J. F. and G. H. THOMSON, 1923. The social and geographical distribution of intelligence in Northumberland (Brit. Journ. Psychol., 14: 192-198).

DUGGAR, B. M., 1936. Biological effects of radiation (2 vols, New York-London, Mc Graw Hill, 1936: 1343 pp.).

DUNCAN, F. N., 1917. Orthodactyly (Journ. Hered., 8: 174-175).

DUNN. L. C. (with P. CHESLEY and S. GLUECKSOHN-SCHOENHEIMER) 1936-1939. The inheritance of taillessness (anury) in the house mouse I-III (Genetics, 21: 525-536; 24: 587-609, 728-731).

DU RIETZ, G. E., 1930. The fundamental units of biological taxonomy (Svensk Botan. Tidskr., 24: 333-428).

DUSTIN, A. P., 1936. La colchicine, réactif de l'imminence caryocinétique (Arch. portug. Sc. biol., 5: 38-43).

DUSTIN, A. P., 1938. L'action des arsénicaux et de la colchicine sur la mitose, la stathmocinèse (C.R. Ass. Anat., 33: 7 pp.).

EAST, E. M., 1926. The physiology of self-sterility in plants (Journ. gen. Physiol., 8: 403-416).

EAST, E. M., 1929a. Self-sterility (Bibliogr. genet., 5: 331-370).

EAST, E. M., 1929b. The concept of the gene (Proc. Int. Congr. Plant. scienes, Ithaca, 1926, 1: 889-895).

EAST, E. M., 1934. The nucleus-plasma problem (Am. Nat., 68: 289-303, 402-439).

EAST, E. M., 1936. Heterosis (Genetics, 21: 375-397).

EAST, E. M. and H. K. HAYES, 1911. Inheritance in maize (Connecticut Agr. Expt. Station, Bull. 167: 141 pp.).

EAST, E. M. and H. K. HAYES, 1912. Heterozygosis in evolution and plantbreeding (U. S. Dept. Agric.

Bur. of Plantind., Bull. 243: 58 pp.).

EAST, E. M. and D. F. JONES, 1919. Inbreeding and outbreeding. Their genetic and sociological signific ance (Philadelphia. London, Lippincott Co., 1919: 285 pp.).

EIGSTI, O. J. and P. DUSTIN JR., 1947-1949. Colchicine bibliography II, III (Lloydia 10: 65-114; 12: 185-207).

EISENTRAUT, M., 1950. Die Eidechsen der Spanischen Mittelmeer Inseln und ihre Rassenaufspaltung im Lichte der Evolution (Berlin, Akademie Verlag, 1950: 225 pp.).

EKMAN, G., 1927. Ueber den Unterschied zwischen Reduktions- und Aequationsteilung (Ann. Soc. zool. bot. Fennica Vanamo, 6: 1-36).

ELOFF, G., 1932. A theoretical and experimental study on the changes in the crossing-over value, their causes and meaning (Genetica, 14: 1-116).

EMERSON, R. A., 1915. Anomalous endosperm development in maize and the problem of bud sports (Zschr. ind. Abst. Vererb. Lehre, 14: 241-259).

EMERSON, R. A., 1921 Genetic evidence of aberrant chromosome behavior in maize endosperm (Am. Journ. Bot., 8: 411-424).

EMERSON, R. A., G. W. BEADLE and A. C. FRASER, 1935. A summary of linkage studies in Maize (Cornell Univ. Agr. exp. Stat. Memoir, 80: 1-83).

EMERSON, S. and A. H. STURTEVANT, 1931. Genetic and cytological studies in Oenothera III. The translocation interpretation (Zschr. ind. Abst. Vererb. Lehre, 59: 395-419).

EPHRUSSI, B., 1939. Génétique physiologique (Actual. Scient. Industr., 789, 21: 39 pp.).

EPHRUSSI, B., 1953. Nucleo-cytoplasmic relations in micro-organisms (Oxford, Clarendon Press, 1953: 127 pp.).

EPHRUSSI, B. et G. W. BEADLE, 1935. La transplantation des disques imaginaux chez la Drosophile (C. R. Ac. Sc. Paris, 201: 98-99).

EPHRUSSI, B. et S. CHEVAIS, 1938. Développement des couleurs des yeux chez la *Drosophile*. Relations entre procudtion, utilisation et libération des substances diffusibles (Bull. Biol. France et Belg., 72: 48-78).

ERNST, A., 1918. Bastardierung als Ursache der Apogamie im Pflanzenreich (Jena, Fischer, 1918: 665 pp.).

ESSEN-MOELLER, E. und E. GEYER, 1938. Die Beweiskraft der Aehnlichkeit im Vaterschaftsnachweis (Mitt. Anthr. Ges. Wien, 68: 9-87).

EULER, H. v., 1929. Chemische Untersuchungen an Chlorophyllmutanten. I (Hereditas, 13: 61-79).

EULER, H. v. u. A., 1931a. Zur chemischen Characterisierung von erblichen Chlorophylldefekten (Zschr. ind. Abst. Vererb. Lehre, 59: 131-152).

EULER, H. v. u. A., 1931b. Vergleichende Versuche über verschiedene Arten von Chlorophylldefekten (Zschr. ind. Abst. Vererb. Lehre, 60: 1-16).

EULER, H. v., u. A., 1933. Ueber die Konstanz des Chlorophyllgehaltes in drei Chlorophyllmutanten (Hereditas, 18: 225-244).

EULER, H. v., u. A., 1935. Konstanz des Chlorophyllgehaltes und Chromatophorendegeneration chlorophyllmutierender Gerstensippen Herreditas, 21: 119-128).

EYSTER, W. H., 1925. Mosaic pericarp in maize (Genetics, 10: 179-196).

EYSTER, W. H., 1928. The mechanism of variegations (Verh. V. Int. Kongr. Vererb. wiss., 1: 666-686).

EYSTER, W. H., 1929. The bearing of variegations on the nature of the gene (Proc. Int. Congr. Plantsciences Ithaca, 1926, 1: 923-941).

EYSTER, W. H., 1934. Genetics of *Zea Mays* (Bibliogr. Genet., 11: 187-392).

FANKHAUSER, G., 1937. The production and development of haploid Salamander larvae (Journ. Hered., 28: 3-15).

FEDERLEY, H., 1913. Das Verhalten der Chromosomen bei der Spermatogenese der Schmetterlinge *Pygaera anachoreta, curtula* und *pigra* sowie einiger ihrer Bastasde (Zschr. f. ind. Abst. u. Vererb. Lehre, 9: 1-110).

FEDERLEY, H., 1920. Die Bedeutung der polymeren Faktoren für die Zeichnung der Lepidopteren (Hereditas, 1: 221-269).

FEDERLEY, H., 1928. Das Inzuchtproblem (Handb. Vererb. wiss. II. I: 42 pp.).

FISCHER, E., 1901. Experimentelle Untersuchungen über die Vererbung erworbener Eigenschaften (Allgem. Zschr. f. Entom., 6: 49-51, 363-365, 377-381).

FISCHER, E., 1907. Zur Physiologie der Aberrationen- und Varietäten-Bildung der Schmetterlinge (Arch. Rassen- und Ges. Biol., 4: 761-793).

FISCHER, I., 1942. Grundriss der Gewebezüchtung (Jena, G. Fischer, 1942: 164 pp.).

FISCHER-WASELS, B., 1938. Die Erblichkeit in der Geschuwlstentwicklung (Fortschr. Erbpathol., 2: 221-262).

FISHER, R. A., 1925. Theory of statistical estimation (Proc. Cambridge Philos. Soc., 22: 700-725).

FISHER, R. A., 1946. Table of Chi-Square (Edinburgh, Oliver and Boyd, 1946).

FISHER, R. A., 1948. Statistical methods for research workers (10th ed. Edinburgh & London, Oliver & Boyd 1948: 356 pp.).

FISHER, R. A., 1949. The theory of inbreeding (Edinburgh-London, Oliver and Boyd, 1949: 120 pp.).

FISHER, R. A., and B. BALKAMUND, 1928. The estimation of linkage from the offspring of selfed heterozygotes (Journ. Genet., 20: 79-92).

FLORKIN, M., 1949. Biochemical evolution (New York, Acad. Press, 1949: 157 pp.).

FOCKE, W. O., 1881. Die Pfanzenmischlinge, ein Beitrag zur Biologie der Gewächse (Berlin, Borntraeger, 1881: 569 pp.).

FÖYN, B., 1932. Geschlechtsgebundene und geschlechtskontrollierte Vererbung (Handb. Vererb. wiss. I. I: 122 pp.).

FRANZ, V., 1920. Die Vervollkommnung in der lebenden Natur. Eine Studie über ein Naturgesetz (Jena, G. Fischer, 1920: 138 pp.).

FRASER ROBERTS, J. A., 1940. An introduction to medical genetics (London, Oxford Univ. Press, 1940: 266 pp.).

FRETS, G. P., 1930. Keimgifte (Arch. Rassen- u. Ges. Biol., 24: 83-96).

FRETS, G.P., 1931 Alcohol and the other germpoisons (Den Haag, M. Nijhoff, 1931: 179 pp.).

FREY-WYSSLING, A., 1948. Submicroscopic morphology of protoplasm and its derivatives (Amsterdam, Elsevier, 1948: 255 pp.).

FRIESEN, H., 1936. Die kosmischen Strahlen und der Mutationprozess (C.R. Ac. Sc. U.R.S.S., 1: 183-186).

FROST, H. B., 1915. The inheritance of doubleness in *Matthiola* and *Petunia*. I. The hypotheses (Am. Nat. 49: 623-635).

FUCHS, W. H. und K. v. ROSENSTIEL, 1941. Physiologische Resistenz (In: Roemer-Rudorf 1941, Handb. d. Pflanzenzüchtung, 1: 265-296).

GAERTNER, C. F. VON, 1826. Nachricht über Versuche, die Befruchtung einiger Gewächse betreffend (Naturwiss. Abhandl. Ver. Vaterl. Naturk. Tübingen, 1: 35-66).

GAERTNER, C. F. VON, 1827. Notices sur des expériences concernant la fécondation de quelques végétaux (Ann. Sc. natur., 10: 113-144).

GAERTNER, C. F. VON, 1838. Verhandeling ter beantwoording der vraag: ,,Wat leert de ondervinding aangaande het ontstaan van nieuwe soorten of bijsoorten van planten door kunstmatige bevruchting van bloemen van de ene met het bloemstof van andere soorten? En welke nieuwe, nuttige of fraaie plantgewassen kunnen op die wijze worden voortgebragt of vermenigvuldigd?'' (Natuurk. Verhand. Holl. Mij. v. Wetensch., 24, 1: 202 pp.).

GAERTNER, C. F. VON, 1849. Versuche und Beobachtungen über die Bastarderzeugung im Pflanzenreich (Stuttgart, Hering und Co., 1849: 790 pp.).

GAGER, C. S., 1908. Effects of the rays of radium on plants (Mem. New-York Bot. Gard., 4: 1-278).

GAGER, C. S., 1911. Cryptomeric inheritance in *Onagra* (Bull. Torrey bot. Club, 28: 461-471).

GAIRDNER, A. E., 1929. Male-sterility in flax. II. A case of reciprocal crosses differing in F_2 (Journ. Genet., 21: 117-124).

GAISER, L., 1926—1933. Chromosome numbers in angiosperms (I. Genetica 8: 401-484; II. Bibliographia genetica 6: 171-466; III. Genetica 12: 161-260; IV. Bibliographia genetica 10: 105-250).

GALIPPE, V., 1905. L'hérédité des stigmates de dégénérescence et les familles souveraines (Paris, Masson Cie., 1905: 455 pp.).

GALTON, F., 1869. Hereditary genius. An inquiry into its laws and consequences (London, Macmillan, 1869: 390 pp., Sec. ed. 1914: 374 pp.).

GALTON, F., 1871. Experiments in pangenesis by breeding from rabbits of a pure variety, into whose circulation blood taken from other varieties had previously been largely transfused (Proceed. Roy. Soc., 19: 393-410).

GALTON, F., 1876. A theory of heredity (Journ. Anthr. Inst. London, 5: 329:348. Revue scientifique, 2me Série 10: 198-205).

GALTON, F., 1877. Typical laws of heredity (Proceed. Roy. Inst. London, 8: 282–301; Nature, 1877: 492-495, 512-514, 532-533; Revue scientifique, 2me Série 13: 385-394).

GALTON, F., 1883. Inquiries into human faculty and its development (London, Macmillan, 1883: 387 pp.).

GALTON, F., 1889. Natural inheritance (London, Macmillan, 1889: 259 pp.).

GALTON, F., 1897a. A new law of heredity (Nature, 56: 235-237).

GALTON, F., 1897b. The average contribution of each several ancestor to the total heritage of the offspring (Proc. Roy. Soc. London, 61: 401-413).

GALTON, F., 1898. A diagram of heredity (illustrating the ,,Ancestral Law'') (Nature, 57: 293).

GALTON, F., 1901. Biometry (Biometrika. I. p. 7-10).

GALTON, F., 1908. Memories of my life (London, Methuen Co, 1908).

GARDNER, I. C. and H. H. NEWMAN, 1940. Mental and physical traits of identical twins reared apart (Journ. Hered., 31: 119-126).

GARROD, A. E., 1909, 1923. Inborn errors of metabolism (2nd ed. London,

Hodder and Stroughton, 1909, 1923: 216 pp.).

GATES, R. R., 1908. A study of reduction in *Oenothera rubrinervis* (Bot. Gazette, 46: 1-34).

GATES, R. R., 1909. The stature and chromosomes of *Oenothera gigas* de Vries (Arch. Zellf., 3: 525-552).

GATES, R. R., 1915a. Heredity and mutation as cell-phenomena (Amer. Journ. Bot., 2: 519-528).

GATES, R. R., 1915b. The mutation-factor in evolution with particular reference to *Oenothera* (London, Macmillan Co., 1915: 353 pp.).

GATES, R. R., 1928. The cytology of *Oenothera* (Bibliogr. genet. 4: 401-492).

GATES, R. R. and C. E. FORD, 1938. Chromosome catenations in *Oenothera* (Tabulae Biologicae, 15: 122-153).

GATES, R. R. and K. M. GOODWIN, 1930. A new haploid *Oenothera*, with some considerations on haploidy in plants and animals (Journ. Genet. 23: 123-156).

GAUTHERET, R. J., 1942. Manuel technique de culture des tissus végétaux (Paris, Masson, 1942: 172 pp.

GEDDES, P. and J. A. THOMSON, 1889. The evolution of sex (London, W. Scott 1889: 342 pp.).

GEITLER, L., 1934. Grundriss der Zytologie (Berlin, Borntraeger, 1934: 295 pp.).

GEITLER, L., 1938. Chromosomenbau (Protoplasma-Monogr., Berlin, Borntraeger, 14: 191 pp.).

GEITLER, L., 1941. Das Wachstum des Zellkerns in tierischen und pflanzlichen Geweben (Ergebn. Biol., 18: 1-54).

GEITLER, L., 1953. Endomitose und endomitotische Polyploidisierung (Protoplasmatologia, VI, C: 89 pp.).

GEPPERT, H., und S. KOLLER, 1938. Erbmathematik. Theorie der Vererbung in Bevölkerung und Sippe (Leipzig, Quelle u. Meyer, 1938: 228 pp.).

GERASSIMOW, J. J., 1902. Die Abhängigkeit der Grösse der Zelle von der Menge ihrer Kernmasse (Zschr. allgem. Physiol., 1: 220-258).

GERASSIMOW, J. J., 1904. Ueber die Grösse des Zellkerns (Beih. Bot. Zbl., 18,1: 45-118).

GILSE, P. H. G. VAN, A. B. HINNEN and A. C. NIEUWENHUYSE, 1942. Heredity in diseases in the field of oto-, rhino- laryngology (Bibliogr. Genet., 13: 301-424).

GINNEKEN, J. v., 1926. De erfelijkheid der klankwetten (Meded. K. A. W. Amst. Afd. Letterk., 61,5: 50 pp.).

GLASS, B., 1949. The genes and the gene action (Survey of biological progress, 1: 15-58).

GLASS, B., 1954. Genetic changes in human populations, especially those due to gene flow and genetic drift (Adv. Genet., 6: 95-140).

GLUECKSOHN-SCHOENHEIMER, S., 1938 The development of two tailless mutants in the house mouse (Genetics, 23: 573-584).

GLUECKSOHN-SCHOENHEIMER, S., 1940 The effect of an early lethal (t°) in the house mouse (Genetics, 25: 391-400).

GLUECKSOHN-WAALSCH, S., 1951. Physiological genetics of the mouse (Adv. Genet., 4: 2-52).

GODRON, D. A., 1844. De l'hybridité dans les végétaux (Nancy, Raybois, 1844: 22 pp.).

GODRON, D. A., 1863a. Des hybrides végétaux, considérés au point de vue de leur fécondité et de la perpétuité ou non-perpétuité de leurs caractères (Ann. Sc. natur. Bot., 4,19: 135-179).

GODRON, D. A., 1863b. Recherches expérimentales sur l'hybridité dans le régne végétale (Nancy, Raybois, 1863: 76 pp.).

GOLDSCHMIDT, R., 1912. Erblichkeitsstudien an Schmetterlingen. I (Zschr. ind. Abst. Vererb. Lehre, 7: 1-62).

GOLDSCHMIDT, R., 1916. Genetic factors and enzyme reactions (Science, N.S. 43: 77-100).

GOLDSCHMIDT, R., 1917. Crossing-over ohne chiasmatypie? (Genetics, 2: 82-97).

GOLDSCHMIDT, R., 1920a. Die quantitative Grundlage von Vererbung und Artbildung (Vortr. u. Aufs. Entwicklungsmech., 24: 163 pp.).

GOLDSCHMIDT, R., 1920b. Untersuchungen über Intersexualität (Zschr. ind. Abst. Vererb. Lehre, 23: 1-199).

GOLDSCHMIDT, R., 1927. Physiologische Theorie der Vererbung (Berlin, Springer, 1927: 247 pp.).

GOLDSCHMIDT, R., 1928a. The gene (Quart. Rev. Biol., 3: 307-324).

GOLDSCHMIDT, R., 1928b. Gen und Aussencharakter (Verh. V. Int. Kongr. f. Vererb. wiss., 1: 223-233).

GOLDSCHMIDT, R., 1931a. Analysis of intersexuality in the gipsy-moth (Quart. Rev. Biol., 6: 125-142).

GOLDSCHMIDT, R., 1931b. Die sexuellen Zwischenstufen (Monogr. Ges. Geb. Phys. Pflanzen und Tiere, 23: 528 pp.).

GOLDSCHMIDT, R., 1931c. Die Entwicklungsphysiologische Erklärung des Falls der sogenannten Treppenallelomorphe des Gens scute von Drosophila (Biol. Zbl., 51: 507-526).

GOLDSCHMIDT, R., 1934. Lymantria (Bibliogr. Genet., 11: 1-186).

GOLDSCHMIDT, R. 1935. Gen und Ausseneigenschaft I. (Zschr. ind. Abst. Vererb. Lehre, 69: 38-69).

GOLDSCHMIDT, R., 1937a. A critical review of some recent work in sex determination. I. Fishes (Quart. Rev. Biol., 12: 426-439).

GOLDSCHMIDT, R., 1937b. Le déterminisme du sexe et l'intersexualité (Paris, Alcan, 1937: 192 pp.).

GOLDSCHMIDT, R., 1938a. Physiological genetics (New York-London, Mc Graw Hill, 1938: 375 pp.).

GOLDSCHMIDT, R., 1938b. The time-law of intersexuality (Genetica, 20: 1-50).

GOLDSCHMIDT, R., 1940. The material basis of evolution (New Haven, Yale Univ. Press. 1940: 436 pp.).

GOLDSCHMIDT, R., 1946. The interpretation of the structure of triploid intersexes in Solenobia (Arch. Julius Klaus Stift., 21: 269-272).

GOLDSCHMIDT, R. und K. KATSUKI, 1927-1931. Erblicher Gynandromorphismus und somatische Mosaïkbildung bei Bombyx mori L. (Biol. Zbl., 47: 45-54; 48: 39-42, 685-699; 51: 58-74).

GOODSPEED, T. H., 1929. Cytological and other features of variant plants produced from X-rayed sex cells of Nicotiana tabacum (Bot. Gaz., 87: 563-582).

GOODSPEED, T. H., 1936. Chromosome alterations (In: Duggar, Biological effects of radiation, 1936: 1281-1296).

GOODSPEED, T. H., 1937. Significance of cytogenetic alterations induced by high frequency radiations in Nicotiana species (Cytologia, Fujii jub. Vol., 2: 961-966).

GOODSPEED, T. H., 1954. The genus Nicotiana (Waltham, Mass., Chronica Botanica Cy, 1954: 536 pp.).

GOODSPEED, T. H. and R. E. CLAUSEN, 1917. Mendelian factor differences versus reaction system contrasts in herdity (Am. Nat., 51: 31-46, 92-101).

GOODSPEED, T. H. and F. M. UBER, 1939. Radiation and plant cytogenetics (Bot.Rev., 5: 1-48).

GORDON, M., 1947. Speciation in fishes. Distribution in time and space of seven dominant multiple alleles in Platypoecilus maculatus (Adv. Genetics, 1: 95-132).

GOSS, J., 1822. On the variation in the colour of peas, occasioned by crossimpregnation (Transact. Hort. Soc. London, 5: 234-236).

GOWEN, J. W., 1927. Milk secretion as influenced by inheritance (Quart. Rev. Biol., 2: 516-531).

GOWEN, J. W., 1933. On the genetic structure of inherited construction for disease resistance (Quart. Rev. Biol., 8: 338-347).

GOWEN, J. W., 1937. Contributions of genetics to understanding of animal disease (Journ. Hered., 28: 233-241).

GOWEN, J. W., ed. 1952. Heterosis (Ames, Iowa State Coll. Press, 1952: 552 pp.).

GRAY, A. P., 1954. Mammalian hybrids A check-list with bibliography (Farnham, Commonw. Agric. Bureaux, Techn. Comm. 10: 144 pp.).

GRÉGOIRE, V., 1905. Les résultats acquis sur les cinèses de maturation dans les deux règnes (La Cellule, 22: 219-376).

GRÉGOIRE, V., 1910. Les cinèses de maturation dans les deux règnes (La Cellule, 26: 221-422).

GREGORY, R. P., 1914. On the genetics of tetraploid plants in Primula sinensis (Proc. Roy. Soc., B. 87: 484-492).

GREGORY, R. P., D. DE WINTON and W. BATESON, 1923. Genetics of *Primula sinensis* (Journ. Genet., 13: 219-253).

GREIS, H., 1940. Vergleichende physiologische Untersuchungen an diploiden und tetraploiden Gersten (Der Züchter, 12: 62-73).

GREIS, H., 1942. Relative Sexualität und Sterilisationsfaktoren bei dem Hymenomyceten *Solenia* (Biol. Zentralbl., 62: 46-92).

GUETT, A. u.A., 1937-1940. Handbuch der Erbkrankheiten (6 vols., Leipzig, Thieme, 1937-1940: 358, 336, 454, 335, 310, 373 pp.).

GUILLIERMOND, A., G. MANGENOT et L. PLANTEFOL, 1933. Traité de cytologie végétale (Paris, E. le François, 1195 pp.).

GUILLIERMOND, A., 1941 The cytoplasm of the plant cell (Waltham, Mass., U.S.A., Chronica botanica Cy., 1941: 247 pp.).

GULICK, A., 1944. The chemical formulation of gene structure and gene action (Adv. Enzymology, 4: 1-39).

GUSTAFSSON, A., 1930. Kastierungen und Pseudogamie bei *Rubus* (Bot. Not., 1930: 477-494).

GUYÉNOT, E., 1935. La détermination du sexe et l'hérédité (Actual. sci. industr., Nr. 258; Paris, Hermann, 1935: 79 pp.).

GYÖRFFI, B., 1941. The physiological and chemical conditions in polyploid plants (Magyar biol. kut. munk., Arb. ungar. biol. Inst. Tihany, 13: 362-446).

HÅKANSSON, A., 1929. Chromosomenringe in *Pisum* und ihre mutmassliche genetische Bedeutung (Hereditas, 12: 1-10).

HÅKANSSON, A., 1938. Zytologische Studien an *Salix*-Bastarden (Hereditas, 24: 1-32).

HÅKANSSON, A., 1942. Zytologische Studien von *Godetia Whitneyi* und verwandter Arten (Lunds Univ. Årsskr. N.F. Avd. 2, 38, 5: 70 pp.).

HÅKANSSON, A., 1950. Spontaneous chromosome variation in the roots of a species hybrid (Hereditas, 36: 39-59).

HAAN, H. DE, 1933. Inheritance of chlorophyll deficiencies (Bibliogr. Genet., 10: 357-416).

HAAN, IZ. DE and C. J. GORTER, 1936. On the differences in longitudinal growth of some varieties of *Pisum sativum* (Rec. Trav. bot. néerl., 33: 434-446).

HAASE BESSELL, G., 1936. Chromatin, Chromosome. Gene (Planta 25: 240-257).

HADORN, E., 1951. Developmental action of lethal factors in *Drosophila* (Adv. Genet., 4: 53-86).

HADORN, E., 1955. Letalfaktoren in ihrer Bedeutung für Erbpathologie der Entwicklung (Stuttgart, Thieme, 1955: 338 pp.).

HADORN, E. und J. NEEL, 1938. Der hormonale Einfluss der Ringdrüse (Corpus allatum) auf die Puparienbildung bei Fliegen (Arch. Entw. Mech., 138: 281-304).

HAECKEL, E., 1866. Generelle Morphologie der Organismen (Berlin, Reimer, 2 vols, 1866: 574 + 462 pp.).

HAECKEL, E., 1868, 1911. Natürliche Schöpfungsgeschichte (Elfte Aufl. Berlin, Reimer, 1911: 832 pp.).

HAECKER, V., 1902. Ueber das Schicksal der elterlichen und grosselterlichen Kernanteile (Jen. Zschr., 37: 297-400).

HAECKER, V., 1912. Der Familientypus der Habsburger (Zschr. ind. Abst. Vererb. Lehre, 6: 61-89).

HAECKER, V., 1917. Die Erblichkeit im Mannesstamm und der vaterrechtliche Familienbegriff (Jena, G. Fischer, 1917: 32 pp.).

HAECKER, V., 1918. Entwicklungsgeschichtliche Eigenschaftsanalyse (Phaenogenetik). Gemeinsame Aufgaben der Entwicklungsgeschichte, Vererbungs- und Rassenlehre (Jena, Fischer, 1918: 344 pp.).

HAECKER, V., 1925. Aufgabe und Ergebnisse der Phaenogenetik. (Bibliogr. genet., 1: 93-314).

HAEMMERLING, J., 1929. Dauermodifikationen (Handb. Vererb. wiss., I. E: 69 pp.).

HAEMMERLING, J., 1934. Ueber Genomwirkungen und Formbildungsfähigkeit bei *Acetabularia* (Arch. Entw. Mech., 132: 424-462).

HAGEDOORN, A. L., 1911. Autokatalytical substances the determinants for the inheritable characters. A biomechanical theory of inheritance and evolution (Vortr. u. Aufs. Entwick-lungsmech., 12: 35 pp.).

HAGEDOORN, A. L., 1946. Animal breeding (Sec. ed., London, Crosby Lockwood, 1946: 304 pp.).

HAGEDOORN, A. L., en A. C., 1919. Het overgeërfde moment bij bacterie-ele ziekten (Nederl. Tijdschr. v. Geneesk. 63,2: 179-182).

HAGEDOORN, A. L. and A. C., 1921. The relative value of the processes causing evolution (The Hague, Nij-hoff, 1921: 294 pp.).

HALDANE, J. B. S., 1930. Theoretical genetics of autopolyploids (Journ. Genet., 22: 359-372).

HALDANE, J. B. S., 1936. A search for incomplete sexlinkage in man (Ann. Eugen., 7: 28-57).

HALDANE, J. B. S., 1939. The theory of the evolution of dominance (Journ. Gen., 37: 365-374).

HALDANE, J. B. S., 1942. New paths in genetics (New York-London, Harper, 1942: 206 pp.).

HAMMARLUND, C., 1923, 1928. Ueber einen Fall von Koppelung und freier Kombination bei Erbsen (Hereditas, 4: 235-238, 10: 303-327).

HAMMARLUND, C. and A. HÅKANSSON, 1930. Parallelism of chromosome-ring formation, sterility and linkage in Pisum (Hereditas, 14: 97-98).

HANNING, E., 1907. Ueber pilzfreies Lolium temulentum (Botan. Ztg., 66: 25-38).

HANSEN, H. P., 1934. Inheritance of resistance to plant diseases caused by Fungi, Bacteria and Vira. A collective review with bibliography (Kgl. Veter. Landbohojskole Kbhvn Aarsskr., 1934: 1-74).

HANSON, F. B. and F. HEYS, 1930. A possible relation between natural (earth) radiation and gene mutations (Science, N.S. 71: 43-44).

HARDER, R., 1927a. Zur Frage nach der Rolle von Kern und Protoplasma im Zellgeschehen und bei der Ueber-tragung von Eigenschaften (Zschr. f. Bot., 19: 337-407).

HARDER, R., 1927b. Ueber mikro-chirurgische Operationen an Hyme-nomyzeten (Zschr. wiss. Mikr. und mikr. Techn., 44: 173-182).

HARDER, R., 1929. Ueber Defektver-suche zur Lösung der Frage nach der Rolle von Zellkern und Plasma bei der Vererbung (Festschr. Techn. Hochsch. Stuttgart, 1829—1929: 172-178).

HARDY, G. H., 1908. Mendelian pro-portions in a mixed population (Science, N.S. 28: 49-50).

HARLAND, S. C., 1936. The genetical conception of the species (Biol. Rev., 11: 83-112).

HARLAND, S. C., 1939. The genetics of cotton (London, Jonathan Cape, 1939: 193 pp.).

HARTMANN, M., 1912. Vererbungsstu-dien. I. Ueber einen experimentel-len Beweis für die Beziehung der Chromosomenreduktion zur Mendel-vererbung (Zool. Suppl., 15,3: 493-500).

HARTMANN, M., 1929. Verteilung, Be-stimmung und Vererbung des Ge-schlechts bei den Protisten und Thal-lophyten (Handb. Vererb. wiss. II. E: 115 pp.).

HARTMANN, M., 1930. Die Sexualität der Protisten und Thallophyten und ihre Bedeutung für eine allgemeine Sexualitätstheorie (Zschr. ind. Abst. Vererb. Lehre, 54: 76-126).

HARTMANN, M., 1932. Neue Ergebnis-se zum Befruchtungs- und Sexuali-tätsproblem (Naturwiss., 20: 567-573).

HARTMANN, M., 1943. Die Sexualität. Das Wesen und die Grundgesetz-lichkeiten des Geschlechts und die Geschlechtsbestimmung im Tier- und Pflanzenreich (Jena, Fischer, 1943: 426 pp.).

HARVEY, E. B., 1916—1920. A review of the chromosome numbers in Me-tazoa. I-II (Journ. Morphol., 28: 1-63, 34: 1-68).

HAYES, H. K. and F. R. IMMER, 1942. Methods of plant breeding (New York-London, McGraw Hill Book Cy, 1942: 432 pp.).

HEBERER, G. u.A., 1943. Die Evolu-tion der Organismen. Ergebnisse und Probleme der Abstammungslehre (Jena, Fischer, 1943: 774 pp.).

HEILBORN, O., 1924. Chromosome numbers and dimensions, species-formation and phylogeny in the genus Carex (Hereditas, 5: 129-216).

HEILBORN, O., 1925. Genetic cytology

and genetics in *Carex* (Bibliogr. Genet., 1: 459-462).

HEILBORN, O., 1930. Temperatur und Chromosomenkonjugation (Svensk Bot. Tidskr., 24: 12-25).

HEILBORN, O., 1932. Aneuploidy and polyploidy in *Carex* (Svensk Botan. Tidskr., 26: 137-146).

HEILBORN, O., 1934. On the origin and preservation of polyploidy (Hereditas, 19: 233-242).

HEITZ, E., 1933. Ueber totale und partielle somatische Heteropyknose sowie strukturelle Geschlechtschromosomen bei *Drosophila funebris* (Zschr. Zellf. mikr. An., 19: 720-742).

HEITZ, E., 1935. Chromosomenstruktur und Gene (Zschr. ind. Abst. Vererb. Lehre, 70: 402-447).

HEITZ, E., 1943. Ueber die Beziehung zwischen Polyploidie und Gemischtgeschlechtlichkeit bei Moosen (Arch. Julius Klaus Stift., 17: 444-448).

HEITZ, E. und H. BAUER, 1933. Beweise für die Chromosomennatur der Kernschleifen in den Knäuelkernen von *Bibio hortulanus* L. (Cytologische Untersuchungen an Dipteren I) (Zschr. Zellf. mikr. An., 17: 67-82).

HENKING, H., 1891. Untersuchungen über die ersten Entwicklungsvorgänge in den Eiern der Insekten. I. Ueber Spermatogenese und deren Beziehung zur Entwicklung bei *Pyrrhocotis apterus* L. (Zschr. wiss. Zool., 51: 280-354).

HERBERT, W., 1819. On the production of hybrid vegetables; with the result of many experiments made in the investigation of the subject (Transact. Hort. Soc. London, 4: 15:50).

HERBERT, W., 1837. On crosses and hybrid intermixtures in vegetables (In his work: Amaryllidaceae; preceded by an attempt to arrange the monocotyledonous ordres, and followed by a treatise on cross-bred vegetables, and supplement. p. 335-380. London, Ridgway sons, 1837: 428 pp.).

HERBST, C., 1907. Auf der Suche nach der Ursache der grösseren oder geringeren Aehnlichkeit der Nachkommen mit einem der beiden Eltern (Arch. Entw. Mech., 24: 185-238).

HERBST, C., 1909. Vererbungsstudien. VI. Die cytologischen Grundlagen der Verschiebung der Vererbungsrichtung nach der mütterlichen Seite (Arch. Entw. Mech., 27: 266-308).

HERBST, C., 1926. Die Physiologie des Kernes als Vererbungssubstanz (Bethe: Handb. norm. path. Physiol. 17: 991-1039).

HERBST, C., 1928–1940. Untersuchungen zur Bestimmung des geschlechtes. I-IX (Sitz. Ber. Heidelb. Ak. Wiss. 1928: 1-19; 1929, 16: 43 pp.; Naturwiss., 20: 375-379; Arch. Entw. Mech., 132: 576-599; 134: 313-330; 135: 178-201; 136: 147-168; 138: 451-464; 139: 282-302; 140: 252-284).

HÉRITIER, Ph. L', 1948. Sensitivity to CO_2 in *Drosophila* — a Review (Heredity 2: 325-348).

HÉRITIER, PH. L' et A. TEISSIER, 1937. Elimination des formes mutantes dans les populations des Drosophiles (C. R. Soc. Biol., 124: 880-884).

HERSKOWITZ, I. H., 1953. Bibliography on the genetics of *Drosophila*. Part two (Commonw. Bur. Anim. Breeding and Genetics, Bibliography 6, 212 pp.).

HERTWIG, G., 1928. Strahleneinwirkung auf Wachstum und Entwicklung (Handb. der ges. Strahlenheilkunde, Biologie, Pathologie und Therapie. 1: 444-458).

HERTWIG, G. und P., 1922. Die Vererbung des Hermaphroditismus bei *Melandrium*. Ein Beitrag zur Frage der Bestimmung und Vererbung des Geschlechts (Zschr. ind. Abst. Vererb. Lehre, 28: 259-294).

HERTWIG, O., 1884. Das Problem der Befruchtung und der Isotropie des Eies. Eine Theorie der Vererbung (Jena ,Fischer, 1884: 68 pp.).

HERTWIG, P., 1927a. Partielle Keimschädigungen durch Radium- und Röntgenstrahlen (Handb. Vererb. wiss. III. C: 48 pp.).

HERTWIG, P., 1927b. Tabellen zur Vererbungslehre (Tab. Biol., 4: 114-214).

HERTWIG, P., 1936. Artbastarde bei Tieren (Handb. Vererb. wiss. IIB, Berlin, Borntraeger, 1936: 140 pp.).

HERTWIG, P., 1937. Allgemeine Erblehre. Teil I. Zytogenetik und Mutationsforschung (Fortschr. Erbpathol. 1: 160-192).

HERTWIG, P., 1939. Allgemeine Erblehre. II. Geschlechtsbestimmung und Entwichlungsphysiologie (Fortschr. Erbpath., 3: 103-150).

HERTWIG, P., 1940. Mutationen bei den Säugetieren und die Frage ihrer Entstehung durch kurzwellige Strahlen und Keimgifte (Handb. Erbb. Menschen, 1: 245-287).

HERWERDEN, M. A. VAN, 1910. Ueber die Kernstruktur in den Speicheldrüsen der Chironomuslarve (Anat. Anz., 36: 193-207).

HERWERDEN, M. A. VAN, 1911, Ueber den Kernfaden und den Nucleolus in den Speicheldrüsenkernen der Chironomuslarve (Anat. Anz., 38: 387-393).

HERWERDEN, M. A. VAN und H. H. LAUGHLIN, 1927. Ein einheitliches System für Stammtafeln und Symbole (Zschr. ind Abst. Vererb. Lehre, 43: 260-262).

HESTON, W. E., 1948. Genetics of cancer (Adv. Genetics, 2: 99-126).

HEUKELS, H. und H. B. BONE 1915. Kreuz- und Selbstbefruchtung und die Vererbungslehre (Rec. Trav. botan. néerl., 12: 278-339).

HIRSZFELD, L., 1936. Ueber die Ausschliessung der Mutterschaft (Dtsch. Zschr. ges. gerichtl. Medizin, 27: 70-74).

HOFACKER, J. D., 1828. Ueber die Eigenschaften, welche sich bei Menschen und Thieren von den Eltern auf die Nachkommen vererben, mit besonderer Rücksicht auf die Pferdezucht (Tübingen, Osiander, 1828: 158 pp.).

HOGBEN, L., 1946. An introduction to mathematical genetics (New York, Norton, 1946: 260 pp.).

HOGE, M. A., 1915. The influence of temperature on the development of a mendelian character (Journ. exper. Zool., 18: 241-298).

HOLFELDER, H. und A. VOGT, 1938. Die Bedeutung der Röntgenstrahlen als Keimgift (Fortschr. Erbpathol., 2: 207-220).

HOLLINGSHEAD, L., 1930. A cytological study of haploid Crepis capillaris plants (Univ. Calif. Publ. Agr. Sc., 6: 107-134).

HOLMES, S. J., 1924. A bibliography of eugenics (Univ. Calif. Publ. Zool., 25: 514 pp).

HONING, J. A., 1909. De tweelingbastaarden van Oenothera Lamarckiana (Diss., Amsterdam, 1909: 104 pp.).

HONING, J. A., 1911. Die Doppelnatur der Oenothera Lamarckiana (Zschr. ind. Abst. Vererb. Lehre, 4: 227-278).

HONING, J. A., 1923. Nicotiana deformis n.sp. und die Enzymtheorie der Erblichkeit (Genetica, 5: 455-476).

HONING, J. A., 1927. Erblichkeitsuntersuchungen an Tabak (Genetica, 9: 1-18).

HONING, J. A., 1930. Nucleus and plasma in the heredity of the need of light for germination in Nicotiana seeds (Genetica, 12: 441-468).

HONING, J. A., 1931. Canna-crosses. III. Plasmatic influences. IV. Canna aureovittata-gigas a vegetative mutation (Meded. Landbouwhoogeschool Wageningen, 35,1: 26 pp.).

HOROWITZ, N. H., 1950. Biochemical genetics of Neurospora (Adv. Genet. 3: 33-72).

HOROWITZ, N. H. and U. LEOPOLD, 1951. Some recent studies bearing on the one gene-one enzyme hypothesis (Cold Spring Harbor Symp. Quant. Biol., 16: 65-74).

HOROWITZ, N. H., a.o. 1945. Genic control of biochemical reactions in Neurospora (Amer. Natur., 79: 304-317).

HUDSON, P. S., 1937. Genetics in its application to plant breeding (Biol. Rev. 12: 285-319).

HUETTIG, W., 1931. Ueber den Einfluss der Temperatur auf die Keimung und Geschlechtsverteilung bei Brandpilzen (Zschr. Botan., 24: 529-577).

HUETTNER, A. F., 1924. Maturation and fertilization in Drosophila melanogaster (Journ. Morph. and Physiol., 39: 249-265).

HULTKRANTZ, J. V. und G. DAHLBERG, 1927. Die Verbreitung eines monohybriden Erbmerkmals in einer Population und in der Verwandtschaft von Merkmalsträgern (Arch. Rassen- u. Ges. Biol., 19: 129-165).

HUNTER, H. and H. M. LEAKE, 1933. Recent advances in agricultural plantbreeding (London, Churchill, 1933: 361 pp.).

HUSKINS, C. L. and S. G. SMITH, 1935. Meiotic chromosome structure in *Trillium erectum* L. (Ann. of Bot., 49: 119-150).

HUTCHINSON, J. B., R. A. SILOW and S. G. STEPHENS, 1947. The evolution of *Gossypium* (Londen, Oxford Univ. Press, 1947: 160 pp.).

HUTCHISON, C. B., 1922. The linkage of certain aleurone and endosperm factors in maize, and their relation to other linkage groups (Mem. Cornell Univ. Agr. Exp. Stat. 60: 1425-1473).

HUTT, F. B., 1949. Genetics of the fowl (New York-London, McGraw Hill Book Cy, 1949: 590 pp.).

HUXLEY, J. S., 1932. Problems of relative growth (New York, Dial Press, 1932: 276 pp.).

HUXLEY, J., 1945. Evolution .The modern synthesis (Fourth impr., London, Allen and Unwin, 1945: 637 pp.).

HUXLEY, J., 1949. Soviet genetics and world science. Lysenko and the meaning of heredity (London, 1949, Chatto and Windus, 1949: 245 pp.).

HUXLEY, J. a.o. 1940. The new systematics (Oxford, Clarendon Press, 1940: 583 pp.).

IKENO, S., 1918. On hybridisation of some species of *Salix* (Journ. Genet., 8: 33-58).

IKENO, S., 1922. On hybridisation of some species of *Salix*. II. (Ann. of Bot., 36: 175-191).

IMAI, Y., 1926. On the rolled leaves and their linked characters in the Japanese morning glory (Zschr. ind. Abst. Vererb. Lehre, 40: 205-231).

IMMER, F. R., 1930. Formulae and tables for calculating linkage intensities (Genetics, 15: 81-98).

IRWIN, .M R., 1946. Antigens, antibodies and genes (Biol. Rev. 21: 93-1001).

IRWIN, M. R., 1947. Immunogenetics (Adv. Genetics 1: 133-160).

IVANOV, M. A., 1938. Experimental production of haploids in *Nicotiana rustica* L. (and a discussion of ha-

ploidy in flowering plants) (Genetica, 20: 285-397).

IVEN, H., 1925. Neuere Untersuchungen über die Wirkungder Röntgenstrahlen auf Pflanzen (Strahlentherapie, 19: 431-461).

JAEGER, G., 1878. Lehrbuch der allgemeinen Zoologie (Leipzig, 1878).

JANSSEN, L. W., 1939. Biosynthesis and the outline of protein structures (Protoplasma, 33: 410-426).

JANSSENS, F. A., 1909. Spermatogénèse dans les Batraciens. 5. La Théorie de la chiasmatypie. Nouvelle interprétation des cinèses de maturation (La Cellule, 25: 387-411).

JANSSENS, F. A., 1924. La chiasmatypie dans les insectes. Spermatogénèse dans 1° *Stethophyma grossum* (L.) et 2° *Chorthippus parallelus* (Zetterstedt) (La Cellule, 34: 135-359).

JEPSEN, G. L., E. MAYR and G. G. SIMPSON, 1949. Genetics, paleontology and evolution (Princeton, Princeton Univ. Press, 1949: 474 pp.).

JOHANNSEN, W., 1903. Ueber Erblichkeit in Populationen und in reinen Linien. Ein Beitrag zur Beleuchtung schwebender Selektionsfragen (Jena, Fischer, 1903: 68 pp.).

JOHANNSEN, W., 1909, 1926. Elemente der exakten Erblichkeitslehre (Jena, Fischer. 1e Aufl. 1909: 515 pp., 3e Aufl. 1926: 735 pp.).

JOHANNSEN, W., 1915. Experimentelle Grundlagen der Deszendenzlehre; Variabilität, Vererbung, Kreuzung, Mutation (Die Kultur der Gegenwart. III. 4. I. Allgemeine Biologie p. 597-661. Leipzig-Berlin. B. G. Teubner, 1915).

JOHANNSEN, W., 1917a. Arvelighed i historisk of experimentel Belysning (Andet oplag, Kjöbenhavn-Kristiana Gyldendalske Boghandel, 1917: 294 pp.).

JOHANNSEN, W., 1917b. Die Vererbungslehre bei Aristoteles und Hippokrates im Lichte heutiger Forschung (Die Naturwissenschaften, 6: 389-397).

JOLLOS, V., 1913. Experimentelle Untersuchungen an Infusorien (Biol. Zbl., 33: 222-236).

JOLLOS, V., 1914. Variabilität und

Vererbung bei Mikro-organismen (Zschr. ind. Abst. Vererb. Lehre, 12: 14-35).

JOLLOS, V., 1921. Experimentelle Protistenstudien. I. Untersuchungen über Variabilität und Vererbung bei Infusorien (Arch. f. Protistenk., 43: 1-222).

JOLLOS, V., 1931a. Genetik und Evolutionsproblem (Verhandl. Deutsch. Zool. Ges., 1931: 252-295).

JOLLOS, V., 1931b. Gerichtete Mutationen und ihre Bedeutung für das Evolutionsproblem (Biol. Zbl., 51: 137-140).

JOLLOS, V., 1932. Weitere Untersuchungen über de experimentelle Auslösung erblicher Veränderungen bei *Drosophila melanogaster* (Zschr. ind. Abst. Vererb. Lehre, 62: 15-23).

JOLLOS, V., 1936. Mutations observed in *Drosophila* stocks taken up into the stratosphere (Nat. Geogr. Soc. Techn. Pap. Stratosphere, 2: 153-157).

JOLLOS, V., 1937-1939. Some attempts to test the rôle of cosmic radiation in the production of mutations in *Drosophila melanogaster* (Genetics, 22: 113-130).

JOLLOS, V., 1939. Grundbegriffe der Vererbungslehre insbesondere Mutation, Dauermodifikation, Modifikation (Hdb. Vererb. Wiss., I.D.: 106 pp.).

JONES, D. F., 1917. Dominance of linked factors as a means of accounting for heterosis (Genetics, 2: 466-479).

JONES, D. F., 1918. The effects of inbreeding and crossbreeding upon development (Conn. Agr. Exp. Stat. Bull., 207: 100 pp.).

JONES, D. F., 1928. Selective fertilization (Chicago Univ. Press, 1928: 163 pp.).

JONES, D. F., 1934. Unisexual maize plants and their bearing in sex differentiation in other plants and animals (Genetics, 19: 552-567).

JONES, D. F., 1936. Segregation of color and growthregulation genes in somatic tissue of maize (Proc. Nat. Ac. Sc., 22: 163-166).

JONES, D. F., 1937. Somatic segregation and its relation to atypical growth (Genetics, 22: 484-522).

JONES, D. F., and P. C. MANGELSDORF, 1925. The improvement of naturally cross-pollinated plants by selection in selffertilized lines. I. The production of inbred strains of corn (Conn. Agr. Exp. Stat. Bull., 266: 353-418).

JONES, W. NEILSON, 1934. Plant chimeras and graft hybrids (London, Methuen, 1934: 136 pp.).

JONES, W. NEILSON, 1937. Chimeras: a summary and some special aspects (Bot. Rev., 3: 545-562).

JULL, M. A., 1946. Poultry breeding (Sec. ed., New York, Wiley, 1946: 484 pp.).

JUST, E. E., 1936. A single theory for the physiology of development and genetics (Amer. Nat., 70: 267-312).

JUST, E. E., a.o., 1937. Symposium on genetics and development (Amer. Nat., 71: 97-142).

JUST, G., 1934. Faktorenkoppelung, Faktorenaustausch und Chromosomenaberrationen beim Menschen, (Nebst einem einleitenden Abschnitt zu Fragen des höheren Mendelismus beim Menschen) (Ergebn. Biol., 10: 566-624).

JUST, G. u.A., 1930. Vererbung und Erziehung (Berlin, Springer, 1930: 333 pp.).

JUST, G. u.A., 1939—1940. Handbuch der Erbbiologie des Menschen (5 vols, Berlin, Springer, 1939—1940: 739, 820, 750, 1272, 1324 pp.).

KAPPERT, H., 1920. Untersuchungen über den Merkmalskomplex glatterunzlige Samenoberfläche bei der Erbse (Zschr. ind. Abst. Vererb. Lehre, 24: 185-210).

KAPPERT, H., 1937. Die Genetik des immerspaltenden Leukojen (*Matthiola incana*) (Zschr. ind. Abst. Vererb. Lehre, 73: 233-281).

KAPPERT, H., 1948. Die vererbungswissenschaftlichen Grundlagen der Pfanzenzüchtung (Berlin-Hamburg, Parey 1948: 244 pp.).

KARCZAG, L., 1928. Die Stereogene als Erbeinheiten. Eine neue Theorie der Vererbungserscheinungen (Zschr. ind. Abst. Vererb. Lehre, 48: 86-144).

KARPECHENKO, G. D., 1927. The production of polyploid gametes in hybrids (Hereditas, 9: 349-368).

KARPECHENKO, G. D., 1928. Polyploid hybrids of *Raphanus sativus* L. × *Brassica oleracea* L. (Zschr. ind. Abst. Vererb. Lehre, 48: 1-85)

KAUFMANN, B. P., 1934. Somatic mitoses of *Drosophila melanogaster* (Journ. Morph., 56: 125-155).

KAUFMANN, B. P., 1936. Chromosome structure in relation to the chromosome cyle (Bot.Rev., 2: 529-553).

KEEBLE, F. and C. PELLEW, 1910. The mode of inheritance of stature and flowering time in peas (*Pisum sativum*) (Journ. Genet., 1: 47-56).

KELLER, K., 1920. Zur Frage des sterilen Zwillingskalbes (Wien. tierärztl. Monatsschr., 7: 146-168).

KELLER, K. und J. TANDLER, 1916. Ueber das Verhalten der Eihäute bei der Zwillingsträchtigkeit des Rindes. Untersuchungen über die Entstehungsursache der geschlechtlichen Unterentwicklung von weiblichen Zwillingskälbern, welche neben einem männlichen Kalbe zur Entwicklung gelangen (Wien. tierärztl. Monatsschr., 3: 513-526).

KELLEY, R. B., 1946. Principles and methods of animal breeding New York, Wiley, 1946: 296 pp.).

KEMP, T., 1942. Statistiske metoder i medecin og biologi (Köbenhavn, 1942, Munksgaard, 1942, 172 pp.).

KEMP, T. und L. MARX, 1936-1937. Beeinflussung von erblichem hypophysärem Zwergwuchs bei Mäusen durch verschiedene Hypophysenauszüge und Tyroxin I-II (Acta pathol. scand., 13: 512-531; 14: 197-227).

KIHARA, H., 1930-1949. Genomanalyse bei *Triticum* und *Aegilops*. I-IX (Cytologia. 1: 263-284; 2: 106-156, 234-255; 3: 384-456; 6: 87-122, 195-216; Mem. Coll. Agric. Kyoto Imp. Univ., 41: 1-61; Cytologia 11: 493-506; 14: 135-144).

KIHARA, H. and T. ONO, 1925. The sex-chromosomes of *Rumex acetosa* (Zschr. in d.Abst. Vererb. Lehre, 39: 1-7).

KIHARA, H. und T. ONO, 1926. Chromosomenzahlen und systematische Gruppierung der *Rumex*-Arten

(Zschr. f. Zellf. mikr. Anat., 4: 475-481).

KIKKAWA, H., 1953. Biochemical genetics of *Bombyx mori* (Silkworm). (Adv. Genet., 5: 107-141).

KING, H. D., 1916. Experimental inbreeding (Journ. Heredity, 7: 70-76).

KING, H. D., 1918. Studies on inbreeding. I-III (Journ. exp. Zool., 26: 1-54, 55-98; 27: 1-35).

KLATT, B., 1927. Entstehung der Haustiere (Handb. Vererbwiss. III.. K: 107 pp.).

KLEINSCHMIDT, O., 1926. Die Formenkreislehre und das Weltwerden des Lebens (Halle, Gebauer-Schwetschke, 1926: 188 pp.).

KNAPP, E., 1935. Untersuchungen über die Wirkung von Röntgenstrahlen an dem Lebermoos *Sphaerocarpus*, mit Hilfe von Tetradenanalyse. I (Zschr. ind. Abst. Vererb. Lehre, 70: 309-349).

KNAPP, E., 1936. Zur Genetik von *Sphaerocarpus* (Tetradenanalytische Untersuchungen) (Ber. Deutsch. bot. Ges., 54: (58)-(69)).

KNAPP, E., 1937a. Crossing-over und Chromosomenreduktion (Zschr. ind. Abst. Vererb. Lehre, 73: 409-418).

KNAPP, E., 1937b. Mutationsauslösung durch ultraviolettes Licht bei dem Lebermoos *Sphaerocarpus Donnelli* Aust. (Zschr. ind. Abst. Vererb. Lehre, 74: 54-69).

KNAYSI, G., 1946. Elements of bacterial cytology (Ithaca, N.Y. Comstock Publ. Cy, 1946: 209 pp.).

KNAYSI, G., 1949. Cytology of bacteria, II. (Bot. Rev. 15: 106-151).

KNIEP, H., 1928. Die Sexualität der niederen Pflanzen (Jena, Fischer, 1928: 544 pp.).

KNIEP, H., 1929. Vererbungserscheinungen bei Pfilzen (Bibliogr. genet., 5: 371-478).

KNIGHT, T. A., 1823. Some remarks on the supposed influences of the pollen in crossbreeding upon the colour of the seed-coats of plants, and the qualities of their fruits (Transact. Hortic. Soc. London, 5: 377-380).

KNIGHT, T. A., 1841. A selection from the physiological and horticultural papers published in the transactions of the Royal and Horticultural Societies, to which is added a sketch of his life (London, 1841: 379 pp.).

KOBOZIEFF, N. et N. A. POMRIASKINSKY-KOBOZIEFF, 1943. Précis de gégétique appliquée à la médecine vétérinaire (Paris, Vigot fréres, 1943: 216 pp.).

KOEHLER, E., 1940. Anbau und Züchtung krankheitsresistenter Sorten (Handb. Pflanzenkrankheiten, 6: 362-406).

KOEHLER, O., 1916. Ueber die Ursachen der Variabilität bei Gattungsbastarden von Echiniden (Zschr. ind. Abst. Vererb. Lehre, 15: 1-163, 177-295).

KOEHLER, W. und W. FELDOTTO, 1937. Morphologische und experimentelle Untersuchungen über Farbe, Form und Struktur der Schuppen von *Vanessa urticae* und ihre gegenseitige Beziehungen (Arch. Entw. Mech., 136: 313-399).

KOELREUTER, J. G., 1761-1766. Vorläufige Nachricht von einigen das Geschlecht der Pflanzen betreffenden Versuchen und Beobachtungen nebst Fortsetzungen 1, 2 und 3. (Leipzig, Gleditsch, 1761-1766: 50, 72, 128, 156 pp. Again edited in: Ostwalds Klassiker d. exakt. Wiss. Nr. 41. Leipzig, Engelmann, 1893).

KOERNICKE, M., 1922. Die Wirkung der Röntgenstrahlen auf die Pflanzen (mit Ausnahme der Bakterien) (Handb. ges. mediz. Anw. Elektrizität, 3: 157-180).

KOLLER, P. C., 1931. The relation of fertility factors to crossing-over in the *Drosophila obscura* hybrid (Zschr. ind. Abst. Vererb. Lehre, 60: 137-151).

KOLLER, S., 1940a. Allgemeine statistische Methoden in speziellem Blick auf die menschliche Erblehre(Handb. Erbbiol. Menschen, herausgeg. v. G. Just, Berlin, Springer, 2: 112-212).

KOLLER, S., 1940b. Methodik der menschlichen Erforschung (mit Ausnahme der Mehrlingsforschung) (Handb. Erbbiol. Menschen, herausgeg. v. G. Just, Berlin, Springer, 2: 249-309).

KOLTZOFF, N. K., 1928. Physikalischchemische Grundlage der Morphologie (Biol. Zbl. 48: 345-369).

KOLTZOFF, N. K., 1935. Physiologie du développement et génétique. (Actual. sci. industr., Nr. 254, Paris, J. Hermann, 1935: 56 pp).

KOMAI, T., 1927. Criteria for distinguishing identical and fraternal twins (Quart. Rev. Biol. 3: 408-418).

KOOIMAN, H. N., 1931. Monograph on the genetics of *Phaseolus* (Bibliogr. genet., 8: 295-413).

KOOPMANS, A., 1951. Cytogenetic studies on *Solanum tuberosum* and some of its relatives (Genetica, 25: 193-337).

KOOPMANS, A., 1952-1955. Changes in sex in the flowers of the hybrid *Solanum rybinii* × *S. chacoense*. I-III (Genetica 26: 359-380; 27: 273-285, 465-471).

KOSSWIG, C., 1939. Die Geschlechtsbestimmungsanalyse bei Zahnkarpfen (Rev. Fac. Sc. Univ. Istanbul, N. S. 4: 32 pp.).

KOSSWIG, C., 1941. Mitteilungen zum Geschlechtsbestimmungsproblem bei Zahnkarpfen (Rev. Fac. Sc. Univ. Istanbul, B. 6: 32 pp.).

KOSTOFF, D., 1929. An androgenic *Nicotiana* haploid (Zschr. f. Zellforsch., 9: 640-642).

KOSTOFF, D., 1941. The problem of haploidy (Cytogenetic studies on *Nicotiana* haploids and their bearings to some other cytogenetic problems) Bibliogr. Genet. 13: 1-148).

KOSTOFF, D., 1942. Wheat phylesis and wheat breeding from a cytogenetic point of view (cytogenetic indices for the rôle of interspecific hybridization in producing valuable wheat forms) (Bibliogr. Genet. 13: 149-224).

KOSTOFF, D., 1943. Haploide *Triticum vulgare* und die Variabilität ihrer diploiden Nachkommenschaften (Züchter 15: 121-125).

KRANZ, H., 1938, 1940. Erbforschung über den angeborenen Schwachsinn (Fortschr. Erbpathol. 1: 281-322; 4: 1-48).

KREDIET, G., 1942. Zoogdierenintersexualiteit (Gorinchem, Noorduyn, 1942: 136 pp.).

KRUMBHOLZ, G., 1925. Untersuchungen über die Scheckung der *Oenotheren* bastardeinsbesondere über die

Möglichkeit der Entstehung der Peri-klinalchimaeren (Jenaïsche Zschr. Naturw., 62: 87-260).

KRYTHE, J. M. and S. J. WELLEN-SIEK, 1942. Five years of colchicine research (Bibliogr. genet., 14: 1-132).

KUCKUCK, H. und A. MUDRA, 1950. Lehrbuch der allgemeinen Pflanzen-züchtung (Zürich, Hirzel, 1950: 280 pp.).

KUEHN, A., 1927. Die Pigmentierung von *Habrobracon juglandis* Ashmed, ihre Prädetermination und ihre Ver-erbung durch Gene und Plasmon (Nachr. Ges. Wiss. Göttingen, math. phys. Kl. 1927: 407-421).

KUEHN, A., 1939. Grundriss der Ver-erbungslehre (Leipzig, Quelle und Meyer, 1939: 164 pp.).

KUESTER, E., 1917. Die Verteilung des Anthocyans bei *Coleus*spielarten (Flora, N. F. 10: 1-33).

KUHN, E., 1930. Pseudogamie und Androgenesis bei Pflanzen (Züch-ter, 2: 124-136).

KUHN, E., 1937. Befruchtungsphysiolo-gische Untersuchungen zum Problem der Vererbung der Blütenfüllung bei *Matthiola* (Zschr. ind. Abst. Ver-erb. Lehre, 72: 387-482).

KUHN, E., 1938. Ueber die Trabanten bei normalen und immerspaltenden Sippen von *Matthiola incana* (Zschr. ind. Abst. Vererb. Lehre, 74: 388-408).

KUHN, E., 1939. Selbstbestäubungen subdioecischer Blütenpflanzen, ein neuer Beweis für die genetische Theorie der Geschlechtsbestimmung (Planta, 30: 457-470).

KUHN, E., 1942. Polyploidie und Ge-schlechtsbestimmung bei zweihäusi-gen Blütenpflanzen (Naturwiss., 30: 189-198).

KUWADA, Y., 1939. Chromosome struc-ture. A critical review (Cytologia, 10: 213-256).

LAIBACH, F., 1925. Das Taubwerden von Bastardsamen und die künst-liche Aufzucht früh absterbender Bastardembryonen (Zschr. Bot., 17: 417-459).

LAIBACH, F., 1931. Ueber Störungen in den physiologischen Beziehungen zwischen Mutter und Embryo bei Bastardierung (Zschr. ind. Abst. Vererb. Lehre, 59: 102-125).

LAM. H. J., 1946. Evolutie. Een poging tot synthese in algemeen begrijpe-lijke vorm (Leiden, Univ. Pres, 1946. 92 pp.).

LAMARCK, J. B. P. DE, 1809. Philoso-phie zoologique (Paris, Dentu, 1809: 428 + 475 pp.; Ed. Charles Mar-tins, Paris, F. Savoy, 1873: 412 + 431 pp.).

LAMBERT, W. V., 1933. The evidence for inheritance of resistance to bac-terial diseases in animals (Quart. Rev. Biol., 8: 331-337).

LANDAUER, W., 1937. Loss of body heat and disease (Amer. Journ. Med. Sc., 194: 667-674).

LANG, A., 1904. Ueber Vorversuche zu Untersuchungen über die Varie-tätenbildung von *Helix hortensis* Müller und *Helix nemoralis* L. (Festschrift f. Haeckel. Jenaïsche Denkschriften 11: 439-506).

LANG, A., 1905. Ueber die Mendelschen Gesetze, Arten und Varietätenbil-dung, Mutation und Variation, ins-besonderen bei unseren Hain- und Gartenschnecken (Verhandl. Schweiz. naturf. Gesellschaft, 1905: 48 pp.).

LANG, A., 1911. Fortgesetzte Verer-bungsstudien (Zschr. ind. Abst. Vererb. Lehre, 5: 97-138).

LANG, A. 1914. Die experimentelle Vererbungslehre in der Zoologie seit 1900. Mit einem Abschnitt: An-fangsgründe der Variation und Kor-relation. Erste Hälfte (Jena, Fischer, 1914: 892 pp.).

LANGLET, O., 1927a. Beiträge zur Zy-tologie der Ranunculazeen (Svensk Botan. Tidskr., 21: 1-17).

LANGLET, O., 1297b. Zur Kenntnis der polysomatischen Zellkerne im Wur-zelmeristem (Svensk Botan. Tidskr., 21: 397-422).

LANGLET, O. und E. SÖDERBERG, 1927. Ueber die Chromosomenzahlen eini-ger Nymphaeaceen (Acta Horti Ber-giani, 9: 85-104).

LAWRENCE. W. J. C. and R. PRINCE, 1940. The genetics and chemistry of flower colour variation (Biol. Rev., 15: 35-58).

LEA, D. E., 1940a. A radiation method for determining the number of genes in the X-chromosome of *Drosophila* (Journ. Genet., 39: 181-188).

LEA, D. E., 1940b. Determination of the size of viruses and genes by radiation methods (Nature, 146: 137).

LEA, D. E., 1946. Actions of radiations in living cells (Cambridge, Univ. Press, 1946: 402 pp.).

LECOQ, H., 1827. Recherches sur la reproduction des végétaux (Paris, 1827: 30 pp.).

LECOQ, H., 1845. De la fécondation naturelle et artificielle des végétaux et de l'hybridation, considérée dans ses rapports avec l'horticulture, l'agriculture et le sylviculture ou études sur les croisements des plantes des principaux genres cultivés dans les jardins d'ornaments, fruitiers et maraichers, sur les végétaux économiques et de grande culture, les arbres forestiers, etc. Contenant les moyens pratiques d'opérer l'hybridation et de créer facilement des variétés nouvelles (Paris, 1845: 287 pp.).

LEDERBERG, J., 1947. Gene recombination and linked segregations in *Escherichia coli* (Genetics, 32: 505-525).

LEHMANN, E., 1914. Art, reine Linie, isogene Einheit (Biol. Cbl., 34: 285-294).

LEHMANN, E., 1918. Ueber reziproke Bastarden zwischen *Epilobium roseum* und *parviflorum* (Zschr. Bot., 10: 497-511).

LEHMANN, E., 1922. Die Theorien der *Oenothera*forschung (Jena, G. Fischer, 1922: 526 pp.).

LEHMANN, E., 1925. Die Gattung *Epilobium* (Bibliogr. genet., 1: 363-418).

LEHMANN, E., 1928. Selbststerilität, Heterostylie (Handb. Vererb. wiss. I. J: 43 pp.).

LEHMANN, E., 1939-1944. Zur Genetik der Entwicklung in der Gattung *Epilobium*. I-IV (Jahrb. wiss. Bot., 87: 625-641; 88: 284-343; 89: 637-686, 686-753; 90: 49-98; 91: 440-502).

LEHMANN, E., (I), G. HINDERER (II), H. GRAZE und G. SCHLENKER (III), 1936. Versuche zur Klärung der reziproken Verschiedenheiten von *Epilobium*-Bastarden I-III (Jahrb. wiss. Bot., 82: 657-668, 669-686, 687-695).

LEHMANN, E. und W. DUPPEL, 1950. Plasmonbegriff und Störungssysteme in der Gattung *Epilobium* (Züchter 20: 103-125).

LEHMANN, E. und J. SCHWEMMLE, 1927. Genetische Untersuchungen in der Gattung *Epilobium* (Biblioth. bot., 95: 156 pp.).

LERNER, I. M., 1950. Population genetics and animal improvement (Cambridge, Univ. Press, 1950: 342 pp.).

LERNER, I. M., 1954. Genetic homoeostasis (Edinburg-London, Oliver and Boyd, 1954: 134 pp.).

LESLEY, M. M., 1925. Chromosomal chimaeras in the tomato (Am. Nat., 59: 570-574).

LESLEY, M. M. and J. W., 1930. The mode of origin and chromosome behaviour in pollen mother cells of a tetraploid seedling tomato (Journ. Genet., 22: 419-424).

LEVAN, A., 1938. The effect of colchicine on root mitoses in *Allium* (Hereditas, 24: 471-486).

LEVAN, A. and G. OESTERGREN, 1943. The mechanism of c-mitotic action. Observations on the naphtaline series (Hereditas, 29: 381-443).

LEWIS, E. B., 1939. Star-recessive, a spontaneous mutation in *Drosophila melanogaster* (Proc. Minn. Ac. Sc., 7: 23-26).

LEWIS, E. B., 1950. The phenomenon of position effect (Adv. Genetics, 3: 73-116).

LEWIS, E. B., 1951. Pseudoallelism and gene evolution. (Cold Spring Harbor Symp. Quant. Biol., 16: 159-174).

LIDFORSS, B., 1905. Studier öfver Artbildningen inom släktet *Rubus* (Ark. Bot., 4,6: 40 pp.).

LIDFORSS, B., 1907. Studier öfver Artbildningen inom släktet *Rubus* II (Ark. Bot., 6,16: 43 pp.).

LIDFORSS, B., 1914. Résumé seiner Arbeiten über *Rubus* (Zschr. ind. Abst. Vererb. Lehre, 12: 1-13).

LILLIE, F. R., 1917. The free-martin: a study of the action of sex hormones in the foetal life of cattle (Journ. exp. Zool., 23: 371-452).

LINDEGREN, C. C., 1945. Mendelian and cytoplasmic inheritance in yeasts (Ann. Missouri Bot. Garden 32: 107-123).

LINDEGREN, C. C., 1946. A new gene theory and an explanation of the phenomenon of dominance to mendelian segregation of the cytogene (Proc. nat. Ac. Sc. Wash., 32: 68-70).

LINDEGREN, C. C., 1949. The yeast cell, its genetics and cytology (St. Louis, Educational Publishers, 1949: 308 pp.).

LINDEGREN, C. C., 1952. Gene conversion in Saccharomyces (Journ. Genet., 51: 625-637).

LINDEGREN, C. C. and S. A. HADDAD, 1953. The control of nuclear and cytoplasmic synthesis by the nueleocytoplasmic ratio in Saccharomyces (Exp. Cell Res., 5: 549-550).

LINDSTROM, E. W., 1936. Genetics of polyploidy (Bot. Rev., 2: 197-215).

LINDSTROM, E. W. and F. GERHARDT, 1926. Inheritance of carbohydrates and fat in crosses of dent and sweetcorn (Iowa Agr. Exp. Stat. Res. Bull. 98: 259-277).

LINDSTROM, E. W. und F. GERHARDT, 1927. Inheritance of chemical characters in maize (Iowa State Coll. Journ. Sc., 2: 9-18).

LINNAEUS, C., 1759. Generatio ambigena. Respondes C. L. Ramström (Amoen. academ., 5).

LINNAEUS, C., 1760. Disquisitio de quaestione academia imper. scient. petropol. proposita: Sexum plantarum argumentis et experimentis novis, praeter adhuc iam cognita, vel corroborare, vel impugnare etc. (Petropol. Acad. Scient. Typogr. 1760: 30 pp.).

LINNAEUS, C., 1762. Fundamentum fructificationis. Respondens J. M. Graberg (Amoen. academ., 6: 279-304).

LITARDIÈRE, R. DE, 1923. Les anomalies de la caryocinèse somatique chez le Spinacia oleracea L. (Rev. Gén. Bot., 35: 369-381).

LITARDIÈRE, R. DE, 1925. Sur l'existence de figures didiploïdes dans le méristème radiculaire du Cannabis sativa L. (La Cellule, 35: 19-26).

LJUNGDAHL, H., 1922. Zur Zytologie der Gattung Papaver (Svensk Bot. Tidskr., 16: 103-114).

LJUNGDAHL, H., 1924. Ueber die Herkunft der in der Meiosis konjugie-

renden Chromosomen bei Papaver-Hybriden (Svensk bot. Tidskr., 18: 279-291).

LOEB, J., 1906. Vorlesungen über die Dynamik der Lebenserscheinungen (Leipzig, Barth, 1906- 324 pp.).

LOEB, J., 1909. Die chemische Entwicklungserregung des tierischen Eies (Berlin, J. Springer, 1909).

LOEB, J., 1916. The organism as a whole (New York, G. B. Putnam Sons, 1916: 279 pp.).

LOCHEM, J. J. V., 1951. Genetische aspecten van bloedgroepen, in het bijzonder van de Rhesus-factor (Vakbl. v. Biol., 31: 2-11).

LONGLEY, A. E., 1941. Chromosome morphology in maize and its relations (Bot. Rev., 7: 263-289).

LORBEER, G., 1927. Untersuchungen über Reduktionsteilung und Geschlechtsbestimmung bei Lebermoosen (Zschr. ind. Abst. Vererb. Lehre, 44: 1-109).

LORBEER, G., 1934. Die Zytologie der Lebermoose mit besonderer Berücksichtigung allgemeiner Chromosomenfragen (Jahrb. wiss. Bot., 89: 567-581).

LORBEER, G., 1941. Struktur und Inhalt der Geschlechtschromosomen (Ber. D. Bot. Ges., 59: 369-418).

LOSSEN, H., 1905. Die Bluterfamilie Mampel in Kirchheim bei Heidelberg (Deutsche Ztg. Chirurgie, 76: 1-18).

LOTSY, J. P., 1917. Over Oenothera Lamarckiana als type van een nieuwe groep van organismen, die der kernchimaeren (Den Haag, Nijhoff, 1917: 52 pp.).

LOTZE, R., 1937. Zwillinge. Einführung in die Zwillingsforschung (Schriften d. deutschen Naturkunde' vereins, N.F., 6: 176 pp.).

LÖVE, A., 1951. Taxonomical evolution of polyploids (Caryologia, 3: 263-284).

LUCAS, P., 1847-1850. Traité philosophique de l'hérédité naturelle dans les états de santé et de maladie du système nerveux (Paris, Baillière, 1 (1847): 626 pp., 2 (1850): 936 pp.).

LUDWIG, W., 1938. Faktorenkoppelung und Faktorenaustausch bei normalen und aberranten Chromosomenbestand (Leipzig, Thieme, 1938: 245 pp.).

LUSH, J. L., 1938. Animal breeding plans (Ames, Coll. Press, 1938: 350 pp)..

LUTZ, A. M., 1912. Triploid mutants in *Oenothera* (Biol. Cbl., 32: 385-435).

LUXENBURGER, H., 1940. Die Zwillingsforschung als Methode der Erbforschung beim Menschen (Handb. Erbbiol. Menschen, herausgeg. v. G. Just, Berlin, Springer, 2: 213-248).

LWOFF, A., 1950. Problems of morphogenesis of ciliates. The kinetosome in development, reproduction and evolution. (New York, Wiley, 1950: 103 pp.).

LYNCH, C. J., 1921. Short ears, an autosomal mutation in the house mouse (Am. Nat., 55: 421-426).

McCLINTOCK, B., 1930. A cytological demonstration of the location of an interchange between two non-homologous chromosomes of *Zea Mays* (Proc. nat. Ac. Sc., 16: 791-796).

McCLINTOCK, B., 1931. Cytological observations of deficiencies involving known genes, translocations and an inversion in *Zea Mays* (Missouri Agr. Exp. Stat. Res. Bull., 163: 30 pp.).

McCLINTOCK, B., 1938. The production of homozygous deficient tissues with mutant characteristics by means of the aberrant mitotic behavior of ring-shaped chromosomes (Genetics, 23: 315-376).

McCLINTOCK, B., 1941. The stability of broken ends of chromosomes in *Zea Mays* (Genetics 26: 234-282).

McCLINTOCK, B., 1944. The relation of homozygous deficiencies to mutations and allelic series in maize (Genetics, 29: 478-502).

McCLINTOCK, B., 1950. The origin and behavior of mutable loci in maize (Proc. nat. Ac. Sc. Wash., 36: 344-355).

McCLINTOCK, B., 1951. Chromosome organization and genetic expression (Cold Spring Harbor Symp. Quant. Biol., 16: 13-47).

MacCLUNG, C. E., 1902a. Note on the accessory chromosome (Anat. Anz., 20: 220-226).

MacCLUNG, C. E., 1902b. The accessory chromosome sex-determinant? (Biol. Bull., 3: 43-84).

MacCLUNG, C. E., 1927. The chiasmatype theory of Janssens (Quart. Rev. Biol., 2: 344-366).

MacCLUNG, C. E., 1940. Chromosome numbers in animals (Tabulac Biologicae, 18: 1-60).

MacDOUGAL, D. T., 1911. Alterations in heredity induced by ovarial treatments (Bot. Gaz., 51: 241-257).

MacDOUGAL, D. T., a.o., 1905. Mutants and hybrids of the *Oenotheras* Carnegie Inst. Wash. Publ., 29: 57 pp.).

MacDOWELL, E. C., 1915. Bristle inheritance in *Drosophila* (Journ. exper. Zool., 19: 61-82).

MacDOWELL, E. C. and E. M. LORD, 1927. Reproduction in alcoholized mice. I. Treated females. II. Treated males (Arch. Entw. Mech., 109: 549-583, 110: 427-449).

MacKELVEY, S. D. and K. SAX, 1933. Taxonomic and cytological relationships of *Yucca* and *Agave* (Journ. Arnold Arbor., 14: 76-81).

MacLEAN EVANS, H. and O. SWEZY, 1929. The chromosomes in man: sex and somatic (Mem.. Univ Calif., 9: 1-65).

MAKINO, S., 1950. An atlas of chromosome numbers in animals (Ames, Iowa State Univ. Press, 1950: 290 pp.).

MANGELSDORF, P. C., 1947. The origin and evolution in maize (Adv. Genetics, 1: 161-209).

MANGOLD, O., 1931. Versuche zur Analyse der Entwicklung des Haftfadens; ein Beispiel für die Induktion artfremder Organe (Naturwiss. 19: 905-911).

MANTON, I., 1950. Problems of cytology and evolution in the Pteridophyta (Cambridge, Univ. Press, 1950: 316 pp.).

MARCHAL, EL. et EM., 1907-1911. Aposporie et sexualité chez les mousses. I-III (Bull. Ac. Roy. Belg., 1907: 765-789; 1909: 1249-1288; 1911: 750-778).

MARCHAL, E., 1920. Recherches sur les variations numériques des chromosomes dans la série végétale (Mém. Ac. Roy. Belg. Cl. d. Sc., 2,4: 108 pp.).

MARCHAL, P., 1904. Recherches sur la biologie et le développement des Hyménoptères parasites. La polyembryonie spécifique ou germinogonie (Arch. Zool. expér., 4,2: 257-336).

MATHER, K., 1933. Interlocking as a demonstration of the occurrence of genetical crossing-over during chiasma-formation (Am. Nat., 67: 476-479).

MATHER, K., 1938a. The measurement of linkage in heredity (New York, Chemical Publishing Co. of New York, 1938: 132 pp.).

MATHER, K., 1938b. Crossing-over (Biol. Rev., 13: 252-292).

MATHER, K., 1941. Variation and selection of polygenic characters (Journ. Genet., 41: 159-193).

MATHER, K., 1943. Polygenic inheritance and natural selection (Biol. Reviews, 18: 32-64).

MATHER, K., 1953. The genetical structure of populations (Symp. Soc. exp. Biol., 7: 66-95).

MATSUURA, H., 1929. A bibliographical monograph on plant genetics (genic analysis) 1900-1925 (Contrib. Cytol. Genet. Tokyo Imp. Univ., 82: 499 pp.).

MATTHEY, R., 1949. Les chromosomes des vertébrés (Lausanne, Rouge, 1949: 356 pp.).

MATTHEY, R., 1953. A propos de la polyploïdie animale: réponse à un article de C. D. Darlington (Rev. Suisse de Zoöl., 60: 466-471).

MAVOR, J. W., 1923. An effect of X-rays on the linkage of Mendelian characters in the first chromosome of Drosophila (Genetics, 8: 355-366).

MAVOR, J. W. and H. K. SVENSON, 1924a. A comparison of the effects of X-rays and temperature on linkage and fertility in Drosophila (Genetics, 9: 588-608).

MAVOR, J. W. and H. K. SVENSON, 1924b. An effect of X-rays on the linkage of Mendelian characters in the second chromosomes of Drosophila melanogaster (Genetics, 9: 70-89).

MAYR, E., 1939. The sex ratio in wild birds (Am. Nat., 73: 156-179).

MAYR, E., 1944. Systematics and the origin of species from the viewpoint of a zoologist (New York, Columbia Univ. Press, 1944: 334 pp.).

MAYR, E., 1948. The bearing of the new systematics on genetical problems. The nature of species (Adv. Genetics 2: 205-239).

MECKEL, J. F., 1812-1818. Handbuch der pathologischen Anatomie (Leipzig, Reclam, 1: 759 pp., 2,1: 499 pp., 2,2: 468 pp.).

MEHLING, E., 1915. Ueber die gynandromorphen Bienen des Eugsterschen Stockes (Verh. physik. med. Ges. Würzburg, N.F., 43: 173-236).

MEISENHEIMER, J., 1921. Geschlecht und Geschlechter im Tierreiche. I. Die natürlichen Beziehungen (Jena, Fischer, 1921: 896 pp.).

MELCHERS, G., 1937. Die Wirkung von Genen, tiefen Temperaturen und blühenden Pfropfpartnern auf die Blühreife von Hyoscyamus niger L. (Biol. Zbl., 57: 568-614).

MELCHERS, G., 1939. Genetik und Evolution (Zschr. ind. Abst. Vererb. Lehre, 76: 229-259).

MELCHERS, G. und A. LANG, 1941. Weitere Untersuchungen zur Frage der Blühhormone (Biol. Zbl., 61: 16-39).

MENDEL, G., 1865. Versuche über Pflanzenhybriden (Verhandl. naturf. Verein Brünn, 4: 3-47. Also in: Ostwalds Klass. exakt. Wiss. Nr. 121. Leipzig, Engelman, 1901).

MENDEL, G., 1869. Ueber einige aus künstlicher Befruchtung gewonnene Hieraciumbastarde (Verhandl. naturf. Verein. Brünn, 8: 26-31. Also in Ostwalds Klassiker No. 121).

MENDEL, G., 1905. Gregor Mendels Briefe an Carl von Naegeli, herausgegeben von Carl Correns (Abh. math. phys. Kl. Kön. Sächs. Ges. Wiss., 29: 189-265).

METZ, C. W., 1927, Chromosome behavior and genetic behavior in Sciara. II. Genetic evidence of selective segregation in S. coprophila (Zschr. ind. Abst. Vererb. Lehre, 44: 184-201).

METZ, C. W., 1938. Chromosome behavior, inheritance and sex determination in Sciara (Am. Nat., 72: 485-520).

METZ, C. W. and E. G. LAWRENCE,

1937. Studies on the organization of the giant gland chromosome of Diptera (Quart. Rev. Biol., 12: 135-151).

MICHAELIS, P., 1929. Ueber den Einfluss von Kern und Plasma auf die Vererbung (Biol. Zbl., 49: 302-316).

MICHAELIS, P., 1932. Ueber die Beziehungen zwischen Kern und Plasma bei den reziprok verschiedenen Epilobium-Bastarden (Zschr. ind. Abst. Vererb. Lehre, 62: 95-102).

MICHAELIS, P., 1938. Ueber die Konstanz des Plasmons (Zschr. ind. Abst. Vererb. Lehre, 74: 435-459).

MICHAELIS, P. a.o., 1931-1936. Entwicklungsgeschichtlich-genetische Untersuchungen an Epilobium. I-VIII (Planta, 14: 566-582; Zschr. ind. Abst. Vererb. Lehre, 65: 1-71, 353-411; Planta, 23: 486-500; Ber. Deutsch. bot. Ges., 53: 143-150; Planta, 23: 604-622; Zschr. ind. Abst. Vererb. Lehre, 70: 138-157; Jahrb. wiss. Bot., 82: 45-64; Planta, 25: 282-301).

MICHAELIS, P. a.o., 1940—1948. Ueber reziprok verschiedene Sippen-Bastarde bei Epilobium hirsutum. I-VIII (Zschr. ind. Abst. Vererb. Lehre, 78: 187-222, 223-237, 295-337; 80: 373-428, 429-453, 454-499; 82: 343-383, 384-414).

MICHAELIS, P., 1949. Ueber Abänderungen des plasmatischen Erbgutes (Zschr. ind. Abst. Vererb. Lehre 83: 36-85).

MICHAELIS, P., 1954. Cytoplasmic inheritance in Epilobium and its theoretical significance (Adv. Genet., 6: 288-402).

MILLARDET, A., 1894. Note sur l'hybridation sans croisement ou fausse hybridation (Mém. Soc. Sc. phys. et nat. Bordeaux, 4,4: 347-372).

MILLER, PH., 1731. The gardeners dictionary, containing the methods of cultivating and improving the kitchen, fruit and flower gardening (London, Printed for the Author and sold by C. Rivington. 1st Ed. 1731. 6th Ed. 1752. 7th Ed. 1759).

MIRSKY, A. E., 1943. Chromosomes and nucleoproteins (Adv. Enzymology, 3: 1-34).

MOEWUS, F., 1938. Carotinoide als Sexualstoffe von Algen (Jahrb. wiss. Bot., 86: 753-783).

MOEWUS, F., 1941-1943. Zur Sexualität der niederen Organismen. I-II Ergebn. Biol., 18: 287-356; 19: 82-142).

MOHR, O. L., 1919. Character changes caused by mutation of an intire region of a chromosome in Drosophila (Genetics, 4: 257-292).

MOHR, O. L., 1921. A case of hereditary brachyphalangy utilized as evidence in forensic medecine (Hereditas, 2: 290-298).

MOHR, O. L., 1926. Ueber Letalfaktoren, mit Berücksichtigung ihres Verhaltens bei Haustieren und beim Menschen (Zschr. ind. Abst. Vererb. Lehre, 41: 59-109).

MOHR, O. L., 1929. Exaggeration and inhibition phenomena encountered in the analysis of an autosomal dominant (Zschr. ind. Abst. Vererb. Lehre, 50: 113-200).

MOHR, O. L., 1931. Letalfaktoren bei Haustieren (Züchtungsk., 4: 105-125).

MOHR, O. L., 1934. Heredity and disease (New York, W. W. Norton, 253 pp.).

MOHR, O. L., 1939. Lethal genes in higher animals and man (Relaz. 4. Congr. Int. Patol. Comp., 1: 247-263).

MOHR, O. L. and CHR. WRIEDT, 1919. A new type of hereditary brachyphalangy in man (Carnegie Instit. Wash., Publ., 295: 64 pp.).

MONNÉ, L., 1948. Functioning of the cytoplasm (Adv. Enzymol. 8: 1-70).

MONTALENTI, G., 1939. I recenti studi sul problema della determinazione del sessa e dei caratteri sessuali secondari negli animali (Genus, 3,3/4: 193-213).

MORGAN, L. V., 1925. Polyploidy in Drosophila melanogaster with two attached X-chromosomes (Genetics, 10: 148-178).

MORGAN, T. H., 1910a. Sex-limited inheritance in Drosophila (Science, N.S. 32: 120-122).

MORGAN, T. H., 1910b. The method of inheritance of two sex-limited characters in the same animal (Proc. Soc. exper. Biol. Med., 8: 17-19).

MORGAN, T. H., 1910c. Chromosomes

and heredity (Am. Nat., 44: 449-496).

MORGAN, T. H., 1913. Heredity and sex (New York, Columbia Univ. Press, 1913: 284 pp.).

MORGAN, T. H., 1914a. Sex-limited and sex-linked inheritance (Am. Nat. 48: 577-583).

MORGAN, T. H., 1914b. Multiple allelomorphs in mice (Am. Nat., 48: 449-458).

MORGAN, T. H., 1916. A critique of the theory of evolution (Princeton. Univ. Press, 1916: 197 pp.).

MORGAN, T. H., 1919. The physical basis of heredity (Philadelphia-London. Lippincott Co., 1919: 303 pp.).

MORGAN, T. H., 1928. The theory of the gene (2nd. ed. New Haven, Yale Univ. Press, 1928: 358 pp.).

MORGAN, T. H., 1934. Embryology and genetics (New York, Columb. Univ. Press, 1934: 258 pp.).

MORGAN, T. H. and C. B. BRIDGES, 1916. Sexlinked inheritance in Drosophila (Carnegie Inst. Wash. Publ. 237: 87 pp.).

MORGAN, T. H., and C. B. BRIDGES 1919. The origin of gynandromorphs (Carnegie Inst. Wash. Publ., 278: 1-122).

MORGAN, T. H., C. B. BRIDGES and A. H. STURTEVANT, 1925. The genetics of Drosophila (Bibliogr. Genet 2: 1-262).

MORGAN, T. H., C. B. BRIDGES and A. H. STURTEVANT, 1928. The constitution of the germinal material in relation to heredity (Carnegie Inst. Yearb., 27: 330-335).

MORGAN, T. H. and C. J. LYNCH, 1912. The linkage of two factors in Drosophiia that are not sex-linked (Biol. Bull., 23: 174-182).

MORGAN, T. H., A. H. STURTEVANT, H. J. MULLER and C. B. BRIDGES, 1915. The mechanism of mendelian heredity (New York, H. Holt Co., 1915: 262 pp.).

MÜNTZING, A., 1935. The evolutionary significance of autopolyploidy (Hereditas 21: 263-378).

MULLER, H. J., 1918. Genetic variability, twinhybrids and constant hybrids, in a case of balanced lethal factors. (Genetics, 3: 422-499).

MULLER, H. J., 1925. The regionally differential effect of X-rays on crossingover in autosomes of Drosophila (Genetics, 10: 470-507).

MULLER, H. J., 1926. Induced crossingover variation in the X-chromosome of Drosophila (Am. Nat., 60: 192-195).

MULLER, H. J., 1928a. The problem of genic modification (Verh. V. Int. Kongr. Vererb. wiss., 1: 234-260).

MULLER, H. J., 1928b. The production of mutations by X-Rays (Proc. nat. Ac. Sc. 14: 714-726).

MULLER, H. J., 1929. The gene as the basis of life (Proc. Intern. Congress on Plant sciences, 1: 897-921).

MULLER, H. J., 1930a. Radiation and genetics (Am. Nat., 64: 220-251).

MULLER, H. J., 1930b. Types of visible variations induced by X-rays in Drosophila (Journ. Genet., 22: 299-334).

MULLER, H. J., 1935. On the dimensions of chromosomes and genes in dipteran salivary glands (Am. Nat., 69: 405-410).

MULLER, H. J., 1936. On the variability of mixed races (Am. Nat., 70: 409-442).

MULLER, H. J., 1939. Bibliography on the genetics of Drosophila (Imp. Bur. Anim. Breed. Genet., 1939: 132 pp.).

MULLER, H. J., 1948. Mutational prophylaxis (Bull. New York Ac. Med. 8: 447-469).

MULLER, H. J., 1950a. Radiation damage to the genetic material (Amer. Scientist 36: 33-59, 126, 399-425).

MULLER, H. J., 1950b. Some present problems in the genetic effects of radiation (Journ. Cell and Comp. Physiol., 35 (Suppl. 1): 9-70).

MULLER, H. J., C. C. LITTLE, L. H. SNYDER, 1947. Genetics, medecine and man. (Ithaca, N.Y., Cornell Univ. Press, 1947: 158 pp.).

MULLER, H. J. and L. M. MOTT-SMITH, 1930. Evidence that natural radioactivity is inadequate to explain the frequency of „natural" mutations (Proc. nat. Ac. Sc., 16: 277-285).

MULLER, H. J. and T. PAINTER, 1932. The differentiation of the sex chromosomes of Drosophila into genetically active and inert regions

(Zschr. ind. Abst. Vererb. Lehre, 62: 316-365).

MULLER, H. J. and A. A. PROKOFYEVA, 1935. The individual gene in relation to the chromomere and the chromosome (Proc. Nat. Ac. Sc. Wash., 21: 16-26).

MULLER, H. J., A. A. PROKOFYEVA and D. RAFFEL, 1935. Minute intergenic rearrangement as a cause of apparent „gene mutation" (Nat., 135: 253-255).

MULSOW, K., 1912. Der Chromosomencyclus bei *Ancyracanthus cystidicola* (Arch. f. Zellf., 9: 63-72).

NABOURS, R. K., 1929. The genetics of the *Tettigidae* (Bibliogr. genet., 5: 27-104).

NABOURS, R. K., 1930. Mutations and allelomorphism in the grouse locusts (*Tettigidae, Orthoptera*) (Proc. nat. Ac. Sc., 16: 350-353).

NACHTSHEIM, H., 1913. Cytologische Studien über die Geschlechtsbestimmung bei der Honigbiene (*Apis mellifica* L.) (Arch. f. Zellf., 11: 169-241).

NACHTSHEIM, H., 1938-1940. Erbpathologie der Haustiere I-II (Fortschr. Erbpathol., 2: 58-104; 4: 49-97).

NACHTSHEIM, H., 1940. Erbpathologie des Stützgewebes der Säugetiere (Handb. Erbb. d. Menschen, 2: 46-104).

NACHTSHEIM, H., 1949. Vom Wildtier zum Haustier (2 Auflage, Berlin, Parey, 1949: 123 pp.).

NAEGELI, C., 1884. Mechanisch-physiologische Theorie der Abstammungslehre (München-Leipzig, R. Oldenbourg, 1884: 822 pp.).

NAUDIN, C., 1855. Réflexions sur l'hybridation dans les végétaux (Revue Horticole 4,4,: 351-360)

NAUDIN, C., 1861. Sur les plantes hybrides (Rev. Hort., 4,10: 245-250).

NAUDIN, C., 1863. Nouvelles recherches sur l'hybridité dans les végétaux (Ann. Sc. nat. Botan., 4,19: 180-203).

NAUDIN, C., 1865a. De l'hybridité considérée comme cause de variabilité dans les végétaux (Ann. Sc. nat. Botan., 5,3: 153-163).

NAUDIN, C., 1865b. Nouvelles recherches sur l'hybridité dans les végé-

taux (Nouv. Arch. Mus. Nat. Paris, 1: 1-176).

NAVASHIN, M., 1931. Spontaneous chromosome alterations in *Crepis tectorum* L. (Univ. Calif. Publ. Agr. Sc., 6: 201-206).

NEBEL, B. R., 1937. Mechanism of polyploidy through colchicine (Nat., 140: 1101).

NEBEL, B. R., 1939. Chromosome structure (Bot. Rev., 5: 563-626).

NEBEL, B. R. and M. L. RUTTLE, 1938. Action of colchicine on mitosis (Genetics, 23: 161-162).

NEBEL, B. R. a.o., 1939. Symposium on chromosome structure (Am. Nat., 73: 289-338).

NEBEL, B. R., a.o., 1941. Symposium on theoretical and practical aspects of polyploidy in crop plants (Amer. Nat., 75: 289-363).

NEEL, J. V., 1947. Genetic effects of the atomic bombs in Hiroshima and Nagasaki. (Science 106: 331-333).

NEWELL, W., 1915. Inheritance in the honey bee (Science, N.S. 41: 168-169).

NEWMAN, H. H., 1917. The biology of twins (Chicago, Univ. Press, 1917: 186 pp.).

NEWMAN, H. H., 1942. Twins and super-twins. (London, Hutchinson's Sc. Techn. Publ., 1942: 164 pp.).

NEWMAN, H. H., F. N. FREEMAN and K. J. HOLZINGER, 1937. Twins. A study of heredity and environment (Chicago, Univ. Chic. Press, 1937: 369 pp.).

NEWMAN, H. H. and J. T. PATTERSON, 1909. A case of normal identical quadruplets in the nine-banded armadillo (Biol. Bull., 17: 181-187).

NEWTON, W. F. C. and C. PELLEW, 1929. *Primula kewesis* and its derivates (Journ. Genet., 20: 405-467).

NILSSON-EHLE, H., 1909. Kreuzungsuntersuchungen an Hafer und Weizen. I (Lunds Univ. Aarsskr. N.F. Afd. 2, 5,2: 122 pp.).

NILSSON-EHLE, H., 1911. Ueber Fälle spontanen Wegfallens eines Hemmungsfaktors beim Hafer (Zschr. ind. Abst. Vererb. Lehre, 5: 1-37).

NILSSON-EHLE, H., 1913. Zur Kenntnis der Erblichkeitsverhältnisse der Eigenschaft Winterfestigkeit beim

Weizen (Zschr. Pflanzenz., 1: 3-12).

NILSSON-EHLE, H., 1915. Den modärna ärftlighetslarna och dess betydelse för växtodlingen (Stockholm, Nordiska Bokhandeln, 1915: 85 pp.).

NILSSON-EHLE, H., 1920. Multiple Allelomorphe und Komplexmutationen beim Weizen (Hereditas, 1: 277-311).

NILSSON, N. HERIBERT, 1915a. Eliminierung der positiven Homozygoten bezüglich der Rotnervigkeit bei Oenothera Lamarckiana (Botan. Not., 1915: 23-25).

NILSSON, N. HERIBERT, 1915b. Die Spaltungserscheinungen der Oenothera Lamarckiana (Lunds Univ. Aarsskr. N.F. Afd. 2., 12,1: 132 pp.).

NILSSON, N. HERIBERT, 1918. Experimentelle Studien über Variabilität, Spaltung, Artbildung und Evolution in der Gattung Salix (Lunds Univ. Aarsskr. N.F. Afd. 2, 14,28: 146 pp.).

NILSSON, N. HERIBERT, 1920a. Zuwachsgeschwindigkeit der Pollenschläuche und gestörte Mendelzahlen bei Oenothera Lamarckiana (Hereditas, 1: 41-67).

NILSSON, N. HERIBERT, 1920b. Kritische Betrachtungen und faktorielle Erklärung der laeta-velutina-Spaltung bei Oenothera (Hereditas, 1: 312-342).

NILSSON, N. HERIBERT, 1928. Salix laurina. Die Entwicklung und die Lösung einer mehr als hundertjährigen phylogenetischen Streitfrage (Lunds Univ. Aarsskr. N.F. Afd. 2, 24,6: 89 pp.).

NILSSON, N. HERIBERT, 1930a. Sind die mutierenden reinen Linien auch rein? (Hereditas, 14: 33-49).

NILSSON, N. HERIBERT, 1930b. Synthetische Bastardierungsversuche in der Gattung Salix (Lunds Univ. Aarsskr. N. F. Afd. 2, 27,4: 97 pp.).

NILSSON, N. HERIBERT, 1931. Ueber das Entstehen eines ganz cinerea-ähnlichen Typus aus dem Bastard Salix viminales × caprea (Hereditas, 15: 309-319).

NILSSON, N. HERIBERT, 1935. Die Analyse der synthetische hergestellten Salix aurina (Hereditas, 20: 339-353).

NILSSON, N. HERIBERT, 1937a Eine Prüfung der Wege und Theorien der Inzucht (Hereditas, 23: 236-256).

NILSSON, N. HERIBERT, 1937b. Ein oktonärer, fertilrt Salix-Bastard und seine Deszendenz (Hereditas, 22: 361-375).

NILSSON, N. HERIBERT, 1953. Synthetische Artbildung. I (Lund, Gleerup, 1953: 734 pp.).

NUSSBAUM, M., 1880. Die Differenzierung des Geschlechts im Tierreich (Arch. mikrosk. Anat, 18:1-121).

OEHLER, E., 1941. Art- und Gattungsbastarde (In: Roemer-Rudorf, Handb. d. Pflanzenzüchtung, 1: 503-541).

OEHLKERS, F., 1937. Die cytologischen Grundlagen des genetischen „crossing-overs" (Ber. Deutsch. bot. Ges., 55: (96)-(118)).

OEHLKERS, F., 1938-1941. Bastardierungsversuche in der Gattung Streptocarpus. I-IV (Zschr. f. Bot., 32: 305-393; Ber. Deutsch. bot. Ges., 58: 76-91; Zschr. f. Bot., 37: 158-182).

OEHLKERS, F., 1940. Meiosis und crossing-over (Biol. Zentralbl., 60: 337-348).

OEHLKERS, F., 1942. Faktorenanalytische Ergebnisse an Artbastarden (Biol. Zbl., 42: 280-289).

OEHLKERS, F. e.a., 1935-1940. Untersuchungen zur Physiologie der Meiosis. I-XI (Zschr. f. Bot., 29: 1-53; 30: 1-57, 253-276, 577-603; 31: 273-328; Jahrb. wiss. Bot., 85: 450-484; Zschr. f. Bot., 32: 225-268; Biol. Zbl., 57: 126-149; Jahrb: wiss. Bot., 85: 706-731; Zschr. f. Bot., 34: 273-310; 36: 161-212).

OEHLKERS, F., 1949. Ueber Erbtraeger ausserhalb des Zellkerns (Ber. Naturf. Ges. Freiburg i. Br., 39:83-121).

OEHLKERS, F., 1952. Neue Ueberlegungen zum Problem der ausserkaryologischen Vererbung (Zschr. ind. Abst. Vererb. Lehre, 84: 213-250).

OESTERGREN, G., 1949. Luzula and the mechanism of chromosome movements (Hereditas 35: 445-468).

OESTERGREN, G., 1950. Considerations on some elementary features of mitosis (Hereditas 36: 1-18).

OFFERMANN, C. A., 1935. The position effect and its bearing on genetics (Bull. Ac. Sc. U.R.S.S., Sc. math. et nat., 1935: 129-140).

OGUMA, K. and S. KAKINO, 1932. A revised check-list of the chromosome number in vertebrata (Journ. Genet. 26: 239-254).

OLIVER, C. P., 1934. Radiation genetics (Quart. Rev. Biol., 9: 381-408).

ONO, T., 1939-1940. Polyploidy and sex determination in *Melandrium* I-III (Bot. Mag. Tokyo, 53: 549-556; 54: 225-230, 348-356).

OORDT, G. J. VAN, 1942. Geslachtsveranderingen bij gewervelde dieren (Gorinchem, Noorduyn, 1942: 134 pp.).

ORTON, W. A., 1911. The development of disease resistant varieties of plants (C. R. IV. Conf. Int. Génét., 1911: 247-265).

OVERBEEK, J. VAN, 1935. The growth hormone and the dwarf type of growth in corn (Proc. Nat. Ac. Sc. Wash., 21: 292-299).

OVERBEEK, J. VAN, 1938. Auxin production in seedlings of dwarf maize (Plant physiology, 13: 587-598).

OWEN, A. R. G., 1950. The theory of genetical recombinations (Adv. Genet., 3: 117-159).

OWEN, F. V., 1928. Inheritance studies in soybeans. III. Seed-coat color and summary of all other Mendelian characters thusfar reported (Genetics, 13: 50-79).

PACKARD, C., 1931. The biological effects of short radiations (Quart.Rev. Biol., 6: 253-280).

PAETAU, K., 1935. Chromosomenmorphologie bei *Drosophila melanogaster* und *Drosephila simulans* und ihre genetische Bedeutung (Naturwiss., 23: 537-543).

PAETAU, K., 1939. Die mathematische Analyse der Evolutionsvorgänge (Zschr. ind. Abst. Vererb. Lehre, 76: 220-228).

PAETAU, K., 1942. Eine neue χ^2 Tafel (Zschr. ind. Abst. Vererb. Lehre, 80: 558-564).

PAINTER, T. S., 1923. Studies in mammalian spermatogenesis. II. The spermatogenesis of man (Journ. exper. Zool., 37: 291-321).

PAINTER, T. S., 1933. A new method for the study of chromosome rearrangements and the plotting of chromosome maps (Science, N.S., 78: 585-586).

PAINTER, T. S., 1934a. Salivary chromosomes and the attack on the gene (Journ. Hered., 25: 465-476).

PAINTER, T. S., 1934b. The morphology of the X-chromosome in salivary glands of *Drosophila melanogaster* and a new type chromosome map for this element (Genetics, 19: 448-469).

PAINTER, T. S. and H. J. MULLER, 1929. Parallel cytology and genetics of induced translocations and deletions in *Drosophila* (Journ. Hered., 20: 287-298).

PARKES, A. S., 1926. The mammalian sex-ratio (Biol. Rev., 2: 1-51).

PARNELL, F. R., 1921. Note on the detection of segregation by examination of the pollen of rice (Journ. Genet., 11: 209-212).

PASCHER, A., 1916. Ueber die Kreuzung einzelliger, haploider Organismen. *Chamydomonas* (Ber. Deutsch. Bot. Ges., 34: 228-242).

PASCHER, A., 1918. Ueber die Beziehung der Reduktionsteilung zur Mendelschen Spaltung (Ber. Deutsch. bot. Ges., 36: 163-168).

PASTEUR, L., 1870. Etudes sur la maladie des vers à Soie (2 vols, Paris, Gauthier-Villars, 1870: 322 + 327 pp.).

PATTERSON, J. T., 1927. Polyembryony in animals (Quart. Rev. Biol., 2: 399-426).

PATTERSON, J. T. and W. S. STONE, 1952. Evolution in the genus *Drosophila* (New York, MacMillan, 1952: 610 pp.).

PAYNE, F., 1920. Selection for high and low bristle number in the mutant strain „reduced" (Genetics, 5: 501-542).

PEARL, R., 1939. The natural history of population (Oxford Univ. Press, 1939: 416 pp.).

PEARL, R. and F. M. SURFACE, 1910a. On the inheritance of the barred color pattern in poultry (Arch. Entw.-Mech., 30: 45-61).

PEARL, R. and F. M. SURFACE, 1910b. Further data regarding the sex-limited inheritance of the barred color pattern in poultry (Science, N. S., 32: 870-874).

PEARSON, K., 1900. On the criterion that a given system of deviation from the probable in the case of a correlated system of variables is such that it can be reasonably supposed to have arisen from random sampling (Phil. Mag. Ser. V., 1: 157-175).

PEARSON, K., 1901. On the fundamental conceptions of biology (Biometrika, 1: 320-344).

PEASE, D. C. and R. F. BAKER, 1949. Preliminary investigations of chromosomes and genes with the electron microscope (Science, 109, p. 8-10, 22).

PELLEW, C., 1929. The genetics of unlike reciprocal hybrids (Biol. Rev., 4: 209-217).

PELLEW, C. and E. RICHARDSON SANSOME, 1931. Genetical and cytological studies on the relations between Asiatic and European varieties of Pisum sativum (Journ. Gen., 25: 25-55).

PETO, F. H., 1935. Associations of somatic chromosomes induced by heat and chloral hydrate treatments (Canad. Journ. Res., 13: 301-314).

PFITZNER, W., 1882. Ueber den feineren Bau der bei der Zelltheilung auftretenden fadenförmigen Differenzierungen des Zellkerns (Morphol. Jahrb., 7: 289-311).

PHILIPTCHENKO, J., 1927. Variabilität und Variation (Berlin, Borntraeger, 1927: 101 pp.).

PHILP, J. and C, L. HUSKINS, 1931. The cytology of Matthiola incana R. Br. especially in relation to the inheritance of double flowers (Journ. Genet., 24: 359-404).

PIRSCHLE, K., 1942. Quantitative Untersuchungen über Wachstum und „Ertrag" autopolyploider Pfanzen; Weitere Untersuchungen über Wachtum und „Ertrag" von Autopolyploiden (2n, 3n, 4n) und ihren Bastarden (Zschr. ind. Abst. Vererb. Lehre, 80: 126-156, 247-270).

PLAGGE, E., 1939. Genabhängige Wirkstoffe bei Tieren (Ergebn. Biol., 17: 105-150).

PLATE. L., 1910. Vererbungslehre und Deszendenztheorie (Festschr. f. R. Hertwig, 2: 537-610).

PLATE, L., 1913, 1932—1938. Vererbungslehre (Leipzig, Engelmann, 1913: 519 pp.; 2nd ed., 3 vols., Jena, Fischer, 1932-1938: 1451 pp.).

PLOUGH, H. H., 1917. The effect of temperature on crossingover in Drosophila (Journ. exper. Zool., 24: 147-209).

PLOUGH, H. H., 1921. Further studies on the effect of temperature on crossingover (Journ. exper. Zool., 32: 187-212).

POLMAN, A., 1950. Anencephaly, Spina bifida and Hydrocephaly (Genetica 25: 29-78).

PONSE, K., 1949. La différenciation du sexe et l'intersexualité chez les vertébrés (Lausanne, Rouge, 1949: 366 pp.).

PONTECORVO, G., 1952. Genetical analysis of cell organization. (Symposia Soc. Exper. Biol. 6: 218-229).

POULSON, D. F., 1940. The effects of certain X-Ray chromosome deficiencies on the embryonic development of Drosophila melanogaster (Journ. Exp. Zool., 83: 271-326).

POULSON, D. F., 1945 Chromosomal control of embryogenesis in Drosophila (Amer. Natur., 79: 340-363).

PUNNETT, R. C., 1913. Reduplication series in sweet peas (Journ. Genet., 3: 77-103).

QUÉTELET, L. A. J., 1871. Anthropométrie ou mesure des différentes facultés de l'homme (Bruxelles, Musquardt; Paris, Baillière, 1871: 479 pp.).

RABAUD, E., 1919. Recherches sur l' hérédité et la variation. Etude expérimentale et théorie physiologique (Bull. Biol. France et Belgique. Supplém. 1: 313 pp.).

RAFFEL, D. and H. J. MULLER, 1940. Position effect and gene divisibility considered in connection with three strikingly similar scute mutations (Genetics, 25: 541-583).

RAJEWSKY, B., A. KREBS und H. ZICKLER, 1936. Mutationen durch Höhenstrahlung (Naturwiss., 24: 619-620).

RASMUSSON, J., 1933. A contribution to the theory of quantitative inheritance (Hereditas, 18: 245-261).

RAVEN, C. P., 1954. An outline of

developmental physiology (London, Pegasus Press, 1954: 216 pp.).

RECK, B., 1936. Untersuchungen über Faktorenaustausch am X-Chromosom von *Drosophila melanogaster* (Zschr. ind. Abst Vererb. Lehre, 72: 138-205).

REINIG, W. F., 1928. Ueber das Manifestieren zweier Genovariationen bei *Drosophila funebris* (Biol. Zbl., 48: 115-125).

REINIG, W. F., 1939. Die genetisch-chorologischen Grundlagen der gerichteten geographischen Variabilität (Zschr. ind. Abst. Vererb. Lehre, 76: 260-308).

REINOEHL, F., 1938. Tierzüchtung (Oehringen, Hohenlohesche Buchh., 1938: 112 pp.).

REITBERGER, A., 1940. Die Cytologie des paedogenetischen Entwicklungszyklus der Gallmücke *Oligarces paradoxus* Mein. (Chromosoma, 1: 391-473).

RENNER, O., 1917. Versuche über die gametische Konstitution der Oenotheren (Zschr. ind. Abst. Vererb. Lehre, 18: 121-294).

RENNER, O., 1918a. Artbastarde und Bastardarten in der Gattung *Oenothera* (Ber. Deutsch. Bot. Ges., 35: (21)-(26)).

RENNER, O., 1918b. *Oenothera Lamarckiana* und die Mutationstheorie (Naturwiss., 46: 37-41, 49-52).

RENNER, O., 1918c. Weitere Vererbungsstudien an Oenotheren (Flora, 111: 641-667).

RENNER, O., 1919a. Ueber Sichtbarwerden der Mendelschen Spaltung in Pollen von *Oenothera*bastarden (Ber. Deutsch. Bot. Ges., 37: 129-135).

RENNER, O., 1919b. Zur Biologie und Morphologie der männlichen Haplonten einiger Oenotheren (Zschr. Bot., 11: 305-380).

RENNER, O., 1920. Mendelsche Spaltung und chemisches Gleichgewicht — Zur Richtigstellung (Biol. Zbl., 40: 268-277, 287-288).

RENNER, O., 1924a. Die Scheckung der Oenotherenbastarde (Biol. Zbl., 44: 309-336).

RENNER, O., 1924b. Vererbung bei Artbastarden (Zschr. ind. Abst. Vererb. Lehre, 33: 317-347).

RENNER, O., 1925. Untersuchungen über die faktorielle Konstitution einiger Komplex-heterozygotischer Oenotheren (Biblioth. genet., 9: 168 pp).

RENNER, O., 1929. Artbastarde bei Pflanzen (Handb. Vererb. wiss. II. A: 161 pp.).

RENNER, O., 1934. Die pflanzlichen Plastiden als selbständige Elemente der genetische Konstitution (Ber. mathem. phys. Kl. Sächs. Ak. Wiss., 86: 241-266).

RENNER, O., 1940. Kurze Mitteilungen zwischen Heterogamie und Embryosackentwicklung und über diplarrhene Verbindungen (Flora, 134: 145-158).

RENNER, O. und W. KUPPER, 1921. Artkreuzungen in der Gattung *Epilobium* (Ber. Deutsch. Bot. Ges., 39: 201-206).

RENSCH, B., 1929. Das Prinzip geographischer Rassenkreise und das Problem der Artbildung (Berlin, Borntraeger, 1929: 206 pp.).

RENSCH, B., 1939. Typen der Artbildung (Biol. Rev., 14: 180-222).

RENSCH, B., 1954. Neuere Probleme der Abstammungslehre. Die transspezifische Evolution (Stuttgart, Enke, 1954: 436 pp.).

RESENDE, F., 1936. Cytologischer Nachweis von Postreduktion bei einer Phanerogame (Planta, 25: 665-666).

RHOADES, M. M. and B. McCLINTOCK, 1935. The cytogenetics of maize (Bot. Rev., 1: 292-325).

RICE, V. A., 1934. Breeding and improvement of farm animals (2nd ed. New York, Mc Graw-Hill, 1934: 516 pp.).

RICHARDSON SANSOME, E., 1929. A chromosome ring in *Pisum* (Nature, 124: 578).

RICHEY, F. D., 1950. Corn breeding (Adv. Genetics 3: 160-192).

RITTER, R., 1939. Die Zigeunerfrage und das Zigeunerbastardproblem (Fortschr. Erbpathol., 3: 2-20).

RITTER, R., 1941. Die Asozialen, ihre Vorfahren und Nachkommen (Fortschr. Erbpathol., 5: 137-155).

ROBBINS, R. B., 1917-1918. Applications of mathematics to breeding problems. I-III (Genetics, 2: 489-504, 3: 73-92, 375-389).

ROBERTIS, E. D. P. DE, W. W. NOWINSKI and F. A. SAEZ, 1953. General cytlogy 2nd impr. Philadelphia-London, W. G. Saunders Cy., 1953; 456 pp.).

ROBERTS, H. F., 1929. Plant hybridization before Mendel (Princeton Univ. Press, 1929: 374 pp.).

ROBSON, G. C., 1928. The species problem (Edinburgh-London. Oliver and Boyd, 1928: 283 pp.).

ROEMER, TH., W. H. FUCHS und K. ISENBECK, 1938. Die Züchtung resistenter Rassen der Kulturpflanzen (Berlin, Parey, 1938: 427 pp.).

ROEMER, TH, and W. RUDORF, 1941 ..... Handbuch der Pflanzenzüchtung (5 vols. Berlin, Parey, 1: 610 pp.; 3: 482 pp., rest not yet complete.

ROSENBERG, O., 1927. Die semiheterotypische Teilung und ihre Bedeutung für die Entstehung verdoppelter Chromosomenzahlen (Hereditas, 8: 305-338).

ROSENBERG, O., 1930. Apogamie und Parthenogenesis bei Pflanzen (Hand Vererb. wiss. II. L: 66 pp.).

ROSS, H., 1941-1948. Ueber die Verschiedenheiten des dissimilatorischen stoffwechsels in reziproken Epilobium-Bastarden und die physiologisch-genetische Ursache der reziproken Unterschiede I-V (Zschr. ind. Abst. Vererb. Lehre, 79: 503-529; Planta, 32: 447-488; 33: 161-184; Zschr. ind. Abst. Vererb. Lehre, 82: 98-129, 187-196).

ROTH, L., 1927. Untersuchungen über die periklinalbunten Rassen von Pelargonien (Zschr. ind. Abst. Vererb. Lehre, 45: 125-159).

RUNDQUIST, E., 1933. Inheritance of spontaneous activity in rats (Journ. comp. Psychol., 16: 415-438).

SACHS, L., 1952. Polyploid evolution and mammalian chromosomes (Herdity, 6: 357-364).

SAGERET, A., 1826. Considérations sur la production des hybrides, des variantes, et des variétés en général et sur celles des Cucurbitacées en particulier (Ann. Sc. natur., 8: 294-314).

SANDERS, J., 1949. Onderzoek en behandeling der gegevens (In: Sirks en Waardenburg, Geneeskunde en erfelijkheid, 3rd edition, Lochem, De Tijdstroom, 1949: 51-72).

SANSOME, F. W., und J. PHILP, 1932. Recent advances in plant genetics. (London, Churchill, 1932: 414 pp.).

SATINA, S., A. F. BLAKESLEE and A. AVERY, 1937. Balanced and unbalanced haploids in Datura (Journ. Hered., 28: 193-202).

SAUNDERS, E. R., 1928. Matthiola (Bibliogr. genet., 4: 141-170).

SAX, K., 1930. Chromosome structure and the mechanism of crossing-over (Journ. Arnold Arbor., 11: 193-220).

SAX, K., 1935. The cytological analysis of species hybrids (Bot. Rev., 1: 100-117).

SAX, K., 1936. Chromosome coiling in relation to meiosis and crossing-over (Genetics, 21: 324-338).

SCHARNAGEL, TH., 1943. Die Bedeutung der Resistenzzüchtung für den Acker- und Pflanzenbau (Prakt. Bl. Pflanzenbau u. Pflanzenschutz, 20 (42): 205-219).

SCHARRER, B. und E. HADORN, 1938. The structure of the ring-gland (corpus allatum) in normal and lethal larvae of Drosophila melanogaster (Proc. Nat. Ac. Sc. Wash., 24: 236-242).

SCHAXEL, J., 1915. Die Leistungen der Zellen bei der Entwicklung der Metazoën (Jena, Fischer, 1915: 336 pp)..

SCHIEMANN, E., 1931, 1937, 1943. Geschlechts- und Artkreuzungen bei Fragaria I-III (Jena, Fischer, 1931: 112 pp.; Zschr. ind. Abst. Vererb. Lehre, 73: 375-390; Flora, 137: 166-192).

SCHIEMANN, E., 1943. Entstehung der Kulturpflanzen (Ergebn. Biol., 19: 409-552).

SCHLEIP, W., 1927. Entwicklungsmechanik und Vererbung bei Tieren (Handb. Vererb. wiss. III A: 81 pp.).

SCHLÖSSER, L. A., 1935. Beitrag zu einer physiologischen Theorie der plasmatischen Vererbung (Zschr. ind. Abst. Vererb. Lehre, 69: 159-192).

SCHLOTTKE, E., 1926. Ueber die Variabilität der schwarzen Pigmentierung und ihre Beeinflussbarkeit durch Temperaturen bei *Habrobracon juglandis* Ashmead (Zschr. vgl. Physiol., 3: 692-736).

SCHMALHAUSEN, J. J., 1949. Factors of evolution. The theory of stabilizing selection (Philadelphia, Blakiston, 1949, 327 pp.).

SCHMIDT, JOHS., 1920. Racial investigations. IV. The genetic behaviour of a secondary sexual character (C. R. Trav. Lab. Carlsberg, 14,8: 1-12).

SCHRADER, F., 1953. Mitosis. The movement of chromosomes in cell-division, (2nd impr., New York, Columbia Univ. Press, 1953: 110 pp.).

SCHRADER, F. and S. H., 1931. Haploidy in metazoa (Quart. Rev. Biol. 6: 411-438).

SCHRÖDINGER, E., 1945. What is life? (Cambridge, Univ. Press, 1945: 91 pp.).

SCHRIJVER, F., 1941. Eugenetische wetgeving (In: Sirks en Waardenburg, Geneeskunde en erfelijkheid, 3rd ed., Lochem, De Tijdstroom, 1941: 441-467).

SCHUBERT, G. und A. PICKHAN, 1938. Erbschädigungen (Leipzig, Thieme, 1938: 164 pp).

SCHUBELER, F. C., 1862. Die Kulturpflanzen Norwegens (Christiania, 1862: 197 pp.).

SCHUBELER, F. C., 1873. Die Pflanzenwelt Norwegens (Christiania, 1863: 201 pp.).

SCHULTZ, J., 1936. Radiation and the study of mutation in animals (In: Duggar, Biological effects of radiation, 1936: 1209-1262).

SCHULTZ, W., 1915. Schwarzfärbung weisser Haare durch Rasur und die Entwicklungsmechanik der Farbmuster von Haaren und Federn (Arch. Entw. Mech. Org., 41: 535-557).

SCHULZ, B., 1936. Methodik der medizinischen Erbforschung unter besonderer Berücksichtigung der Psychiatrie (Leipzig, Thieme, 1936: 189 pp.).

SCOTT-MONCRIEFF, R., 1936. A biochemical survey of some Mendelian factors for flower colour (Journ. Genet., 32: 117-170).

SEARS, E. R., 1948. The cytology and genetics of the wheats and their relatives (Adv. Genet., 2: 240-270).

SEILER, J., 1922. Geschlechtschromosomen-Untersuchungen an Psychiden III. Chromosomenkoppelungen bei *Solenobia pineti*, Z. Eine zytologische Basis für die Faktorenaustauschhypothese (Archiv f. Zellf., 16: 171-216).

SEILER, J., 1924. Die crossing-over-Studien der Schule Morgan (Naturwiss., 12: 677-685).

SEILER, J., 1925. Ergebnisse aus Kreuzungen von Schmetterlingsrassen mit verschiedener Chromosomenzahl. Ein Beweis für das Mendeln der Chromosomen (Arch. Julius Klaus Stift., 1: 63-117).

SEILER, J. 1926. Die Chiasmatypie als Ursache des Faktorenaustausches (Zschr. ind. Abst. Vererb. Lehre, 41: 259-284).

SEILER, J., 1942. Resultate aus der Kreuzung parthenogenetischer und zweigeschlechtlicher Schmetterlinge (Arch. Julius Klaus Stift., 17: 513-528).

SEILER, J., 1946. Bemerkungen zu Goldschmidts Interpretation der intersexen Solenobien (Arch. Julius Klaus Stift., 21: 273-275).

SEILER, J., 1949. Das Intersexualitätsphaenomen (Experientia, 5: 425-438).

SEILER, J. und C. B. HANIEL, 1921. Das verschiedene Verhalten der Chromosomen in Eireifung und Samenreifung von *Lymantria monacha* L. (Zschr. ind. Abst. Vererb. Lehre, 27: 81-103).

SEMON, R., 1904. Die Mneme als erhaltendes Prinzip im Wechsel des organischen Geschehens (Leipzig, Engelmann 1904: 420 pp.).

SEMON, R., 1909. Die mnemischen Empfindungen in ihren Beziehungen zu den Originalempfindungen. Erste Fortsetzung der „Mneme" (Leipzig, Engelmann, 1909: 392 pp.).

SEMON, R., 1911. Der Stand der Frage nach der Vererbung erworbener Eigenschaften (Fortschr. d. Naturwiss. Forschung, 2: 1-82).

SENGBUSCH, R. v., 1937. Pflanzenzüchtung und Rohstoffversorgung (Leipzig, Thieme, 1937: 131 pp.).

SENN, H. A., 1938. Chromosome number relationships in the Leguminosae (Bibliogr. Genet., 12: 175-345).

SEREBROVSKY, A. S., 1930. Untersuchungen über Treppenallelomorphismus. IV. Transgenation scute-6 und ein Fall des Nicht-Allelomorphismus von Gliedern eines Allelomorphenreie bei *Drosophila melanogaster* (Arch. Entw.-mech. Org., 122: 88-104).

SEREBROVSKY, A. S. and N. P. DUBININ, 1930. X-ray experiments with *Drosophila* (Journ. Hered., 21: 259-265).

SERRA, J. A., 1942. Relations entre la chimie et la morphologie nucléaire (Bol. Soc. Broteriana, 2e Série, 16: 83-135).

SERRA, J. A., 1943. Sur la nature des mélanines et la mélanogenèse (Genetica, 23: 300-314).

SETON, A., 1822. On the variation in the colours of peas from cross-impregnation (Transact. Hort. Soc. London, 5: 236-237).

SHARP, L. W., 1934. Introduction to cytology (3rd ed. New York, Mac Graw Hill, 1934: 567 pp.).

SHARP, L. W., 1943. Fundamentals of cytology. (Sixth impr., MacGraw Hill, New York, 1943: 270 pp).

SHEAR, C. L. and B. O. DODGE, 1927. Life histories and heterothallism of the red bread-mould fungi of the *Monilia sitophila* group (Journ. Agr. Res., 34: 1019-1042).

SHIMOTOMAI, N., 1938. Cytogenetische Untersuchungen über *Chrysanthemum* (Bibliogr. Genet., 12: 161-174).

SHIWAGO, P. I. und A. H. ANDRES, 1932. Die Geschlechtschromosomen in der Spermatogenese des Menschen (Zschr. f. Zellf. u. mikr. Anat., 16: 413-431).

SHRODE, R. R. and J. L. LUSH, 1947. The genetics of cattle (Adv. Genetics 1: 210-263).

SHULL, G. H., 1910. Inheritance of sex in *Lychnis* (Bot. Gaz., 2: 110-125).

SHULL, G. H., 1911a. The genotypes of maize (Am. Nat., 45: 234-252).

SHULL, G. H., 1911b. Reversible sex-mutants in *Lychnis dioica* (Bot. Gaz., 52: 329-368).

SHULL, G. H., 1912. Phenotype and clone (Science, N.S. 35: 182-183).

SHULL, G. H., 1914a. Duplicate genes for capsule-form in *Bursa bursa-pastoris* (Zschr. ind. Abst. Vererb. Lehre, 12: 97-149).

SHULL, G. H., 1914b. Sex-limited inheritance in *Lychnis dioica* L. (Zschr. ind. Abst. Vererb. Lehre, 12: 265-302).

SIEMENS, H. W., 1924. Die Zwillingspathologie, ihre Bedeutung, ihre Methodik, ihre bisherigen Ergebnisse (Berlin, Springer, 1924: 103 pp.).

SIMPSON, G. G., 1944. Tempo and mode in evolution (New York, Columbia Univ. Press, 1944: 237 pp.)

SIMPSON, G. G., 1949. The meaning of evolution (New Haven, Yale Univ. Press, 1949: 364 pp.).

SINNOTT, E. W. and L. C. DUNN, 1935. The effect of genes on the development of use and form (Biol. Rev., 10: 123-151).

SINNOTT, E. W., L. C. DUNN and TH. DOBZHANSKY, 1950. Principles of genetics (McGraw-Hill, New York, 1950: 506 pp.).

SIRKS, M. J., 1915. Waren die *Salix*-hybriden Wichuras wirklich konstant? (Zschr. ind. Abst. Vererb. Lehre, 15: 164-166).

SIRKS, M. J., 1923. Die Verschiebung genotypischer Verhältniszahlen innerhalb Populationen laut mathematischer Berechnung und exprimenteller Prüfung (Meded. Landbouwhoogeschool Wageningen, 26,4: 40 pp.).

SIRKS, M. J., 1925. The inheritance of the seedweight in the garden bean. I (Genetica, 7: 118-169).

SIRKS, M. J. 1926a. Mendelian factors in *Datura* I. Certation (Genetica, 8: 485-500).

SIRKS, M. J., 1926b. Multiple allelomorphs versus multiple factors (Proc Intern. Congr. Plantsciences, 1926, 1: 803-814).

SIRKS, M. J. 1929a. Mendelian factors in *Datura*. III. Separate factors for certation and their differential value (Genetica, 11: 257-266).

SIRKS, M. J., 1929b. The interrelations of some anthocyane-factors in the potato (Genetica, 11: 239-328).

SIRKS, M. J., 1931. Beiträge zu einer genotypischen Analyse der Ackerbohne, *Vicia Faba* (Genetica, 13: 209-632).

SIRKS, M. J., 1938a. A case of bud-variation in *Phaesolus* caused by a transitory plasmatic change (Genetica, 20 : 121-158).

SIRKS, M. J., 1938b. Plasmatic inheritance (Bot. Rev., 4 : 113-131).

SIRKS, M. J., 1942. De ontwikkeling der biologie (Noorduyns Wetensch. Reeks 2, Gorinchem, Noorduyn, 1942 : 183 pp.).

SIRKS, M. J., 1946. Genetik und Chemie (Arch. Julius Klaus Stift. 21 : 377-396).

SIRKS, M. J., 1948. Bevolking en qualiteit (In: Methorst en Sirks, Het bevolkingsvraagstuk, Amsterdam, Scheltema en Holkema, 1948: 88-230).

SIRKS, M. J., 1949a. De dynamiek van populaties (Natuurwet. Tijdschrift, Gent 31 : 227-242).

SIRKS, M. J., 1949b. De variabiliteit van het soortsbgerip (Dodonaea, 16 : 176-194).

SIRKS, M. J., 1949c Het geslacht. Uitingen en oorzaken (2e dr., Gorinchem, Noorduyn, 1949: 176 pp.).

SIRKS, M. J., 1951. The scope of genetics in problems of evolution (Rep. Brisbane meeting A.N.Z.Z.Z.S., 28: 170-177).

SIRKS, M. J. en J. BIJHOUWER, 1919. Onderzoekingen over de eenheid der Linneaansche soort *Chrysantemum leucanthemum* L. (Genetica, 1: 401-442).

SIRKS, M. J. en P. J. WAARDENBURG, 1949. Geneeskunde en erfelijkheid (3e dr., Lochem, Tijdstroom, 1949, 484 pp.).

SITOWSKI, L., 1910. Experimentelle Untersuchungen über vitale Färbung des Mikrolepidopterenraupen (Bull. Acad. Sc. Cracovie, 1910 : 775-790).

SITUATION, 1949. The situation in Biological Sciences. (Proc. Lenin Acad. Agric. Sc. U.S.S.R. Session July 31-Aug. 7, 1948. Verbatim Report (Moscow, Foreign Lang. Publ. House, 1949: 631 pp.)).

SKALINSKA, M., 1930. On the significance of the cytoplasm in matrocli-nous hybrids of *Aquilegia* (Acta biol. exp. 5 : 1-18).

SMITH, G., 1906. Rhizocephala (Flora und Fauna d. Golfes von Neapel. Monographie 29).

SMITH, P. E. and E. C. McDOWELL, 1930. An hereditary anterior-pituitary deficiency in the mouse (Anat. Rec., 46 : 249-257).

SMITH, P. E. and E. C. McDOWELL, 1931. The differential effect of hereditary mouse dwarfism on the anterior pituitary hormones (Anat. Rec., 50 : 85-93).

SMITH, S. G., 1932. Cytology of *Anchusa* and its relation to the taxonomy of the genus (Bot. Gaz., 94: 394-403).

SOLMS-LAUBACH, H., 1907. Ueber unsere Erdbeeren und ihre Geschichte (Bot. Ztg., 65,1 : 45-76).

SOMMERMEYER, K. und U. DEHLINGER 1939. Beiträge zur Diskussion eines Gen-Modells (Physik. Zschr., 40 : 67-70).

SONNEBORN, T. M., 1945. Gene action in *Paramaecium*. (Ann. Missouri Bot. Garden, 32 : 213-221).

SONNEBORN, T. M., 1947. Recent Advances in the Genetics of *Paramaecium* and *Euplotes* (Adv. Genetics 1 : 264-358).

SONNEBORN, T. M., 1949. Beyond the gene (Amer. Scient., 37 : 33-59).

SONNEBORN, T. M., 1950. The cytoplasma in heredity (Heredity 4 : 11-36).

SONNEBORN, T. M., 1953 (1954). Patterns of nucleocytoplasmic integration in *Paramaecium* (Proc. 9th Intern. Congr. Genet. Bellagio, 1953: 30 pp.).

SORSBY, A., 1951. Genetics in ophthalmology. (London, Butterworth, 1951 251 pp.).

SPEMANN, H., 1924. Vererbung und Entwicklungsmechanik (Zschr. ind. Abst. Vererb. Lehre 33 : 272-293).

SPEMANN, H., 1932. Ueber xenoplastische Transplantation als Mittel zur Analyse der embryonalen Induktion (Naturwiss., 20 : 463-467).

SPEMANN, H., 1936. Experimentelle Beiträge zu einer Theorie der Entwicklung (Dtsch. Ausgabe Silliman Lectures 1933, Berlin, J. Springer, 1936 : 296 pp.).

SPENCER, H., 1864, 1898. The principles of biology (London, Williams and Norgate, 1864. Rev. Ed., 1898: 706 pp. + 660 pp.).

SPENCER, H., 1893a. The inadequacy of „natural selection" (Contemporary Review, Febr.-March 1893: 45 pp.).

SPENCER, H., 1893b. Professor Weismanns theories (Contemporary Review, May 1893 : 24 pp.).

SPENCER, H., 1893c. A rejoinder to Professor Weismann (Contemporary Review, Dec. 1893 : 29 pp.).

SPENCER, H., 1894. Weismannism once more (Contemporary Review, Oct. 1894 : 23 pp.).

SPENCER, W. P., 1947. Mutations in wild populations in *Drosophila* (Adv. Genetics 1 : 359-402).

SPIEGELMANN, S., 1945. The physiology and genetic significance of enzymic adaptation (Ann. Missouri Bot. Garden, 32 : 139-163).

SPILLMAN, W. J., 1908. Spurious allelomorphism: Results of some recent investigations (Am. Nat., 42 : 610-615).

SPILLMAN, W. J., 1909. Barring in barred Plymouth Rocks (Poultry, 5 : 7-8).

SPRENGEL, C. K., 1793. Das entdeckte Geheimniss der Natur im Blau und in der Befruchtung der Blumen (Berlin, Vieweg, 1793: 443 pp.).

SRB, A. M. and R. D. OWEN, 1953. General Genetics (San Francisco, W. H. Freeman Cy, 1953: 560 pp.).

STADLER, L. J., 1928. Genetic effects of X-rays in maize (Proc. Nat. Ac. Sc. Wash., 14: 69-75).

STADLER, L. J., 1936. Induced mutations in plants (In: Duggar, Biological effects of radiations, 1936: 1263-1280).

STARK, M. B., 1918. An hereditary tumor in the fruit fly, *Drosophila* (Journ. Cancer Res., 3: 279-300).

STARK, M. B., 1919. An hereditary tumor (Journ. exper. Zool., 27: 509-522).

STEBBINS, G. L., 1947. Types of polyploids: their classification and significance (Adv. Genetics 1: 413-430).

STEBBINS, G. L., 1950. Variation and evolution in plants (New York, Columbia Univ. Press, 1950: 643 pp.).

STEIN, E., 1926. Untersuchungen über die Radiomorphosen von *Antirrhinum* (Zschr. ind. Abst. Vererb. Lehre, 43: 1-87).

STEIN, E., 1930. Weitere Mitteilung über die durch Radium-bestrahlung induzierte Gewebe-Entartungen in *Antirrhinum* (Phytocarcinome) und ihr erbliches Verhalten (Somatische Induktion und Erblichkeit) (Biol. Zbl., 50: 129-158).

STEIN, E., 1931. Zur Entstehung und Vererbung der durch Radiumbestrahlung erzeugten Phytocarcinome (Zschr. ind. Abst. Verb. Lehre, 62: 1-13).

STEIN, E., 1936. Erbliche durch Radiumbestrahlung erzeugte Zell- und Gewebe-Entartung beim Löwenmaul (*Antirrhinum majus*) ˙(Naturwiss., 24: 337—342).

STEPHENS, S. G., 1947. Cytogenetics of *Gossypium* and the problem of the origin of New World cottons (Adv. Genetics 1: 431-442).

STERN, C., 1926a. An effect of temperature and age on crossing-over in the first chromosome of *Drosophila melanogaster* (Proc. nat. Ac. Sc., 12: 530-532).

STERN, C., 1926b. Eine neue Chromosomenaberration von *Drosophila melanogaster* und ihre Bedeutung für die Theorie der linearen Anordnung der gene (Biol. Zbl., 46: 505-508).

STERN, C., 1926c. Vererbung im Y-Chromosom von *Drosophila melanogaster* (Biol. Zbl., 46: 344-348.

STERN, C., 1927a. Die Chromosomen elimination bei der Taufliege (Naturwiss., 15: 740-746).

STERN, C., 1927b. Die genetische Analyse der Chromosomen (Naturwiss., 15: 465-473).

STERN, C., 1928. Fortschritte der Chromosomentheorie der Vererbung (Ergebn. Biol., 4: 205-359).

STERN, C., 1929a. Die Bedeutung von *Drosophila melanogaster* für die genetische Forschung (Züchter, 1: 237–243).

STERN, C., 1929b. Untersuchungen über Aberrationen des Y-Chromosoms von *Drosophila melanogaster* (Zschr. ind. Abst. Vererb. Lehre, 51: 253-353).

STERN, C., 1929c. Ueber additive Wirkung multipeler Allelen (Biol. Zbl.,
49: 261–290).

STERN, C., 1930a. Konversionstheorie
und Austauschtheorie. (Biol. Zbl.,
50: 608–624).

STERN, C., 1930b. Multiple Allelie
(Handbuch der Vererbungswissen-
schaft. I. G: 147 pp.).

STERN, C., 1931. Zytologisch-geneti-
sche Untersuchungen als Beweise
für die Morgansche Theorie des
Faktorenaustausches (Biol. Zbl.,
51: 547-587).

STERN, C., 1932. Ueber die Konversionstheorie (Biol. Zbl., 52: 367-
379).

STERN, C., 1933. Faktorenkoppelung
und Faktorenaustausch (Handb.
Vererb. Wiss., I, H: 331 pp).

STERN, C., 1936. Somatic crossing-
over and segregation in Drosophila
melanogaster (Genetics, 21: 625-
730).

STERN, C., 1949. Principles of human
genetics (San Francisco, Freeman,
1949: 617 pp).

STOUT, A. B., 1915. The establishment of varieties in Coleus by the
selection of somatic variations (Carnegie Instit. Wash., Publ. 218: 80
pp.).

STRASBURGER, E., 1884. Neue Untersuchungen über den Befruchtungs-
vorgang bei den Phanerogamen als
Grundlage für eine Theorie der Zeu-
gung (Jena, Fischer, 1884: 176 pp)..

STRASBURGER, E., 1907. Apogamie
bei Marsilia (Flora, 97: 123-191).

STRAUB, J., 1941. Ergenbisse und Probleme der Polyploidieforschung
(Forschungsdienst, 12: 318-324).

STROHMAYER, W., 1937. Die Vererbung des Habsburger Familientypus. Eine erbphysiognomische
Betrachtung auf genealogischer
Grundlage (Nova Acta Leopoldina,
5: 217-296).

STUBBE, H., 1930-1938. Untersuchungen über experimentelle Auslösung von Mutationen bei Antirrhinum majus I-VI (Zschr. ind. Abst.
Vererb. Lehre, 56: 1-38, 202-232;
60: 474-513; 64: 181-204; 67:
152-172; 72: 378-386; 75- 341-351).

STUBBE, H., 1933. Labile Gene (Bibliogr. Genet., 10: 299-356).

STUBBE, H., 1937. Spontane und
strahleninduzierte Mutabilität (Leipzig, Thieme, 1937: 190 pp.).

STUBBE, H., 1938. Genmutation. I.
Allgemeiner Teil (Handb. Vererb.
Wiss. II F., Berlin, Borntraeger,
1938: 429 pp.).

STUMPFL, F., 1941. Psychopathien und
Kriminalität (Fortschr. Erbpathol.,
5: 33-116).

STURTEVANT, A. H., 1912. An experiment dealing with sex-linkage in
fowls (Journ. exper. Zool., 12: 499-
518).

STURTEVANT, A. H., 1913. The linear
arrangement of six sex-linked factors in Drosophila, as shown by
their mode of association (Journ.
exper. Zool., 14: 43-60

STURTEVANT, A. H., 1915. The behavior of the chromosomes as studied
through linkage (Zschr. ind. Abst.
Vererb. Lehre, 13: 234-287).

STURTEVANT, A. H., 1919. Inherited
linkage variations in the second
chromosome (Carnegie Inst. Wash.
Publ. 278: 305-341).

STURTEVANT, A. H., 1920. The vermilion gene and gynandromorphism
(Proc. Soc. exper. Biol. Med., 17:
70-71).

STURTEVANT, A. H., 1921. A case of
rearrangement of genes in Drosophila (Proc. nat. Ac. Sc., 7: 235-
237).

STURTEVANT, A. H., 1923. Inheritance
of direction of coiling in Limnaea
(Science, N. S. 58: 269-270).

STURTEVANT, A. H., 1925. The effects
of unequal crossing-over at the bar
locus in Drosophila (Genetics, 10:
117-147).

STURTEVANT, A. H., 1926. A cross-
over reducer in Drosophila melanogaster due to inversion of a section
of the third chromosome (Biol. Zbl.,
46: 697-702).

SUOMALAINEN, E., 1950. Parthenogenesis in animals (Adv. Genetics
3: 193-254).

SUTTON, W. S., 1903. The chromosomes in heredity (Biol. Bull., 4:
231-248).

SVÄRDSON, G., 1945. Chromosome
studies in Salmonidae. (Medd. Sta-
tens undersöknings- och försökanst.
f. söttvattensfisket 23; 151 pp.).

SYMPOSIUM, 1941a. Cytology, genetics and evolution (Philadelphia, Univ. Penns. Press, 1941: 168 pp.).

SYMPOSIUM 1941b. Genes and chromosomes. Structure and organization (Cold Spring Harbor Symp. Quant. Biol. 9: 315 pp.).

SYMPOSIUM, 1942. The relation of hormones to development (Cold Spring Harbor Symp. Quant. Biol. 10: 167 pp.).

SYMPOSIUM, 1945a. Gene action in micro-organisms (Ann. Missouri Bot. Garden, 32: 107-263).

SYMPOSIUM, 1945b. Mammary tumors in mice (Washington, Amer. Ass. Adv. Sc., 1945: 223 pp.).

SYMPOSIUM, 1946. Heredity and variation in micro-organisms (Cold Spring Harbor Symp. Quant. Biol. 11: 314 pp.).

SYMPOSIUM, 1947a. Nucleic acid (Symp Soc. Exp. Biol. 1: 290 pp.).

SYMPOSIUM, 1947b. Nucleic acids and nucleoproteins. (Cold Spring Harbor Symp. Quant. Biol. 12: 279 pp.).

SYMPOSIUM 1948a. On radiation genetics. Oak Ridge National Laboratory. (Journ. Cell. and Comp. Physiology, 35, Suppl. 1, 1950: 210 pp.).

SYMPOSIUM, 1948b. The Rh factor in the clinic and the laboratory. (New York, Grave and Stratton, 1948: 192 pp.).

SYMPOSIUM, 1949a. Acidi nucleici, proteine e differenziamento normale e patologice. (Torino, Rosenberg e Sellier, 1949: 127 pp.).

SYMPOSIUM, 1949b. Natural Selection and adaptation (Proc. Amer. Philos. Soc. 93: 459-519).

SYMPOSIUM 1949c. Sui fattori ecologici e genetici della speciazione negli animali (La Ricerca Scient. 19, Suppl.: 134 pp.).

SYMPOSIUM 1949d. Unités biologiques données de continuité génétique (Coll. internat. Centre nation. Rech. Scient. Paris, 8: 205 pp.).

SYMPOSIUM, 1950a. Aminoacids and proteins (Cold Spring Harbor Symp. Quant. Biol., 14: 217 pp.).

SYMPOSIUM, 1950b. Biochemical aspects of genetics (Biochem. Soc. Symposia nr. 4.). (Cambridge Univ. Press, 1950: 66 pp.).

SYMPOSIUM, 1950c. Gli agenti mutageni. (Publ. Staz. Zoologica Napoli, 22, Suppl.: 200 pp.).

SYMPOSIUM, 1955. On genetic recombination (Oak Ridge Nation. Lab.; Journ. Cell. Comp. Phys., 45, Suppl. 2: 321 pp.).

TÄCKHOLM, G., 1922. Zytologische Studien über die Gattung *Rosa* (Acta Horti Bergiani, 7: 97-381).

TAHARA, M., 1921. Cytologische Studien an einigen Kompositen (Journ. Coll. Sc. Imp. Univ. Tokyo, 43,7: 53 pp.).

TAHARA, M. und N. SHIMOTOMAI, 1927. Bastardierung als eine Ursache für die Entstehung von Chromosomenpolyploïdie. I. Bastarde zwischen *Chrysanthemum marginatum* und *C. lavandulaefolium* (Sci. Rep. Tohoku Imp. Univ., 4,2: 293-299).

TAMMES, T., 1914. Die Erklärung einer scheinbaren Ausnahme der Mendelschen Spaltungsregel (Rec. Trav. botan. neerl., 54-69).

TAMMES, T., 1926. Dominanzwechsel bei *Dianthus barbatus* (Genetica, 8: 513-517).

TAMMES, T., 1930. The use of symbols for indicating the history of individuals or groups of individuals in genetic investigations (Genetica, 12: 145-150).

TANAKA, Y., 1917. Genetic studies on the silkworm (Journ. Coll. Agr. Tohoku Imp. Univ. Sapporo, 7: 129-255).

TANDLER, J. und K. KELLER, 1911. Ueber das Verhalten des Chorions bei verschieden-geschlechtlicher Zwillingsgravidität des Rindes und über die Morphologie der Genitalen der weiblichen Tiere, welche einer solchen Gravidität entstammen (Dtsch. tierärztl. Woch., 19: 148-149).

TATUM, E. L. and J. LEDERBERG, 1947. Gene recombination in the bacterium, *Escherichia coli* (Journ. Bacter. 53: 673-684).

TATUNO, S., 1934-1938. Geschlechtschromosomen bei Lebermoosen I-IX (Jap. Journ. Gen., 9: 95-96; Journ. Sc. Hirosh. Univ. Ser. B. Div. 2, 3: 1-9; Bot. Mag. Tokyo, 50: 401-405, 526-531; 51: 812-819, 860-866; 52: 374-379, 650-654).

TAYLOR, J. W. and C. E. LEIGHTY,

1931. Inheritance in a „constant" hybrid between *Aegilops ovata* and *Triticum dicoccum* (Journ. Agr. Res., 43: 661-679).

TERMAN, L. M., 1925. Genetic studies of genius. I. Mental and physical traits of a thousand gifted children (Stanford Univ. Press, 1925: 648 pp.).

THERMAN, E. and S. TIMONEN, 1951. Inconstancy of the human somatic chromosome complement (Hereditas, 37: 266-279).

TIMOFEEFF-RESSOVSKY, N. W., 1927. Studies on the phenotypic manifestation of hereditary factors. I. On the phenotypic manifestation of the genovariation radius incompletus in *Drosophila funebris* (Genetics, 12: 128–198).

TIMOFEEFF-RESSOVSKY, N. W., 1931. Gerichtetes Variïeren in der phänotypischen Manifestierung einiger Genovariationen von *Drosophila funebris* (Naturwiss. 19: 493-497).,

TIMOFEEFF-RESSOVSKY, N. W., 1934a. The experimental production of mutations (Biol. Rev., 9: 411-457).

TIMOFEEFF-RESSOVSKY, N. W., 1934b. Ueber die Vitalität einiger Genmutationen und ihrer Kombinationen bei *Drosophila funebris* und ihre Abhängigkeit vom „genotypischen" und vom äusseren Milieu (Zschr. ind. Abst. u. Vererb. Lehre, 66: 319-344).

TIMOFEEFF-RESSOVSKY, N. W., 1934c. Ueber den Einfluss des genotypischen Milieus und der Aussenbedingungen auf die Realisation des Genotypus. Genmutation v.t.i. (venae transversae incompletae) bei *Drosophila funebris* (Nachr. Ges. Wiss. Göttingen. Math. phys. Kl. Biologie, N.F. I, 53-106).

TIMOFEEFF-RESSOVSKY, N. W., 1937. Experimentelle Mutationsforschung in der Vererbungslehre. Beeinflussung der Erbanlagen durch Strahlung und andere Faktoren (Leipzig, Steinkopff, 1937: 181 pp.).

TIMOFEEFF-RESSOVSKY, N. W., 1939. Genetik und Evolution (Bericht eines Zoologen) (Zschr. ind. Abst. Vererb. Lehre, 76: 158-219).

TIMOFEEFF-RESSOVSKY, N. W., 1940a. Allgemeines über die Entstehung neuer Erbanlagen (Handb. Erbb. des Menschen, 1: 139-244).

TIMOFEEFF-RESSOVSKY, N. W., 1940b. Eine biophysikalische Analyse des Mutationsvorganges (Nova Acta Leopoldina, N.F., 9: 209-240).

TIMOFEEFF-RESSOVSKY, N. W., 1940c. Der Positionseffekt der Gene (Hndb. Erbb. d. Menschen, 1: 181-190).

TIMOFEEFF-RESSOVSKY, N. W., K. G. ZIMMER und M. DELBRUECK, 1935. Ueber die Natur der Genmutation und der Genstruktur (Nachr. Ges. Wissensch. Göttingen, Math, phys, Kl. Biologie, N.F., 1: 189-245).

TISCHLER, G., 1925. Die cytologischen Verhältnisse bei pflanzlichen Bastarden (Bibliogr. Genet., 1: 39-68).

TISCHLER, G., 1927-1938. Pflanzliche Chromosomenzahlen (Tab. biol., 4: 1-83; 7: 109-226; 11: 281-304; 12: 57-115; 16: 162-218).

TISCHLER, G., 1950. Die Chromosomenzahlen der Gefässpflanzen Mitteleuropas (Den Haag, Junk, 1950: 263 pp.).

TISCHLER, G., 1951. Allgemeine Pflanzenkaryologie 2 Hälfte: Kernteilung und Kernverschmelzung (2te Aufl., Handb. d. Pflanzenanatomie. II; Berlin, Borntraeger, 1951: 1040 pp.).

TJEBBES, K., 1931. Polymerism (Bibliogr. Genet., 8: 227-268).

TRIMBLE, H. C. and C. E. KEELER, 1938. The inheritance of „high uric acid excretion" in dogs (Journ. Hered., 29: 280-289).

TROLAND, L. T., 1917. Biological enigmas and the theory of enzyme action (Am. Nat., 51: 321-350).

TSCHERMAK, E., 1900a. Ueber künstliche Kreuzung bei *Pisum sativum* (Ber. Deutsch. bot. Ges., 18: 232-239).

TSCHERMAK, E., 1900b. Ueber künstliche Kreuzung bei *Pisum sativum* (Zschr. landw. Versuchsw. Oesterreich, 3: 465-555).

TSCHERMAK, E., 1902. Der gegenwärtige Stand der Mendelschen Lehre (Zschr. landw. Versuchsw. Oesterreich, 5: 28-70).

TSCHERMAK, E., 1904a. Weitere Kreuzungsstudien an Erbsen, Levkojen und Bohnen (Zschr. landw. Versuchsw. Oesterreich, 8: 533-638).

TSCHERMAK, E., 1904b. Die Theorie der Kryptomerie und des Kryptohybridismus (Beih. bot. Cbl., 16: 11-35).

TSCHERMAK, E., 1914. Notiz über den Begriff der Kryptomerie (Zschr. f. ind. Abst. Vererb. Lehre, 11: 183-191).

TSCHERMAK, E. und H. BLEIER, 1926. Ueber fruchtbare *Aegilops*-Weizenbastarde (Ber. Deutsch. Bot. Ges., 44: 110-132).

TURESSON, G., 1925. The plant species in relation to habitat and climate. Contribution to the knowledge of genecological units (Hereditas, 6: 147-236).

TURESSON, G., 1929. Zur Natur und Begrenzung der Arteinheiten (Hereditas, 12: 323-334).

UBISCH, L. v., 1939. Keimblattchimärenforschung an Seeigellarven (Biol. Rev., 14: 88-103).

UMBGROVE, J. H. F., 1943. Leven en materie (Den Haag, Nyhoff, 1943: 132 pp.).

VAARAMA, A., 1949. Spindle abnormalities and variation in chromosome number in *Ribes nigrum* (Hereditas, 35: 136-162).

VANDEL, A., 1937. Chromosome number, polyploidy and sex in the animal kingdom (Proc. Zool. Soc. London, Ser. A., 107: 519-542).

VERSCHUER, O. v., 1930. Erbpsychologische Untersuchungen an Zwillingen (Zschr. ind. Abst. Vererb. Lehre 54: 280-285).

VERSCHUER, O. v., 1932. Die biologischen Grundlagen der menschlichen Mehrlingsforschung (Zschr. ind. Abst. Vererb. Lehre, 61: 147-205).

VERSCHUER, O. v., 1937. Erbpathologie (2. Aufl., Dresden, Steinkopff, 1937: 244 pp).

VERSCHUER, O. v., 1941, 1943. Leitfaden der Rassenhygiene (Leipzig, Thieme, 1941: 260 pp., 2 Aufl., 1943: 274 pp.).

VILMORIN, PH. DE, 1911. Fixité des races de froment (C. R. IV Conf. génétique, 1911: 312-316).

VOGT, O., 1926. Psychiatrisch wichtige Tatsachen der zoologisch-botanischen Systematik (Zschr. f. ges. Neurol. u. Psych., 101: 805-832).

VRIES, H. DE, 1889. Intracelluläre Pan-genesis (Jena, Fischer, 1889: 212 pp.).

VRIES, H. DE, 1899. Ueber Curvenselektion bei *Chrysanthemum segetum* (Ber. Deutsch. bot. Ges., 17: 84-98).

VRIES, H. DE, 1900. Das Spaltungsgesetz der Bastarde (Ber. Deutsch. bot. Ges., 18: 83-90).

VRIES, H. DE, 1901-1903. Die Mutationstheorie. I. Die Entstehung der Arten durch Mutation. II. Elementare Bastardlehre (Leipzig, Veit, 1901-1903: 648 + 752 pp).

VRIES, H. DE, 1907. On twin hybrids (Bot. Gaz., 44: 401-407).

VRIES, H. DE, 1908. Ueber die Zwillingsbastarde von *Oenothera nanella* (Ber. Deutsch. bot. Ges., 26a: 667-676).

VRIES, H. DE, 1909. On triple hybrids (Bot. Gaz., 47: 1-8).

VRIES, H. DE, 1911. Ueber doppeltreziproke Bastarde von *Oenothera biennis* L. und *O. muricata* L. (Biol. Cbl., 31: 97-104).

VRIES, H. DE, 1913. Gruppenweise Artbildung (Berlin, Borntraeger, 1913: 265 pp.).

WAARDENBURG, P. J., 1927. De biologische achtergrond van aanleg, milieu en opvoeding (Groningen, P. Noordhoff, 1: 108 pp.; 2: 186 pp.).

WAARDENBURG, P. J., 1932. Das menschliche Auge und seine Erbanlagen (Bibliogr. Genet., 7: 631 pp.).

WAARDENBURG, P. J., 1937. Het rassenvraagstuk in onzen tijd. Een biologische toelichting (Arnhem, v. Loghum Slaterus, 1937: 228 pp.).

WAARDENBURG, P. J., 1940. De biologische achtergrond van de criminaliteit. I-III (Afkomst en Toekomst, 6: 129-144, 193-214, 293-306).

WADA, Bt., 1933. Mikrodissektion der Chromosomen von *Tradescantia reflexa* (Cytologia, 4: 222-227).

WADDINGTON. C. H., 1947. Organisers and genes (Cambridge, Univ. Press, 1947: 160 pp.).

WAGNER, R. P. and H. K. MITCHELL. 1955. Genetics and metabolism (New York, Wiley Sons, 1955: 444 pp.).

WAHLUND, S., 1928. Zusammensetzung von Populationen und Korrelationserscheinungen vom Standpunkt der Vererbungslehre aus betrachtet (Hereditas, 11: 65-106).

WARMKE, H. E. and A. F. BLAKESLEE, 1939. Sex mechanism in polyploids of *Melandrium* (Science, N.S., 89: 391-392).

WARNER, M. F., M. A. SHERMAN and E. M. COLVIN, 1934. A bibliography of plant genetics (U.S. Dept. AGRIC., Misc. Publ. 164: 552 pp.).

WEBBER, H. J., 1903. New horticultural and agricultural terms (Science, N.S. 18: 501-503).

WEBER, E., 1935. Einführung in die Variations- und Erblichkeits-Statistik (München, Lehmann, 1935: 255 pp.).

WEIJER, J., 1945. A genetical investigation into the td-locus of *Neurospora crassa* (Genetica, 27: 173-252).

WEISMANN, A., 1883. Ueber die Vererbung. Ein Vortrag (Jena, Fischer, 1883. Also in: Aufsätze über Vererbung und verwandte biologische Fragen. (Jena, Fischer, 1892: 73-122).

WEISMANN, A., 1885. Die Continuität des Keimplasmas als Grundlage einer Theorie der Vererbung (Jena, Fischer, 1885. Also in: Aufsätze etc. 1892: 191-302).

WEISMANN, A., 1891. Amphimixis oder die Vermischung der Individuen (Jena, Fischer, 1891. Also in: Aufsätze etc., 1892: 673-826).

WEISMANN, A., 1892. Das Keimplasma. Eine Theorie der Vererbung (Jena, Fischer, 1892: 628 pp.).

WEISS, F. E., 1930. The problem of graft hybrids and chimaeras and Addendum (Biol. Rev., 5: 231-271, 6: 132).

WELLENSIEK, S. J., 1925. Genetic monograph on *Pisum* (Bibliogr. Genet., 3: 343-476).

WELLENSIEK, S. J., 1927. Methods for calculating the actual gametic F_2-series from a given zygotic series (Genetica, 9: 329-340).

WELLENSIEK, S. J., 1943. Grondslagen der algemene plantenveredeling (Haarlem, H. D. Tjeenk Willink, 1943: 492 pp.).

WENRICH, D. H., 1916. The spermatogenesis of *Phrynotettix magnus*, with special reference to synapsis and the individuality of the chromosomes (Bull. Mus. Comp. Zool. Harvard, 60: 57-135).

WENTWORTH, E. N. and R. M. REMICK 1916. Some properties of the generalised Mendelian population (Genetics, 1: 608-616).

WESTERGAARD, M., 1936. On the satellites in the eversporting *Matthiola* races (C. R. Trav. Lab. Carlsberg, Série Physiol., 21: 195-204).

WESTERGAARD, M., 1940. Studies on cytology and sex determination in polyploid forms of *Melandrium album* (Dansk bot. Ark., 10,5: 131 pp.).

WETTSTEIN, F. v., 1923-1924a. Kreuzungsversuche mit multiploiden Moosrassen. I-II (Biol. Zbl., 43: 71-83; 44: 145-169).

WETTSTEIN, F. v., 1924b-1928. Morphologie und Physiologie des Formwechsels der Moose auf genetischer Grundlage. I-II (Zschr. ind. Abst. Vererb. Lehre, 33: 1-236; Bibliotheca genet., 10: 216 pp.).

WETTSTEIN, F. v., 1925. Genetische Untersuchungen an Moosen (Musci und Hepaticae) (Bibliogr. Genet., 1: 1-38).

WETTSTEIN, F. v., 1926-1930. Ueber plasmatische Vererbung, sowie Plasma- und Genwirkung .I-II (Nachr. Ges. Wiss. Göttingen, math. phys. Klasse, 1926: 250-281; 1930. 109-118).

WETTSTEIN, F. v., 1927. Die Erscheinungen der Heteroploidie, besonders im Pflanzenreich (Ergebn. Biol., 2: 311-356).

WETTSTEIN, F. v., 1936. Gesichertes und Problematisches zur Geschlechts-bestimmung (Ber. Deutsch. Bot. Ges., 54: (23)-(38).

WETTSTEIN, F. v., 1937. Die genetische und entwicklungsphysiologische Bedeutung der Cytoplasmas (Zschr. ind. Abst. Vererb. Lehre, 73: 349-366).

WETTSTEIN, F. v. (und J. STRAUB), 1937-1942. Experimentelle Untersuchungen zum Artbildungsproblem I-III (Zschr. ind. Abst. Vererb. Lehre, 74: 34-53, Ber. D. Bot. Ges., 58: 374-388, Zschr. ind. Abst. Vererb. Lehre, 80: 271-280).

WETTSTEIN, F. v. und K. PIRSCHLE,

1938. Ueber die Wirkung hetero-
plastischer Pfropfungen und die
Uebertragung eines genbedingten
Stoffes durch Pfropfung bei *Petunia*
(Biol. Zbl., 58: 123-142).

WHELDALE, M., 1914. Our present
knowledge of the chemistry of the
Mendelian factors for flower colours
(Journ. Genet., 4: 109-129, 369-
376).

WHELDALE, M. and H. L. BASSETT,
1914. The chemical interpretation
of some mendelian factors for flo-
wercolours (Proc. Roy. Soc. B. 87:
300-311).

WHITE, M. J. D., 1940. Evidence for
polyploidy in the hermaphrodite
group of animals (Nature 146: 132-
133).

WHITE, M. J. D., 1945, 1954. Animal
cytology and evolution (Cambridge,
Univ. Press, 1945: 374 pp.; Sec. ed.
1954: 454 pp.).

WHITE, P. R., 1936-1946. Plant
tissue cultures, (Bot. Rev. 2: 419-
437; 12: 521-529).

WHITE, P. R., 1943. A handbook of
tissue culture (Lancaster, Cattell,
1943).

WHITING, P. W., 1935. Sex determination in bees and wasps (Journ.
Hered., 26: 263-278).

WHITING, P. W. and A. R., 1927.
Gynandromorphs and other irregu-
lar types in *Habrobracon* (Biol. Bull.
Woods Hole, 52: 89-120).

WICHLER, G., 1913. Untersuchungen
über den Bastard *Dianthus armeria*
× *Dianthus deltoides* nebst Bemerkungen über einige andere Artkreuzungen der Gattung *Dianthus* (Zsch.
ind. Abst. Vererb. Lehre, 10: 177-
232).

WICHURA, M., 1865. Die Bastardbe-
fruchtung im Pflanzenreich, erläu-
tert an den Bastarden der Weiden
(Breslau, Morgenstern, 1865: 94
pp.)

WIEGMANN, A. F., 1828. Ueber Bastarderzeugung im Pflanzenreich
(Braunschweig, Vieweg, 1828: 40
pp.).

WILLIER, B. H., 1939. Embryonic development of sex (In: E. Allen,
1939. Sex and internal secretions,
2nd ed., London, Baillière, Tindall
& Cox, 1939: 64-144).

WILSON, E. B., 1896, 1928, 1937. The
cell in development and heredity
(First ed. 1896, 3rd ed. with corrections; New York, MacMillan,
1937: 1232 pp.).

WILSON, E. B. and T. H. MORGAN,
1920. Chiasmatype and crossing-
over (Am. Nat., 54: 193-219).

WINGE, Ö., 1917. The chromosomes,
their numbers and general significance (C. R. Trav. Lab. Carlsberg,
13: 131-275).

WINGE, Ö., 1923a. Crossing-over between the X- and the Y- chromosome in *Lebistes* (Journ. Genet., 13:
201-217).

WINGE, Ö., 1923b. On sex chromosomes, sex determination, and preponderance of females in some
dioecious plants (C. R. Trav. Lab.
Carlsberg, 15,5: 1-27).

WINGE, Ö., 1927a. On a Y-linked gene
in *Melandrium* (Hereditas, 9: 274-
283).

WINGE, Ö., 1927b. The location of
eighteen genes in *Lebistes reticulatus*
(Journ. Genet., 18: 1-43).

WINGE, Ö., 1932a. On the origin of
constant species-hybrids (Svensk
Bot. Tidskr., 26: 107-122).

WINGE, Ö., 1932b. The inheritance of
double flowers and other characters
in *Matthiola* (Zschr. f. Pflanzenz.,
17: 118-135).

WINGE, Ö., 1936. Linkage in *Pisum*
(C. R. Trav. Lab. Carlsberg, Série
physiol, 21: 271-393).

WINGE, Ö., 1949. Inheritance of enzymatic characters in yeasts (Proc.
8th Intern. Congr. Genet., Stockholm, 1919:520-529).

WINGE, Ö., 1950. Inheritance in dogs.
With special reference to hunting
breeds (Ithaca. N. Y., Comstock,
1950: 153 pp.).

WINGE, Ö., 1952. The genetic situation
concerning fermentation in yeast
(Heredity, 6 : 263-269).

WINGE, Ö. and O. LAUSTSEN, 1940. On
a cytoplasmatic effect of inbreeding
in homozygous yeast (C. R. Lab.
Carslberg, Série physiol., 23: 17-
39).

WINGE, Ö. and C. ROBERTS, 1954.
Causes of deviations from 2: 2
segregations in the tetrad of monohybrid yeasts (C. R. Trav. Lab.

Carlsberg, Série physiol., 25: 285-329).

WINKLER, H., 1907. Ueber Pfropfbastarde und pflanzliche Chimaeren (Ber. Deutsch. bot. Ges., 25: 568-576).

WINKLER, H., 1916. Ueber die experimentelle Erzeugung von Pflanzen mit abweichenden Chromosomenzahlen (Zschr. Bot., 8: 417-528).

WINKLER, H., 1920. Verbreitung und Ursache der Parthenogenesis im Pflanzen- und Tierreiche (Jena, Fischer, 1920: 231 pp.).

WINKLER, H., 1930. Die Konversion der Gene (Jena, Fischer, 1930: 116 pp.).

WINKLER, H., 1938. Ueber einen Burdonen von *Solanum nigrum* (Planta, 27: 680-707).

WISSELINGH, C. v., 1919. Ueber Variabilität und Erblichkeit (Zschr. ind. Abst. Vererb. Lehre, 22: 65-126).

WITSCHI, E., 1929. Bestimmung und Vererbung des Geschlechts bei Tieren (Handb. Vererb. wiss. II. D: 115 pp.).

WITSCHI, E., 1934. Genes and inductors of sex differentiation in Amphibians (Biol. Rev., 9: 460-488).

WITSCHI, E., 1939. Modification of development of sex in lower vertebrates and in mammals (In: E. Allen, 1939. Sex and internal secretions, 2nd ed., London, Baillière, Tindall & Cox, 1939: 145-226).

WOESS, F. v., 1941. Experimentelle Untersuchungen zum Artbildungsproblem an *Arenaria serpyllifolia* und *Arenaria Marschlinsi* (Zschr. ind. Abst. Vererb. Lehre, 79: 444-472).

WOLFF, C. F., 1759-1774. Theoria generationis (Halae, 1759: 146 pp.; ed. nova Halae 1774: 231 pp.).

WOOD, T. B., 1909. The inheritance of horns and face colour in sheep (Journ. agr. Sc., 3: 145-154).

WOODGER, J. H., 1930-1931. The „concept of organism" and the relation between embryology and genetics (Quart. Rev. Biol., 5: 1-22, 438-463; 6: 178-207).

WOUTERS, W., 1948. Contribution à l'étude taxonomique et caryologique du genre *Gossypium* et application à l'amélioration du cotonnier au Congo belge (Public. I.N.E.A.C., Série scient., 34: 394 pp.).

WRIEDT, CHR., 1931. Die Vererbung quantitativer Eigenschaften bei den Wirbeltieren (Zschr. ind. Abst. u. Vererb. Lehre, 57: 211-225).

WRIGHT, S., 1921. Systems of mating (Genetics, 6: 111-178).

WRIGHT, S., 1922. The effects of inbreeding and crossbreeding in guinea pigs. I. Decline in vigor. II. Differentiation among inbred families. III. Crosses between highly inbred families. (U. S. Deptm. Agriculture. Bull. 1090: 63 pp. en Bull. 1121: 60 pp.).

WRIGHT, S., 1931. Evolution in mendelian populations (Genetics, 16: 97-159).

WRIGHT, S., 1939a. Statistical genetics in relation to evolution (Act. Sc. Industr. Exp. Biom. et Stat. biol., 13: 64 pp.).

WRIGHT, S., 1939b. The distribution of self-sterility alleles in populations (Genetics, 24: 538-552).

WRIGHT, S., 1941. The physiology of the gene (Physiol. Rev., 21: 487-527).

WRIGHT, S., 1945a. Genes as physiological agents (Amer. Natur., 79: 289-303).

WRIGHT, S., 1945b. Physiological aspects of genetics (Ann. Review Physiol., 7: 75-106.).

WRIGHT, S., 1951. The genetical structure of populations (Ann. Eugen., 15: 323-354).

WRIGHT, S. and O. N. EATON, 1929. The persistence of differentiation among inbred families of guinea-pigs (U. S. Deptm. Agric. Techn. Bull. 103: 45 pp.).

YANOFSKY, C., 1952. The effects of gene change on tryptophan desmolase formation (Proc. Nat. Ac. Sc., 38: 215-226).

Yearbook, 1936, 1937. U. S. Dept. of Agriculture (1936: 119-1141; 1937: 117-1477).

ZELENY, C., 1920. A change in the bar gene of *Drosophila melanogaster* involving further decrease in facet number further decrease in facet number and increase in dominance (Journ. Exp. Zool., 30: 293-324).

ZIEGLER, H. E., 1910. Die Streitfrage

der Vererbungslehre (Lamarckismus oder Weismannismus) (Naturwiss. Wochenschr., N. F. 9: 193-200).

ZIMMER, K. G., 1934. Ein Beitrag zur Frage nach der Beziehung zwischen Röntgenstrahlendosis und dadurch ausgelöster Mutationsrate (Strahlentherapie, 51: 179-184).

ZIMMER, K. G., 1937. Strahlungen. Wesen, Erzeugung und Mechanismus der biologischen Wirkung (Leipzig, Thieme, 1937: 72 pp.).

ZIMMERMAN, W., 1938. Vererbung „erworbener Eigenschaften" und Ausleze (Jena, G. Fischer, 345 pp.).

ZIRKLE, C., 1935. The beginnings of plant hybridization (Philadelphia, Univ. Penn. Press., 1935: 231 pp.).

INDEX